T0305831

EPIGENETICS AND DERMATOLOGY

EPIGENETICS AND DERMATOLOGY

QIANJIN LU
Professor and Director, Hunan Key Laboratory of Medical Epigenetics,
Department of Dermatology, The 2nd Xiangya Hospital,
Central South University, Changsha, China

CHRISTOPHER C. CHANG
Professor of Medicine and Associate Director, Allergy and Immunology
Fellowship Program, Division of Rheumatology, Allergy and Clinical Immunology,
University of California at Davis, California, USA

BRUCE C. RICHARDSON
Professor of Medicine, Epigenetic Research Team Leader, Division of Rheumatology,
Department of Internal Medicine,
University of Michigan, Ann Arbor, Michigan, USA

AMSTERDAM • BOSTON • HEIDELBERG • LONDON
NEW YORK • OXFORD • PARIS • SAN DIEGO
SAN FRANCISCO • SINGAPORE • SYDNEY • TOKYO

Academic Press is an imprint of Elsevier

Academic Press is an imprint of Elsevier
32 Jamestown Road, London NW1 7BY, UK
525 B Street, Suite 1800, San Diego, CA 92101-4495, USA
225 Wyman Street, Waltham, MA 02451, USA
The Boulevard, Langford Lane, Kidlington, Oxford OX5 1GB, UK

Copyright © 2015 Elsevier Inc. All rights reserved.

No part of this publication may be reproduced or transmitted in any form or by any
means, electronic or mechanical, including photocopying, recording, or any information
storage and retrieval system, without permission in writing from the publisher. Details on
how to seek permission, further information about the Publisher's permissions policies and
our arrangements with organizations such as the Copyright Clearance Center and the
Copyright Licensing Agency, can be found at our website: www.elsevier.com/permissions.

This book and the individual contributions contained in it are protected under copyright
by the Publisher (other than as may be noted herein).

Notices
Knowledge and best practice in this field are constantly changing. As new research
and experience broaden our understanding, changes in research methods, professional
practices, or medical treatment may become necessary.

Practitioners and researchers must always rely on their own experience and knowledge in
evaluating and using any information, methods, compounds, or experiments described
herein. In using such information or methods they should be mindful of their own safety
and the safety of others, including parties for whom they have a professional responsibility.

To the fullest extent of the law, neither the Publisher nor the authors, contributors,
or editors, assume any liability for any injury and/or damage to persons or property as
a matter of products liability, negligence or otherwise, or from any use or operation of
any methods, products, instructions, or ideas contained in the material herein.

ISBN: 978-0-12-800957-4

British Library Cataloguing-in-Publication Data
A catalogue record for this book is available from the British Library.

Library of Congress Cataloging-in-Publication Data
A catalog record for this book is available from the Library of Congress.

For Information on all Academic Press publications
visit our website at http://store.elsevier.com/

Typeset by MPS Limited, Chennai, India
www.adi-mps.com

Printed and bound in the United States of America

Working together
to grow libraries in
developing countries

www.elsevier.com • www.bookaid.org

Dedication

To our patients who suffer from skin diseases. May this book be a seed for future research and development of novel treatments to help alleviate dermatological illness of all forms, from allergic diseases to autoimmune skin diseases and cancer. We hope that epigenetics will provide potential cures and personalized approaches for many of these diseases.

QL, CC, and BR

Contents

1

BIOLOGICAL AND HISTORICAL ASPECTS OF EPIGENETICS

1. Introduction to Epigenetics
YU LIU AND QIANJIN LU

2. Laboratory Methods in Epigenetics
YU LIU, JIEYUE LIAO, AND QIANJIN LU

3. Keratinocyte Differentiation and Epigenetics
JEUNG-HOON LEE

4. Epigenetics and Fibrosis: Lessons, Challenges, and Windows of Opportunity
BIN LIU, XIN SHENG WANG, HUI-MIN CHEN, QIANJIN LU, M. ERIC GERSHWIN, AND PATRICK S.C. LEUNG

5. Epigenetic Modulation of Hair Follicle Stem Cells
HAIJING WU AND QIANJIN LU

6. Epigenetics and the Regulation of Inflammation
CHRISTIAN M. HEDRICH

7. Malignant Transformation and Epigenetics
YIXING HAN, JIANKE REN, WEISHI YU, MINORU TERASHIMA, AND KATHRIN MUEGGE

8. Epigenetic Mechanisms of Sirtuins in Dermatology
ALEXANDER LO, MELISSA SERRAVALLO, AND JARED JAGDEO

9. MicroRNAs in Skin Diseases

MARIANNE B. LØVENDORF AND LONE SKOV

2

IMMUNOLOGIC SKIN DISEASES

10. Systemic Lupus Erythematosus

BRUCE C. RICHARDSON

11. Epigenetics in Psoriasis

KUAN-YEN TUNG, FU-TONG LIU, YI-JU LAI, CHIH-HUNG LEE, YU-PING HSIAO,
AND YUNGLING LEO LEE

12. Epigenetics and Systemic Sclerosis
NEZAM ALTOROK AND AMR H. SAWALHA

13. Epigenetics of Allergic and Inflammatory Skin Diseases
NINA POLIAK AND CHRISTOPHER CHANG

14. Epigenetics and Other Autoimmune Skin Diseases
MING ZHAO, RUIFANG WU, AND QIANJIN LU

3

NONIMMUNOLOGIC SKIN DISEASES

15. Epigenetics and Infectious Skin Disease
JACK L. ARBISER AND MICHAEL Y. BONNER

4

APPLICATIONS OF EPIGENETICS

List of Contributors

Nezam Altorok Division of Rheumatology, Department of Internal Medicine, University of Toledo Medical Center, Toledo, OH

Jack L. Arbiser Department of Dermatology, Emory School of Medicine, Winship Cancer Institute, Atlanta, GA; Department of Dermatology, Atlanta Veterans Affairs Medical Center, Decatur, GA

Michael Y. Bonner Department of Dermatology, Emory School of Medicine, Winship Cancer Institute, Atlanta, GA

Wesley H. Brooks Department of Chemistry, University of South Florida, Tampa, FL

Christopher Chang Division of Rheumatology, Allergy and Clinical Immunology, University of California, Davis, CA

Jessica Charlet Department of Urology, Keck School of Medicine, University of Southern California, Los Angeles, CA

Frederic L. Chedin Department of Molecular and Cellular Biology, University of California, Davis, CA

Hui-Min Chen Department of Molecular and Cellular Biology, University of California, Davis, CA; Division of Rheumatology, Allergy and Clinical Immunology, University of California, Davis, CA

Suresh de Silva Center for Retrovirology Research, Department of Veterinary Biosciences, The Ohio State University, Columbus, Ohio

Pierre Gazeau EA2216, INSERM ESPRI, ERI29, European University of Brittany and Brest University, Brest, France; SFR ScInBioS, LabEx IGO "Immunotherapy Graft Oncology," and "Réseau Épigénétique du Cancéropole Grand Ouest," France; Laboratory of Immunology and Immunotherapy, CHU Morvan, Brest, France

M. Eric Gershwin Division of Rheumatology, Allergy and Clinical Immunology, University of California, Davis, CA

Yixing Han Mouse Cancer Genetics Program, Center for Cancer Research, National Cancer Institute, Frederick, MD

Christian M. Hedrich Pediatric Rheumatology and Immunology, Children's Hospital Dresden, University Medical Center "Carl Gustav Carus," Technische Universität Dresden, Dresden, Germany

Yu-Ping Hsiao Department of Medical Education, Taichung Veterans General Hospital, Taichung, Taiwan; Institute of Medicine, Chung Shan Medical University, Taichung, Taiwan

Jared Jagdeo Department of Dermatology, SUNY Downstate Medical Center, Brooklyn, NY; Department of Dermatology, University of California at Davis, Sacramento, CA; Dermatology Service, Sacramento VA Medical Center, Mather, CA

Yi-Ju Lai Institute of Biomedical Sciences, Academia Sinica, Taipei, Taiwan

Christelle Le Dantec EA2216, INSERM ESPRI, ERI29, European University of Brittany and Brest University, Brest, France; SFR ScInBioS, LabEx IGO "Immunotherapy Graft Oncology," and "Réseau Épigénétique du Cancéropole Grand Ouest," France

Chih-Hung Lee Department of Dermatology, Kaohsiung Chang Gung Memorial Hospital, Kaohsiung, Taiwan

Jeung-Hoon Lee Department of Dermatology, College of Medicine, Chungnam National University, Daejeon, South Korea

Yungling Leo Lee Institute of Biomedical Sciences, Academia Sinica, Taipei, Taiwan; Institute of Epidemiology and Preventive Medicine, National Taiwan University, Taipei, Taiwan

Patrick S.C. Leung Division of Rheumatology, Allergy and Clinical Immunology, University of California, Davis, CA

Gangning Liang Department of Urology, Keck School of Medicine, University of Southern California, Los Angeles, CA

Jieyue Liao Department of Dermatology, Second Xiangya Hospital of Central South University, Hunan Key Laboratory of Medical Epigenetics, Changsha, Hunan, PR China

Bin Liu Department of Rheumatology and Immunology, The Affiliated Hospital of Medical College Qingdao University, Qingdao City, Shandong Province, PR China; Division of Rheumatology, Allergy and Clinical Immunology, University of California, Davis, CA

Fu-Tong Liu Institute of Biomedical Sciences, Academia Sinica, Taipei, Taiwan

Yu Liu Department of Dermatology, Second Xiangya Hospital of Central South University, Hunan Key Laboratory of Medical Epigenetics, Changsha, Hunan, PR China

Alexander Lo SUNY Downstate College of Medicine, Brooklyn, NY

Marianne B. Løvendorf Department of Dermato-Allergology, Gentofte Hospital, University of Copenhagen, Hellerup, Denmark

Qianjin Lu Department of Dermatology, The Second Xiangya Hospital of Central South University, Hunan Key Laboratory of Medical Epigenetics, Changsha, Hunan, PR China

Anjali Mishra Comprehensive Cancer Center and Division of Dermatology, Department of Internal Medicine, The Ohio State University, Columbus, Ohio

Kathrin Muegge Basic Science Program, Leidos Biomedical Research, Inc., Mouse Cancer Genetics Program, Frederick National Laboratory for Cancer Research, Frederick, MD; Mouse Cancer Genetics Program, Center for Cancer Research, National Cancer Institute, Frederick, MD

Sreya Mukherjee Department of Chemistry, University of South Florida, Tampa, FL

Nina Poliak Division of Allergy and Immunology, Nemours/AI duPont Hospital for Children, Wilmington, DE

Pierluigi Porcu Comprehensive Cancer Center and Division of Hematology, Department of Internal Medicine, The Ohio State University, Columbus, Ohio

Jianke Ren Mouse Cancer Genetics Program, Center for Cancer Research, National Cancer Institute, Frederick, MD

Yves Renaudineau EA2216, INSERM ESPRI, ERI29, European University of Brittany and Brest University, Brest, France; SFR ScInBioS, LabEx IGO "Immunotherapy Graft Oncology," and "Réseau Épigénétique du Cancéropole Grand Ouest," France; Laboratory of Immunology and Immunotherapy, CHU Morvan, Brest, France

Bruce C. Richardson Division of Rheumatology, Department of Internal Medicine, University of Michigan, Ann Arbor, MI

Sabita N. Saldanha Department of Biological Sciences, Alabama State University, Montgomery, AL

Amr H. Sawalha Center for Computational Medicine and Bioinformatics, University of Michigan, Ann Arbor, MI; Division of Rheumatology, Department of Internal Medicine, University of Michigan, Ann Arbor, MI

Melissa Serravallo Department of Dermatology, SUNY Downstate Medical Center, Brooklyn, NY

Lone Skov Department of Dermato-Allergology, Gentofte Hospital, University of Copenhagen, Hellerup, Denmark

Minoru Terashima Mouse Cancer Genetics Program, Center for Cancer Research, National Cancer Institute, Frederick, MD

Shannon Doyle Tiedeken Department of Pediatrics, Thomas Jefferson University, Nemours/A.I. duPont Hospital for Children, Wilmington, DE

Kuan-Yen Tung Institute of Biomedical Sciences, Academia Sinica, Taipei, Taiwan; Institute of Epidemiology and Preventive Medicine, National Taiwan University, Taipei, Taiwan

Xin Sheng Wang Department of Urology, The Affiliated Hospital of Medical College Qingdao University, Qingdao City, Shandong Province, PR China

Louis Patrick Watanabe Department of Biology, University of Alabama at Birmingham, Birmingham, AL

Henry K. Wong Comprehensive Cancer Center and Division of Dermatology, Department of Internal Medicine, The Ohio State University, Columbus, Ohio

Haijing Wu Department of Dermatology, The Second Xiangya Hospital of Central South University, Hunan Key Laboratory of Medical Epigenomics, Changsha, Hunan, PR China

Li Wu Center for Retrovirology Research, Department of Veterinary Biosciences, The Ohio State University, Columbus, Ohio

Ruifang Wu Department of Dermatology, The Second Xiangya Hospital of Central South University, Hunan Key Laboratory of Medical Epigenomics, Changsha, Hunan, PR China

Weishi Yu Mouse Cancer Genetics Program, Center for Cancer Research, National Cancer Institute, Frederick, MD

Ming Zhao Department of Dermatology, The Second Xiangya Hospital of Central South University, Hunan Key Laboratory of Medical Epigenomics, Changsha, Hunan, PR China

Preface

Epigenetics—the word epigenetics has been used since the 1940s, when Dr. Charles Waddington used the term to describe how gene regulation impacts development. In those days, before we even knew the structure of DNA, Dr. Waddington also coined the term *chreode*, to describe the cellular developmental process which leads to the paths that cells take toward development, a sort of cellular destiny. Now, some 70 plus years later, the term *epigenetics* has taken on a different meaning, though not necessarily a discordant philosophy, and is used to describe the study of how genes are regulated without a change in DNA sequence.

The concept of epigenetics embodies a broad range of cellular and biological phenomena, but the premise is based on the fact that gene expression may be altered in the absence of mutations or deletions, or other changes in DNA sequence, leading to different states of health and disease. How this is achieved is through the mechanisms of epigenetics, which includes DNA methylation and alterations in histone structure. MicroRNAs, which are short sequences of noncoding RNA that bind to promoter regions of genes to affect translation, have also been classified by some as an epigenetic phenomenon, but this is not without controversy.

The skin is the largest organ in the body. It is a dynamic, living, immunologic structure that possesses many functions, serving as a protective barrier to the outside world and a homeostatic system to support life. It is also an immune organ, and while it protects us from the dangers of microbes, pollutants, and toxins, it also participates in how we identify safety from hazardous exposures, thus acting as a medium for the development of tolerance. The systems in the skin are complex, involving numerous cell types and signaling molecules, and the pathways that govern the regulation of skin function add an additional layer of complexity. Thus, much can go wrong. Therefore, diseases of the skin range from neoplasms to infections to autoimmune diseases and allergic conditions. Solving the mysteries of skin function will help us find new ways to restore skin "health" or "normalcy." Epigenetics will no doubt play a significant role in these endeavors.

The first application of epigenetics was in cancer diagnosis and treatment. Interestingly, research scientists, pharmacologists, and physicians have been using products that act by impacting epigenetics for many years without knowing it. For example, many herbal products were found to be efficacious in the treatment of some diseases, and were therefore widely used, and though we did not know it at the time, some of these herbal products actually act through epigenetic mechanisms. We are gradually recognizing that epigenetics is involved in many aspects of diseases, and the acquisition of data on how these processes work will help guide us in the development of novel, epigenetic treatment modalities that promise to help diagnose, treat, or even cure diseases in the coming future.

This book is divided into three sections. The first includes chapters addressing the basic science of epigenetics in various skin cell types. The second describes the role of epigenetics in dermatological conditions, and the third touches upon more general epigenetic diagnostic and therapeutic concepts and discusses the future of epigenetics and skin diseases.

It is the hope of us, the editors, that this book on epigenetics in dermatology will benefit readers from many disciplines, including but not limited to dermatologists, rheumatologists, biologists, allergists, immunologists, and oncologists. We hope that the reader will enjoy the discussions on all the various aspects by which epigenetics can impact skin function and diseases.

Qianjin Lu
Christopher C. Chang
Bruce C. Richardson

Acknowledgments

The editors of this book thank all the authors for their tireless contribution to their respective chapters. They also thank Elsevier for the opportunity to communicate this important topic to our readers, especially Catherine Van Der Laan, Lisa Eppich, and Graham Nisbet. The editors also thank their families for their sacrifices in order that they could spend hours on weekends and weeknights working on bringing this book to fruition.

BIOLOGICAL AND HISTORICAL ASPECTS OF EPIGENETICS

Introduction to Epigenetics

Yu Liu and Qianjin Lu

Department of Dermatology, Second Xiangya Hospital of Central South
University, Hunan Key Laboratory of Medical Epigenetics, Changsha,
Hunan, PR China

The human genome project has been one of the most important scientific achievements in modern history. It has ushered in a new era in the field of life science research. However, among the project's many great discoveries, surprising findings such as only particular subsets of genes being able to be expressed at a particular location and time, led to the realization that knowledge of DNA sequences is insufficient to understand phenotypic manifestations. The mechanism by which DNA, or the genetic code, is translated into protein sequences is not merely dependent on the sequence itself but also on a sophisticated regulatory system that interplays between genetic and environmental factors. These mechanisms comprise the science of epigenetics, and the control of genes through various chemical interactions for the basis of at least part of the regulatory system overseeing the expression of the genetic code [1].

Epigenetics is defined as heritable changes in gene expression without changes in the DNA sequence. The prefix epi- is derived from the Greek preposition ἐπί, meaning above, on, or over. The term was first coined in 1942 by C.H. Waddington to denote a phenomenon that conventional genetics could not explain [2]. Since then, epigenetics has evolved into a branch of science that studies biological pathways and systems with well-understood molecular mechanisms. Simplistically, epigenetic mechanisms may involve modifications to DNA and surrounding structures such as DNA methylation, chromatin modification, and noncoding RNA (ncRNA).

DNA methylation is a stable and inheritable epigenetic mark. This genetically programmed modification is almost exclusively found on the $5'$ position of the pyrimidine ring of cytosines (5mC) adjacent to a guanine. These sites are referred to as CpG sites, and the modification is

© 2015 Elsevier Inc. All rights reserved.

mediated by specific enzymes called DNA methyltransferases (DNMTs). Transcription is generally repressed by hypermethylation of active promoters associated with CpG-rich sequences [3]. DNA methylation-based imprinting disorders play an important role in skin diseases such as systemic lupus erythematosus (SLE) [4], psoriasis vulgaris [5], primary Sjögren's syndrome [6], and other diseases. In addition, aberrations in the function of DNMTs and methyl-CpG-binding proteins (MBDs) can also contribute to skin diseases [7]. Recently, another modified form of cytosine, 5-hydroxymethylcytosine (5hmC), has been identified and is now recognized as the "sixth base" in the mammalian genome, following 5mC (the "fifth base") [8]. 5mC can be converted to 5hmC by the ten−eleven translocation (Tet) family proteins, which can further oxidize 5hmC to 5-formylcytosine (5fC) and 5-carboxycytosine (5caC) to achieve active DNA demethylation [9]. Emerging evidence has indicated that 5hmC-mediated DNA demethylation and Tet family proteins may play essential roles in diverse biological processes including development and diseases, as illustrated by the critical function of 5hmC in the development of melanoma [10].

The other main mechanism in epigenetics involves changes to non-DNA gene components. DNA is tightly compacted by histone proteins. Posttranslational modifications on the tails of core histones, including lysine acetylation, lysine and arginine methylation, serine and threonine phosphorylation, and lysine ubiquitination, and sumoylation are important epigenetic modifications that regulate gene transcription. Abnormalities in these modifications, especially acetylation and deacetylation, can alter the structure of chromatin and perturb gene transcription, which can then contribute to disease development and progression. Histone acetylation status is reversibly regulated by two distinct competing families of enzymes, histone acetyltransferases (HATs) and histone deacetylases (HDACs). Until now, four classes of HDACs have been identified (including Class I, Class II, and Class IV). HDACs are zinc-dependent proteases consisting of HDAC1−11, and Class III, also known as sirtuins (SIRT1−7), which require the cofactor NAD+ for their deacetylase function [11].

Another widely studied histone modification is methylation. Methylation of lysine or arginine in histone proteins alters the compaction or relaxation of chromatin depending on the position of amino acid and the number of methyl groups; for example, histone 3 tri-methylated at lysine 4 promotes gene transcription, while histone 3 tri-methylated at lysine 9 inhibits gene transcription [3]. Increasing evidence indicates the critical role of histone modifications in skin diseases including immune-mediated skin diseases, infectious diseases, and cancer [12−14].

It is debatable whether or not the role of ncRNAs constitutes an epigenetic phenomenon. There are some who will claim that ncRNAs such as microRNAs (miRNAs) are a fundamental part of nature and do not satisfy the definition of epigenetics. However, others feel that since miRNAs do affect regulation of genes, they are a bona fide mechanism of epigenetic change.

The family of ncRNAs is diverse and complex. It can be divided into eight groups: ribosomal RNAs, transfer RNAs, miRNAs, long noncoding RNAs (lncRNAs), small nucleolar RNAs, small interfering RNAs, small nuclear RNAs, and piwi-interacting RNAs. ncRNAs are important epigenetic regulators in development and disease, especially miRNAs and lncRNAs. miRNAs are short ncRNA sequences (19–25 nucleotides) that regulate gene expression by binding to complementary sequences in the 3′ UTR of multiple target mRNAs, leading to translational repression (imperfect sequence match) or mRNA cleavage (perfect match) [15]. Since the first miRNA *lin-4* was characterized in 1993, an increasing number of miRNAs have been identified. Altered expression profiles of miRNAs in patients revealed a crucial role of miRNAs in cellular events and the development of diseases [16].

lncRNAs are functional ncRNAs, each exceeding 200 nucleotides in length and lacking functionally open reading frames. lncRNAs regulate gene expression through different molecular mechanisms. They can mediate the activity of proteins involved in chromatin remodeling and histone modification, or act as an RNA decoy or sponge for miRNAs. They can also bind to specific protein partners to modulate the activity of that particular protein [17]. Recent advancements in technology to identify ncRNAs using microarrays provide a great bulk of novel data from genomewide studies, and have revealed potential use of ncRNAs as diagnostic and prognostic biomarkers in various human disorders including skin diseases [18].

The role of genetics in disease is indisputable. But environmental exposures have also been demonstrated to play an essential role in the pathogenesis of skin diseases. Many diseases are now believed to occur as a result of a combination of genetic and environmental factors, but how do these two opposing forces interact? Epigenetic mechanisms may play a role in linking genetic and environmental factors, adding an additional element to the mechanism of disease.

Epigenetic regulation is generally accepted to play a key role in cellular processes. Aberrations of epigenetic modifications contribute to the pathogenesis of human diseases. With a growing knowledge of epigenetic mechanisms, we are confident that epigenetic markers can be applied as sensitive and specific biomarkers in disease diagnosis, evaluation, and prognosis. Moreover, epigenetic interventions may become an important supplement to traditional therapeutic approaches in the near future. The specific role of epigenetics in the pathogenesis, clinical phenotypes, and treatment of skin diseases is rapidly expanding as we continually increase our understanding of the mechanisms of epigenetics.

References

[1] Lu Q. The critical importance of epigenetics in autoimmunity. J Autoimmun 2013; 41:1−5.

[2] Choudhuri S. From Waddington's epigenetic landscape to small noncoding RNA: some important milestones in the history of epigenetics research. Toxicol Mech Methods 2011;21(4):252−74.

[3] Liu Y, Li H, Xiao T, Lu Q. Epigenetics in immune-mediated pulmonary diseases. Clin Rev Allergy Immunol 2013;45(3):314−30.

[4] Zhang Y, Zhao M, Sawalha AH, Richardson B, Lu Q. Impaired DNA methylation and its mechanisms in CD4(+) T cells of systemic lupus erythematosus. J Autoimmun 2013;41:92−9.

[5] Zhang P, Zhao M, Liang G, et al. Whole-genome DNA methylation in skin lesions from patients with psoriasis vulgaris. J Autoimmun 2013;41:17−24.

[6] Yu X, Liang G, Yin H, et al. DNA hypermethylation leads to lower FOXP3 expression in CD4+ T cells of patients with primary Sjogren's syndrome. Clin Immunol 2013;148(2):254−7.

[7] Lei W, Luo Y, Lei W, et al. Abnormal DNA methylation in CD4+ T cells from patients with systemic lupus erythematosus, systemic sclerosis, and dermatomyositis. Scand J Rheumatol 2009;38(5):369−74.

[8] Ye C, Li L. 5-Hydroxymethylcytosine: a new insight into epigenetics in cancer. Cancer Biol Ther 2014;15(1):10−15.

[9] Sun W, Guan M, Li X. 5-Hydroxymethylcytosine-mediated DNA demethylation in stem cells and development. Stem Cells Dev 2014;23(9):923−30.

[10] Lian CG, Xu Y, Ceol C, et al. Loss of 5-hydroxymethylcytosine is an epigenetic hallmark of melanoma. Cell 2012;150(6):1135−46.

[11] Shi BW, Xu WF. The development and potential clinical utility of biomarkers for HDAC inhibitors. Drug Discov Ther 2013;7(4):129−36.

[12] Trowbridge RM, Pittelkow MR. Epigenetics in the pathogenesis and pathophysiology of psoriasis vulgaris. J Drugs Dermatol 2014;13(2):111−18.

[13] Liang Y, Vogel JL, Arbuckle JH, et al. Targeting the JMJD2 histone demethylases to epigenetically control herpesvirus infection and reactivation from latency. Sci Transl Med 2013;5(167):167ra5.

[14] Rangwala S, Zhang C, Duvic M. HDAC inhibitors for the treatment of cutaneous T-cell lymphomas. Future Med Chem 2012;4(4):471−86.

[15] Hauptman N, Glavac D. MicroRNAs and long non-coding RNAs: prospects in diagnostics and therapy of cancer. Radiol Oncol 2013;47(4):311−18.

[16] Thamilarasan M, Koczan D, Hecker M, Paap B, Zettl UK. MicroRNAs in multiple sclerosis and experimental autoimmune encephalomyelitis. Autoimmun Rev 2012;11 (3):174−9.

[17] Katsushima K, Kondo Y. Non-coding RNAs as epigenetic regulator of glioma stem-like cell differentiation. Front Genet 2014;5:14.

[18] Jinnin M. Various applications of microRNAs in skin diseases. J Dermatol Sci 2014;74 (1):3−8.

2

Laboratory Methods in Epigenetics

Yu Liu, Jieyue Liao, and Qianjin Lu

Department of Dermatology, Second Xiangya Hospital of Central
South University, Hunan Key Laboratory of Medical Epigenetics,
Changsha, Hunan, PR China

2.1 INTRODUCTION

Epigenetic changes occur during cell differentiation, and serve to activate or suppress genes once the cells have reached terminal differentiation. Thus, epigenetics builds a bridge between genetics and environmental stimuli. Gene expression is up- or downregulated through epigenetic mechanisms in response to environmental changes. Abnormalities of epigenetic marks, such as DNA methylation, histone modifications, and aberrant expression of microRNAs (miRNAs), lead to the development of diseases. Mapping of the human epigenome is one of the most exciting and promising endeavors in terms of increasing our understanding of the etiology of diseases, and of developing new treatment strategies. Recent advances in technology have made it possible to interpret parts of the "epigenetic code." In this chapter, we summarize the classical strategies used in epigenetic studies and give a description of technological advancement in detection methodology.

2.2 DNA METHYLATION ANALYSIS

DNA methylation is an important epigenetic mark and a widely studied epigenetic change. The developments of DNA methylation studies keep pace with the advancements of detection technology. Over the past three decades, a large number of different methods have been applied in DNA methylation analysis. From the initial Southern blot analysis using methylation-sensitive restriction endonucleases to the current availability of microarray-based epigenomics, the technology used for DNA methylation analysis has been revolutionized [1]. Here, we discuss methods to distinguish 5-methylcytosine (5mC) from cytosine as well as methods that can distinguish 5-hydroxymethylcytosine (5hmC) from 5mC. Different methodologies available for analyzing DNA methylation are discussed, with a comparison of their relative strengths and limitations.

2.2.1 Methods to Distinguish 5-Methylcytosine from Cytosine

There are four major methods to distinguish 5-methylcytosine from cytosine. Many additional DNA methylation analysis techniques have been developed based on these primary methods (Figure 2.1).

2.2.1.1 Restriction Endonuclease-Based Analysis

2.2.1.1.1 Southern Blot

Southern blot analysis using methylation-sensitive restriction endonucleases is one of the classical and initial methods utilized in the measurement of DNA methylation in particular sequences. The two most commonly used pairs of isoschizomers are *Hpa*II-*Msp*I, which recognize CCGG, and *Sma*I-*Xma*I, which recognize CCCGGG. Neither *Hpa*II nor *Sma*I can digest methylated cytosine [2]. Although this method is relatively inexpensive and the interpretation of results is straightforward, it is limited by the availability of restriction enzyme sites in the target DNA. Other

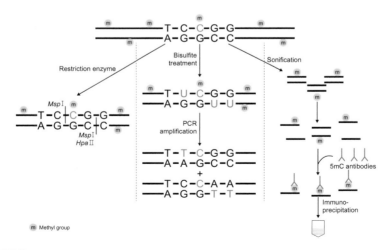

FIGURE 2.1   Principles to distinguish 5-methylcytosine from cytosine.

limitations include large amounts of high-quality DNA and problems with incomplete digestions. These disadvantages render this method time-consuming with relatively low resolution. Thus, it is not widely applicable.

2.2.1.1.2 Methylation-Sensitive Amplified Polymorphism

The methylation-sensitive amplified polymorphism (MSAP) method is based on digestion with methylation-sensitive restriction endonucleases followed by amplification of restriction fragments [3]. MSAP is a simple and relatively inexpensive genome-wide method for the identification of putative changes in DNA methylation. Unlike methods based on bisulfite modification or immunoprecipitation, MSAP is independent on the availability of genome sequence information, but the choice of the particular restriction enzymes may lead to ambiguous interpretation of MSAP data [4].

2.2.1.2 Bisulfite Conversion Technique and Derivatives

The bisulfite conversion technique is a revolutionary mark that has accelerated the study of DNA methylation. Treatment of the DNA with sodium bisulfite can convert unmethylated cytosine into uracil, while methylated cytosine remains unchanged. During the following polymerase chain reaction (PCR) process, uracil is then converted to thymidine. This chemical modification in the DNA sequence can be detected by using a variety of methods [5].

2.2.1.2.1 Bisulfite Sequencing PCR

Bisulfite sequencing PCR (BSP), which is regarded as the "gold standard" of DNA methylation analysis, is an unbiased and sensitive alternative to the use of restriction enzymes. This method combines the bisulfite treatment of genomic DNA with PCR amplification and sequencing analyses [6]. PCR products can be sequenced directly or as single clones. The latter is much more popular as it enables mapping of methylated sites at single-base-pair resolution. To acquire this high-quality data, the bisulfite-treated amplified DNA is usually cloned into bacterial cells with subsequent isolation of plasmids from numerous bacterial clones to be sequenced to determine the extent of methylation within the DNA sequence of interest; this is a process which is quite time-consuming and labor-intensive [7].

2.2.1.2.2 Pyrosequencing

Pyrosequencing is an attractive alternative to the conventional BSP. Pyrosequencing detects luminescence from the release of pyrophosphate on nucleotide incorporation into the complementary strand. Pyrosequencing studies also require the coupling of bisulfite treatment of genomic DNA with PCR amplification of the target sequence, but the advantage of pyrosequencing is that quantitative DNA methylation data can be obtained from direct sequencing of PCR products without requiring cloning into bacterial expression vectors and sequencing a large number of clones [8]. On the other hand, the quality of the data decreases with the distance of the CpG from the 3′ end of the forward primer, thus the number of bases that can be analyzed in a single sequencing reaction is limited [9].

2.2.1.2.3 Combined Bisulfite and Restriction Analysis

Bisulfite treatment of DNA can lead to the creation of new methylation-dependent restriction sites or the maintenance of restriction sites in a methylation-dependent manner. Based on this property, a quantitative method termed "combined bisulfite restriction analysis" (COBRA) was developed which merged the bisulfite and restriction analysis protocols. The use of COBRA is again limited by the availability of restriction enzyme recognition sites in the target DNA. This method is relatively labor-intensive but is cost-effective [10].

2.2.1.2.4 Methylation-Sensitive Single-Nucleotide Primer Extension and SnuPE Ion Pair Reversed-Phase High Performance Liquid Chromatography

Methylation-sensitive single-nucleotide primer extension (Ms-SNuPE) assay analyzes methylation status at individual CpG sites in a quantitative way and with the capability of multiple analyses. This method couples bisulfite treatment with strand-specific PCR which is performed to generate a DNA template. Subsequently, an internal primer that terminates immediately 5′ of the single nucleotide to be assayed is extended with a DNA polymerase that uses ^{32}P-labeled dCTP or dTTP [11]. This protocol can be carried out using multiple internal primers in a single primer-extension reaction; thus a relatively high throughput is possible. However, Ms-SNuPE assay is usually labor-intensive and requires radioactive substrates. To overcome this restriction, several variants which omitted radioactive labeling were developed, such as SNaPshot technology from Applied Biosystems (ABI) [12], SNuPE ion pair reversed-phase HPLC (SIRPH) and matrix-assisted laser desorption/ionization mass spectrometry (MALDI-MS) [13].

2.2.1.2.5 Methylation-Sensitive Melting Curve Analysis

Based on the principle that the higher GC (base pair of guanine and cytosine) content of DNA sequence makes it more resistant to melting, a new approach to DNA methylation analysis, methylation-sensitive melting curve analysis (MS-MCA), was developed. This method detects sequence difference between methylated and unmethylated DNA obtained after sodium bisulfite treatment by continuous monitoring of the change of fluorescence as a DNA duplex melts while the temperature is increased. If equal proportions of fully methylated and fully unmethylated molecules are amplified, two distinct melting peaks are observed, and interpretation is easy. If the target sequence is heterogeneously methylated, a complex melting will result in a pattern which is difficult to interpret [14].

2.2.1.2.6 Methylation-Sensitive High-Resolution Melting

The principle behind methylation-sensitive high-resolution melting (MS-HRM) is the same as for MS-MCA, but MS-HRM possesses some methodological advantages. First, the HRM approach acquires more data points so that it is more sensitive to detecting subtle differences within the amplicons. Second, the temperature variations produced with HRM instrumentation are generally extremely small. Third, the data obtained in HRM are more stable and reliable because most of the software provided with the instruments allows normalization for end-level fluorescence, temperature shifting, and use of internal oligonucleotide calibrators [14]. This technique requires the use of double-stranded DNA-binding dyes that can be used at saturating concentrations without inhibiting PCR amplification. Both MCA and HRM are semiquantitative measurements that cannot offer detailed information about the methylation of single cytosines within the sequence of interest, but they can distinguish fully and partially methylated samples, which may enable early detection of diseases [15].

2.2.1.2.7 MethyLight

MethyLight technology is a sensitive, sodium-bisulfite-dependent, fluorescence-based real-time PCR technique that quantitatively analyzes DNA methylation. Execution of MethyLight requires the designation of methylation-specific primers and fluorogenic probes [16]. The MethyLight method has major advantages. First, it is a relatively simple assay procedure, without the need to open the PCR tubes after the reaction has ended, thereby reducing the risk of contamination and the handling errors associated with manual manipulation. Second, only small amounts and modest quality of DNA template are required, making the method compatible with plasma samples and small biopsies. Third, it has the potential ability to be used as a rapid screen tool and is uniquely well suited for detection of low-frequency DNA methylation biomarkers as evidence of disease. However, the drawback of MethyLight technology is that it is not designed to offer high-resolution methylation information [17,18].

2.2.1.3 Immunoprecipitation-Based Methods

Immunoprecipitation-based methods utilize methylation-binding proteins such as MeCP2 and MBD2, or 5mC-specific antibodies to enrich the methylated fraction of the genome. Different strategies using this approach have been successfully applied for the analysis of DNA methylation information. The two most commonly used methods are methylated DNA immunoprecipitation (MeDIP) and methyl-CpG immunoprecipitation (MCIp). MeDIP is an adaptation of the chromatin immunoprecipitation (ChIP) technique and uses 5mC-specific antibodies to immunoprecipitate methylated DNA. MCIp uses a recombinant protein that contains the methyl-CpG-binding domain and the Fc fraction of the human IgG1 to directly bind and enrich methylated DNA. These methods are relatively straightforward without either digestion of genomic DNA or bisulfite treatment and the results are relatively easier to analyze and interpret. However, immunoprecipitation-based methods do not provide DNA methylation information at single-nucleotide resolution [19].

2.2.1.3.1 Methylated-CpG Island Recovery Assay

The methyl-CpG island recovery assay (MIRA) is based on the fact that methyl-CpG-binding domain protein-2 (MBD2) has the capacity to bind specifically to methylated DNA sequences and this interaction is enhanced by the methyl-CpG-binding domain protein 3-like-1 (MBD3L1) protein. DNA isolated from cells or tissue is sonicated and incubated with a matrix containing glutathione-S-transferase-MBD2b and MBD3L1. Then, specifically bound DNA is eluted from the matrix and gene-specific PCR reactions are performed to detect CpG island methylation. The MIRA procedure can detect DNA methylation using 1 ng of DNA or 3000 cells. It is quite specific, sensitive, and labor-saving [20].

2.2.1.3.2 Methyl-Binding-PCR

Methyl-binding (MB)-PCR relies on a recombinant, bivalent polypeptide with high affinity for CpG-methylated DNA. This polypeptide is coated onto the walls of a PCR vessel and can selectively capture methylated DNA fragments from a mixture of genomic DNA. Then, the degree of methylation of a specific DNA fragment is detected in the same tube by gene-specific PCR. MB-PCR is particularly useful to screen for methylation levels of candidate genes. Given the enormous amplification capability and specificity of PCR, MB-PCR provides a quick, simple, and extremely sensitive technique that can reliably detect the methylation degree of a specific genomic DNA fragment from <30 cells [21].

2.2.1.4 Mass Spectrometry-Based Methods

Mass spectrometry is recognized as an extremely useful and reliable measurement for acquiring molecular information. The principle of mass spectrometry is that a charged particle passing through a magnetic field is deflected along a circular path on a radius that varies with the mass-to-charge ratio (m/z). One adapted mass spectrometry platform for DNA methylation analysis is MassARRAY EpiTYPER, which uses base-specific enzymatic cleavage coupled to MALDI-TOF (matrix-assisted laser desorption ionization time-of-flight mass spectrometry) mass spectrometry analysis. Although the limited throughput and high cost restrict this approach in becoming a genome-wide technology, it is an excellent tool to analyze DNA methylation for its fast and accurate analysis power and its multichannel analysis capability [22].

2.2.1.4.1 MALDI-TOF Mass Spectrometry with Base-Specific Cleavage

The base-specific cleavage strategy involves amplification of bisulfite-treated DNA followed by *in vitro* transcription, and subsequent base-specific RNA cleavage by an endoribonuclease to produce different cleavage patterns. Bisulfite treatment of genomic DNA converts unmethylated cytosine into uracil and it appears as a thymidine (T) in the PCR products while the methylated cytosine remains unchanged. These C/T appear as G/A variations in the reverse strand. In the subsequent base-specific RNA cleavage reaction, methylated regions are cleaved at every C to create fragments containing at least one CpG site each. But both methylated and unmethylated regions are cleaved at every T to produce fragments in the T-cleavage reaction. G/A variations in the cleaved products generated from the reverse strand show a mass difference of 16 Da per CpG site. In MALDI-TOF analysis, the relative amount of methylated sequence can be calculated by comparing the signal intensity between the mass signals of methylated and unmethylated templates to generate quantitative results. This approach is recommended for purposes requiring the analysis of larger regions of unknown methylation content [23].

2.2.1.4.2 MALDI-TOF Mass Spectrometry with Primer Extension

The primer-extension strategy requires the designation of a primer that anneals immediately adjacent to the CpG site under investigation in a post-PCR primer-extension reaction. The primer is then extended with a mixture of four different terminators and the extension reaction will terminate on different nucleotides depending on the methylation status of the CpG site. Therefore, distinct signals are generated for MALDI-TOF mass spectrometry analysis. This approach should be used in routine analyses of a relatively small number of well-characterized informative CpG sites [14].

2.2.2 Genome-Scale DNA Methylation Analysis

Given the importance of DNA methylation, it is not surprising that many researchers have taken advantage of array- and sequencing-based technologies that have become available in recent years to perform genome-scale association studies which will provide valuable new information with high throughput and lower cost.

2.2.2.1 Microarray-Based Analysis of DNA Methylation Changes

2.2.2.1.1 Sample Preparation

There are three basic techniques applied to sample preparation in microarray platforms: digesting the DNA with methylation-sensitive or methylation-insensitive restriction endonucleases, sodium bisulfite conversion of unmethylated cytosine into uracil, and affinity purification by applying antibodies binding to methylated cytosines. Coupled with these techniques, a wide range of microarray platforms have evolved to enable genome-scale DNA methylation analysis [22].

2.2.2.1.2 Microarray Used in DNA Methylation Profiling

The initially applied microarray platform was a CpG island microarray used to identify genomic loci that exhibited differential methylation. CGI microarrays used clones from libraries in which CpG-rich fragments had been enriched by MeCP2 columns [24]. However, these arrays have low resolution and limited methylome coverage. Therefore, microarrays made of short oligonucleotides are now commercially available to overcome the drawbacks of CGI microarrays [25]. These oligonucleotide arrays, such as a promoter array, can reach a high resolution, can be easily configured according to the user's need and often contain a high density of probes spanning each CGI [26]. The first "complete" high-resolution DNA methylome profile of a living organism (*Arabidopsis thaliana*) was generated using a tiling array platform [27]. This approach involves up to several million oligonucleotides and has greater methylome coverage than promoter and CGI microarrays. It has allowed researchers to study DNA methylation in noncoding areas in addition to regulatory regions of genes [28,29]. However, to cover the entire human genome, more array slides and a relatively larger amount of genomic DNA are required. The single-nucleotide polymorphism (SNP) arrays combine the use of methylation-specific endonucleases with an SNP-ChIP. This approach can provide an integrated genetic and epigenetic profiling and allows allele-specific methylation analysis at heterozygous loci [30,31]. Besides the methods cited above, microarrays based on methyl-sensitive restriction enzymes, methylation-dependent restriction enzymes, bisulfite conversion, or immunoprecipitation are widely used in epigenomic studies [32]. These microarray-based technologies show differences in terms of resolution, coverage, and sample preparation; therefore, it is necessary to determine the advantages and disadvantages of each specific technique (Table 2.1) [33–41].

TABLE 2.1 Comparison of Microarray Assays in DNA Methylation Detection

Microarray platform	Resolution (bp)	Coverage of CpGs	Principles	Advantages	Limitations	References
SNP arrays	1	10^4	Restriction endonuclease	Identify an integrated genetic and epigenetic profiling and allow allele-specific methylation analysis at heterozygous loci	Limit to restriction enzymes digested sites	[30,31]
HELP (dual-adapter approach)	50–200	10^6	Restriction endonuclease	Positive display of hypomethylated loci	Limit to restriction enzymes digested sites and relatively low resolution	[33,34]
CHARM	50–600	Better than HELP (lack exact data)	Restriction endonuclease	Detect hypermethylated CpG sites in CpG island core and CGI "shore" regions	Limit to restriction enzymes digested sites and relatively low resolution	[35,36]
Bead array (Infinium/GoldenGate)	1	10^4	Bisulfite conversion	Low sample input and low cost		[37–39]
MeDIP	1000	10^6	Immunoprecipitation	Cost-effective and independent on a specific restriction site	Less sensitive to CpG-poor sites	[32,40]
MIRA	100	10^6	Immunoprecipitation	More sensitive to CpG-poor sites than MeDIP and do not require DNA to be denatured to single strands	Depending on MBD-binding ability	[32,41]

HELP, HpaII tiny fragment enrichment by ligation-mediated PCR; CHARM, comprehensive high-throughput arrays for relative methylation; MeDIP, methylated DNA immunoprecipitation; MIRA, methylated-CpG island recovery assay; CGI shores, stretches of ~2 kb bordering CpG islands.

2.2.2.2 *Next-Generation Sequencing Techniques*

The evolution of sequencing technologies marked by the first massively parallel DNA sequencing platforms in 2005, has revolutionized research in biological science and ushered in a new era of next-generation sequencing (NGS). There are three major NGS platforms, namely Roche/454, Illumina/Solexa, and Life Technologies/SOLiD, and each of them has different features (Table 2.2) [42−47]. Compared with microarray-based methodologies, NGS offers higher resolution and a larger coverage, and is independent of knowledge of the reference genome or genomic features. Most importantly, NGS methods allow for assessment of DNA methylation in interspersed repetitive genomic regions that are inaccessible using microarrays [22]. However, sequencing-based methods would produce a dramatically large number of bioinformatics data, which leads to extreme difficulties in downstream data management. Thus, the selection of bioinformatics software tools is particularly critical for efficient and appropriate data processing [48]. NGS techniques also include methylation-sensitive restriction enzymes (MRE-seq), affinity-based methods (such as MeDIP-seq, MBD-seq), and bisulfite conversion approaches (e.g., reduced-representation bisulfite sequencing (RRBS)). MRE-seq methods evaluate relative rather than absolute methylation levels through incorporating parallel digestions with three to five restriction endonucleases. With single CpG resolution and the ability to assay a more significant portion of the methylome including most CGIs, MRE-seq becomes a relatively simple and accurate method to analyze DNA methylation [49]. Compared with MRE or bisulfite-based sequencing, MeDIP-seq shows lower resolution, but an important advantage of MeDIP over restriction enzyme methods is that it lacks bias for a specific nucleotide sequence, other than CpGs [49]. RRBS can assess absolute quantification of methylation of more than 1 million CpG sites at single base-pair resolution, which prevails over other sequencing methods [50]. However, all the methods cited above can generate largely comparable methylation calls, but differ in CpG coverage, resolution, quantitative accuracy, efficiency, and cost [51].

2.3 TECHNIQUES USED FOR 5hmC MARK DETECTION

5′-hmC is an oxidative metabolite of 5′mC catalyzed by ten−eleven translocation dioxygenases (TET). It is widely distributed among tissues, especially in embryonic stem cells and Purkinje neurons, but depleted in cancer cell lines, which indicates that 5hmC might serve biologically important roles [52]. Traditional methods for detecting 5mC cannot differentiate 5mC from 5hmC. Methylation-sensitive restriction enzymes

TABLE 2.2 Parameters of NGS Platforms in Profiling DNA Methylome

	Sequencing	Amplification	Read length	Throughput	Advantages	Limitations	References
Roche/454	Pyrosequencing	Emulsion PCR	1000 bp	400–600 Mb	Sequences long reads	Higher costs and a relatively high error rate for calling insertions and deletions (indels) in homopolymers	[42–44]
Illumina/Solexa	Sequencing-by-synthesis with reversible terminator	Bridge PCR	35–150 bp	10–95 Gb	Ultra high-throughput and cost-effective	Short-read sequencing and GC-related bias	[42,45,47]
Life Technologies/SOLiD	Sequencing by ligation	Emulsion PCR	~50 bp	10–15 Gb	Presents the lowest error rate	Short-read sequencing and GC-related bias	[42,46,47]

are equally blocked by 5mC or 5hmC. Analogously to 5mC, 5hmC remains unconverted after bisulfite treatment [53]. The anti-5mC antibody and the methylated CpG-binding proteins (such as MBD1, MBD2b, MBD3L, MBD4, and the MBD domain of MeCP2) cannot recognize the oxidized base [54]. Therefore, exploration of special methods for 5hmC detection is necessary. To overcome the limitation of traditional bisulfite sequencing, two methods of base-resolution hydroxymethylome mapping were developed: oxidative bisulfite sequencing (oxBS-seq) and Tet-assisted bisulfite sequencing (TAB-seq) (Figure 2.2). The oxidative bisulfite sequencing strategy uses potassium perruthenate ($KRuO_4$) to oxidase 5hmC to 5fC (5-formylcytosine), which could be converted into uracil (U) in the subsequent bisulfite treatment, while 5mC remains as Cs. This allows determination of the amount of 5hmC at a particular nucleotide position by subtraction of this readout from a BS-seq one [55]. The TAB-seq protocol comprises two steps. First, a glucose moiety is introduced onto the hydroxyl group of 5hmC by using β-glucosyl-transferase (β-GT) to generate 5ghmC (β-glucosyl-5hmC), in order to prevent further oxidization by Tet1 protein in the next step. After blocking of 5hmC, all 5mC is converted to 5caC (5-carboxylcytosine) by oxidation with excess recombinant Tet1 protein. Then, in the followed bisulfite sequencing, all Cs and 5caCs (from 5mCs) read as Ts, while 5ghmCs (from 5hmCs) are sequenced as Cs. This approach allows for the detection of 5hmC at single-base resolution in both genome-wide and loci-specific studies [56,57]. For genome-wide studies of 5hmC, 5hmC-specific antibodies (hMeDIP) or chemical affinity tags to enrich

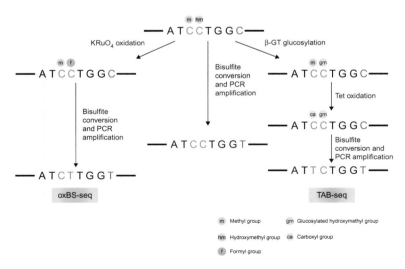

FIGURE 2.2   Methods to detect 5-hydroxymethylcytosine.

5hmC-containing DNA are applied [58−60]. These studies yielded many insightful observations, but were limited by resolution. Recently, Sun et al. reported a genome-wide high-resolution method for hydroxy-methylome study which utilizes a 5hmC-dependent restriction endonu-clease, AbaSI, that recognizes 5ghmC with high specificity when compared to 5mC and C. This AbaSI-coupled sequencing (Aba-seq) allows researchers to determine the exact 5hmC locations [61].

2.4 HISTONE MODIFICATION ANALYSIS

Modification of histone proteins is an essential component of the regulation of gene activity. Due to the extra complexity and multivari-ate nature of histone modifications, it has been suggested that the his-tone modifications may be assimilated in the form of a "histone code." Advances have been made in techniques to assess histone modifica-tions on specific residues as well as genome-wide analysis. Of the many assays used to assess the histone modification status, the most fundamental technique is ChIP. The coupling of ChIP with DNA microarray (ChIP-on-chip) and high-throughput sequencing (ChIP-seq) has significantly increased the scope of histone modification analysis.

2.4.1 Chromatin Immunoprecipitation

The ChIP assay is a powerful and versatile technique used for prob-ing interactions between specific proteins or modified forms of proteins and a genomic DNA region. The first use of the ChIP assay was by Gilmour and colleagues in 1984 to monitor the association of RNA poly-merase II with transcribed and poised genes in *Drosophila* cells. Now, this assay is widely used to monitor the presence of particular histone modifications at specific genomic locations. In addition, the ChIP assay can be used to analyze binding of transcription factors, DNA replication factors, and DNA repair proteins [62,63].

When performing the ChIP assay, the initial step is the cross-linking of protein−DNA in live cells with formaldehyde. After cross-linking, cells are lysed and chromatin is harvested and fragmented using either sonication or enzymatic digestion. Proteins together with cross-linked DNA are subsequently subjected to immunoprecipitation using antibo-dies specific to a particular protein or histone modification. Any DNA fragments that are connected with the protein or histone modification of interest will co-precipitate as part of the cross-linked protein−DNA complexes. After immunoprecipitation, the protein−DNA cross-links

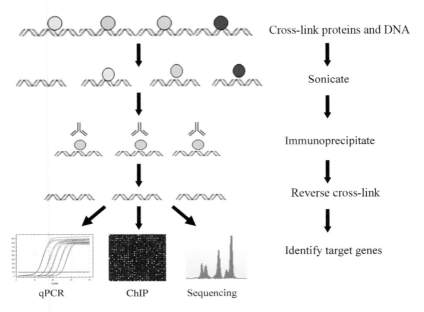

FIGURE 2.3   Overview of chip experiment.

are reversed and the DNA is purified. The enrichment of a particular DNA sequence can then be detected by agarose gel electrophoresis or more commonly by quantitative PCR (qPCR). Alternatively, the ChIP assay can be combined with genomic tiling microarray (ChIP-on-chip) techniques, sequencing (ChIP-seq), or cloning strategies, which allow for genome-wide mapping of protein–DNA interactions and histone modifications (Figure 2.3).

2.4.2 ChIP-on-Chip

ChIP-on-chip is a technique that combines ChIP with microarray technology. In ChIP-on-chip, immunoprecipitated material is labeled with fluorescent dyes and hybridized to DNA microarrays containing several hundred thousand to several million probes [64]. In the first step, ChIP is performed on cross-linked chromatin as described above. After purification and amplification of the DNA, the samples are labeled with a fluorescent tag such as Cy5 or Alexa 647. Labeled IP and input samples are hybridized onto the DNA microarray at 67°C for 24 h. Whenever a labeled fragment finds a complementary fragment on the array, they will hybridize and form a double-stranded DNA fragment. Using the ChIP-on-chip technique, global patterns of different histone modifications can be observed.

The biological significance may be derived from identifying genomic regions where ChIP-on-chip probes show significant enrichment. Often normalized signals and P values are employed for each degree of genomic probe-enrichment. As the microarray has unbiased, high-throughput capabilities, the ChIP-on-chip technique has several significant advantages. First, it allows the identification of histone modification and permits discovery of unanticipated sites of protein-binding DNA on a genome-wide basis instead of a limited number of loci selected by researchers. Second, the ChIP-on-chip technique often has an optimized commercially available platform, and therefore it is time-saving compared with large-scale quantitative PCR assays. Third, the parallel analysis of thousands of genes allows one to parse the data into distinct classes of genes based on different binding distributions or behaviors, and permits statistical comparisons between classes.

2.4.3 ChIP-seq

Owing to the rapid progress in NGS technology, ChIP followed by sequencing (ChIP-seq) can be used to accurately survey interactions between protein, DNA, and RNA.

2.4.3.1 Workflow of ChIP-seq

ChIP-seq typically begins with the process of ChIP and yields several to a few hundred nanograms of DNA as 75- to 300-bp fragments surrounding histone mark locations. High-throughput sequencing is then used to read the ChIP-DNA fragments on a genome-wide scale. The key steps of ChIP-seq are summarized as follows: cross-link protein and shear DNA; add protein-specific antibody; immunoprecipitate and purify complexes; reverse cross-links, purify DNA and prepare for sequencing; sequence DNA fragment and map to genome.

2.4.3.2 Analysis Pipeline of ChIP-seq Data

A number of NGS technology platforms have been developed that are suitable for ChIP-seq, including Illumina/Solexa, ABI/SOLiD, 454/Roche, and Helicos. On the Illumina/Solexa Genome Analyzer, clusters of clonal sequences are generated by bridge PCR, and sequencing is performed by synthesis of single-molecule arrays with reversible terminators. Illumina/Solexa is capable of generating 35-bp reads and producing at least 1 Gb of sequence per run in 2—3 days. The 454/Roche platform is capable of generating 80—120 Mb of sequence in 200- to 300-bp reads in a 4-h run. The technology used in 454/Roche platform is based on emulsion PCR and pyrosequencing in high-density picoliter reactors. The approach of ABI SOLiD platform is massively parallel sequencing by hybridization—ligation. ABI/SOLiD is capable of producing 1—3 Gb of sequence data in 35-bp reads per an 8-day run. On a single-molecule sequencing platform such as Helicos, fluorescent nucleotides incorporated into templates can be imaged at the level of single molecules, which makes clonal amplification unnecessary [65,66].

The Illumina/Solexa and the ABI/SOLiD sequencers produce shorter reads but provide tens of millions of reads per sample lane, whereas the 454/Roche sequencer allows for longer yet fewer sequencing reads per run [67]. However, most of the ChIP-seq data are generated through the Illumina Genome Analyzer. The typical workflows associated with the analysis pipeline of ChIP-seq data include read aligner, peak calling, and motif finding.

2.4.3.2.1 Read Aligner

The short DNA sequence reads are first aligned to the genome using alignment algorithms. DNA sequence reads that are uniquely aligned to a genome can be viewed as a custom track in the University of California Santa Cruz (UCSC) genome browser. Due to the large number of reads, the conventional alignment algorithms are time-consuming. A new generation of aligners include Eland, Maq, and Bowtie. Eland, a fast and efficient aligner for short reads, was developed by Illumina. Mapping and Assembly with Qualities (Maq), a widely used aligner, is based on a straightforward but effective strategy called spaced seed indexing. Maq has excellent capabilities for detecting SNPs. Bowtie is an extremely fast mapper based on an algorithm which was originally developed for file compression.

2.4.3.2.2 Peak Calling

Peak calling, using data from the ChIP profile and a control profile (which is usually created from input DNA), generates a list of enriched regions that are ordered by false discovery rate as a statistical measure. The analysis of ChIP-seq data critically depends on this step and many statistical and computational methods have been developed to support the analysis of the massive data sets from these experiments. There are softwares available for the large quantities of data generated by ChIP-seq, including MACS and Avadis [68]. Model-based Analysis of ChIP-seq (MACS) is an open-source solution available from http://liulab.dfci.harvard.edu/MACS Galaxy. MACS consists of four steps: removing redundant reads, adjusting the read position based on fragment size distribution, calculating peak enrichment using local background normalization, and estimating the empirical false discovery rate by exchanging ChIP-seq and control samples.

2.4.3.2.3 Motif Finding

Motifs are conserved regions in multiple sequences and they often correspond to structurally or functionally important residues. Such information is useful for classifying diverse sequences. The motif discovery algorithm programs are employed to look for motifs with statistical significance. The popular motif-finding algorithm programs include MEME, MDScan, Weeder, and WebMOTIFS. MEME works by searching for repeated, ungapped motifs that occur in the DNA or protein sequences provided by the user and it is not suited to whole genome transcription factor binding site (TFBS) motif discovery. MDScan discovers motifs efficiently by the combination of word enumeration motif search strategies and position-specific weight matrix updating motif search strategies. Weeder is user-friendly and requires very little prior knowledge about the motifs to be found. WebMOTIFS provides a web interface and identifies biologically relevant motifs from diverse data in several species [69].

2.4.3.3 Advantages of ChIP-seq

Different methods often have complementary strengths, and the choice of which method to use depends on the nature of the search being conducted. Therefore, a thorough understanding of the technological variation between ChIP-on-chip and ChIP-seq is of great importance. Compared to microarray analysis (ChIP-on-chip), ChIP-seq has lots of advantages (Table 2.3). ChIP-seq requires no hybridization probes and therefore generally produces profiles with greater spatial resolution, dynamic range, and genomic coverage. In addition, any species can be studied with ChIP-seq since it is not constrained by the availability of a species-specific microarray. Furthermore, ChIP-seq can reveal binding events located in repetitive regions in the mammalian genome. Another advantage of ChIP-seq over ChIP-on-chip is the smaller amount of sample material needed. All of these advantages result in a higher sensitivity and specificity for ChIP-seq technology [70].

TABLE 2.3   Comparison of ChIP-on-Chip and ChIP-seq

	ChIP-seq	ChIP-on-chip
Sample quantity	10−50 ng DNA	4−5 µg DNA
Coverage	Genome-wide coverage	Limited by the repertoire of probe sequence fixed on the array
Resolution	Single base-pair resolution	30−100 bp resolution, which is limited by probe spacing
Sensitivity	More sensitive	Less sensitive
Mapping	Mapping reads	Mapping signals
Platforms	454 Life Science, Illumina, SOLiD system	Affymetrix, Agilent, NimbleGen

2.4.4 Challenges for Histone Modification Analysis

Researchers have identified many histone modifications and several medium- and large-scale epigenomic efforts have already been initiated. These practices will certainly provide new insights for different biological processes such as basic gene regulatory processes, cellular differentiation, reprogramming, and the role of epigenetic regulation in aging and disease development. Over the last decade, epigenetic research has seen a shift from site-specific studies to a genome-wide assessment. However, the histone code is just beginning to be uncovered and a number of questions are yet to be fully answered.

First, the time and temperature at which the DNA-binding protein is cross-linked to DNA with formaldehyde must be optimized. If cross-linking of proteins and DNA is too severe, it is more difficult to efficiently shear the chromatin into small fragments and to reverse the cross-links before sequencing. Fragments of ~100−300 bp range are generally required for NGS. A good starting point for further optimization is a 10-min incubation of 1% formaldehyde at 37°C. However, in ChIP experiments, micrococcal nuclease (mNase) digestion without cross-linking is most often used to fragment the chromatin.

Second, both ChIP-on-chip and ChIP-seq experiments demand an antibody that recognizes the protein of interest with high affinity, high specificity, and low background. It is a good practice to employ ChIP-validated antibodies. In order to test the antibody specificity, Western blots or immunofluorescence staining can be performed. If only one band can be detected at the specific sites or only the nuclei are stained, the antibody may be considered to be specific. In addition, in order to get comprehensive epigenomic maps, it is necessary to assay every modification. However, this is also limited by the lack of specific antibodies. An alternative strategy is suggested: to identify a set of key histone modifications such as H3K4me1, H3K4me3, H3K9ac, H3K9me3, H3K27me3, and H3K36me3 (Figure 2.4).

Third, the volume of data generated by a large-scale epigenomics project is great and substantial challenges exist in efficient data access and processing. For example, epigenomic maps from different cell types need to be compared; integrative analysis that integrates individual data sets—including genomic, epigenomic, transcriptomic, and proteomic information—has become an essential part of determining how errors lead to disease; quality control and statistically correct normalization measures must be performed followed by sequencing. It is expected that this comparative analysis will lead to the first epigenome-wide association studies.

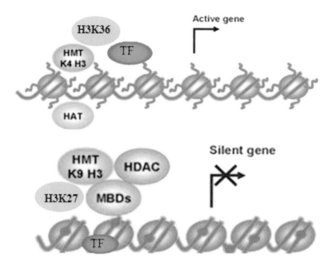

FIGURE 2.4   Histone modifications and gene expression.

2.5 miRNA ANALYSIS

miRNAs are small noncoding RNAs that derive from different transcription units. miRNAs act as key regulators of development, cell proliferation, differentiation, and the cell cycle. This regulation is exerted by base-pairing to the target mRNA, which directs translational repression or posttranscriptional silencing. Significant progress has been made in miRNA research since the first discovery of lin-4 RNA in 1993 [71]. The classic strategies for miRNA research are as follows (Figure 2.5).

The integral first step in miRNA research is detection of the miRNA by microarray analysis and deep sequencing. Northern blot and quantitative reverse transcriptase-PCR (RT-PCR) analysis on the original starting material can serve to verify the microarray data.

Each miRNA is believed to regulate multiple genes. miRNA targets can be obtained through a number of freely available programs such as TargetScan, PicTar, DIANA-microT, miRBase, and so on. Lastly, it is important to confirm these predictions using miRNA target validation techniques. Usually, synthetic miRNA inhibitors/mimics and luciferase reporter vectors are employed in this step [72].

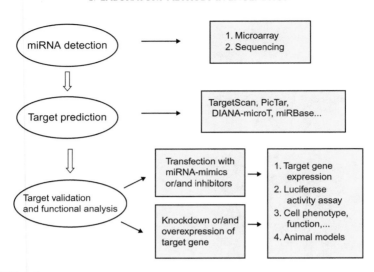

FIGURE 2.5   General analysis pipeline for miRNA research.

2.5.1 miRNA Detection

miRNA expression levels can be studied using several different methods, including real-time PCR, Northern blots, Microarray analysis, and NGS [73].

2.5.1.1 Microarray

miRNA microarray is a tool based on nucleic acid hybridization to explore the expression profiling of miRNAs. The ready-to-use miRNA microarray consists of glass slides immobilized with 5′ amine-modified oligonucleotide probes which are antisense to miRNAs. The isolated miRNAs are labeled with fluorescent dye and then hybridized with the miRNA microarray. The biotinylated miRNAs are then captured on the microarray at different positions by oligonucleotide probes in hybridization. Consequently, the specific miRNAs and their relative quantities can be evaluated by analyzing the fluorescence signal data [74].

The small size of miRNAs poses difficulties using the above methodology. New microarray platforms based on locked nucleic acid (LNA)-modified, Tm-normalized capture probes spotted onto N-hydroxy-succinamide (NHS)-coated glass slides have been successfully introduced into miRNA profiling microarray detection.

2.5.1.2 Next-Generation Sequencing

NGS technology has been employed to survey the expression of miRNAs over the past few years. Profiling of miRNAs by NGS measures allows for the discovery of novel miRNAs or other small RNA species. Analysis of profiling data from deep sequencing can be carried out using publicly available tools, such as miRDeep, CD-miRNA, MiRank, miRCat, and others. One challenge will be to effectively integrate these various types of data with each other, in order to get useful information for further research [75,76].

2.5.1.3 RT-PCR

Another popular method to quantitate miRNA expression is RT-PCR, which can detect miRNA in real time. In this method, an artificial tail is added to miRNA and then reverse transcribed by using a universal primer. The synthesized cDNA is then used as a template for qPCR with one miRNA-specific primer and a second universal primer. To monitor the miRNA expression in real time, different approaches can be used. The basic principle is the detection of a fluorescent reporter molecule whose signal intensity correlates with the amount of DNA present in each cycle of amplification. For example, a fluorescent probe such as SYBR Green I, which intercalates double-stranded DNA, can be added. Also, a dual-labeled probe, containing a fluorescent reporter and quencher upon adjacent nucleotides can be added, which is cleaved by polymerase.

2.5.1.4 Northern Blot Analysis

Northern blot analysis is a widely used method to assess accumulation levels of miRNAs of interest. It is usually a readily available method for laboratories and does not require special equipment and technical knowledge. The sample containing miRNA is run on agarose gel electrophoresis. Next, the miRNA is transferred to a positively charged nylon membrane, followed by soaking in a hybridization solution containing a fluorescent or radiolabeled oligonucleotide probe complementary to part of or the entire target sequence. After unhybridized probe has been removed by washing in several changes of buffer, the miRNA target can be detected. However, the major drawback of this technique is its low sensitivity when using the traditional DNA oligonucleotide probes. It is not feasible when monitoring expression of low-abundant miRNAs. In order to solve the sensitivity problem, LNA-modified oligonucleotide probes have been used. It has significantly improved sensitivity and high specificity of miRNAs detection [77].

2.5.1.5 Others

Other miRNA analysis methods include electrochemical detection in which the miRNA is directly labeled with a redox active and electrocatalytic moiety; fluorescence correlation spectroscopy methodology in which two organic fluorophore-labeled oligonucleotides are added to miRNA targets; and surface-enhanced Raman scattering (SERS) platform based on oblique angle deposition (OAD)-fabricated silver nanorod arrays which can be used to classify miRNA patterns with high accuracy.

Most of the miRNA detection methods are based on hybridization and require a label for detection. However, not all of the methods are perfect. Therefore, there is a great need for the development of rapid and sensitive miRNA profiling methods for detection and identification of miRNAs.

2.5.2 Target Prediction

It has been further confirmed that many 3′ untranslated region (3′ UTR) elements that mediate mRNA decay and translational repression have predicted targets for the 5′ region of different miRNAs. Based on this principle, a number of prediction softwares including TargetScan, PicTar, DIANA-microT, and so on have been designed and are available on the Web (Table 2.4). A large class of miRNA targets can be confidently detected. However, these systems have a high degree of overlap because they now all require stringent seed pairing and they are not 100% identical.

2.5.2.1 Target Scan

The TargetScan software is available for download at http://genes. mit.edu/targetscan. TargetScan predicts biological targets of miRNAs by identifying mRNAs with conserved complementarity to the 5′ region of the miRNA, known as the miRNA seed [78]. As an option, targeting can also be detected in nonconserved sites. Less than 2% identified are

TABLE 2.4   Tools for miRNA Target Prediction

Tools	Prediction criteria	Ranking criteria	Website
TargetScan	Stringent miRNA seed pairing	Site efficacy scores	http://genes.mit.edu/targetscan
PicTar	Search for nearly but not fully complementary regions of seeds	Overall predicted pairing stability	http://pictar.mdc-berlin.de
DIANA-MicroT	Single site prediction	Weighted sum of conserved and unconserved miRNA recognition elements	http://www.microrna.gr/microT
miRBase	Moderately stringent seed pairing	Overall pairing	http://microrna.sanger.ac.uk

sites with mismatches in the seed region that are compensated by the 3' end of miRNAs. In mammals, predictions are ranked based on the site efficacy scores [79].

2.5.2.2 PicTar

PicTar is an algorithm for the identification of miRNA targets and the online service is available at http://pictar.mdc-berlin.de. First, PicTar searches for nearly, but not fully, complementary regions of conservative 3' UTRs. Then PicTar uses a maximum likelihood method to compute the likelihood that sequences are miRNA target sites when comparing to the 3' UTR background. A hidden Markov model is used to score the 3' UTR that has at least one aligned conserved predicted binding site for an miRNA.

2.5.2.3 DIANA-microT

DIANA-microT searches sites with canonical central bulge and it is mainly constructed for single site prediction. The comprehensive predicted score of a miRNA—target gene interaction is the weighted sum of conserved and unconserved miRNA recognition elements (MREs) of a gene. It provides a unique signal-to-noise ratio (SNR) for the evaluation of its specificity. SNR is defined as the average SNR between a total of predicted targets by a real miRNA in 3' UTR and a total of predicted targets by mock miRNA with randomized sequence in searched 3' UTR. DIANA-microT-coding sequences (CDS), also known as DIANA-microT web server v5.0, is the updated version of the microT algorithm. The new DIANA-microT web server is freely available at http://www.microrna.gr/webServer. It is specifically trained on a set of miRNA recognition elements (MREs) located in both the 3' UTR and CDS. The new DIANA-microT web server increases the target prediction accuracy and usability of the server interface, helping users to perform advanced multistep functional miRNA analyses [80].

2.5.2.4 Others

Beyond the above algorithms, miRBase, RNA22, and PITA have also been successfully used to predict miRNA targets in mammals [81]. Unlike TargetScan, PicTar, DIANA-microT, and miRBase, the algorithms RNA22 and PITA do not rely on cross-species conservation. PITA predicts miRNA targets using a parameter-free model that is based on target-site accessibility and minimum free energy. RNA22 uses a pattern-based approach to identify miRNA-binding sites and their corresponding heteroduplexes. RNA22 can be applied to find putative miRNA-binding sites without requiring the identification of the targeting miRNA. This indicates that RNA22 can recognize binding sites even if the targeting miRNA is not among those currently known [82].

However, it is important to keep in mind that even the most recent miRNA target prediction may also produce a huge number of false-positive and an unknown number of false-negative results. To extract high confidence targets, it is better to retrieve results of several algorithms.

2.5.3 Target Validation and Functional Analysis

2.5.3.1 Luciferase Reporter Assays

Several studies have attempted to further investigate the conservation of predicted miRNA—mRNA regulatory relationships [83]. The luciferase assay is the biochemical method most widely used to identify miRNA targets. A fragment sequence from the 3′ UTR of the target gene is cloned into the 3′ UTR of the luciferase gene contained in a plasmid (test plasmid). A control plasmid is then constructed by generating reporter constructs with mutations in the 3′ UTR of the target gene. The test plasmid or control plasmid is then transfected into a cell line with or without the miRNA overexpression vector, and luciferase assays are carried out in parallel. Firefly luciferase activity is measured using the Dual-Luciferase reporter assay system and luminometer. When the candidate is an authentic target, luciferase activity is lower in miRNA over-expressing cells containing the test plasmid compared with those containing the control plasmid.

2.5.3.2 Gain-of-Function and Loss-of-Function Experiments

Specific miRNA function can be explored by up- and downregulating specific miRNA levels. Gain-of-function experiments are performed by transfecting a plasmid containing a constitutive promoter (e.g., cytomegalovirus (CMV)) to overexpress a pri-miRNA or a pre-miRNA sequence. Viral vectors can also be used, or the pre-miRNA itself can be transfected. Usually, the associated companies offer the pre-miRNA precursor molecule, a miRNA mimic that is chemically synthesized as a modified double-stranded oligonucleotide [84]. At the same time, miRNA functional analysis can also be examined by using synthetic miRNA inhibitors.

2.6 CONCLUSION

The identification of epigenetic modifications and the understanding of their roles in the regulation of gene expression ushered in a new era for the field of life science research. The emerging picture of epigenetic regulation in humans is far more complicated than previously imagined. However, the technological advances have revolutionized our understanding and appreciation of the importance of epigenetic changes in phenotypes and in the etiology of diseases. This chapter outlines the available methods used for DNA methylation detection, histone modification analysis, and miRNA profiling and target validation, and discusses the advantages and drawbacks of each technology. Specifically, NGS technologies are of great importance and will be widely used in epigenetic studies, especially in the analysis of DNA methylation that cannot be identified using regular DNA sequencing which does not enrich the methylated genomic compartment. Ultimately, with the increasing interest in epigenetic mechanisms, further developments in methodology are certain to emerge in the coming years and many more related findings will come to light in the near future to provide valuable new insights into the developmental basis of diseases.

List of Abbreviations

β-GT	β-glucosyltransferase
5caC	5-carboxylcytosine
5fC	5-formylcytosine
5ghmC	β-glucosyl-5hmC
5hmC	5-hydroxymethylcytosine
5mC	5-methylcytosine
BSP	bisulfite sequencing PCR
CGI	CpG island
ChIP	chromatin immunoprecipitation
COBRA	combined bisulfite and restriction analysis
MACS	Model-Based Analysis of ChIP-seq
MALDI-TOF	matrix-assisted laser desorption ionization time-of-flight mass spectrometry
Maq	Mapping and Assembly with Qualities
MBD	methyl-CpG-binding domain protein
MCIp	methyl-CpG immunoprecipitation
MeDIP	methylated DNA immunoprecipitation
MIRA	methylated-CpG island recovery assay
miRNA	microRNA
mNase	micrococcal nuclease
MRE	miRNA recognition element
MRE-seq	methylation-sensitive restriction enzymes
MSAP	methylation-sensitive amplified polymorphism
MS-HRM	methylation-sensitive high-resolution melting
MS-MCA	methylation-sensitive melting curve analysis
Ms-SNuPE	methylation-sensitive single-nucleotide primer extension
NGS	next-generation sequencing
NHS	*N*-hydroxysuccinamide
oxBS-seq	oxidative bisulfite sequencing
PCR	polymerase chain reaction
SERS	surface-enhanced Raman scattering
SIRPH	SnuPE ion pair reversed-phase HPLC

SNP	single-nucleotide polymorphism
SNR	signal-to-noise ratio
TAB-seq	Tet-assisted bisulfite sequencing
Tet	ten−eleven translocation dioxygenases

References

[1] Shames DS, Minna JD, Gazdar AF. Methods for detecting DNA methylation in tumors: from bench to bedside. Cancer Lett 2007;251(2):187−98.

[2] Yaish MW, Peng M, Rothstein SJ. Global DNA methylation analysis using methyl-sensitive amplification polymorphism (MSAP). Methods Mol Biol 2014;1062:285−98.

[3] Albertini E, Marconi G. Methylation-sensitive amplified polymorphism (MSAP) marker to investigate drought-stress response in Montepulciano and Sangiovese grape cultivars. Methods Mol Biol 2014;1112:151−64.

[4] Fulnecek J, Kovarik A. How to interpret methylation sensitive amplified polymorphism (MSAP) profiles? BMC Genet 2014;15(1):2.

[5] Sasaki M, Anast J, Bassett W, Kawakami T, Sakuragi N, Dahiya R. Bisulfite conversion-specific and methylation-specific PCR: a sensitive technique for accurate evaluation of CpG methylation. Biochem Biophys Res Commun 2003;309(2):305−9.

[6] Frommer M, McDonald LE, Millar DS, et al. A genomic sequencing protocol that yields a positive display of 5-methylcytosine residues in individual DNA strands. Proc Natl Acad Sci USA 1992;89(5):1827−31.

[7] Fraga MF, Esteller M. DNA methylation: a profile of methods and applications. BioTechniques 2002;33(3): 632, 4, 6−49.

[8] Reed K, Poulin ML, Yan L, Parissenti AM. Comparison of bisulfite sequencing PCR with pyrosequencing for measuring differences in DNA methylation. Anal Biochem 2010;397(1):96−106.

[9] Dupont JM, Tost J, Jammes H, Gut IG. De novo quantitative bisulfite sequencing using the pyrosequencing technology. Anal Biochem 2004;333(1):119−27.

[10] Xiong Z, Laird PW. COBRA: a sensitive and quantitative DNA methylation assay. Nucleic Acids Res 1997;25(12):2532−4.

[11] Gonzalgo ML, Liang G. Methylation-sensitive single-nucleotide primer extension (Ms-SNuPE) for quantitative measurement of DNA methylation. Nat Protoc 2007;2(8):1931−6.

[12] Kaminsky Z, Petronis A. Methylation SNaPshot: a method for the quantification of site-specific DNA methylation levels. Methods Mol Biol 2009;507:241−55.

[13] Hu J, Zhang CY. Single base extension reaction-based surface enhanced Raman spectroscopy for DNA methylation assay. Biosens Bioelectron 2012;31(1):451−7.

[14] Kristensen LS, Hansen LL. PCR-based methods for detecting single-locus DNA methylation biomarkers in cancer diagnostics, prognostics, and response to treatment. Clin Chem 2009;55(8):1471−83.

[15] Zhang C, Shao Y, Zhang W, et al. High-resolution melting analysis of ADAMTS9 methylation levels in gastric, colorectal, and pancreatic cancers. Cancer Genet Cytogenet 2010;196(1):38−44.

[16] He Q, Chen HY, Bai EQ, et al. Development of a multiplex MethyLight assay for the detection of multigene methylation in human colorectal cancer. Cancer Genet Cytogenet 2010;202(1):1−10.

[17] Trinh BN, Long TI, Laird PW. DNA methylation analysis by MethyLight technology. Methods 2001;25(4):456−62.

[18] Campan M, Weisenberger DJ, Trinh B, Laird PW. MethyLight. Methods Mol Biol 2009;507:325−37.

[19] Down TA, Rakyan VK, Turner DJ, et al. A Bayesian deconvolution strategy for immunoprecipitation-based DNA methylome analysis. Nat Biotechnol 2008;26(7):779—85.

[20] Rauch T, Pfeifer GP. Methylated-CpG island recovery assay: a new technique for the rapid detection of methylated-CpG islands in cancer. Lab Invest 2005;85(9):1172—80.

[21] Gebhard C, Schwarzfischer L, Pham TH, Andreesen R, Mackensen A, Rehli M. Rapid and sensitive detection of CpG-methylation using methyl-binding (MB)-PCR. Nucleic Acids Res 2006;34(11):e82.

[22] Gupta R, Nagarajan A, Wajapeyee N. Advances in genome-wide DNA methylation analysis. Biotechniques 2010;49(4):iii—xi.

[23] van den Boom D, Ehrich M. Mass spectrometric analysis of cytosine methylation by base-specific cleavage and primer extension methods. Methods Mol Biol 2009;507:207—27.

[24] Yan PS, Chen CM, Shi H, Rahmatpanah F, Wei SH, Huang TH. Applications of CpG island microarrays for high-throughput analysis of DNA methylation. J Nutr 2002;132(8 Suppl.):2430—4S.

[25] Shi H, Maier S, Nimmrich I, et al. Oligonucleotide-based microarray for DNA methylation analysis: principles and applications. J Cell Biochem 2003;88(1):138—43.

[26] Mund C, Beier V, Bewerunge P, Dahms M, Lyko F, Hoheisel JD. Array-based analysis of genomic DNA methylation patterns of the tumour suppressor gene p16INK4A promoter in colon carcinoma cell lines. Nucleic Acids Res 2005;33(8):e73.

[27] Zhang X, Yazaki J, Sundaresan A, et al. Genome-wide high-resolution mapping and functional analysis of DNA methylation in *Arabidopsis*. Cell 2006;126(6):1189—201.

[28] Zilberman D, Gehring M, Tran RK, Ballinger T, Henikoff S. Genome-wide analysis of *Arabidopsis thaliana* DNA methylation uncovers an interdependence between methylation and transcription. Nat Genet 2007;39(1):61—9.

[29] Schob H, Grossniklaus U. The first high-resolution DNA "methylome." Cell 2006;126 (6):1025—8.

[30] Yuan E, Haghighi F, White S, et al. A single nucleotide polymorphism chip-based method for combined genetic and epigenetic profiling: validation in decitabine therapy and tumor/normal comparisons. Cancer Res 2006;66(7):3443—51.

[31] Kerkel K, Spadola A, Yuan E, et al. Genomic surveys by methylation-sensitive SNP analysis identify sequence-dependent allele-specific DNA methylation. Nat Genet 2008;40(7):904—8.

[32] Fouse SD, Nagarajan RO, Costello JF. Genome-scale DNA methylation analysis. Epigenomics 2010;2(1):105—17.

[33] Oda M, Greally JM. The HELP assay. Methods Mol Biol 2009;507:77—87.

[34] Oda M, Glass JL, Thompson RF, et al. High-resolution genome-wide cytosine methylation profiling with simultaneous copy number analysis and optimization for limited cell numbers. Nucleic Acids Res 2009;37(12):3829—39.

[35] Irizarry RA, Ladd-Acosta C, Carvalho B, et al. Comprehensive high-throughput arrays for relative methylation (CHARM). Genome Res 2008;18(5):780—90.

[36] Irizarry RA, Ladd-Acosta C, Wen B, et al. The human colon cancer methylome shows similar hypo- and hypermethylation at conserved tissue-specific CpG island shores. Nat Genet 2009;41(2):178—86.

[37] Bibikova M, Fan JB. GoldenGate assay for DNA methylation profiling. Methods Mol Biol 2009;507:149—63.

[38] Bibikova M, Lin Z, Zhou L, et al. High-throughput DNA methylation profiling using universal bead arrays. Genome Res 2006;16(3):383—93.

[39] Huang YW, Huang TH, Wang LS. Profiling DNA methylomes from microarray to genome-scale sequencing. Technol Cancer Res Treat 2010;9(2):139—47.

[40] Pelizzola M, Koga Y, Urban AE, et al. MEDME: an experimental and analytical methodology for the estimation of DNA methylation levels based on microarray derived MeDIP-enrichment. Genome Res 2008;18(10):1652—9.

[41] Rauch TA, Wu X, Zhong X, Riggs AD, Pfeifer GP. A human B cell methylome at 100-base pair resolution. Proc Natl Acad Sci USA 2009;106(3):671–8.

[42] Xuan J, Yu Y, Qing T, Guo L, Shi L. Next-generation sequencing in the clinic: promises and challenges. Cancer Lett 2013;340(2):284–95.

[43] Wommack KE, Bhavsar J, Ravel J. Metagenomics: read length matters. Appl Environ Microbiol 2008;74(5):1453–63.

[44] Gilles A, Meglecz E, Pech N, Ferreira S, Malausa T, Martin JF. Accuracy and quality assessment of 454 GS-FLX Titanium pyrosequencing. BMC Genomics 2011;12:245.

[45] Metzker ML. Sequencing technologies—the next generation. Nat Rev Genet 2010;11 (1):31–46.

[46] Harismendy O, Ng PC, Strausberg RL, et al. Evaluation of next generation sequencing platforms for population targeted sequencing studies. Genome Biol 2009;10(3):R32.

[47] Ratan A, Miller W, Guillory J, Stinson J, Seshagiri S, Schuster SC. Comparison of sequencing platforms for single nucleotide variant calls in a human sample. PLOS ONE 2013;8(2):e55089.

[48] Huss M. Introduction into the analysis of high-throughput-sequencing based epigenome data. Brief Bioinform 2010;11(5):512–23.

[49] Stevens M, Cheng JB, Li D, et al. Estimating absolute methylation levels at single-CpG resolution from methylation enrichment and restriction enzyme sequencing methods. Genome Res 2013;23(9):1541–53.

[50] Wang J, Xia Y, Li L, et al. Double restriction-enzyme digestion improves the coverage and accuracy of genome-wide CpG methylation profiling by reduced representation bisulfite sequencing. BMC Genomics 2013;14:11.

[51] Harris RA, Wang T, Coarfa C, et al. Comparison of sequencing-based methods to profile DNA methylation and identification of monoallelic epigenetic modifications. Nat Biotechnol 2010;28(10):1097–105.

[52] Li Y, O'Neill C. 5'-Methylcytosine and 5'-hydroxymethylcytosine each provide epigenetic information to the mouse zygote. PLOS ONE 2013;8(5):e63689.

[53] Lee DH, Tran DA, Singh P, et al. MIRA-SNuPE, a quantitative, multiplex method for measuring allele-specific DNA methylation. Epigenetics 2011;6(2):212–23.

[54] Jin SG, Kadam S, Pfeifer GP. Examination of the specificity of DNA methylation profiling techniques towards 5-methylcytosine and 5-hydroxymethylcytosine. Nucleic Acids Res 2010;38(11):e125.

[55] Booth MJ, Branco MR, Ficz G, et al. Quantitative sequencing of 5-methylcytosine and 5-hydroxymethylcytosine at single-base resolution. Science 2012;336(6083):934–7.

[56] Yu M, Hon GC, Szulwach KE, et al. Base-resolution analysis of 5-hydroxymethylcytosine in the mammalian genome. Cell 2012;149(6):1368–80.

[57] Yu M, Hon GC, Szulwach KE, et al. Tet-assisted bisulfite sequencing of 5-hydroxymethylcytosine. Nat Protoc 2012;7(12):2159–70.

[58] Sui WG, Tan QP, Yan Q, et al. Genome-wide analysis of DNA 5-hmC in peripheral blood of uremia by hMeDIP-chip. Ren Fail 2014;36(6):937–45.

[59] Pastor WA, Pape UJ, Huang Y, et al. Genome-wide mapping of 5-hydroxymethylcytosine in embryonic stem cells. Nature 2011;473(7347):394–7.

[60] Song CX, Szulwach KE, Fu Y, et al. Selective chemical labeling reveals the genome-wide distribution of 5-hydroxymethylcytosine. Nat Biotechnol 2011;29(1):68–72.

[61] Sun Z, Terragni J, Borgaro JG, et al. High-resolution enzymatic mapping of genomic 5-hydroxymethylcytosine in mouse embryonic stem cells. Cell Rep 2013;3(2):567–76.

[62] Radpour R, Haghighi MM, Fan AX, et al. High-throughput hacking of the methylation patterns in breast cancer by *in vitro* transcription and thymidine-specific cleavage mass array on MALDI-TOF silico-chip. Mol Cancer Res 2008;6(11):1702–9.

[63] Liu XS, Brutlag DL, Liu JS. An algorithm for finding protein–DNA binding sites with applications to chromatin-immunoprecipitation microarray experiments. Nat Biotechnol 2002;20(8):835–9.

[64] Johnson DS, Li W, Gordon DB, et al. Systematic evaluation of variability in ChIP-chip experiments using predefined DNA targets. Genome Res 2008;18(3):393–403.

[65] Schadt EE, Linderman MD, Sorenson J, Lee L, Nolan GP. Computational solutions to large-scale data management and analysis. Nat Rev Genet 2010;11(9):647–57.

[66] Marguerat S, Wilhelm BT, Bahler J. Next-generation sequencing: applications beyond genomes. Biochem Soc Trans 2008;36(Pt 5):1091–6.

[67] Choudhuri S. From Waddington's epigenetic landscape to small noncoding RNA: some important milestones in the history of epigenetics research. Toxicol Mech Methods 2011;21(4):252–74.

[68] Feng J, Liu T, Qin B, Zhang Y, Liu XS. Identifying ChIP-seq enrichment using MACS. Nature Protoc 2012;7(9):1728–40.

[69] Pavesi G, Mereghetti P, Mauri G, Pesole G. Weeder Web: discovery of transcription factor binding sites in a set of sequences from co-regulated genes. Nucleic Acids Res 2004;32(Web Server issue):W199–203.

[70] Gilchrist DA, Fargo DC, Adelman K. Using ChIP-chip and ChIP-seq to study the regulation of gene expression: genome-wide localization studies reveal widespread regulation of transcription elongation. Methods 2009;48(4):398–408.

[71] Lee RC, Feinbaum RL, Ambros V. The *C. elegans* heterochronic gene lin-4 encodes small RNAs with antisense complementarity to lin-14. Cell 1993;75(5):843–54.

[72] van Rooij E. The art of microRNA research. Circ Res 2011;108(2):219–34.

[73] Huang Y, Zou Q, Wang SP, Tang SM, Zhang GZ, Shen XJ. The discovery approaches and detection methods of microRNAs. Mol Biol Rep 2011;38(6):4125–35.

[74] Li W, Ruan K. MicroRNA detection by microarray. Anal Bioanal Chem 2009;394 (4):1117–24.

[75] Cissell KA, Deo SK. Trends in microRNA detection. Anal Bioanal Chem 2009;394 (4):1109–16.

[76] Friedlander MR, Chen W, Adamidi C, et al. Discovering microRNAs from deep sequencing data using miRDeep. Nat Biotechnol 2008;26(4):407–15.

[77] Varallyay E, Burgyan J, Havelda Z. MicroRNA detection by northern blotting using locked nucleic acid probes. Nat Protoc 2008;3(2):190–6.

[78] Lewis BP, Burge CB, Bartel DP. Conserved seed pairing, often flanked by adenosines, indicates that thousands of human genes are microRNA targets. Cell 2005;120 (1):15–20.

[79] Ritchie W, Flamant S, Rasko JE. Predicting microRNA targets and functions: traps for the unwary. Nat Methods 2009;6(6):397–8.

[80] Friedman RC, Farh KK, Burge CB, Bartel DP. Most mammalian mRNAs are conserved targets of microRNAs. Genome Res 2009;19(1):92–105.

[81] Griffiths-Jones S, Saini HK, van Dongen S, Enright AJ. miRBase: tools for microRNA genomics. Nucleic Acids Res 2008;36(Database issue):D154–8.

[82] Kiriakidou M, Nelson PT, Kouranov A, et al. A combined computational-experimental approach predicts human microRNA targets. Genes Dev 2004;18 (10):1165–78.

[83] Thomson DW, Bracken CP, Goodall GJ. Experimental strategies for microRNA target identification. Nucleic Acids Res 2011;39(16):6845–53.

[84] Du T, Zamore PD. Beginning to understand microRNA function. Cell Res 2007;17 (8):661–3.

Keratinocyte Differentiation and Epigenetics

Jeung-Hoon Lee

Department of Dermatology, College of Medicine, Chungnam National University, Daejeon, South Korea

3.1 INTRODUCTION

The skin is the largest organ of the body; it covers an area of approximately 1.4–2.0 m^2 and its weight reaches around 4 kg. Skin plays a variety of functional roles, including protection, preservation, temperature regulation, and vitamin D synthesis. Skin is composed of many cell types, among which keratinocytes, melanocytes, and fibroblasts are regarded as the most abundant and functionally important. Developmentally, keratinocytes originate from ectoderm and eventually participate in the formation of epidermal structures such as epidermis, hair, nails, and sebaceous glands. Specifically, keratinocytes constitute a majority of the epidermis and form cornified layers that help to contain body fluids and provide barrier protection from the environment [1]. Melanocytes are derived from the neural crest and located in the lower epidermis and hair follicles, where they generate melanin to provide coloration and protection from ultraviolet damage [2]. Fibroblasts are the major cells of the dermal portion. They originate from mesenchyme and synthesize essential extracellular matrix (ECM) components to provide structural support [3].

The characteristics of each skin cells are determined by cell type-specific gene expression patterns. By comparing the respective gene expression profiles of keratinocytes, melanocytes, and fibroblasts, many signature genes for each cell type can be identified (Figure 3.1) [4]. To understand their specified roles in establishing each type of skin cell, it is necessary to study the functions of each signature gene. In addition, the regulatory mechanism underlying differential expression of signature genes in a cell

 © 2015 Elsevier Inc. All rights reserved.

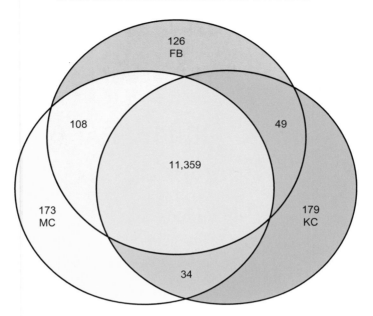

FIGURE 3.1   **The number of specific genes identified by cDNA microarray.** A total of 12,028 genes were obtained for which expression was detected in at least one of three cell types. KC, keratinocytes; FB, fibroblasts; MC, melanocytes.

type-specific manner should be investigated. It is believed that gene expression is regulated in a cell type-specific manner by a number of signaling pathways and transcription factors [5]. Although the precise mechanisms governing cell type-specific gene expression are still under investigation, one important regulatory mechanism, epigenetic modulation, has been emerging as a critical player in gene expression control [6].

Keratinocytes provide the rigid stratified structure through a highly complicated and tightly regulated process of differentiation [7]. The keratinocyte progenitor cells in the basal layer proliferate and move upward; the differentiation process begins in the suprabasal layers and culminates in fully differentiated dead cells on the surface. During this process, keratinocytes show several distinctive morphological changes and can be divided into several stratified layers such as basal, spinous, granular, and cornified layers. Given that this process takes place along a pathway that leads to cell cycle arrest and terminal differentiation, specific gene sets should show precise spatiotemporally regulated expression. Many differentiation-related genes, including those encoding transglutaminases 1 and 3, involucrin, cornifin, loricrin, filaggrin, and small proline-rich proteins (SPRs), have been shown to be expressed in a temporally regulated manner in keratinocyte differentiation [8–12]. Several factors have been suggested to play roles in keratinocyte differentiation *in vivo* and *in vitro*, including calcium, vitamin D, retinoic acid, and 12-*O*-tetradecanoylphorbol-13-acetate (TPA) [13−16]. Calcium is the best characterized of these factors as a differentiating agent for keratinocytes, and has well been chosen as a differentiating agent to investigate the molecular events involved in the keratinocyte differentiation process. In the epidermis, it takes 14 days for keratinocytes to undergo terminal differentiation and become dead corneocytes. In a previous study, the temporal gene expression patterns during keratinocyte differentiation were investigated using microarray analysis [17]. Four time points (1, 3, 7, and 14 days after calcium treatment) revealing the specific situation of differentiating keratinocytes were chosen and total RNAs were isolated. Calcium treatment resulted in marked changes in the gene expression profile in a time-dependent manner. The time-dependent expression patterns of the genes over four time points were analyzed by hierarchical clustering (average linkage clustering). The TreeView image of clustering clearly indicated that calcium triggered distinctive yet well-ordered changes in gene expression patterns in a time-dependent manner (Figure 3.2A). The differentially expressed genes were clustered into six groups according to their expression pattern using self-organizing map analysis and showed the global feature of function-related regulation (Figure 3.2B). The genes related to keratinocyte differentiation were markedly upregulated by calcium treatment. In addition, a unique pattern of increase was seen in the expression of genes related to ribosomal proteins. On the other hand, transcripts involved in metabolism, DNA repair, transcription, and translation were generally downregulated. These results demonstrate the complexity of the gene expression profile that contributes to the spatiotemporal regulation of keratinocyte differentiation. They provide important clues on which to base further investigations of the molecular events controlling the process of keratinocyte differentiation, including signal transduction events, transcription factors, and epigenetic modulation. Investigations into gene expression can lead to further study regarding epigenetic modulation, including the variability of the regulatory regions to DNA−protein interactions, histone acetylation linked to activation of transcription, and chromatin remodeling by polycomb complexes [18].

3.2 GENE EXPRESSION IN KERATINOCYTE DIFFERENTIATION

During embryogenesis, stratified epithelium is established by a highly organized keratinocyte differentiation process, in which many differentiation-related genes are coordinately expressed. Interestingly, many differentiation-related genes are clustered in several genomic loci,

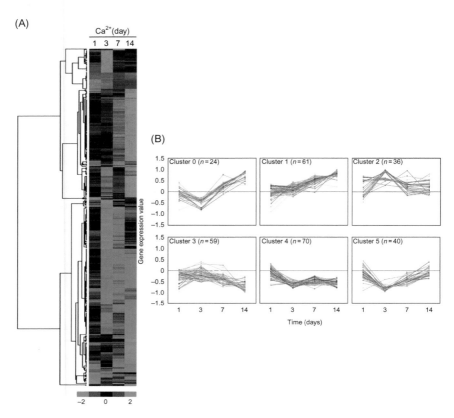

FIGURE 3.2   **Analysis of calcium-inducible genes in keratinocytes.** (A) Hierarchical clustering of gene expression profiles in calcium-induced keratinocyte differentiation. Hierarchical clustering was performed using Cluster version 2.12 and TreeView version 1.50. (B) Self-organizing map analysis of the differentially expressed genes. Two hundred and ninety genes were selected as differentially expressed genes, the expression level of which showed changes greater than twofold on at least one occasion over four time points on microarray analysis. The numbers in each graph show the number of genes belonging to the corresponding cluster.

e.g., keratin type I and type II loci, epidermal differentiation complex (EDC), and keratin-associated protein loci [19,20]. The EDC cluster consists of at least 43 genes that primarily make up the structural elements of the cornified layer, e.g., involucrin, filaggrin, loricrin, late cornifying envelope (LCE) genes, and SPRR genes [21,22]. The coordinated expression of EDC genes is required for proper keratinocyte terminal differentiation and epidermal function. In fact, comparative genomics study demonstrates that a network of interspersed *cis*-regulatory elements in EDC participates in coordinated gene expression during mammalian epidermal differentiation [19]. And it appears that multiple EDC genes are involved in the pathogenesis of chronic skin diseases including two common inflammatory skin disorders that are barrier-impaired, atopic dermatitis and psoriasis. Therefore, the understanding of the mechanisms of EDC gene regulation would help to identify potential therapeutic targets for such skin disorders.

3.3 EPIGENETIC MODULATION IN KERATINOCYTE DIFFERENTIATION

Epigenetic modulation is defined as a mechanism in which heritable changes in the genome functions occur without alterations to the DNA sequence [23]. It occurs by various methods, such as DNA methylation, histone modification, and microRNA (miRNA) interference [24]. First, DNA methylation is regarded as the most important method for modifying gene expression and cell functions without DNA sequence change. DNA methylation is catalyzed by the enzyme DNA methyltransferase (DNMT), which adds the methyl group to cytosine (C) and guanine (G) dinucleotides [6]. In the human genome, there are regions with stretches of up to a few kilobases in length in which CpG dinucleotides are rich. These regions are called CpG islands. The human genome contains about 30,000 CpG islands, 50−60% of which occur in the promoter region of genes [25]. It is thought that methylation of CpG in the promoter region usually results in the repression of gene expression [26]. In addition to CpG islands, recent genome-wide DNA methylation studies reveal that non-CpG methylation in both the intragenic and intergenic sites is also important for regulation of differential gene expression. For example, CpG island shores (low CpG density regions that lie up to 2 kb from CpG islands) show highly variant methylation [27]. Second, histone modification is also an important epigenetic modulatory mechanism. Histone acetylation and deacetylation are the major events regulating gene expression. Usually histone acetylation activates the transcription of certain genes, while histone deacetylation is associated with the transcriptionally inactive status of a chromosome [28]. Since histone modification and DNA methylation have similar roles in gene expression, it is conceivable that there is significant interplay between these two epigenetic regulatory mechanisms. Histone modifications appear more dynamic, while DNA methylation functions as a more stable regulatory mechanism [29]. Third, the influence of the miRNA on gene expression is regarded as an important epigenetic mechanism. The miRNA-coding genes may control the expression of 30% of all protein-coding genes [30].

Epigenetic modulation has been recognized as an important regulatory mechanism in various biological events. For example, parental imprinting is regulated by epigenetic changes such as DNA methylation and histone modifications, by which a single allele is differentially expressed depending on the sex of the parent transmitting the allele [31]. Another example includes the maintenance of cellular differentiation. Certain tissue is comprised of specific differentiated cells showing their distinct pattern of gene expression, although these cells are genetically identical to other tissues' cells. The maintenance of a specific phenotype from one cell to its daughter cells in a differentiated tissue is regulated by epigenetic changes [32].

The epidermis is a multilayered epithelium, which is maintained by stem cells residing in the basal layer. To maintain tissue homeostasis, progeny of epidermal stem cells must choose either to self-renew or to differentiate. Epigenetic modulation plays a critical role to ensure appropriate gene expression associated with fate decision. For example, an epigenetic network, in which at least five complexes with chromatin remodeling activities are involved, is required for maintenance of epidermal stem cells in an undifferentiated state [33].

The onset of keratinocyte differentiation involves cell cycle withdrawal and detachment from the basement membrane. Differentiation is processed by downregulation of keratins 5 and 14, whereas it is accompanied by upregulation of EDC genes. DNA methylation and histone modification play important roles in this process. For example, primary cultured keratinocytes grown under low calcium conditions show DNA hypomethylation of the keratin 5 locus [34]. With regard to histone modification, a histone methylase JMJD3 is recruited to the promoter regions of EDC genes during keratinocyte differentiation and removes the methyl groups from trimethylated Lys 27 of histone 3 (H3K27me3). Erasure of H3K27me3 results in loss of repressive effects and then upregulation of the differentiation-related genes. In contrast, depletion of JMJD3 leads to enrichment of H3K27me3 at the promoters of epidermal differentiation genes, resulting in repression of transcription and impaired epidermal differentiation [35]. It is likely that methylation change occurs at numerous genetic loci in both positive and negative ways; methylation represses expression of genes related to proliferation, while hypomethylation is linked to differentiation-related genes.

The methyltransferase EZH2 controls the proliferative potential of basal progenitors by repressing the INK4A−INK4B locus and tempers the developmental rate of differentiation by preventing premature recruitment of AP1 transcriptional activator to the structural genes that are required for epidermal differentiation [36]. In addition, transcription factor GRHL2 inhibits recruitment of histone demethylase JMJD3 to the

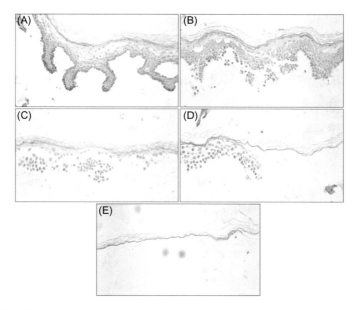

FIGURE 3.3   **Histologic analysis of an epidermal sample.** (A) H&E stained section of the entire epidermis after thermolysin incubation and removal of the dermis. Epidermal fragments remaining after the first, second, and third trypsin treatments, respectively, constitute the T1 (B), T2 (C), and T3 fractions (D). Fragments shown in (E) are mainly composed of the cornified layer and constitute the T4 fraction.

EDC gene promoters and enhances the level of histone 3 Lys 27 trimethylation enrichment at these promoters, leading to inhibition of gene expression [37]. Another important histone modifier is a histone deacetylase (HDAC). HDAC removes acetylation from histone protein and other transcription factors such as p53, which is critical for epidermal proliferation, differentiation, and hair follicle formation [38]. Many factors can affect gene expression during keratinocyte differentiation by epigenetic modulation. UVA induces specific histone acetylation in the promoter region of the p16 (*INK4a*) gene [39], and vitamin D3 regulates cathelicidin and CD14 gene expression in keratinocytes through histone modulation [40].

To investigate whether epigenetic modulation is involved in keratinocyte differentiation-specific gene regulation, we used trypsin for epidermal fragmentation with successive short-term enzyme incubation to progressively detach cells from the deep layers and to purify the cells (Figure 3.3). Incubations were performed at 4°C to stop cellular metabolic activity and to preserve the mRNA pool from degradation. This point is crucial as many growth factors, cell cycle regulators, and transcription factors are encoded by short-lived mRNAs. We performed

TABLE 3.1   DNA Methylation Microarray Data

Gene name	Symbol	Accession no.	Map
(A)			
Protocadherin gamma subfamily A, 3	*PCDHGA3*	NM_018916	5q31
F-box and leucine-rich repeat protein 17	*FBXL17*	NM_022824	5q21.3
Spastic paraplegia 20, spartin (Troyer syndrome)	*SPG20*	NM_015087	13q13.3
Dihydrouridine synthase 3-like (*S. cerevisiae*)	*DUS3L*	NM_020175	19p13.3
Mitochondrial ribosomal protein L36	*MRPL36-NDUFS6*	NM_032479	5p15.3
Protein kinase D3	*PRKD3*	NM_005813	2p21
Pleckstrin homology, Sec7, and coiled-coil domains 2 (cytohesin-2)	**PSCD2**	NM_017457	19q13.3
Chromosome 2 open reading frame 3	*C2orf3*	NM_003203	2p11.2-p11.1
Hypothetical protein LOC151278	*FLJ32447*	NM_153038	2q36.1
(B)			
Zinc finger, ZZ-type with EF-hand domain 1	*ZZEF1*	NM_015113	17p13.2
Zinc finger, HIT type 3	*ZNHIT3*	NM_001033577	17q12
Zinc finger, AN1-type domain 2A	*ZFAND2A*	NM_182491	7p22.3
Zinc finger and BTB domain containing 11	*ZBTB11*	NM_014415	3q12.3
Tyrosine 3-monooxygenase/tryptophan 5-monooxygenase activation protein, theta polypeptide	**YWHAQ**	NM_006826	2p25.1
WD repeat domain 54	*WDR54*	NM_032118	2p13.1
Ventral anterior homeobox 1	*VAX1*	NM_199131	10q26.1
Thioredoxin domain containing 9	*TXNDC9-EIF5B*	NM_005783	2q11.2
Signal peptidase complex subunit 2 homolog (*S. cerevisiae*)	*SPCS2*	NM_014752	11q13.4

Hypermethylated genes (A) and hypomethylated genes (B) from differentiated keratinocytes.
S. cerevisiae, Saccharomyces cerevisiae.

methylation DNA microarray analysis with genomic DNA isolated from the basal (T1) and cornified layers (T4). A methylated CpG-assisted microarray analysis was performed, and genes showing hyper- and/or hypomethylation were verified (Table 3.1). Most of these genes had not previously been associated with keratinocyte differentiation. *PSCD2* (pleckstrin homology, Sec7, and coiled-coil domain 2) was the most significant gene in the hypermethylation field, and *YWHAQ* (tyrosine 3-monooxygenase/tryptophan 5-monooxygenase activation protein, theta polypeptide) was the most reliable gene in the hypomethylation field. The *PSCD2* gene functions to promote activation of adenosine diphosphate-ribosylation factor (ARF) through replacement of GDP with GTP. Members of this family have an identical structural organization that consists of an N-terminal coiled-coil motif, a central Sec7 domain, and a C-terminal pleckstrin homology domain [41]. The functions of this family include mediating regulation of protein sorting and membrane trafficking [42]. Although there are reports of a regulatory role in development of neuronal processes [43], the role of *PSCD2* during keratinocyte differentiation has not yet been investigated. *YWHAQ* is a gene associated with an adapter protein that is implicated in the regulation of a signaling pathway. *YWHAQ* is directly involved in cellular processes, such as cytokinesis, cell-contact inhibition, anchorage-independent growth, and cell adhesion, processes that often become deregulated in diseases like cancer [44,45]. As this is a preliminary study, more work is necessary to determine whether changes in the methylation status of these candidate genes actually control keratinocyte differentiation. Although extensive work in this field is clearly needed, these preliminary findings highlight the importance of epigenetic modulation in keratinocyte differentiation-specific gene regulation [46].

Recently, a highly advanced technique, "next-generation sequencing," has been emerging, with the cost for high-resolution assessment of DNA methylation decreasing. Parallel to accumulating knowledge on epigenetic modulation with the advance of technology, there is also a growing interest in development of drugs that can reverse methylation abnormalities. Agents known to inhibit DNA methylation, for example 5-azacytidine (5AC), and histone deacetylation, for example sodium butyrate (NaB), are able to inhibit cell growth and to promote keratinocyte differentiation [47,48]. Currently, the HDAC inhibitors vorinostat and romidepsin are approved for treatment of cutaneous T-cell lymphomas by the U.S. Food and Drug Administration (FDA) [49].

3.4 EPIGENETICS AND SKIN DISEASES

Epigenetic modulation is an important regulatory mechanism not only for the development and cell differentiation of normal skin but also for the pathogenesis of many skin diseases. Although genetics is important in the etiology of many diseases, other exogenous factors including diet and lifestyle also affect susceptibility, in part through epigenetic changes. Specifically, the importance of DNA methylation is demonstrated by the growing number of diseases that occur when methylation is not properly established or maintained in cells [50]. A pivotal role for altered methylation has well been established in various cancers. Cancer cells are usually hypomethylated, and the loss of genomic methylation is an early event in cancer development. The methylation status often correlates with disease severity and metastatic potential. In contrast, genes involved in apoptosis, cell cycle regulation, DNA repair, cell signaling, and transcription have been shown to be silenced by hypermethylation. Thus, DNA methylation profiles may be used as biomarkers for the detection and classification of various cancers. In addition, epigenetic changes can be adopted for prediction of treatment response. For example, hypermethylation and loss of expression of O-6-methylguanine-DNA methyltransferase (MGMT) in glioblastoma multiforme is tightly correlated with enhanced survival ratio [51].

In common skin cancers, aberrant methylation of tumor suppressor gene promoters is associated with their transcriptional inactivation, and carcinogens such as ultraviolet (UV) radiation may act through epigenetic mechanisms. UV radiation triggers direct DNA damage through the formation of cyclobutane pyrimidine dimers, which is responsible for the majority of mutations induced by UV. In addition to these harmful mutations, promoter hypermethylation of tumor suppressor genes is an important mechanism for switching off transcriptional regulation to provide selective cell growth, and is associated with a variety of skin cancers (Table 3.2). For example, hypermethylation of the p53 gene promoter can result in the increase of DNA methyltransferase 1 (DNMT1), because p53 negatively regulated *DNMT1* expression by forming a complex with specificity protein 1 (Sp1) and chromatin modifiers on the *DNMT1* promoter [53]. Therefore, inactivation of p53 in keratinocytes via UV radiation may lead to a considerable increase in DNA methylation, which affects the cells producing a cancer-driving event.

It has also been reported that inorganic arsenic, a potent human carcinogen, can induce malignant transformation of the keratinocytes by decreasing the level of let-7c through its hypermethylation, and thereby activation of the Ras/NF-κB signal pathway [54]. In addition to hypermethylation, histone modification and miRNA interference may have an important role in the pathogenesis of skin cancer, especially in malignant melanoma [55,56].

Epigenetic factors play an important role in the pathogenesis of systemic autoimmune diseases, such as systemic lupus erythematosus (SLE), and scleroderma. In contrast to cancer, promoter hypomethylation is associated with SLE. SLE patients show usually hypomethylated DNA and a decrease in the enzymatic activity of DNMTs, which may be the mechanism to explain the DNA hypomethylation [57,58]. Promoter

TABLE 3.2   Selected Genes That Show Promoter Hypermethylation in Skin Cancer [52]

Gene with promoter methylation	Function of protein or RNA inactivated by methylation	Skin cancers associated
14–13–3s	p53-regulated cell cycle inhibitor	BCC
Cadherin-13	Cell adhesion in epithelial tissues	BCC
FHIT	Tumor suppressor	BCC, SCC
p16	Inhibitor of cell cycle	SCC, MM, CTCL, CBCL
p14ARF	Inhibitor of cell cycle	SCC, MM, CTCL
E-cadherin	Cell adhesion in epithelial tissues	SCC, MM
Cadherin-1	Cell adhesion in epithelial tissues	SCC
IGF-BP3	Growth factor binding protein	SCC
Thrombospondin-1	Antiangiogenic action	SCC
RASSF1A	Inhibitor of cell cycle	SCC, MM
Calcyclin	Calcium binding	SCC
MGMT	DNA repair protein	SCC, MM, CTCL
APC	Wnt signaling	MM
RARb2	Retinoid signaling	MM
Necdin	Tumor suppressor	MM
microRNA-34a	Inhibitor of cell cycle	MM
p15	Inhibitor of cell cycle	CTCL, CBCL
p73	Tumor suppressor	CTCL
MLH1	DNA repair protein	SCC, CTCL
BCL7a	Tumor suppressor	CTCL
PTPRG	Tumor suppressor	CTCL
SHP1	Hematopoiesis	CTCL

BCC, basal cell carcinoma; CBCL, cutaneous B-cell lymphoma; CTCL, cutaneous T-cell lymphoma; MM, malignant melanoma: SCC, squamous cell carcinoma.

hypomethylation results in induction of gene transcription and this could lead to the overreactivity of the immune system seen in SLE.

The histone modification in SLE is less well understood compared to DNA methylation. Aberrant histone acetylation patterns were identified in CD4$^+$ T cells of SLE patients. The degree of histone H3 acetylation correlated negatively with increased disease activity in lupus patients [59].

In addition, the HDAC inhibitor trichostatin A significantly reverses the skewed expression of multiple genes implicated in the immunopathogenesis of SLE [60]. Thus, it can be suggested that histone modification may be an important pathogenic factor in SLE.

Epigenetic modulation may also be relevant in the pathogenesis of inflammatory skin diseases. For example, psoriasis is a skin disease that occurs in genetically predisposed individuals. T-cell activation is important to pathogenesis of psoriasis, leading to a burst of cytokine cascade that results in keratinocyte proliferation [61]. A study by Zhang et al. identified a subset of differentially methylated CpG sites involving 12 sites mapping EDC and methylated regions covering almost the entire genome with sufficient depth and high resolution in involved psoriatic skin lesions [62]. The antiapopototic protein p16 is increased in psoriatic skin, and the p16 gene promoter is methylated in these patients. This epigenetic change correlates with disease severity [63]. The miRNA may be also important in the pathogenesis of psoriasis. miR-146a and miR-203 are increased in psoriatic epidermis, and these miRNAs regulate the TNF-α pathway that is implicated in the pathogenesis of psoriasis [30]. The study on epigenetic changes associated with psoriasis will provide new insight into the pathogenesis of this complex disease.

In lichen simplex chronicus (LSC), an inflammatory skin disorder classified as an endogenous eczema, alterations in the LINE-1 methylation pattern have been demonstrated in keratinocytes from patients. Because the alteration patterns were very distinctive and highly sensitive and specific for detecting LSC, the authors suggested that alterations in the LINE-1 methylation pattern could be used as the biomarkers for this disease [64].

3.5 CONCLUSION

Epigenetic modulation affects a variety of cellular events, including embryonic development, transcription, genomic imprinting, chromatin structure, and chromosome stability. There is growing information regarding epigenetic mechanisms related to pathogenesis of skin disease, especially in the fields of skin cancers and autoimmune and inflammatory diseases. The findings will be applicable for prediction of the treatment response and lead to the development of therapies which target epigenetic modulation. In some fields, inhibitors for methylation and histone deacetylation are already used to treat diseases. Although the information regarding skin diseases is still inadequate, epigenetic modulation will surely be an important subject on which to base further investigation. The findings will provide useful information for future development of novel therapeutics and preventive tools for many skin diseases associated with abnormalities in keratinocyte differentiation.

References

[1] Fartasch M, Ponec M. Improved barrier structure formation in air-exposed human keratinocyte culture systems. J Invest Dermatol 1994;102:366–74.

[2] Sturm RA, Box NF, Ramsay M. Human pigmentation genetics: the difference is only skin deep. Bioessays 1998;20:712–21.

[3] Takeda K, Gosiewska A, Peterkofsky B. Similar, but not identical, modulation of expression of extracellular matrix components during *in vitro* and *in vivo* aging of human skin fibroblasts. J Cell Physiol 1992;153:450–9.

[4] Lee JS, Kim DH, Choi DK, Kim CD, Ahn GB, Yoon TY, et al. Comparison of gene expression profiles between keratinocytes, melanocytes and fibroblasts. Ann Dermatol 2013;25:36–45.

[5] Blanpain C, Fuchs E. Epidermal homeostasis: a balancing act of stem cells in the skin. Nat Rev Mol Cell Biol 2009;10:207–17.

[6] Jaenisch R, Bird A. Epigenetic regulation of gene expression: how the genome integrates intrinsic and environmental signals. Nat Genet 2003;33(Suppl.):245–54.

[7] Kalinin AE, Kajava AV, Steinert PM. Epithelial barrier function: assembly and structural features of the cornified cell envelope. Bioessays 2002;24:789–800.

[8] Rice RH, Green H. The cornified envelope of terminally differentiated human epidermal keratinocytes consists of cross-linked protein. Cell 1977;11:417–22.

[9] Fuchs E. Epidermal differentiation and keratin gene expression. J Cell Sci 1993;17:197–208.

[10] Steinert PM, Marekov LN. The proteins elafin, filaggrin, keratin intermediate filaments, loricrin, and small proline-rich proteins 1 and 2 are isodipeptide cross-linked components of the human epidermal cornified cell envelope. J Biol Chem 1995;270:17702–11.

[11] Steinert PM, Marekov LN. Direct evidence that involucrin is a major early isopeptide cross-linked component of the keratinocyte cornified cell envelope. J Biol Chem 1997;272:2021–30.

[12] Nemes Z, Steinert PM. Bricks and mortar of the epidermal barrier. Exp Mol Med 1999;31:5–19.

[13] Yuspa SH, Hennings H, Tucker RW, Jaken S, Kilkenny AE, Roop DR. Signal transduction for proliferation and differentiation in keratinocytes. Ann N Y Acad Sci 1988;548:191–6.

[14] Pillai S, Bikle DD, Elias PM. Vitamin D and epidermal differentiation: evidence for a role of endogenously produced vitamin D metabolites in keratinocyte differentiation. Skin Pharmacol 1988;1:149–60.

[15] Gibbs S, Backendorf C, Ponec M. Regulation of keratinocyte proliferation and differentiation by all-trans-retinoic acid, 9-cis-retinoic acid and 1,25-dihydroxy vitamin D3. Arch Dermatol Res 1996;288:729–38.

[16] Goldyne ME, Evans CB. 12-O-tetradecanoylphorbol-13-acetate and the induction of prostaglandin E2 generation by human keratinocytes: a re-evaluation. Carcinogenesis 1994;15:141–3.

[17] Seo EY, Namkung JH, Lee KM, Lee WH, Im M, Kee SH, et al. Analysis of calcium-inducible genes in keratinocytes using suppression subtractive hybridization and cDNA microarray. Genomics 2005;86:528–38.

[18] Botchkarev VA, Gdula MR, Mardaryev AN, Sharov AA, Fessing MY. Epigenetic regulation of gene expression in keratinocytes. J Invest Dermatol 2012;132:2505–21.

[19] Bazzi H, Fantauzzo KA, Richardson GD, Jahoda CA, Christiano AM. Transcriptional profiling of developing mouse epidermis reveals novel patterns of coordinated gene expression. Dev Dyn 2007;236:961–70.

[20] Segre JA. Epidermal differentiation complex yields a secret: mutations in the cornification protein filaggrin underlie ichthyosis vulgaris. J Invest Dermatol 2006;126:1202−4.

[21] Hoffjan S, Stemmler S. On the role of the epidermal differentiation complex in ichthyosis vulgaris, atopic dermatitis and psoriasis. Br J Dermatol 2007;157:441−9.

[22] Kypriotou M, Huber M, Hohl D. The human epidermal differentiation complex: cornified envelope precursors, S100 proteins and the "fused genes" family. Exp Dermatol 2012;21:643−9.

[23] Berger SL, Kouzarides T, Shiekhattar R, Shilatifard A. An operational definition of epigenetics. Genes Dev 2009;23:781−3.

[24] Barter MJ, Bui C, Young DA. Epigenetic mechanisms in cartilage and osteoarthritis: DNA methylation, histone modifications and microRNAs. Osteoarthritis Cartilage 2012;20:339−49.

[25] Costello JF, Plass C. Methylation matters. J Med Genet 2001;38:285−303.

[26] Bird AP. The relationship of DNA methylation to cancer. Cancer Surv 1996;28:87−101.

[27] Maunakea AK, Nagarajan RP, Bilenky M, Ballinger TJ, D'Souza C, Fouse SD, et al. Conserved role of intragenic DNA methylation in regulating alternative promoters. Nature 2010;466:253−7.

[28] Bönisch C, Nieratschker SM, Orfanos NK, Hake SB. Chromatin proteomics and epigenetic regulatory circuits. Expert Rev Proteomics 2008;5:105−19.

[29] Cedar H, Bergman Y. Linking DNA methylation and histone modification: patterns and paradigms. Nat Rev Genet 2009;10:295−304.

[30] Sonkoly E, Ståhle M, Pivarcsi A. MicroRNAs: novel regulators in skin inflammation. Clin Exp Dermatol 2008;33:312−15.

[31] Kota SK, Feil R. Epigenetic transitions in germ cell development and meiosis. Dev Cell 2010;19:675−86.

[32] Gudjonsson JE, Krueger G. A role for epigenetics in psoriasis: methylated Cytosine−Guanine sites differentiate lesional from nonlesional skin and from normal skin. J Invest Dermatol 2012;132:506−8.

[33] Mulder KW, Wang X, Escriu C, Ito Y, Schwarz RF, Gillis J, et al. Diverse epigenetic strategies interact to control epidermal differentiation. Nat Cell Biol 2012;14:753−63.

[34] Cheng JB, Cho RJ. Genetics and epigenetics of the skin meet deep sequence. J Invest Dermatol 2012;132:923−32.

[35] Sen GL, Webster DE, Barragan DI, Chang HY, Khavari PA. Control of differentiation in a self-renewing mammalian tissue by the histone demethylase JMJD3. Genes Dev 2008;22:1865−70.

[36] Ezhkova E, Pasolli HA, Parker JS, Stokes N, Su IH, Hannon G, et al. Ezh2 orchestrates gene expression for the stepwise differentiation of tissue-specific stem cells. Cell 2009;136:1122−35.

[37] Chen W, Liu ZX, Oh JE, Shin KH, Kim RH, Jiang M, et al. Grainyhead-like 2 (GRHL2) inhibits keratinocyte differentiation through epigenetic mechanism. Cell Death Dis 2012;3:e450.

[38] LeBoeuf M, Terrell A, Trivedi S, Sinha S, Epstein JA, Olson EN, et al. Hdac1 and Hdac2 act redundantly to control p63 and p53 functions in epidermal progenitor cells. Dev Cell 2010;19:807−18.

[39] Chen IP, Henning S, Faust A, Boukamp P, Volkmer B, Greinert R. UVA-induced epigenetic regulation of P16(INK4a) in human epidermal keratinocytes and skin tumor derived cells. Photochem Photobiol Sci 2012;11:180−90.

[40] Schauber J, Oda Y, Büchau AS, Yun QC, Steinmeyer A, Zügel U, et al. Histone acetylation in keratinocytes enables control of the expression of cathelicidin and CD14 by 1,25-dihydroxyvitamin D3. J Invest Dermatol 2008;128:816−24.

[41] Torii T, Miyamoto Y, Sanbe A, Nishimura K, Yamauchi J, Tanoue A. Cytohesin-2/ARNO, through its interaction with focal adhesion adaptor protein paxillin, regulates preadipocyte migration via the downstream activation of Arf6. J Biol Chem 2010;285:24270—81.

[42] White DT, McShea KM, Attar MA, Santy LC. GRASP and IPCEF promote ARF-to-Rac signaling and cell migration by coordinating the association of ARNO/cytohesin 2 with Dock180. Mol Biol Cell 2010;21:562—71.

[43] Hernández-Deviez D, Mackay-Sim A, Wilson JM. A role for ARF6 and ARNO in the regulation of endosomal dynamics in neurons. Traffic 2007;8:1750—64.

[44] Yaffe MB, Rittinger K, Volinia S, Caron PR, Aitken A, Leffers H, et al. The structural basis for 14-3-3:phosphopeptide binding specificity. Cell 1997;91:961—71.

[45] Tian Q, Feetham MC, Tao WA, He XC, Li L, Aebersold R, et al. Proteomic analysis identifies that 14-3-3zeta interacts with beta-catenin and facilitates its activation by Akt. Proc Natl Acad Sci USA 2004;101:15370—5.

[46] Back SJ, Im M, Sohn KC, Choi DK, Shi G, Jeong NJ, et al. Epigenetic modulation of gene expression during keratinocyte differentiation. Ann Dermatol 2012;24:261—6.

[47] Okada N, Steinberg ML, Defendi V. Re-expression of differentiated properties in SV40-infected human epidermal keratinocytes induced by 5-azacytidine. Exp Cell Res 1984;153:198—207.

[48] Elder JT, Zhao X. Evidence for local control of gene expression in the epidermal differentiation complex. Exp Dermatol 2002;11:406—12.

[49] Glass E, Viale PH. Histone deacetylase inhibitors: novel agents in cancer treatment. Clin J Oncol Nurs 2013;17:34—40.

[50] Robertson KD. DNA methylation and human disease. Nat Rev Genet 2005;6:597—610.

[51] Hegi ME, Diserens AC, Gorlia T, Hamou MF, de Tribolet N, Weller M, et al. MGMT gene silencing and benefit from temozolomide in glioblastoma. N Engl J Med. 2005;352(10):997—1003.

[52] Millington GW. Epigenetics and dermatological disease. Pharmacogenomics 2008;9:1835—50.

[53] Lin RK, Wu CY, Chang JW, Juan LJ, Hsu HS, Chen CY, et al. Dysregulation of p53/Sp1 control leads to DNA methyltransferase-1 overexpression in lung cancer. Cancer Res 2010;70:5807—17.

[54] Jiang R, Li Y, Zhang A, Wang B, Xu Y, Xu W, et al. The acquisition of cancer stem cell-like properties and neoplastic transformation of human keratinocytes induced by arsenite involves epigenetic silencing of let-7c via Ras/NF-κB. Toxicol Lett 2014;227:91—8.

[55] Yamamoto S, Yamano T, Tanaka M, Hoon DS, Takao S, Morishita R, et al. A novel combination of suicide gene therapy and histone deacetylase inhibitor for treatment of malignant melanoma. Cancer Gene Ther 2003;10:179—86.

[56] Zhang L, Huang J, Yang N, Greshock J, Megraw MS, Giannakakis A, et al. microRNAs exhibit high frequency genomic alterations in human cancer. Proc Natl Acad Sci USA 2006;103:9136—41.

[57] Sanders CJ, Van Weelden H, Kazzaz GA, Sigurdsson V, Toonstra J, Bruijnzeel-Koomen CA. Photosensitivity in patients with lupus erythematosus: a clinical and photobiological study of 100 patients using a prolonged phototest protocol. Br J Dermatol 2003;149:131—7.

[58] Luo Y, Li Y, Su Y, Yin H, Hu N, Wang S, et al. Abnormal DNA methylation in T cells from patients with subacute cutaneous lupus erythematosus. Br J Dermatol 2008;159:827—33.

[59] Hu N, Qiu X, Luo Y, Yuan J, Li Y, Lei W, et al. Abnormal histone modification patterns in lupus CD4[+] T cells. J Rheumatol 2008;35:804—10.

[60] Mishra N, Brown DR, Olorenshaw IM, Kammer GM. Trichostatin A reverses skewed expression of CD154, interleukin-10, and interferon-gamma gene and protein expression in lupus T cells. Proc Natl Acad Sci USA 2001;98:2628–33.

[61] Lebwohl M. Psoriasis. Lancet 2003;361:1197–204.

[62] Zhang P, Zhao M, Liang G, Yin G, Huang D, Su F, et al. Whole-genome DNA methylation in skin lesions from patients with psoriasis vulgaris. J Autoimmun 2013;41:17–24.

[63] Elias AN, Barr RJ, Nanda VS. p16 expression in psoriatic lesions following therapy with propylthiouracil, an antithyroid thioureylene. Int J Dermatol 2004;43(12):889–92.

[64] Yooyongsatit S, Ruchusatsawat K, Supiyaphun P, Noppakun N, Mutirangura A, Wongpiyabovorn J. Alterations in the LINE-1 methylation pattern in patients with lichen simplex chronicus. Asian Pac J Allergy Immunol 2013;31:51–7.

4

Epigenetics and Fibrosis: Lessons, Challenges, and Windows of Opportunity

Bin Liu[1,2], Xin Sheng Wang[3], Hui-Min Chen[1,5], Qianjin Lu [4], M. Eric Gershwin[1], and Patrick S.C. Leung[1]

[1]Division of Rheumatology, Allergy and Clinical Immunology, University of California, Davis, CA [2]Department of Rheumatology and Immunology, The Affiliated Hospital of Medical College Qingdao University, Qingdao City, Shandong Province, PR China [3]Department of Urology, The Affiliated Hospital of Medical College Qingdao University, Qingdao City, Shandong Province, PR China [4]Department of Dermatology, Second Xiangya Hospital of Central South University, Hunan Key Laboratory of Medical Epigenetics, Changsha, Hunan, PR China [5]Department of Molecular and Cellular Biology, University of California, Davis, CA

4.1 INTRODUCTION

Fibrotic diseases are comprised of a wide array of clinical entities, accounting for almost one-third of naturally occurring deaths worldwide [1]. Fibrosis is defined as the overgrowth, hardening, and/or scarring of tissues, which results from excessive deposition of extracellular matrix in normal tissue architecture and ultimately obliterates the functions of affected tissues. Development of fibrosis results from a variety of factors, including persistent infections, environmental exposures, connective tissue diseases, drug toxicity, radiation, and trauma [1,2]. Organs affected by fibrosis include lung, liver, heart, kidney, bone marrow, skin, and intestines [3]. Although fibrosis can affect virtually any organ and can be triggered by a variety of insults, it is believed that there are similar, if not identical, core pathways underlying this pathological phenomenon [4]. Further knowledge of the physiological and molecular mechanisms in the regulation of normal tissue growth and the progression of pathological fibrosis development will be essential in the clinical management of fibrotic diseases [5,6]. To date, therapeutic regimens for fibrotic diseases are mainly symptomatic treatment, since there are currently no effective antifibrotic agents available.

Epigenetics, which includes gene expression regulation by DNA methylation and histone modification, has been reported to be mechanistically involved in multiple human diseases such as cancer and immunological disorders [7–10]. Recently, epigenetics has also been shown to play important roles in inducing a heritable profibrotic phenotype in cells involved in scar tissue formation. Thus, knowledge on the epigenetic control of cellular phenotype in the context of tissue injury and aberrant tissue remodeling may lead to the development of epigenetic biomarkers and novel therapeutic regimens [11–13].

4.2 INCIDENCE AND PREVALENCE OF FIBROSIS

Fibrosis can affect virtually any organ, the most commonly affected being the lung, liver, and skin. Pulmonary fibrosis is an incurable disease characterized by scarring of the lungs. It is the leading cause of nonproductive cough on exertion and progressive exertional dyspnea [14,15]. Chronic pulmonary inflammation and fibrosis frequently develop without an identifiable reason and are commonly referred to as idiopathic pulmonary fibrosis (IPF) [16]. The incidence and prevalence of IPF vary between studies due to different diagnostic criteria, case definitions, study populations, and designs. The prevalence of IPF has been estimated to be between 14.0 and 42.7 per 100,000 persons in the United States. The incidence of IPF is approximately 50,000 newly diagnosed patients each year [17]. Data from studies involving 27 European Union countries indicated that the incidence of IPF ranges from 4.6 to 7.4 per 100,000 people, suggesting that approximately 30,000–35,000 new IPF cases will be diagnosed annually. The incidence and mortality of IPF are on the rise [17–19]. Unfortunately, it is associated with a poor prognosis with a median survival time of 3–5 years [20,21].

Liver fibrosis is a significant health issue worldwide with a mortality of approximately 1.5 million deaths per year [22,23], mainly due to cirrhosis and primary liver cancer. Cirrhosis is the last stage of fibrosis, which consists of excessive accumulation of extracellular matrix proteins including collagen. Cirrhosis is secondary to chronic viral hepatitis, alcohol abuse, and other causes. Ascites is the most common complication of cirrhosis and is associated with a lower quality of life, increased risk of infection, and poor long-term prognosis. Other potentially life-threatening complications are hepatic encephalopathy and bleeding from esophageal varices. Cirrhosis can also result in hepatic insufficiency, which will lead to fatal liver failure [24]. Liver cirrhosis mortality accounts for 1.8% of all deaths in Europe with the highest rates observed in southeastern and northeastern Europe [25]. Moreover, cirrhosis is an important cause of morbidity and mortality in the United States and ranks as the 12th leading cause of death, claiming 30,000 lives annually according to the National Center for Health Statistics at the Centers for Disease Control and Prevention. Patients with cirrhosis face a reduced life expectancy. Among a population aged 45−54 years, cirrhosis is the fourth leading cause of death [26−28]. Established cirrhosis has a 10-year mortality with a rate of 34−66% [29].

Systemic sclerosis (SSc; scleroderma) is a highly heterogeneous autoimmune disease characterized by small vessel vasculopathy, presence of autoantibodies, and fibroblast dysfunction, leading to increased deposition of extracellular matrix [30−32]. The resulting tissue fibrosis commonly leads to organ failure, a major cause of morbidity and mortality in patients with SSc [33,34]. SSc is clinically manifested as thickening and tightening of the skin, inflammation, and scarring of various parts of the body, leading to dysfunction in lung, kidney, heart, and digestive system. Organ involvement occurs in approximately 15% of patients with SSc [35−37]. Common complications in SSc include Raynaud's phenomenon, gastroesophageal reflux disease, dysphagia, and skin involvement. Digital ulcer is seen in more than half of the patients with SSc [38,39]. Pulmonary involvement in SSc patients is a major cause of morbidity and mortality. Pulmonary fibrosis occurs in approximately 80% of the patients, while pulmonary hypertension is observed in approximately 50% [40−43]. Clinical manifestations of cardiac disease are present in 15−35% of SSc patients [44,45], whereas 75% of SSc patients have subclinical cardiac dysfunction [46]. Renal crisis occurs in 5−10% of patients with SSc. These patients may present with an abrupt onset of hypertension, acute renal failure, headache, fever, malaise, hypertensive retinopathy, encephalopathy, and pulmonary edema [47].

SSc is a rare, chronic, and systemic autoimmune disease with an incidence of 0.002% [48]. In the past three decades, there has been a significant increase in the incidence of SSc, up to 20 per million, possibly due to improved diagnosis. In the United States, approximately 1 in 1000 is affected [49−51]. Five-year survival rates published in a Detroit and Spanish study were 78% and 84%, respectively, while 10-year survival rates were 55% and 65%, respectively [50−52]. However, the 9-year survival rate of patients with internal organ involvement is only 39% [48,53].

4.3 BIOLOGY OF EPIGENETICS

Epigenetics is defined as the study of regulations in gene function that are inheritable and do not entail a change in DNA sequence. The epigenome refers to DNA methylation, histone modifications, and chromatin accessibility throughout the genome. Each cell type possesses a unique epigenome, which defines its regulatory program [54,55]. The classical epigenetic field includes DNA methylation and histone modifications [56].

4.3.1 DNA Methylation

DNA methylation is important in the fine-tuning of chromatin structure and histone modifications to regulate gene expression at various stages of differentiation and development [56]. DNA methylation usually occurs at clusters of CpG dinucleotides called CpG islands, which are regions of more than 200 base pairs of a G + C content of at least 50% and a ratio of observed to statistically expected CpG frequencies of at least 0.6. CpG islands are generally in promoter regions and often associated with transcriptional inactivation of the target genes [57]. CpG islands in the gene promoters are mostly hypomethylated in transcriptionally active genes in somatic tissues [54]. In mammalian cells, DNA methylation is catalyzed by DNA methyltransferases (DNMTs), with *S*-adenosyl methionine as the methyl donor transferring a methyl group onto the C5 position of cytosine to form 5-methylcytosine. DNA methylation can inhibit gene expression by various mechanisms. Methylated DNA can promote the recruitment of methyl-CpG-binding domain (MBD) proteins; MBD proteins in turn recruit histone-modifying and chromatin-remodeling complexes to methylated sites to alter the chromatin structure toward a compact, silent state [58,59]. DNA methylation can also in some instances directly inhibit transcription by precluding the recruitment of DNA-binding proteins from their target sites [60].

The DNMT family is composed of four members: DNMT1, DNMT3A, DNMT3B, and DNMT3L [61]. The catalytic members of the DNMT family are classified into *de novo* and maintenance enzymes. DNMT3A and DNMT3B are involved in the *de novo* methylation of unmethylated and hemimethylated sites that occurs during embryonic development and germline reprogramming. DNMT3L is expressed during gametogenesis; it acts as a general stimulatory factor for DNMT3A and DNMT3B and interacts with them in the nucleus [61]. DNMT1 is the most abundant DNMT enzyme within the cell and is transcribed mostly during the S phase in the cell cycle. DNMT1, together with its conserved partner UHRF1 [62], localizes to the DNA replication fork, where they ensure that hemimethylated sites generated during semiconservative DNA replication are restored back to full methylation on both DNA strands [63,64]. The combined action of the *de novo* and maintenance functions of DNMTs ensures that DNA methylation patterns are faithfully copied from mother to daughter cells at each generation, therefore representing a form of mitotically stable epigenetic memory [65]. DNA methylation can, in principle, be counterbalanced by DNA demethylation. DNA demethylation can be passive, i.e., generated upon DNA replication in the absence of DNA maintenance methylation and can be active as well, which is thought to be enzyme-mediated DNA demethylation. Multiple mechanisms involving an oxidative pathway mediated by the ten−eleven translocation (Tet) family of enzymes and/or the deamination of methylated residues, followed by the repair of these modified bases, have been proposed [66−68].

4.3.2 Histone Modifications

Modifications of histone proteins play an important role in transcriptional regulation, DNA repair, DNA replication, alternative splicing, and chromosome condensation. Histones are conserved proteins that pack DNA into structural units and are classified as core (H2A, H2B, H3, and H4) and linker (H1 and H5) histones. The core histones aggregate into two H2A−H2B dimers and one H3−H4 tetramer to form a histone octamer. A 147-bp segment of DNA is wrapped in 1.65 turns around the histone octamer to make up one nucleosome, the basic unit of chromatin. Neighboring nucleosomes are separated by a linker DNA. The linker histones bind to linker DNA by sealing off the nucleosome at the location where DNA enters and leaves [69,70]. Several posttranslational modifications occur in histone tails, such as acetylation, methylation, phosphorylation, ubiquitination, SUMOylation, and ADP-ribosylation [71]. These modifications induce changes of the chromatin structure and thereby affect the accessibility of DNA to transcriptional factors and enzymes, resulting in gene activation or repression [72].

The best studied histone modification is acetylation. Histone acetylation neutralizes positive charges on lysine and weakens the interactions between DNA and histone. As a result, histone acetylation is usually associated with upregulated transcriptional activity of the associated gene. On the other hand, histone deacetylation is thought to stabilize the local chromatin structure by restoring the positive charges in lysine residues. Therefore, histone deacetylases are mainly considered as transcriptional repressors [73]. Histone acetylation and deacetylation are catalyzed by histone acetyltransferases (HATs) and histone deacetylases (HDACs), respectively [74,75]. Unlike acetylation, histone methylation does not alter the charge of histone proteins. Histone methylation mainly occurs at lysine and arginine residues of histones H3 and H4 [73]. Depending on the position and the degree of methylation, histone methylation can be associated with gene activation or repression [76–79].

4.4 EPIGENETICS AND FIBROSIS

Extensive evidence indicates that epigenetic mechanisms are involved in the pathogenesis of fibrotic diseases (Table 4.1). Herein, we discuss the current literature findings on epigenetics in fibrosis of the lung, liver, and skin [2,80,81].

4.4.1 Epigenetics and Lung Fibrosis

4.4.1.1 DNA Methylation and Lung Fibrosis

It is widely believed that aberrantly activated alveolar epithelial cells (AECs) induce the formation of fibroblastic and myofibroblastic foci. Activated myofibroblasts consequently secrete exaggerated amounts of extracellular matrix molecules and contribute to the development of fibrosis [106]. Environmental factors and genetic transmission are the

TABLE 4.1   Aberrant Epigenetic Patterns in Fibrotic Diseases

Disease	DNA methylation	Histone modification
IPF	Thy-1 [82,83]	TGF-β1 [84]
		H3K9Me3, HDAC-2, HDAC-4 [85]
		Thy-1 [86]
Liver fibrosis	PTCH1 [87,88]	Wnt [91]
	PTEN [89]	H3K4me1, H3K4me2, H3K4me3, H3K9me2, and H3K9me3
	RASAL1 [90]	SET7/9 [92]
		ASH1 [93]
SSc	DKK1 and SFRP1 [94,95]	HDAC-7 [102]
	Wnts [94]	HDAC-3 [102]
	Fli1 [53,96,97]	p300 [103]
	CD40L [98]	H3K27me3 [104]
	CD70 [99]	H4 and H3K9 [105]
	DNMT1 [100,101]	

most important risk factors for aberrant AEC activation [107]. Among these two aspects, epigenetic mechanisms act as the link between environmental factors and gene expression. Environmental factors can cause epigenetic modifications, affecting gene expression patterns without changes in the nucleotide sequence [108] (Figure 4.1). Epigenomic studies of DNA methylation profiles in IPF have identified several targets responsible for the pathogenesis. Thy-1 is a membrane protein expressed on multiple cells including fibroblasts and has been shown to suppress fibroblast differentiation into myofibroblasts. In IPF, the proliferating myofibroblasts within the fibroblastic foci are Thy-1(−), whereas normal lung fibroblasts are predominantly Thy-1(+) [109]. Downregulated Thy-1 in human lung tissues through hypermethylated Thy-1 promoters has been found in fibroblastic foci in IPF. Furthermore, treatment with DNMT inhibitor restores Thy-1 expression in Thy-1(−) fibroblasts possibly by reversing the epigenetically imposed suppression. Hypoxia

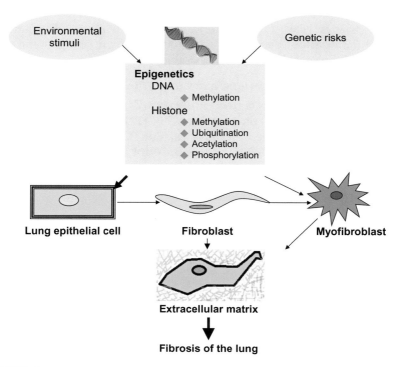

FIGURE 4.1 Epigenetic regulations in fibrosis of the lung. Interactions between environmental factors and genetic predisposition contribute to the modulations of epigenetic mechanisms on epithelial cells, fibroblasts, and myofibroblast in the lung. Aberrantly activated AECs induce the formation of fibroblastic and myofibroblastic foci which lead to excessive deposits of extracellular matrix and subsequently fibrosis.

has been implicated in the pathogenesis of pulmonary fibrosis. Robinson et al. [110] reported an increase in global hypermethylation patterning in human pulmonary fibroblasts cultured under hypoxic conditions and examined the methylation status of Thy-1 of normal human fibroblasts cultured under hypoxic conditions by methylation-specific PCR. Their data showed that Thy-1 is suppressed in hypoxia fibroblasts and that treatment with a DNMT inhibitor (5-aza2dC) reversed the hypoxic suppression of Thy-1. Altogether, these data indicate that hypermethylation of the Thy-1 promoter region will lead to Thy-1 suppression, and it may contribute to the pathogenesis of IPF [82,83].

Comparison of global methylation patterns associated with Agilent Human CpG Islands Microarrays between lung tissues from 12 patients with IPF with adenocarcinomas and normal lung tissues obtained from the same 10 lung cancer patients, showed that 625 CpG islands were differentially methylated in IPF lung tissues. Furthermore, it was interesting to note that the methylation pattern of IPF is more similar to that of lung cancer tissues, with an altered methylation pattern overlapping in 65% of the CpG islands [111]. Further work on the methylated CpG islands in gene regulation should be aimed at examining how the epigenome modulates the pathological development of IPF.

4.4.1.2 Histone Modifications and Lung Fibrosis

Abnormal expression of transforming growth factor-β1 (TGF-β1) plays an important role in the pathogenesis of IPF [112]. Nuclear factor (NF)-κB and AP-1 enhance histone acetylation at the TGF-β1 promoter site and subsequently activate TGF-β1 gene transcription. Using ELISA-based binding and chromatin immunoprecipitation assays, Lee et al. demonstrated that IL-1β-stimulated transcription of TGF-β1 is temporally regulated by ubiquitous inflammatory transcription factors NF-κB and AP-1 through histone hyperacetylation at distinct promoter sites [84]. In the bleomycin-induced mouse model of lung disease, the Fas promoter region in fibroblasts showed diminished histone acetylation and increased histone 3 lysine 9 trimethylation (H3K9Me3). This was associated with increased histone deacetylases HDAC-2 and HDAC-4 expression. The researchers also found that treatment with HDAC inhibitors would increase the Fas expression and restore the susceptibility of Fas-mediated apoptosis. Moreover, fibroblasts from patients with IPF likewise exhibited a reduction in histone acetylation and an elevation in H3K9Me3 in the Fas promoter. In the presence of an HDAC inhibitor, the expression of Fas was upregulated as well. These findings demonstrated the critical role of histone modifications in the development of resistance to apoptosis in fibroblasts both in the murine model and in patients with IPF and suggested novel therapies in progressive fibroproliferative disorders [85]. One study using fibroblasts from rat lungs has shown that in addition to epigenetic regulation by DNA methylation, Thy-1 silencing is regulated by histone modifications [86]. Another study also demonstrated that epigenetic regulation of Thy-1 gene expression is involved in lipopolysaccharide (LPS)-induced lung fibroblast proliferation [83]. However, to date, no epigenome-wide histone modification profile has been conducted in IPF.

4.4.2 Epigenetics and Liver Fibrosis

4.4.2.1 DNA Methylation and Liver Fibrosis

Activation of hepatic stellate cells (HSCs) is the pivotal step in fibrino-genesis and is mediated through methylation or demethylation of Patched1 (PTCH1), phosphatase and tension homolog (PTEN), and Ras GTPase activating-like protein1 (RASAL1) (Figure 4.2). The RASAL1 gene is located on chromosome 12q23−24 and is a member of the RAS−GAP family, which allows Ras inactivation by binding to GTP−Ras [113]. HSC is activated through the hypermethylation and MeCP2 regulation of PTCH1 and RASAL1 as well as by the demethylation of PTEN. Aberrant

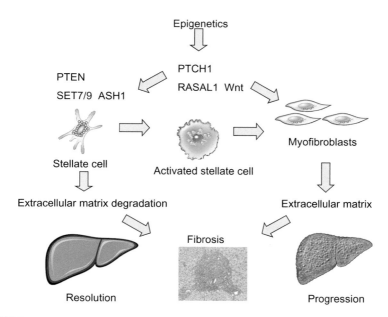

FIGURE 4.2 **Epigenetic regulation in HSC in the development of liver fibrosis.** Epigenetics plays a major role in the transcriptional regulation, conformitying progression and resolution processes. Epigenetic mechanisms regulate HSCs activation, differentiation of HSCs to myofibroblasts, and liver fibrogenesis. On the other hand, expression of other sets of epigenetic factors in HSCs leads to a decrease in collagen production and resolution of fibrosis.

hypermethylation in the PTCH1 gene promoter contributes to increased Gli1 and Smad3 activity during HSC activation. Similarly, the expression of RASAL1 mediated by DNA methylation and MeCP2 may provide molecular mechanisms for HSCs activation and liver fibrosis [90]. DNMT1-mediated PTEN hypermethylation would lead to the loss of PTEN expression, followed by the activation of the PI3K/AKT and ERK pathways, and subsequently HSCs activation [89].

4.4.2.2 Histone Modifications and Liver Fibrosis

Myofibroblastic transdifferentiation (MTD) is a physiological homeostatic response to chronic injury in the liver characterized by transdifferentiation of HSC and fibroblasts. The well-orchestrated, differential expression of the molecular events of MTD in liver clearly demonstrates the contribution of epigenetic regulation. Studies of epigenetic events regarding transcriptional activation of adiopogenic genes showed that DNA demethylation, histone H3K4 methylation, histone H3K9 demethylation, histone deacetylase downregulation, and histone acetylation are involved in the formation of euchromatin and the transcription of the members of the nuclear receptor superfamily, such as peroxisome proliferator-activated receptor gamma (PPAR-γ) [114–116]. On the other hand, epigenetic silencing of adipogeneic genes is mediated by histone deacetylation and di- or trimethylation of histone H3K27 and H3K9, leading to heterochromatin formation [117].

It is known that Wnt signaling plays a role in human fibrotic diseases, such as IPF, renal fibrosis, and liver fibrosis. Wnt signaling promotes liver fibrosis by enhancing HSCs activation and survival [91]. The necdin–Wnt10 pathway has also been reported to play an epigenetic role in suppressing PPAR-γ thus possessing an antiadiogenic effect on HSC MTD [118]. Necdin is a maternally imprinted gene that functions to inhibit adiopocyte differentiation [119] and is induced in HSC MTD in animal models of liver fibrosis. Necdin binds to the Wnt10 promoter to induce Wnt10b transactivation. Blocking canonical Wnt signaling in myofibroblastic HSC with co-receptor antagonist, Dickkopf-1, abrogates these epigenetic mechanisms and restores PPAR-γ expression as well as HSC differentiation [120,121]. In the bile duct ligation (BDL)-induced cholestatic rat model, TGF-β1 gene expression was significantly upregulated in liver tissues. In this model, Chen et al. examined the contribution of histone modifications and showed that during cholestasis, levels of H3K9 methylation decreased while levels of H3K4 methylation increased, as evidenced by an increase in active chromatin marks (H3K4me1, H3K4me2, and H3K4me3) and a decrease in the levels of repressive marks (H3K9me2 and H3K9me3) in the TGF-β1 promoter. Furthermore, the levels and recruitment of SET7/9, a H3K4 methytransferase, were also elevated. Notably, SET7/9 gene knockdown by siRNAs significantly reduced

BDL-induced TGF-β1 gene expression, serum enzymes (aspartate transaminase and alanine transaminase), and liver collagen content. SET7/9 gene knockdown also correlates with the increase of H3K4 methylation and the decrease of H3K9. These findings suggest that SET7/9 can potentially become a therapeutic target for cholestasis and perhaps fibrosis [92].

A recent study has reported that H3K4 methyltransferase contributes to fibrogenesis via the activation of profibrogenic genes [93]. Comparison of the expression profiles of epigenetic regulators between activated and quiescent HSCs revealed that numerous histone methyltransferases of H3K4 are activated during HSC MTD. In particular, using cross-linked chromatin immunoprecipitation, it was shown that ASH1, a histone–lysine N-methyltransferase, directly bound to the regulatory regions of alpha smooth muscle actin (αSMA), collagen I, tissue inhibitor of metalloproteinase-1 (TIMP1), and TGF-β1 in activated HSCs. Depletion of ASH1 caused broad suppression of fibrogenic gene expression. Perugorria et al. also discovered that MeCP2 positively regulated ASH1 expression and therefore identified ASH1 as a key transcriptional activator component of the MeCP2 epigenetic relay pathway that orchestrates coordinated induction of multiple profibrogenic genes [93].

4.4.3 Epigenetics and Systemic Sclerosis (Scleroderma)

4.4.3.1 DNA Methylation and SSc

Although both genetic and environmental factors are believed to play a role in the disease process of SSc, the etiology of SSc remains elusive. HLA genes including *HLA-A, -B, -C, -DR, -DP,* and *-DQ* are most closely linked to SSc susceptibility. In addition, a large number of non-HLA genes have also been identified to be associated with SSc [122,123]. Recent advances in molecular genetics further indicated that epigenetic modifications play an important role in the development and pathogenesis of SSc (Figure 4.3) [48,53,122]. In patients with SSc, DNMT1 expression was significantly upregulated in microvascular endothelial cells and fibroblasts isolated from the skin while the activity of DNA demethylase was significantly diminished [100,101]. The ability for Wnt to induce fibroblast activation and progenitor cell differentiation *in vitro* shows that this developmental pathway plays an important role in SSc pathogenesis. The promoter regions of *DKK1* and *SFRP1*, which are endogenous Wnt antagonists, were hypermethylated in fibroblasts and peripheral blood mononuclear cells (PBMC) from patients with SSc and possibly contributed to aberrant Wnt signaling in SSc pathogenesis. Inhibition of DNMT could effectively reduce Wnt signaling in cultured fibroblasts and treatment with DNMT inhibitors ameliorated bleomycin-induced dermal fibrosis [94,95]. Altogether, these data

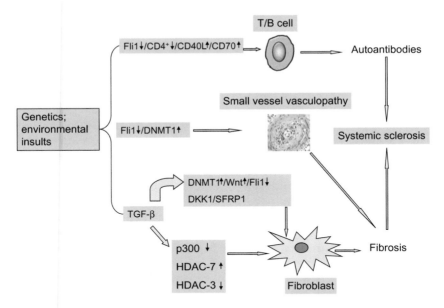

FIGURE 4.3   **Epigenetic aberrations leading to extracellular matrix deposition in SSc.**
Epigenetic aberrations play major roles in the development of small vessel vasculopathy,
production of autoantibodies, and fibroblast dysfunction, leading to increased deposition
of extracellular matrix in SSc.

suggested a link between epigenetic alterations and activated Wnt signaling in the pathogenesis of SSc via promoter hypermethylation and
subsequently silencing of Wnt antagonists. Blocking Wnt-β-catenin
signaling with currently available drugs, such as the PPAR-γ ligand
thiazolidinediones, the vitamin D derivative paricalcitol, the antihel-
minthic drug pyrvinium, or novel small molecules, represents potential
antifibrotic therapeutic approaches awaiting further investigation
[94,95,124,125]. The transcription factor Fli1 (Friend leukemia integra-
tion-1) is expressed in fibroblasts, endothelial cells, and immune cells
with a considerable role in the activation, differentiation, development,
and survival of these cells. In SSc patients, Fli1 expression was attenuated in fibroblasts, regulated by epigenetic mechanisms. It is known
that Fli1 deficiency in fibroblasts and endothelial cells contributes to the
histopathologic features of fibrosis and vasculopathy in SSc, respectively
[53,96,97].

Reduced DNA methylation and aberrant expression of methylation-
related genes in CD4$^+$ T cells are also correlated with SSc [126].
Demethylation of CD40L regulatory elements on the inactive X chromosome contributes to CD40L over-expression in female SSc patients, and

may partly account for the female predilection seen with SSc [98]. Demethylation of the CD70 promoter region contributes to an over-expression of CD70 in $CD4^+$ T cells and may lead to autoimmune responses in SSc [99].

4.4.3.2 *Histone Modifications and SSc*

It has been reported that knockdown of the *HDAC-7* gene in SSc fibroblasts leads to the decrease of type I and type III collagens on both mRNA and protein levels, which suggests that targeting *HDAC-7* may be an effective antifibrotic therapy for SSc [102]. Likewise, acetyltrans-ferase p300 and TGF-β1 play important roles in skin fibrosis. A recent study by Ghosh et al. formulated a link between fibrosis and aberrant p300 expression in patients with SSc. They reported that p300 expres-sion was highly upregulated in skin biopsies from SSc patients. Furthermore, p300 upregulation could be induced by TGF-β1 and was correlated with histone hyperacetylation in fibroblasts. TGF-β1 enhanced p300 recruitment and *in vivo* histone H4 acetylation at the col-lagen type I, α2 locus. Hence, p300-mediated histone acetylation could be a fundamental epigenetic mechanism in the fibrogenesis of SSc [103]. In contrast with other histone methylation markers, histone H3 on lysine 27 (H3K27me3) counterregulated the upregulation of the profi-brotic genes in SSc fibroblasts by repressing the expression of fra-2 [104]. Global histone H4 and H3K9 hypomethylation were observed in peripheral B cells from patients with severe SSc, indicating that aberrant histone acetylation/methylation and histone modifier gene expression contribute to the pathogenesis of SSc [105].

4.5 AUTOIMMUNITY AND FIBROSIS

Autoimmune diseases (AIDs) are thought to result from a combina-tion of genetic susceptibility and environmental exposures, leading to a breakdown in immune tolerance mechanisms [127]. Multiple indepen-dent studies have illustrated that innate immune sensors can stimulate fibrosis through a variety of mechanisms [128]. Epigenetic mechanisms are essential for normal development and function of the immune sys-tem; imbalance of epigenetic homeostasis in the immune responses can lead to aberrant gene expression and contribute to immune dysfunction [129]. Specifically, extensive research has shown that Toll-like receptor signals regulate inflammation, matrix remodeling, and ultimately acti-vate fibrosis. Cytokines such as TGF-β1, interleukin-1 (IL-1), IL-4, IL-6, IL-13, and IFNs appear to be particularly important in regulating the profibrotic phases in innate immune activation. These mechanisms are critical in fibrotic diseases and have an impact on multiple organ systems, including lung, liver, kidney, and skin [129,130]. The differentiation of T-helper cells is known to be regulated at the epigenetic level. Indeed, Th1 cells have an epigenetic profile with a demethylated IFN-γ promoter and with repressive histone H3K27me3 at the IL-4 and IL-13 locus while the opposite was observed in IL-4, IL-13, and IFN-γ promoters in Th2 cells [131]. Studies in animal models showed that environmental factors can either induce or exacerbate AIDs and this can serve as a valuable window to understand the influence of environmental insults regarding epigenetic regulation in AIDs [132].

It is difficult to identify the agent responsible for a specific disease, since the same agent can induce multiple diseases, and different factors can induce and/or exacerbate the same clinical features. Therefore, new criteria are being developed for environmentally associated AIDs [133]. According to findings from a National Institute of Environmental Health Sciences Expert Panel Workshop, solvent exposure contributes to the development of SSc [134]. Current consensus indicates that environmental agents could induce aberrant self antigen modification which then triggers a series of immune reactions and breaks immune tolerance [134]. Epigenetics is receiving unprecedented attention from clinicians and researchers in the study of SSc and other AIDs. The expectation is that in the near future, the expanding knowledge in epigenetics and AIDs will bring about new therapeutics, means for early interventions, and perhaps prevention modalities [54].

4.6 CONCLUSION

There is increasing evidence that shows fibrosis is a dynamic and reversible process. The strongest evidence for the reversal of fibrosis involving parenchymal internal organs was provided by liver data from several clinical studies, which reported resolution of fibrosis upon successful treatment of underlying hepatitis infections [135,136]. Numerous studies have also reported possible regression of established fibrosis in animal models of fibrogenesis [137–139].

Thus far, no drug has been approved by the Food and Drug Administration with antifibrotic effect. One of the major challenges for the development of antifibrotic drugs is the lack of disease-specific biomarkers, which can be used to identify patients who may benefit from a specific therapy. Therefore, future research for identification of novel and specific biomarkers for early diagnosis and prognosis is necessary to improve the clinical management of patients with fibrotic diseases. Fibrotic research is also expected to provide new insights into disease pathogenesis as well as therapeutic approaches [140–142]. In conclusion, knowledge gathered on epigenetic regulation in fibrosis will provide a novel avenue for effective antifibrotic therapies.

Acknowledgments

This work is supported in part by NIH grant DK 39588, Natural Sciences Foundation of China (No. 81241094), and Science and Technology program of basic research projects of Qingdao city (No. 13-1-4-138-jch). There are no financial or commercial conflicts of interest.

List of Abbreviations

αSMA	alpha smooth muscle actin
AEC	alveolar epithelial cell
AID	autoimmune disease
ATII	alveolar epithelial type II
DNMT	DNA methyltransferase
Fli1	Friend leukemia integration-1
HAT	histone acetyltransferase
HDAC	histone deacetylase
HSC	hepatic stellate cell
IFN	interferon
IL-1	interleukin-1
IPF	idiopathic pulmonary fibrosis
LPS	lipopolysaccharide
MBD	methyl-CpG-binding domain
MeCP2	methyl-CpG-binding protein 2
MTD	myofibroblast transdifferentiation
NF-κB	nuclear factor-κB
PBMC	peripheral blood mononuclear cell
PPAR-γ	peroxisome proliferator-activated receptor gamma
PTCH1	patched 1
PTEN	phosphatase and tensin homolog
RASAL1	Ras GTPase activating-like protein 1
SAM	S-adenosyl methionine
SSc	systemic sclerosis
TGF-β1	transforming growth factor-β1
Th cell	T-helper cell
TIMP1	tissue inhibitor of metalloproteinase-1

References

[1] Wynn TA. Cellular and molecular mechanisms of fibrosis. J Pathol 2008;214:199–210.
[2] Robinson CM, Watson CJ, Baugh JA. Epigenetics within the matrix: a neo-regulator of fibrotic disease. Epigenetics 2012;7:987–93.
[3] Mann J, Mann DA. Epigenetic regulation of wound healing and fibrosis. Curr Opin Rheumatol 2013;25:101–7.
[4] Mehal WZ, Iredale J, Friedman SL. Scraping fibrosis: expressway to the core of fibrosis. Nat Med 2011;17:552–3.

[5] Friedman SL, Bansal MB. Reversal of hepatic fibrosis—fact or fantasy? Hepatology 2006;43:S82—8.

[6] Zeisberg M, Neilson EG. Mechanisms of tubulointerstitial fibrosis. J Am Soc Nephrol 2010;21:1819—34.

[7] Rogatsky I, Chandrasekaran U, Manni M, Yi W, Pernis AB. Epigenetics and the IRFs: a complex interplay in the control of immunity and autoimmunity. Autoimmunity 2014;47:242—55.

[8] Pillai S. Rethinking mechanisms of autoimmune pathogenesis. J Autoimmun 2013;45:97—103.

[9] Wang L, Wheeler DA. Genomic sequencing for cancer diagnosis and therapy. Annu Rev Med 2014;65:33—48.

[10] Tarayrah L, Chen X. Epigenetic regulation in adult stem cells and cancers. Cell Biosci 2013;3:41.

[11] Hardie WD, Glasser SW, Hagood JS. Emerging concepts in the pathogenesis of lung fibrosis. Am J Pathol 2009;175:3—16.

[12] Wynn TA, Ramalingam TR. Mechanisms of fibrosis: therapeutic translation for fibrotic disease. Nat Med 2012;18:1028—40.

[13] Rosenbloom J, Mendoza FA, Jimenez SA. Strategies for anti-fibrotic therapies. Biochim Biophys Acta 2013;1832:1088—103.

[14] American Thoracic Society/European Respiratory Society International Multidisciplinary Consensus Classification of the Idiopathic Interstitial Pneumonias. This joint statement of the American Thoracic Society (ATS), and the European Respiratory Society (ERS) was adopted by the ATS board of directors, June 2001 and by the ERS Executive Committee, June 2001. Am J Respir Crit Care Med 2002;165:277—304.

[15] Raghu G, Collard HR, Egan JJ, Martinez FJ, Behr J, Brown KK, et al. An official ATS/ERS/JRS/ALAT statement: idiopathic pulmonary fibrosis: evidence-based guidelines for diagnosis and management. Am J Respir Crit Care Med 2011;183:788—824.

[16] Meltzer EB, Noble PW. Idiopathic pulmonary fibrosis. Orphanet J Rare Dis 2008;3:8.

[17] Raghu G, Weycker D, Edelsberg J, Bradford WZ, Oster G. Incidence and prevalence of idiopathic pulmonary fibrosis. Am J Respir Crit Care Med 2006;174:810—16.

[18] Gribbin J, Hubbard RB, Le Jeune I, Smith CJ, West J, Tata LJ. Incidence and mortality of idiopathic pulmonary fibrosis and sarcoidosis in the UK. Thorax 2006;61: 980—5.

[19] Navaratnam V, Fleming KM, West J, Smith CJ, Jenkins RG, Fogarty A, et al. The rising incidence of idiopathic pulmonary fibrosis in the UK. Thorax 2011;66:462—7.

[20] Collard HR, King Jr. TE, Bartelson BB, Vourlekis JS, Schwarz MI, Brown KK. Changes in clinical and physiologic variables predict survival in idiopathic pulmonary fibrosis. Am J Respir Crit Care Med 2003;168:538—42.

[21] Soares Pires F, Caetano Mota P, Melo N, Costa D, Jesus JM, Cunha R, et al. Idiopathic pulmonary fibrosis—clinical presentation, outcome and baseline prognostic factors in a Portuguese cohort. Rev Port Pneumol 2013;19:19—27.

[22] World Health Organization. Revised global burden of disease 2002 estimates. <http://www.who.int/healthinfo/global_burden_disease/en/index.html> [accessed 27.10.08].

[23] Fleming KM, Aithal GP, Solaymani-Dodaran M, Card TR, West J. Incidence and prevalence of cirrhosis in the United Kingdom, 1992—2001: a general population-based study. J Hepatol 2008;49:732—8.

[24] Pauly MP, Ruebner BH. Hepatic fibrosis and cirrhosis in tropical countries (including portal hypertension). Baillieres Clin Gastroenterol 1987;1:273—96.

[25] Zatonski WA, Sulkowska U, Manczuk M, Rehm J, Boffetta P, Lowenfels AB, et al. Liver cirrhosis mortality in Europe, with special attention to Central and Eastern Europe. Eur Addict Res 2010;16:193—201.

[26] Minino AM. Death in the United States, 2009. NCHS Data Brief 2011:1−8.

[27] Minino AM. Death in the United States, 2011. NCHS Data Brief 2013:1−8.

[28] Asrani SK, Larson JJ, Yawn B, Therneau TM, Kim WR. Underestimation of liver-related mortality in the United States. Gastroenterology 2013;145:375−82, e1−2.

[29] Sorensen HT, Thulstrup AM, Mellemkjar L, Jepsen P, Christensen E, Olsen JH, et al. Long-term survival and cause-specific mortality in patients with cirrhosis of the liver: a nationwide cohort study in Denmark. J Clin Epidemiol 2003;56:88−93.

[30] Lonzetti LS, Joyal F, Raynauld JP, Roussin A, Goulet JR, Rich E, et al. Updating the American College of Rheumatology preliminary classification criteria for systemic sclerosis: addition of severe nailfold capillaroscopy abnormalities markedly increases the sensitivity for limited scleroderma. Arthritis Rheum 2001;44:735−6.

[31] Nadashkevich O, Davis P, Fritzler MJ. A proposal of criteria for the classification of systemic sclerosis. Med Sci Monit 2004;10:CR615−21.

[32] Pattanaik D, Brown M, Postlethwaite AE. Vascular involvement in systemic sclerosis (scleroderma). J Inflamm Res 2011;4:105−25.

[33] Jacobsen S, Halberg P, Ullman S. Mortality and causes of death of 344 Danish patients with systemic sclerosis (scleroderma). Br J Rheumatol 1998;37:750−5.

[34] Jimenez SA, Derk CT. Following the molecular pathways toward an understanding of the pathogenesis of systemic sclerosis. Ann Intern Med 2004;140:37−50.

[35] Ringel RA, Brick JE, Brick JF, Gutmann L, Riggs JE. Muscle involvement in the scleroderma syndromes. Arch Intern Med 1990;150:2550−2.

[36] Sjogren RW. Gastrointestinal motility disorders in scleroderma. Arthritis Rheum 1994;37:1265−82.

[37] Muangchan C, Canadian Scleroderma Research Group, Baron M, Pope J. The 15% rule in scleroderma: the frequency of severe organ complications in systemic sclerosis. A systematic review. J Rheumatol 2013;40:1545−56.

[38] Di Ciaula A, Covelli M, Berardino M, Wang DQ, Lapadula G, Palasciano G, et al. Gastrointestinal symptoms and motility disorders in patients with systemic scleroderma. BMC Gastroenterol 2008;8:7.

[39] Nitsche A. Raynaud, digital ulcers and calcinosis in scleroderma. Reumatol Clin 2012;8:270−7.

[40] Kanarek DJ, Mark EJ. Case records of the Massachusetts General Hospital. Weekly clinicopathological exercises. Case 20-1989. A 33-year-old woman with exertional dyspnea and Raynaud's phenomenon. N Engl J Med 1989;320:1333−40.

[41] Bolster MB, Silver RM. Lung disease in systemic sclerosis (scleroderma). Baillieres Clin Rheumatol 1993;7:79−97.

[42] van den Hoogen F, Khanna D, Fransen J, Johnson SR, Baron M, Tyndall A, et al. 2013 classification criteria for systemic sclerosis: an American College of Rheumatology/European League against Rheumatism collaborative initiative. Arthritis Rheum 2013;65:2737−47.

[43] van den Hoogen F, Khanna D, Fransen J, Johnson SR, Baron M, Tyndall A, et al. 2013 classification criteria for systemic sclerosis: an American College of Rheumatology/European League against Rheumatism collaborative initiative. Ann Rheum Dis 2013;72:1747−55.

[44] Steen VD, Medsger Jr. TA. Severe organ involvement in systemic sclerosis with diffuse scleroderma. Arthritis Rheum 2000;43:2437−44.

[45] Ferri C, Valentini G, Cozzi F, Sebastiani M, Michelassi C, La Montagna G, et al. Systemic sclerosis: demographic, clinical, and serologic features and survival in 1,012 Italian patients. Medicine (Baltimore) 2002;81:139−53.

[46] Hachulla AL, Launay D, Gaxotte V, de Groote P, Lamblin N, Devos P, et al. Cardiac magnetic resonance imaging in systemic sclerosis: a cross-sectional observational study of 52 patients. Ann Rheum Dis 2009;68:1878−84.

[47] Denton CP, Lapadula G, Mouthon L, Muller-Ladner U. Renal complications and scleroderma renal crisis. Rheumatology (Oxford) 2009;48(Suppl. 3):iii32—5.

[48] Nikpour M, Stevens WM, Herrick AL, Proudman SM. Epidemiology of systemic sclerosis. Best Pract Res Clin Rheumatol 2010;24:857—69.

[49] LeRoy EC, Medsger Jr. TA. Criteria for the classification of early systemic sclerosis. J Rheumatol 2001;28:1573—6.

[50] Mayes MD, Lacey Jr. JV, Beebe-Dimmer J, Gillespie BW, Cooper B, Laing TJ, et al. Prevalence, incidence, survival, and disease characteristics of systemic sclerosis in a large US population. Arthritis Rheum 2003;48:2246—55.

[51] Arias-Nunez MC, Llorca J, Vazquez-Rodriguez TR, Gomez-Acebo I, Miranda-Filloy JA, Martin J, et al. Systemic sclerosis in northwestern Spain: a 19-year epidemiologic study. Medicine (Baltimore) 2008;87:272—80.

[52] Steen VD, Medsger TA. Changes in causes of death in systemic sclerosis, 1972—2002. Ann Rheum Dis 2007;66:940—4.

[53] Barnes J, Mayes MD. Epidemiology of systemic sclerosis: incidence, prevalence, survival, risk factors, malignancy, and environmental triggers. Curr Opin Rheumatol 2012;24:165—70.

[54] Antequera F, Bird A. Number of CpG islands and genes in human and mouse. Proc Natl Acad Sci USA 1993;90:11995—9.

[55] Wu C, Morris JR. Genes, genetics, and epigenetics: a correspondence. Science 2001;293:1103—5.

[56] Skinner MK. Role of epigenetics in developmental biology and transgenerational inheritance. Birth Defects Res C Embryo Today 2011;93:51—5.

[57] Herman JG, Baylin SB. Gene silencing in cancer in association with promoter hypermethylation. N Engl J Med 2003;349:2042—54.

[58] Esteller M. Epigenetic gene silencing in cancer: the DNA hypermethylome. Hum Mol Genet 2007;16(Spec. No. 1):R50—9.

[59] Lopez-Serra L, Esteller M. Proteins that bind methylated DNA and human cancer: reading the wrong words. Br J Cancer 2008;98:1881—5.

[60] Kuroda A, Rauch TA, Todorov I, Ku HT, Al-Abdullah IH, Kandeel F, et al. Insulin gene expression is regulated by DNA methylation. PLOS ONE 2009;4:e6953.

[61] Chedin F. The DNMT3 family of mammalian de novo DNA methyltransferases. Prog Mol Biol Transl Sci 2011;101:255—85.

[62] Sharif J, Muto M, Takebayashi S, Suetake I, Iwamatsu A, Endo TA, et al. The SRA protein Np95 mediates epigenetic inheritance by recruiting Dnmt1 to methylated DNA. Nature 2007;450:908—12.

[63] Law JA, Jacobsen SE. Establishing, maintaining and modifying DNA methylation patterns in plants and animals. Nat Rev Genet 2010;11:204—20.

[64] Portela A, Esteller M. Epigenetic modifications and human disease. Nat Biotechnol 2010;28:1057—68.

[65] Tsukada Y. Hydroxylation mediates chromatin demethylation. J Biochem 2012;151: 229—46.

[66] Fritz EL, Papavasiliou FN. Cytidine deaminases: AIDing DNA demethylation? Genes Dev 2010;24:2107—14.

[67] Kohli RM, Zhang Y. TET enzymes, TDG and the dynamics of DNA demethylation. Nature 2013;502:472—9.

[68] Pastor WA, Aravind L, Rao A. TETonic shift: biological roles of TET proteins in DNA demethylation and transcription. Nat Rev Mol Cell Biol 2013;14:341—56.

[69] Daujat S, Zeissler U, Waldmann T, Happel N, Schneider R. HP1 binds specifically to Lys26-methylated histone H1.4, whereas simultaneous Ser27 phosphorylation blocks HP1 binding. J Biol Chem 2005;280:38090—5.

[70] Kouzarides T. Chromatin modifications and their function. Cell 2007;128:693—705.

[71] Cohen I, Poreba E, Kamieniarz K, Schneider R. Histone modifiers in cancer: friends or foes? Genes Cancer 2011;2:631—47.

[72] Dieker J, Muller S. Epigenetic histone code and autoimmunity. Clin Rev Allergy Immunol 2010;39:78—84.

[73] Bannister AJ, Kouzarides T. Regulation of chromatin by histone modifications. Cell Res 2011;21:381—95.

[74] de Ruijter AJ, van Gennip AH, Caron HN, Kemp S, van Kuilenburg AB. Histone deacetylases (HDACs): characterization of the classical HDAC family. Biochem J 2003;370:737—49.

[75] Roth SY, Denu JM, Allis CD. Histone acetyltransferases. Annu Rev Biochem 2001;70:81—120.

[76] Hansen JC. Linking genome structure and function through specific histone acetylation. ACS Chem Biol 2006;1:69—72.

[77] Joanna F, van Grunsven LA, Mathieu V, Sarah S, Sarah D, Karin V, et al. Histone deacetylase inhibition and the regulation of cell growth with particular reference to liver pathobiology. J Cell Mol Med 2009;13:2990—3005.

[78] Tambaro FP, Dell'aversana C, Carafa V, Nebbioso A, Radic B, Ferrara F, et al. Histone deacetylase inhibitors: clinical implications for hematological malignancies. Clin Epigenetics 2010;1:25—44.

[79] Pandey P, Houben A, Kumlehn J, Melzer M, Rutten T. Chromatin alterations during pollen development in *Hordeum vulgare*. Cytogenet Genome Res 2013;141:50—7.

[80] Hinz B, Phan SH, Thannickal VJ, Prunotto M, Desmouliere A, Varga J, et al. Recent developments in myofibroblast biology: paradigms for connective tissue remodeling. Am J Pathol 2012;180:1340—55.

[81] Ghosh AK, Quaggin SE, Vaughan DE. Molecular basis of organ fibrosis: potential therapeutic approaches. Exp Biol Med (Maywood) 2013;238:461—81.

[82] Sanders YY, Pardo A, Selman M, Nuovo GJ, Tollefsbol TO, Siegal GP, et al. Thy-1 promoter hypermethylation: a novel epigenetic pathogenic mechanism in pulmonary fibrosis. Am J Respir Cell Mol Biol 2008;39:610—18.

[83] He Z, Wang X, Deng Y, Li W, Chen Y, Xing S, et al. Epigenetic regulation of Thy-1 gene expression by histone modification is involved in lipopolysaccharide-induced lung fibroblast proliferation. J Cell Mol Med 2013;17:160—7.

[84] Lee KY, Ito K, Hayashi R, Jazrawi EP, Barnes PJ, Adcock IM. NF-kappaB and activator protein 1 response elements and the role of histone modifications in IL-1beta-induced TGF-beta1 gene transcription. J Immunol 2006;176:603—15.

[85] Huang SK, Scruggs AM, Donaghy J, Horowitz JC, Zaslona Z, Przybranowski S, et al. Histone modifications are responsible for decreased Fas expression and apoptosis resistance in fibrotic lung fibroblasts. Cell Death Dis 2013;4:e621.

[86] Sanders YY, Tollefsbol TO, Varisco BM, Hagood JS. Epigenetic regulation of thy-1 by histone deacetylase inhibitor in rat lung fibroblasts. Am J Respir Cell Mol Biol 2011;45:16—23.

[87] Bian EB, Zhao B, Huang C, Wang H, Meng XM, Wu BM, et al. New advances of DNA methylation in liver fibrosis, with special emphasis on the crosstalk between microRNAs and DNA methylation machinery. Cell Signal 2013;25:1837—44.

[88] Yang JJ, Tao H, Huang C, Shi KH, Ma TT, Bian EB, et al. DNA methylation and MeCP2 regulation of PTCH1 expression during rats hepatic fibrosis. Cell Signal 2013;25:1202—11.

[89] Bian EB, Huang C, Ma TT, Tao H, Zhang H, Cheng C, et al. DNMT1-mediated PTEN hypermethylation confers hepatic stellate cell activation and liver fibrogenesis in rats. Toxicol Appl Pharmacol 2012;264:13—22.

[90] Tao H, Huang C, Yang JJ, Ma TT, Bian EB, Zhang L, et al. MeCP2 controls the expression of RASAL1 in the hepatic fibrosis in rats. Toxicology 2011;290:327—33.

[91] Miao CG, Yang YY, He X, Huang C, Huang Y, Zhang L, et al. Wnt signaling in liver fibrosis: progress, challenges and potential directions. Biochimie 2013;95:2326−35.

[92] Sheen-Chen SM, Lin CR, Chen KH, Yang CH, Lee CT, Huang HW, et al. Epigenetic histone methylation regulates transforming growth factor beta-1 expression following bile duct ligation in rats. J Gastroenterol 2014;49:1285−97.

[93] Perugorria MJ, Wilson CL, Zeybel M, Walsh M, Amin S, Robinson S, et al. Histone methyltransferase ASH1 orchestrates fibrogenic gene transcription during myofibroblast transdifferentiation. Hepatology 2012;56:1129−39.

[94] Dees C, Schlottmann I, Funke R, Distler A, Palumbo-Zerr K, Zerr P, et al. The Wnt antagonists DKK1 and SFRP1 are downregulated by promoter hypermethylation in systemic sclerosis. Ann Rheum Dis 2014;73:1232−9.

[95] Bayle J, Fitch J, Jacobsen K, Kumar R, Lafyatis R, Lemaire R. Increased expression of Wnt2 and SFRP4 in Tsk mouse skin: role of Wnt signaling in altered dermal fibrillin deposition and systemic sclerosis. J Invest Dermatol 2008;128:871−81.

[96] Wang Y, Fan PS, Kahaleh B. Association between enhanced type I collagen expression and epigenetic repression of the FLI1 gene in scleroderma fibroblasts. Arthritis Rheum 2006;54:2271−9.

[97] Asano Y, Bujor AM, Trojanowska M. The impact of Fli1 deficiency on the pathogenesis of systemic sclerosis. J Dermatol Sci 2010;59:153−62.

[98] Lian X, Xiao R, Hu X, Kanekura T, Jiang H, Li Y, et al. DNA demethylation of CD40l in CD4$^+$ T cells from women with systemic sclerosis: a possible explanation for female susceptibility. Arthritis Rheum 2012;64:2338−45.

[99] Hedrich CM, Rauen T. Epigenetic patterns in systemic sclerosis and their contribution to attenuated CD70 signaling cascades. Clin Immunol 2012;143:1−3.

[100] Kahaleh B, Wang W. SS.1.1 Decrease activity of DNA demethylase in SSC fibroblast and microvascular endothelial cells: a possible mechanism for persistent SSC phenotype. Rheumatology 2012;51(Suppl. 2):ii5−6.

[101] Bujor AM, Haines P, Padilla C, Christmann RB, Junie M, Sampaio-Barros PD, et al. Ciprofloxacin has antifibrotic effects in scleroderma fibroblasts via downregulation of Dnmt1 and upregulation of Fli1. Int J Mol Med 2012;30:1473−80.

[102] Hemmatazad H, Rodrigues HM, Maurer B, Brentano F, Pileckyte M, Distler JH, et al. Histone deacetylase 7, a potential target for the antifibrotic treatment of systemic sclerosis. Arthritis Rheum 2009;60:1519−29.

[103] Ghosh AK, Bhattacharyya S, Lafyatis R, Farina G, Yu J, Thimmapaya B, et al. p300 is elevated in systemic sclerosis and its expression is positively regulated by TGF-beta: epigenetic feed-forward amplification of fibrosis. J Invest Dermatol 2013;133:1302−10.

[104] Kramer M, Dees C, Huang J, Schlottmann I, Palumbo-Zerr K, Zerr P, et al. Inhibition of H3K27 histone trimethylation activates fibroblasts and induces fibrosis. Ann Rheum Dis 2013;72:614−20.

[105] Wang Y, Yang Y, Luo Y, Yin Y, Wang Q, Li Y, et al. Aberrant histone modification in peripheral blood B cells from patients with systemic sclerosis. Clin Immunol 2013;149:46−54.

[106] Habiel DM, Hogaboam C. Heterogeneity in fibroblast proliferation and survival in idiopathic pulmonary fibrosis. Front Pharmacol 2014;5:2.

[107] King Jr. TE, Pardo A, Selman M. Idiopathic pulmonary fibrosis. Lancet 2011;378:1949−61.

[108] Kabesch M, Adcock IM. Epigenetics in asthma and COPD. Biochimie 2012;94:2231−41.

[109] Hagood JS, Prabhakaran P, Kumbla P, Salazar L, MacEwen MW, Barker TH, et al. Loss of fibroblast Thy-1 expression correlates with lung fibrogenesis. Am J Pathol 2005;167:365−79.

[110] Robinson CM, Neary R, Levendale A, Watson CJ, Baugh JA. Hypoxia-induced DNA hypermethylation in human pulmonary fibroblasts is associated with Thy-1 promoter methylation and the development of a pro-fibrotic phenotype. Respir Res 2012;13:74.

[111] Rabinovich EI, Kapetanaki MG, Steinfeld I, Gibson KF, Pandit KV, Yu G, et al. Global methylation patterns in idiopathic pulmonary fibrosis. PLOS ONE 2012;7:e33770.

[112] Rafii R, Juarez MM, Albertson TE, Chan AL. A review of current and novel therapies for idiopathic pulmonary fibrosis. J Thorac Dis 2013;5:48−73.

[113] Bernards A, Settleman J. Loss of the Ras regulator RASAL1: another route to Ras activation in colorectal cancer. Gastroenterology 2009;136:46−8.

[114] Musri MM, Gomis R, Parrizas M. Chromatin and chromatin-modifying proteins in adipogenesis. Biochem Cell Biol 2007;85:397−410.

[115] Mann J, Oakley F, Akiboye F, Elsharkawy A, Thorne AW, Mann DA. Regulation of myofibroblast transdifferentiation by DNA methylation and MeCP2: implications for wound healing and fibrogenesis. Cell Death Differ 2007;14:275−85.

[116] Mann J, Chu DC, Maxwell A, Oakley F, Zhu NL, Tsukamoto H, et al. MeCP2 controls an epigenetic pathway that promotes myofibroblast transdifferentiation and fibrosis. Gastroenterology 2010;138:705−14, 714. e1−4.

[117] Okamura M, Inagaki T, Tanaka T, Sakai J. Role of histone methylation and demethylation in adipogenesis and obesity. Organogenesis 2010;6:24−32.

[118] Zhu NL, Wang J, Tsukamoto H. The Necdin-Wnt pathway causes epigenetic peroxisome proliferator-activated receptor gamma repression in hepatic stellate cells. J Biol Chem 2010;285:30463−71.

[119] Tseng YH, Butte AJ, Kokkotou E, Yechoor VK, Taniguchi CM, Kriauciunas KM, et al. Prediction of preadipocyte differentiation by gene expression reveals role of insulin receptor substrates and necdin. Nat Cell Biol 2005;7:601−11.

[120] Passino MA, Adams RA, Sikorski SL, Akassoglou K. Regulation of hepatic stellate cell differentiation by the neurotrophin receptor p75NTR. Science 2007;315:1853−6.

[121] Tsukamoto H, Zhu NL, Asahina K, Mann DA, Mann J. Epigenetic cell fate regulation of hepatic stellate cells. Hepatol Res 2011;41:675−82.

[122] Favalli E, Ingegnoli F, Zeni S, Fare M, Fantini F. [HLA typing in systemic sclerosis]. Reumatismo 2001;53:210−14.

[123] Broen JC, Coenen MJ, Radstake TR. Genetics of systemic sclerosis: an update. Curr Rheumatol Rep 2012;14:11−21.

[124] Konigshoff M, Kramer M, Balsara N, Wilhelm J, Amarie OV, Jahn A, et al. WNT1-inducible signaling protein-1 mediates pulmonary fibrosis in mice and is upregulated in humans with idiopathic pulmonary fibrosis. J Clin Invest 2009;119:772−87.

[125] Bhattacharyya S, Wei J, Varga J. Understanding fibrosis in systemic sclerosis: shifting paradigms, emerging opportunities. Nat Rev Rheumatol 2012;8:42−54.

[126] Lei W, Luo Y, Lei W, Luo Y, Yan K, Zhao S, et al. Abnormal DNA methylation in CD4 + T cells from patients with systemic lupus erythematosus, systemic sclerosis, and dermatomyositis. Scand J Rheumatol 2009;38:369−74.

[127] Hewagama A, Richardson B. The genetics and epigenetics of autoimmune diseases. J Autoimmun 2009;33:3−11.

[128] Lafyatis R, Farina A. New insights into the mechanisms of innate immune receptor signalling in fibrosis. Open Rheumatol J 2012;6:72−9.

[129] Strickland FM, Richardson BC. Epigenetics in human autoimmunity. Epigenetics in autoimmunity—DNA methylation in systemic lupus erythematosus and beyond. Autoimmunity 2008;41:278−86.

[130] Powell EE, Edwards-Smith CJ, Hay JL, Clouston AD, Crawford DH, Shorthouse C, et al. Host genetic factors influence disease progression in chronic hepatitis C. Hepatology 2000;31:828−33.

[131] Baguet A, Bix M. Chromatin landscape dynamics of the Il4-Il13 locus during T helper 1 and 2 development. Proc Natl Acad Sci USA 2004;101:11410–15.

[132] Selmi C, Lu Q, Humble MC. Heritability versus the role of the environment in autoimmunity. J Autoimmun 2012;39:249–52.

[133] Brooks WH, Le Dantec C, Pers JO, Youinou P, Renaudineau Y. Epigenetics and autoimmunity. J Autoimmun 2010;34:J207–19.

[134] Germolec D, Kono DH, Pfau JC, Pollard KM. Animal models used to examine the role of the environment in the development of autoimmune disease: findings from an NIEHS Expert Panel Workshop. J Autoimmun 2012;39:285–93.

[135] Farci P, Roskams T, Chessa L, Peddis G, Mazzoleni AP, Scioscia R, et al. Long-term benefit of interferon alpha therapy of chronic hepatitis D: regression of advanced hepatic fibrosis. Gastroenterology 2004;126:1740–9.

[136] Hui CK, Leung N, Shek TW, Yao H, Lee WK, Lai JY, et al. Sustained disease remission after spontaneous HBeAg seroconversion is associated with reduction in fibrosis progression in chronic hepatitis B Chinese patients. Hepatology (Baltimore, MD) 2007;46:690–8.

[137] Ismail MH, Pinzani M. Reversal of liver fibrosis. Saudi J Gastroenterol 2009;15:72–9.

[138] Kisseleva T, Brenner DA. Anti-fibrogenic strategies and the regression of fibrosis. Best Pract Res Clin Gastroenterol 2011;25:305–17.

[139] Friedman SL. Fibrogenic cell reversion underlies fibrosis regression in liver. Proc Natl Acad Sci USA 2012;109:9230–1.

[140] Asano Y. Future treatments in systemic sclerosis. J Dermatol 2010;37:54–70.

[141] Beyer C, Distler O, Distler JH. Innovative antifibrotic therapies in systemic sclerosis. Curr Opin Rheumatol 2012;24:274–80.

[142] Tampe B, Zeisberg M. Contribution of genetics and epigenetics to progression of kidney fibrosis. Nephrol Dial Transplant 2014;29(Suppl. 4):iv72–9.

5

Epigenetic Modulation of Hair Follicle Stem Cells

Haijing Wu and Qianjin Lu

Department of Dermatology, The Second Xiangya Hospital of Central South University, Hunan Key Laboratory of Medical Epigenomics, Changsha, Hunan, PR China

5.1 INTRODUCTION

Unlike cold-blooded animals, most mammals rely on hair to keep warm and avoid injury. The hair follicle (HF) is the particular organ that produces hair. During the production of hair, the follicle undergoes synchronized and dynamic cycles. It starts from a growing phase (anagen) to a remodeling phase (catagen), followed by a quiescent phase (telogen), and starts over again. There are two key elements that are responsible for hair regeneration and pigmentation in a cyclical manner. One is the follicular epithelial stem cell, and the other is the specialized mesenchymal cell. Both are found in a region of the follicle outer root sheath (ORS) known as the bulge [1,2]. In an HF cycle, epithelial cells divide extremely rapidly within follicles and the latter executes precisely timed differentiation programs during the growing phase (anagen). Next, as growth stops and the cycle enters the catagen phase, most mature follicles return to immature and developing status, after which they step into a period of mitotic quiescence (telogen). Stem cells are responsible for the control of these dynamic events that consist of a burst of activity and further morphogenetic remodeling as the follicles start to grow again in a new anagen phase. HF stem cells (HF-SCs) dominate the dynamic changes of this cycle. Meanwhile, cell—cell communication and autocrine or paracrine cytokines determine the fate of stem cells through an epigenetic regulatory network [3]. The theory of epigenetic regulation has been used increasingly to explain the influence of environmental factors on

© 2015 Elsevier Inc. All rights reserved.

the pluripotency of stem cells. The extent to which similar mechanisms affect HF-SCs, however, has been inadequately examined. Against this backdrop, this chapter focuses on epigenetic regulation in HF-SCs.

5.1.1 Identification of HF-SCs

As a fundamental element of HF-SCs, epithelial stem cells act as slow cycling cells but they perform with great proliferative capacity under certain conditions [4]. Some of these newly divided cells escape from the stem cell niche and become transit amplifying (TA) cells with a limited proliferative capacity, and they become postmitotic when this potential is exhausted. There is no reliable marker, however, to distinguish slow cycling stem cells from rapid proliferative TA cells, and researchers are now using cell kinetics to identify slow cycling stem cells [5]. BrdU and 3H-thymidine are widely used reagents to label synthesized DNA. As TA cells divide rapidly, diluted labeled DNA defines these cells. Detectable levels of these labels can identify slow cycling stem cells [6,7]. After a proliferating period, most rapidly proliferating TA cells lose their labels due to dilution. Only the slow cycling stem cells retain their labels, which make them identifiable by investigators, under the name of label-retaining cells (LRCs) [1]. Another method to label infrequently cycling stem cells involves H2BGFP (green fluorescent protein) fusion protein incorporated into nucleosomes, combined with proliferation-associated markers Ki67, phosphorylated histone H3 (p-H3), and basonuclin (BSN), as well as their unique location [8,9]. Despite a lack of specific markers for HF-SCs, as technology develops, quiescent HF-SCs are found to preferentially express certain kinds of molecules, such as keratin 15, CD34, $\alpha 3\beta 1$, Lgr5 integrins, CD71, CD200, nestin, and their transcription factors: Lhx2, Sox9, Nfatc1, Tcf3, and Tcf4 [2]. These markers could be used as additional markers for HF-SCs identification, and the specific methods of detecting HF-SCs vary across different studies.

5.1.2 Characteristics of HF-SCs

Similarly to other stem cells, HF-SCs exhibit several critical features. The first is the high potential of growth. Cells from the upper region of follicles demonstrate greatest ability of growth *in vitro*, in contrast to cells from other regions of follicles and the epidermis [10]. The second feature of HF-SCs is that these bulge cells are relatively undifferentiated, meaning that they are multipotent [11] with the capacity of differentiating into not only all the cell types of the hair, but also the epidermis and the sebaceous glands. Last but not least, HF-SCs show the ability of returning quiescence [11]. In addition, the well-defined developmental biology of HF-SCs, as well as their location and unique cycling, makes them intriguing and probably the best model system with which to study adult stem cells [12].

5.1.3 Applications of HF-SCs

HF-SCs, of course, have been intensively studied for hair regeneration, a cure for hair loss. Until recently there has been no valid treatment for patients. Traditional medicines have promoted hair growth but have failed to generate new HFs in the bald scalp to achieve the ultimate goal of therapy [13]. Therefore, the regenerative power of HF-SCs provides patients with the hope of generating new HFs and being cured.

In addition to the potential in treating hair loss, HF-SCs with nestin expression are found to be multipotent in that the cells can differentiate into keratinocytes, neurons, glia, smooth muscle cells, and melanocytes *in vitro*. They also contribute to the recovery of peripheral nerve and spinal cord injury, indicating that HF-SCs provide an essentially accessible, autologous source of adult stem cells with the potential for use in regenerative medicine [14]. Moreover, recent studies have suggested that adult tissue-derived follicular stem cells can be used as a potential bioengineered organ replacement therapy to treat damaged organs [15]. Therefore, a well-established understanding of HF-SCs and their differentiation potential may provide a novel strategy for treatments of inherited skin diseases, injury, hair loss, and even neuromuscular and hematopoietic diseases.

5.2 EPIGENETIC REGULATION

The term "epigenetics" was first proposed by Conrad H. Waddington in 1942 with the literal meaning of "interactions of genes with their environment that bring the phenotype into being" [16]. Nowadays this concept generally refers to studies of long-term, stable, heritable or not, alternations in gene activities that do not result from changes in the DNA sequence, and occur during cell development and proliferation [17]. Epigenetic regulations are vital for cell differentiation and development, as well as in mature mammals, either by chance or under the influence of environmental factors [18]. DNA methylation was introduced in 1995 as the first and key epigenetic process, which might be responsible for the stable maintenance of certain types of gene expression [19]. However, two decades on, epigenetics has started to draw increasing attention from the public, and numerous publications have now acknowledged the essential role of epigenetics in the

TABLE 5.1   The Summary of Current Knowledge of Epigenetic Regulation of
HF-SCs

Types	Regulators	Effects on HF-SCs	References
DNA methylation	DNMT1	Preserving the progenitor state, triggering the exit from the niche	[29,30]
	UHRF	Maintaining proliferation, inhibiting premature differentiation	[31]
Histone modifications	H3K4me3	Initiating transcription	[30]
	H3K79me2	Activating transcription	[30]
	H3K27me3	Absent transcriptional activity	[30]
	Jarid2	Regulating HF-SC proliferation	[32]
MiRNAs	miR-125b	Regulating differentiation	[33]
	miR-18b	Suppressing smooth muscle cells differentiation from HF-SCs	[34]

pathogenesis of cancer and autoimmune disorders and suggested novel therapeutic potential. Following DNA methylation, histone modifications were recognized as an important mechanism of epigenetic regulation [20]. Currently, microRNAs (miRNAs), constituting a class of small 22-nucleotide-long endogenous RNAs that negatively regulate coding-gene expression [21], have been revealed to be involved in various diseases and have aroused extensive research interest [22−24]. Whereas a large number of articles have been published on epigenetics and increasing numbers of studies are reporting the findings of epigenetic regulation in stem cells [25−28], with the aim of exploring the potential therapeutic benefit of manipulating stem cell differentiation, little has been done on epigenetic regulation in HF-SCs. Therefore, the following section reviews current findings and knowledge of epigenetic regulation of HF-SCs from the aspects of DNA methylation, histone modification, and miRNAs regulation, respectively (Table 5.1).

5.2.1 DNA Modification

the As one of the first identified means of epigenetic regulation, DNA methylation has been intensively studied and found to play a role in stem cell self-renewal and differentiation [35], especially among epidermal stem cells. Over the past few decades, DNA methyltransferase 1 (DNMT1) has been proven to be essential for functions of epidermal stem cells by preserving the progenitor state and triggering the process of exit from the niche [29]. Moreover, UHRF, a regulator for DNMT1, is found to facilitate DNMT1 function, maintain proliferation, and inhibit premature differentiation [31]. However, the understanding of DNA methylation in regulating HF-SC action is far from clear. Although easy purification of HS-SCs allows people to investigate the prevalence of DNA alterations, there is no solid evidence to support DNA methylation in regulating HF-SCs differentiation and function. Only one relevant study has been conducted on investigating DNMT1 in a mouse model [30]. The results showed that DNMT1 was critical in sustaining epidermal lineage precursor cells. The genes that regulate differentiation were demethylated upon differentiation. To examine whether DNMT1 is involved in the regenerative cycling of HFs, the authors crossed a K14-cre line with a floxed DNMT1 line to create a conditional knockout of DNMT1 in the K14-expressing epidermis. It was found that hair numbers decreased gradually from birth, with proliferation being reduced and apoptosis enhanced in matrix TA cells. K15-positive stem cells also displayed impaired ability to proliferate and form a hair germ. These findings have reflected, though indirectly, the importance of DNA methylation in regulating HF-SC development and regeneration [30].

5.2.2 Histone Modifications

It is believed that chromatin modifications are important in maintaining the pluripotency of embryonic stem cells (ESCs). However, their effects on adult HF-SCs are far from having been elucidated. On this score, recent evidence was published by Lien and colleagues in 2011, delineating the histone methylation patterns in HF-SCs, as well as their unique markers [36]. As described in previous chapters, histone modifications affect chromatin structure and alter transcription factor activity. In general, trimethylation of histone H3 on lysine 4 (H3K4me3) in promoter regions initiates transcription, while the broadening of H3K4me3 marks and dimethylation of histone H3 at lysine 79 (H3K79me2) characterizes active transcription. On the other hand, trimethylation of histone H3 lysine 27 (H3K27me3) characterizes certain genes with relatively little or no transcriptional activity [37]. In addition to the genes marked by H3K4me3 or H3K27me3, several thousand "bivalent" genes have been found to show both H3K4me3 and H3K27me3 marks in mammal ESC populations [38]. Furthermore, Hong et al. proposed recently that bivalency existed in ESC populations, thus reflecting their underlying heterogeneity and plasticity [39]. The bivalent genes can be also detected in other progenitor cells, but the patterns and functional relevance of histone methylation in adult stem cell populations remain unknown. To explore this question, Lien and colleagues investigated the role of chromatin modifications in mature stem cells in a well-characterized HF system. They carried out chromatin immunoprecipitation (ChIP)-seq for

H3K27me3 and H3K4me3, using cells from quiescent bulge population, early anagen bulge, and anagen matrix. It was found that in resting bulge stem cells, Polycomb group (PcG) repressive marks were absent from transcription factor genes, which mediate vital bulge cell functions, and hair differentiation lineage regulators and nonskin regulators were PcG-repressed. Moreover, quiescent bulge stem cells showed only a small amount of bivalently marked genes compared to ESCs. As mentioned above, HF-SCs are capable of transiting from the quiescence stage to the activated stage, and then returning to inactivated status, which are capabilities very different from those of other stem cells. The findings that activated bulge cells gain only few alternations in PcG marks and obtain H3K79me2 marks at non-PcG regulated cell cycle regulatory genes may partially explain this phenomenon. Thus, these findings are significant in that they demonstrate chromatin modifications as an important regulator in balancing HF-SC quiescence. However, only few bulge cells are activated during anagen, and this activated population may not in reality be isolated from quiescent cells. In this sense, the results might be contaminated with quiescent cells and thus spurious. Another piece of evidence to support the important role of histone modifications came from the work published, in the same year, by Stefania and colleagues [32]. Their research reported that Jarid2, which was required for recruitment of the polycomb repressive complex-2 in ESCs, was critical for H3K27 trimethylation and regulated HF-SC proliferation, thus echoing the findings in ESCs. Furthermore, the fact that bulge stem cells have limited bivalently marked genes indicates the limitation of developmental potential, which is consistent with the finding that these cells contribute little within the skin. On the other hand, it is observed by others and has been described previously that HF-SCs are multipotent and show great capacity of differentiating *in vitro* under certain conditions. This means that there might be unknown epigenetic regulation, particularly a histone methylation profile, similar to ESCs. It would be fascinating to discover whether other epigenetic regulators, such as miRNAs and long noncoding RNAs, participate in HF-SC fate.

5.2.3 MicroRNAs

miRNAs are tiny noncoding RNA molecules (containing about 22 nucleotides), which negatively modulate gene expression by binding to the 3' untranslated region (UTR) of target messenger RNAs (mRNAs). This process causes degradation or translational repression of their target mRNAs. Since the discovery of miRNAs in 1993, they have provided a new insight in numerous gene regulations and pathways that are involved in cell development, stem cell regulation, and pathogenesis of human diseases. The role of miRNAs has been well documented in stem cell differentiation. However, their function in maintaining stem cell and self-renewal is still unclear. In 2006, Yi and colleagues discovered more than 100 miRNAs that are secreted by epidermis and HF differentially. It is well known that the enzyme Dicer is necessary in the processing of miRNAs, and in the Dicer mutant, degenerated follicles lacking stem cell markers suggest the vital role of miRNAs in HF development [40]. Unfortunately, only two publications have described the relationship between miRNAs and HF-SC actions. In HF-SCs, miR-125b was found to directly target either Blimp1, which is a key transcription factor of these baceous gland lineages, or VDR (vitamin D receptor), a key regulator required for HF differentiation. This mechanism pushed stem cells into the differentiated lineages [33]. The other relevant work by Liu et al. illustrated that miR-18b suppressed smooth muscle cells differentiation from HF-SCs by targeting SMADs [34], implying a novel way to manipulate the differentiation of HF-SCs for therapy.

5.3 PROSPECTS

Epigenetic modifications in various types of stem cells have been well documented. Therefore, it is not beyond our expectation that they play a role in HF-SC differentiation, proliferation, self-renewal, and functions. More than 90% of epidermal stem cells reside in HFs and some research has claimed that HF-SCs represent epidermal ones. But the latter are distinct from HF-SCs since their differentiation is restricted to epidermis lineages [41]. Although most relevant studies were focused on epidermal stem cells, HF-SCs share some common features with them, suggesting that the actions of HF-SCs are also regulated by epigenetic modifications. Indeed, previous studies, though limited in number and preliminary in their conclusions, have shown that DNMT1, chromatin methylation, and Dicer are involved in HF-SC development. However, further research needs to be conducted to explore how epigenetic factors regulate HF-SCs and how to manipulate their differentiation and functions for therapeutic purpose. In contrast to DNMT1, Tet family members serve as enzymes that activate DNA demethylation by modifying 5-methylcytosine (5mC) to 5-hydroxymethylcytosine (5hmC). 5hmC is lost during differentiation and reported to be enriched in ESCs, neurons, and epidermal stem cells [42,43], implying that high 5hmC levels are related to a pluripotent cell state [44]. Therefore, further investigation of the levels of 5hmC and Tet family members in HF-SCs may provide a novel avenue to manipulate their development. Again, since more than 100 miRNAs have been identified to be released from skin and follicles, some of them may participate in the actions of HF-SCs. In addition, other histone modifications that regulate epidermal stem cells may also have an impact on HF-SCs [45]. Therefore, as an optimal model for adult stem cell studies and potential therapies for hair loss, skin injury, and organ damage, HF-SCs have displayed remarkable biomedical significance. Better utilization of HF-SCs in medical treatments necessitates a more comprehensive understanding of epigenetic modification of these cells.

References

[1] Lavker RM, et al. Hair follicle stem cells. J Invest Dermatol Symp Proc 2003;8 (1):28−38.
[2] Blanpain C, Fuchs E. Epidermal homeostasis: a balancing act of stem cells in the skin. Nat Rev Mol Cell Biol 2009;10(3):207−17.
[3] Shimojo H, Ohtsuka T, Kageyama R. Oscillations in notch signaling regulate maintenance of neural progenitors. Neuron 2008;58(1):52−64.
[4] Lavker RM, Sun TT. Epidermal stem cells: properties, markers, and location. Proc Natl Acad Sci USA 2000;97(25):13473−5.
[5] Lehrer MS, Sun TT, Lavker RM. Strategies of epithelial repair: modulation of stem cell and transit amplifying cell proliferation. J Cell Sci 1998;111(Pt 19):2867−75.
[6] Morris RJ, Potten CS. Highly persistent label-retaining cells in the hair follicles of mice and their fate following induction of anagen. J Invest Dermatol 1999;112 (4):470−5.
[7] Taylor G, et al. Involvement of follicular stem cells in forming not only the follicle but also the epidermis. Cell 2000;102(4):451−61.
[8] Kanda T, Sullivan KF, Wahl GM. Histone-GFP fusion protein enables sensitive analysis of chromosome dynamics in living mammalian cells. Curr Biol 1998;8(7): 377−85.
[9] Tumbar T, et al. Defining the epithelial stem cell niche in skin. Science 2004;303 (5656):359−63.
[10] Kobayashi H. Effect of c-kit ligand (stem cell factor) in combination with interleukin-5, granulocyte-macrophage colony-stimulating factor, and interleukin-3, on eosinophil lineage. Int J Hematol 1993;58(1−2):21−6.
[11] Amoh Y, et al. Multipotent nestin-expressing hair follicle stem cells. J Dermatol 2009;36(1):1−9.
[12] Waters JM, Richardson GD, Jahoda CA. Hair follicle stem cells. Semin Cell Dev Biol 2007;18(2):245−54.
[13] Price VH. Treatment of hair loss. N Engl J Med 1999;341(13):964−73.
[14] Amoh Y, Hoffman RM. Isolation and culture of hair follicle pluripotent stem (hfPS) cells and their use for nerve and spinal cord regeneration. Methods Mol Biol 2010;585:401−20.
[15] Toyoshima KE, et al. Fully functional hair follicle regeneration through the rearrangement of stem cells and their niches. Nat Commun 2012;3:784.
[16] Waddington CH. The epigenotype. 1942. Int J Epidemiol 2012;41(1):10−13.
[17] Jaenisch R, Bird A. Epigenetic regulation of gene expression: how the genome integrates intrinsic and environmental signals. Nat Genet 2003;33(Suppl.):245−54.
[18] Issa JP. CpG-island methylation in aging and cancer. Curr Top Microbiol Immunol 2000;249:101−18.
[19] Riggs AD. X inactivation, differentiation, and DNA methylation. Cytogenet Cell Genet 1975;14(1):9−25.

[20] Wiekowski M, et al. Changes in histone synthesis and modification at the beginning of mouse development correlate with the establishment of chromatin mediated repression of transcription. J Cell Sci 1997;110(Pt 10):1147–58.

[21] Bartel DP. MicroRNAs: genomics, biogenesis, mechanism, and function. Cell 2004;116(2):281–97.

[22] Ha TY. MicroRNAs in human diseases: from autoimmune diseases to skin, psychiatric and neurodegenerative diseases. Immune Netw 2011;11(5):227–44.

[23] Galeazzi M, et al. MicroRNAs in autoimmune rheumatic diseases. Reumatismo 2012;64(1):7–17.

[24] Thamilarasan M, et al. MicroRNAs in multiple sclerosis and experimental autoimmune encephalomyelitis. Autoimmun Rev 2012;11(3):174–9.

[25] Zhao L, et al. The dynamics of DNA methylation fidelity during mouse embryonic stem cell self-renewal and differentiation. Genome Res 2014. PMID: 24835587.

[26] Sheaffer KL, et al. DNA methylation is required for the control of stem cell differentiation in the small intestine. Genes Dev 2014;28(6):652–64.

[27] Farifteh F, et al. Histone modification of embryonic stem cells produced by somatic cell nuclear transfer and fertilized blastocysts. Cell J 2014;15(4):316–23.

[28] Kamat V, et al. MicroRNA screen of human embryonic stem cell differentiation reveals miR-105 as an enhancer of megakaryopoiesis from adult CD34[+] cells. Stem Cells 2014;32(5):1337–46.

[29] Tsumura A, et al. Maintenance of self-renewal ability of mouse embryonic stem cells in the absence of DNA methyltransferases Dnmt1, Dnmt3a and Dnmt3b. Genes Cells 2006;11(7):805–14.

[30] Li J, et al. Progressive alopecia reveals decreasing stem cell activation probability during aging of mice with epidermal deletion of DNA methyltransferase 1. J Invest Dermatol 2012;132(12):2681–90.

[31] Zhang J, et al. S phase-dependent interaction with DNMT1 dictates the role of UHRF1 but not UHRF2 in DNA methylation maintenance. Cell Res 2011;21(12):1723–39.

[32] Mejetta S, et al. Jarid2 regulates mouse epidermal stem cell activation and differentiation. EMBO J 2011;30(17):3635–46.

[33] Yi R, Fuchs E. MicroRNAs and their roles in mammalian stem cells. J Cell Sci 2011;124(Pt 11):1775–83.

[34] Liu X, et al. miR-18b inhibits TGF-beta1-induced differentiation of hair follicle stem cells into smooth muscle cells by targeting SMAD2. Biochem Biophys Res Commun 2013;438(3):551–6.

[35] Sen GL, et al. DNMT1 maintains progenitor function in self-renewing somatic tissue. Nature 2010;463(7280):563–7.

[36] Lien WH, et al. Genome-wide maps of histone modifications unwind in vivo chromatin states of the hair follicle lineage. Cell Stem Cell 2011;9(3):219–32.

[37] Suganuma T, Workman JL. Signals and combinatorial functions of histone modifications. Annu Rev Biochem 2011;80:473–99.

[38] Fisher CL, Fisher AG. Chromatin states in pluripotent, differentiated, and reprogrammed cells. Curr Opin Genet Dev 2011;21(2):140–6.

[39] Hong SH, et al. Cell fate potential of human pluripotent stem cells is encoded by histone modifications. Cell Stem Cell 2011;9(1):24–36.

[40] Andl T, et al. The miRNA-processing enzyme dicer is essential for the morphogenesis and maintenance of hair follicles. Curr Biol 2006;16(10):1041–9.

[41] Watt FM, Jensen KB. Epidermal stem cell diversity and quiescence. EMBO Mol Med 2009;1(5):260–7.

[42] Szulwach KE, et al. Integrating 5-hydroxymethylcytosine into the epigenomic landscape of human embryonic stem cells. PLOS Genet 2011;7(6):e1002154.

[43] Davis T, Vaisvila R. High sensitivity 5-hydroxymethylcytosine detection in Balb/C brain tissue. J Vis Exp 2011;(48):pii 2661.

[44] Ruzov A, et al. Lineage-specific distribution of high levels of genomic 5-hydroxymethylcytosine in mammalian development. Cell Res 2011;21(9):1332−42.

[45] Shen Q, Jin H, Wang X. Epidermal stem cells and their epigenetic regulation. Int J Mol Sci 2013;14(9):17861−80.

Epigenetics and the Regulation of Inflammation

Christian M. Hedrich

**Pediatric Rheumatology and Immunology, Children's Hospital Dresden,
University Medical Center "Carl Gustav Carus,"
Technische Universität Dresden, Dresden, Germany**

6.1 INTRODUCTION

Inflammation is part of the complex response of cells, tissues, and organs to certain intrinsic or extrinsic stimuli, such as oxidative stress, damaged or apoptotic cells, debris, or pathogens. The purpose of inflammation is activating the immune system in order to remove injurious agents and to initiate repair processes. Activated complement components, cytokines and chemokines, and antibodies or immune complexes are major contributors to the onset and maintenance of inflammation. Thus, inflammation is the common denominator of the innate and the adaptive immune system. Though inflammation is a somewhat stereotyped response, a multitude of cells and tissues contribute in a receptor- and tissue-specific manner. Impaired activation or control of pro- and anti-inflammatory signals may cause infection, inflammatory conditions, allergies, or cancer [1−4].

Epigenetic mechanisms are regulatory events that govern gene expression without altering the underlying DNA sequence. Epigenetic patterns determine the accessibility and subsequently the expression of genes in a signal-, tissue-, and organ-specific manner. It has become increasingly clear that a multitude of factors contribute to the timely and regionally regulated accessibility of genomic information. Epigenetics comprises the combined action of a number of tightly regulated factors that will be discussed here: DNA methylation, histone modifications, and noncoding transcripts, including micro RNAs (miRNAs) [2−5].

85 © 2015 Elsevier Inc. All rights reserved.

i. *DNA methylation* is achieved through the addition of a methyl-group to the 5′ carbon position of cytosine in so-called cytosine-phosphate-guanosine dinucleotides (CpGs). DNA methylation reduces the accessibility of regulatory regions to transcription factors, transcriptional co-activators, and RNA polymerases. DNA methylation is mediated by certain enzymes, which are referred to as DNA methyltransferases (DNMTs). Historically, two DNMT classes have been distinguished: maintenance (e.g., DNMT1) and *de novo* (e.g., DNMT3a and DNMT3b) DNMT enzymes. Maintenance DNMTs have been claimed as being responsible for re-methylation of the second (or daughter) DNA strand during cell division, while *de novo* DNMTs have been claimed to be responsible for DNA methylation independent of cell division. However, the situation appears to be more complex, with the classical grouping oversimplifying the situation [2,3,5,6].

ii. In the nucleus, *histone proteins* organize in clusters with two copies of each H2A, H2B, H3, and H4. Resulting histone octamers are responsible for DNA packing in three-dimensional structures. To control the three-dimensional DNA structure, histone proteins undergo posttranslational modifications at amino acid residues. These result in a variable electric charge, mediating an "open" or "closed" chromatin conformation, thus regulating accessibility to transcriptional activators and RNA polymerases. The most common histone modifications include acetylation, citrullination, phosphorylation, and methylation. One very central histone mark is histone 3 lysine 18 acetylation (H3K18ac), which mediates an "open" chromatin structure and allows active transcription of genes. Histone 3 lysine 9 (H3K9me3) and/or lysine 27 (H3K27me3) trimethylation, however, mediate epigenetic "closure" and are therefore marks of silenced genes. A constantly growing number of enzymes and multiprotein complexes deposit or remove these epigenetic marks [2,3,5].

iii. The transcription of *noncoding RNAs* from intronic or intergenic regions is a potentially important, yet poorly understood mechanism influencing gene expression at the interface between transcription of adjacent genes, chromatin remodeling, and the translation of RNA into protein products. A correlation between chromatin structure and the presence of intergenic transcripts has been established and is considered a mechanism to solidify and extend an "open" chromatin conformation. Intergenic transcription could therefore allow interactions between core promoters and proximal or distal enhancer elements within a gene cluster or even over longer ranges, potentially between chromosomes. Furthermore, noncoding RNAs from certain regions can be further processed by the nuclear ribonuclease Drosha and the cytoplasmatic enzyme Dicer into so-called miRNAs, which function as translational regulators of target genes [2–4,7].

In the following, we discuss the contribution of the aforementioned epigenetic events to the regulation of inflammation, with the main focus on T cells and B cells. Using the example of antigen receptor assembly, we discuss a group of novel mechanisms contributing to the formation of DNA loops, three-dimensional long-distance interactions that promote the co-expression or rearrangement of functionally related genes [8]. Finally, we discuss the contribution of an impaired epigenetic regulation of X chromosomal genes to gender-specific inflammatory responses, and the influence of selected environmental factors on inflammation [9,10]. Given the constantly growing body of literature and the vast number of cells and tissues involved in inflammatory responses, this text serves as an introduction to the most central concepts and epigenetic mechanisms involved in inflammatory responses.

6.2 EPIGENETIC MECHANISMS DURING T-CELL DIFFERENTIATION

Tightly controlled signal-, receptor-, and time-specific gene expression patterns define the phenotype of immune cells [2–4]. Lineage-defining proteins include surface receptors, and (co-)stimulatory proteins, master transcription factors, and chemokines, or cytokines. Regulatory T cells modulate the immune system, maintain self-tolerance, and abrogate autoimmunity [11–13]. Effector T cells on the other hand induce and maintain cellular immune reactions, thus being essential for pathogen clearance. To exert these functions, effector T cells require tightly controlled cytokine expression profiles, allowing the immune system to be effective but not harmful for the host. The situation *in vivo*, however, significantly differs from *in vitro* settings. In a live system, effector T cells frequently share characteristics of several Th cell subsets defined in *in vitro* settings [11–18].

Though the expression of lineage-defining master regulatory factors and signature cytokines or chemokines have proven useful, they oversimplify the situation, since most T cells express more than only one master regulator and cytokine. Another oversimplification is the currently applied differentiation between regulatory and effector T cells (Figure 6.1), and the subdivision of effector T-cell populations into the various T helper cell subsets, based on their cytokine expression profiles (e.g., Th0, Th1, Th2, Th17) [11–16]. Regulatory and effector T cells exist in both the CD4$^+$ and the CD8$^+$ compartment [19,20].

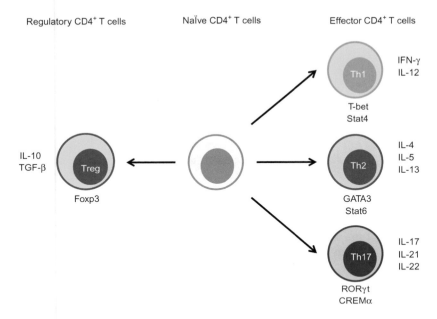

FIGURE 6.1   **Differentiation of regulatory and effector CD4$^+$ T cells.** In response to activation, naïve CD4$^+$ T cells differentiate into specialized regulatory (Treg) or effector CD4$^+$ T-cell subsets. Effector T cells can be subdivided into distinct T helper subsets, including Th1, Th2, and Th17 cells. Those T-cell subsets are characterized by signature transcription factors (displayed under each T-cell subset) and cytokine expression profiles (displayed next to each T-cell subset).

One more recently identified T-cell lineage is that of T-cell receptor (TCR)-$\alpha\beta^+$CD3$^+$CD4$^-$CD8$^-$ T cells, which are frequently referred to as "double negative" (DN) T cells. We recently demonstrated that DN T cells may originate from CD8$^+$ T cells and exert effector functions, including the expression of IL-17A, a potent effector cytokine [21–25].

Regardless of all the aforementioned oversimplifications, the currently applied differentiation between regulatory and effector T cells, based on their gene expression patterns, is the most practicable system. Based on this classification, epigenetic mechanisms have been proven to play a central role for the determination of a cell's phenotype, granting stability while still allowing some plasticity. During differentiation or through a changed (micro-)environmental situation, whenever T cells need to adjust their phenotype and/or cytokine expression patterns, a combination of key transcription factors control epigenetic changes to regulatory regions. This is mediated by transcription factor binding to previously unoccupied regulatory regions and the subsequent recruitment and activation of epigenetic modifiers, resulting in active *cis*-regulatory elements, such as promoters and enhancers [3,11–13,15,16,20,26].

6.2.1 CD4$^+$ T-Cell Phenotype Determination

CD4$^+$, CD8$^+$, and CD4$^-$CD8$^-$ (DN) T cells are critical players during host defense mechanisms. Most data on the contribution of epigenetic regulation on lineage determination and inflammation are available for the CD4$^+$ helper T-cell subset [11−13,15,16,20,22]. Thus, in this manuscript we mainly focus on CD4$^+$ T cells. During the process of subset determination, naïve CD4$^+$ T cells differentiate into distinct regulatory or effector T cells. *Trans*-regulatory and epigenetic factors control the expression of lineage-defining molecules. A further critical aspect during the determination of T-cell phenotypes, however, is the silencing of lineage-inappropriate genes [1,3,5,11−13,16,20]. In this context, the expression of master regulatory factors is crucial. The expression of master transcription factors largely depends on the activation of surface receptors through the extracellular environment, which entails antigens, co-stimulatory molecules, adhesion molecules, and cytokines. As a result, lineage-defining intrinsic programs are induced in naïve CD4$^+$ T cells. Master transcription factors achieve their impact on gene expression not only through direct regulation of effector genes, but also through positive and negative regulation of collaborating and/or antagonistic factors, such as epigenetic modifiers [1,3,5,11−13,16,20,27].

It is a well-established concept that CD4$^+$ T cells specialize in response to (usually) microbial challenges. The first T-cell subsets reported were *Th1* and *Th2 cells*. Their classification was primarily based on the expression of distinct cytokine expression patterns (Figure 6.1) [11−13,16,20,27].

The *T helper cell subset (Th)1* [28−30] is defined by the expression of interferon (IFN)-γ and IL-12. Th1 cells have been proven critical for the clearance of many—particularly intracellular—pathogens, including *Leishmania* spp., mycobacteria, viruses and *Toxoplasma gondii*. Th1 cells have also been discussed in terms of contributing to autoimmune disorders. The Th1 phenotype is determined by the expression of the master regulatory factor T-bet in the presence of the transcription factor signal transducer and activator of transcription (Stat)4. Transcription factors, however, do not act isolated from additional, co-activating molecules and the surrounding chromatin structure. As are many other genes and gene clusters, the *IFN/Ifn* genes are tightly regulated by a series of distal and proximal enhancer elements. The human and murine *IFN/Ifn* genes in Th1 cells have been demonstrated to undergo DNA demethylation in order to allow transcriptional activation [11−13,16,20,28,29,31]. Furthermore, it has been demonstrated that Stat4, whenever co-localizing with T-bet, acts pervasively on Th1-specific enhancers through the recruitment of the transcriptional co-activator p300. p300, through its enzymatic histone acetyltransferase (HAT) activity, mediates epigenetic remodeling. Conversely, lineage-inappropriate genes remove p300 from regulatory elements through Stat4-dependent mechanisms. Furthermore, the Th1 signature cytokine gene *Ifng* exhibits unopposed permissive histone marks, namely H3K4 trimethylation, in murine Th1 cells. Conversely, the murine *Ifng* gene is epigenetically silenced by repressive histone marks, namely H3K27me3, in such subsets that do not express IFN-γ. However, not only cytokines but also lineage-promoting transcription factor genes undergo epigenetic remodeling. Of note, key transcription factor genes generally exhibit even more complicated epigenetic marks with the coexistence of both accessibility- and repression-associated modifications. This implies a complex need for time- and tissue-appropriate expression of these key mediators during the different steps T-cell differentiation [11−13,16,20,27−30,32].

Th2 cells are characterized by their ability to express IL-4, IL-5, and IL-13. Th2 cells have proven important for the elimination of helminths, such as *Schistosoma mansoni*. Uncontrolled Th2 responses, and a recently identified Th2 subset named Th9 cells (based on the expression of the cytokine IL-9) centrally contribute to allergies and asthma. In analogy to *Ifn* in Th1 cells, multiple enhancers and silencers, which undergo subset-specific epigenetic remodeling through DNA methylation and histone modifications, regulate the human and the murine *IL4* gene [31,33−37]. Epigenetic remodeling allows or prevents transcription factor recruitment and transcriptional activation. Deletion of epigenetic modifiers provided valuable insight into epigenetic control of lineage-defining cytokine genes. Deficiency of DNMT1 resulted in inappropriate expression of Th1 or Th2-specific cytokines in the opposed T-cell subset, providing evidence for the central role of DNA methylation during subset determination. In agreement with these findings, deletion of the H3K9 methylase results in inappropriate expression of Th1 lineage-defining cytokines in Th2 cells. The master transcription factor for Th2 cell differentiation is GATA-binding factor (GATA)3. IL-4-induced Stat6, however, is also necessary for Th2 generation and it became evident that simultaneous recruitment of GATA3 and Stat6 are necessary for epigenetic remodeling and successful "opening" of the Th2 locus, spanning the *IL4*, *IL5*, and *IL13* genes [27−30,32].

Over the past decade, evidence accumulated suggesting that physiological mechanisms, such as pathogen clearance, but also autoimmune phenomena, cannot be reduced to the balance between Th1 and Th2 cells and cytokines. The identification of a series of "new" cytokines and a novel, distinct CD4$^+$ T helper cell subset, referred to as *Th17 cells*, started a "small revolution" concerning the understanding of T-cell physiology, autoimmune, and allergic disease [3,5,14,38,39]. Th17 cells produce the effector cytokine family IL-17, including the two main members IL-17A and IL-17F. IL-17 cytokines, for example, induce neutrophil recruitment to sites of inflammation, and affect smooth muscle tonus, thus contributing to immune and nonimmune-mediated effects of inflammation. IL-22, another cytokine produced by Th17 cells, helps to protect barrier functions of various epithelia. Some Th17 cells produce IL-9, IL-10, and IL-21, allowing the definition of various Th17 subsets based on cytokine expression patterns. Th17 differentiation requires stimulation of the TCR complex. Furthermore, a series of cytokines promote Th17 generation, including IL-23, IL-6, and IL-21, all of which induce the transcription factor Stat3. Deletion experiments suggested that Stat3 is necessary for Th17 generation, while forced expression of Stat3 sufficiently induced IL-17 production [14,15,32,38,39]. State-of-the-art approaches in mice allowed determining that Stat3 binds many promoters and enhancer elements of cytokine (e.g., *IL21*, *IL23*) and cytokine receptor (*IL23R*) genes involved in Th17 generation and the *IL17* gene cluster itself. Furthermore, Stat3 regulates other transcriptional regulators involved in Th17 generation, including the master regulators retinoid acid receptor (RAR)-related orphan receptor (ROR)γ, IFN regulatory factor (IRF)4, and Basic leucine zipper transcription factor (Batf). As aforementioned, it has been proposed that Stat transcription factors induce long-range epigenetic remodeling during T-cell subset differentiation, thus allowing the recruitment of additional, so-called master regulatory factors. During Th17 differentiation, the murine *IL17* cluster undergoes epigenetic remodeling with increased histone H3 acetylation and H3K4me3. These changes allow for transcription factor recruitment. Interestingly, histone acetylation is particularly high in such promoter and intergenic enhancer regions recruiting both Stat3 and the transcriptional co-activator p300. As aforementioned, p300 was suggested to co-recruit with Stat transcription factors (Stat4 during Th1 differentiation), acting as a "pioneer factor" promoting epigenetic remodeling that allows for recruitment of the so-called lineage-specific master regulators (e.g., RORgt) [14–16,27,29,30,32,40]. We recently demonstrated that the activating transcription factor (ATF) family transcription factor cAMP response element modulator-α (CREMα) promotes epigenetic remodeling in T cells from patients with systemic lupus erythematosus (SLE). SLE is a systemic autoimmune disorder characterized by an accumulation of autoreactive T cells that express increased amounts of IL-17A while failing to express IL-2. In T cells from SLE patients, CREMα induces epigenetic remodeling (Figure 6.2A and B). Through its interactions with histone deacetylase (HDAC)1 and DNMT3a, CREMα promotes DNA methylation and histone H3K9me3 of the entire *IL2* gene, resulting in a "closure" and transcriptional silencing. Conversely, at the *IL17* cluster, CREMα does not recruit HDAC1 or DNMT3a, but mediates DNA demethylation in a yet to be determined manner. It is likely that the distinctly different transcription factor

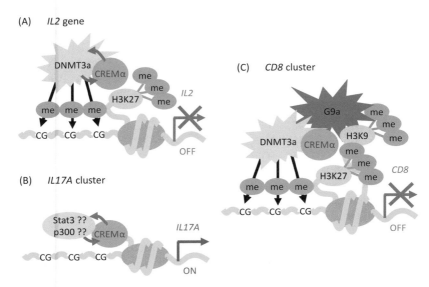

FIGURE 6.2 **Epigenetic remodeling of T-cell lineage-specific genes by CREMα.** CREMα mediates epigenetic remodeling of cytokine genes in T cells. (A) At the *IL2* gene, CREMα mediates transcriptional silencing through *trans*-repression and the recruitment of DNMT3a. (B) At the *IL17A* gene, CREMα mediates *trans*-activation and epigenetic "opening" of the locus in a yet to be determined manner. Unpublished findings indicate that interactions between CREMα and p300, potentially in conjunction with Stat3, may be a central step. (C) During the generation of DN T cells from CD8⁺ T cells, CREMα occupies cAMP response element (CRE)-binding elements, resulting in *trans*-repression and epigenetic silencing through the recruitment of DNMT3a, and the histone methyltransferase G9a. Thus, CREMα may act as a "pioneer factor" during T-cell differentiation, preparing the scene for so-called master regulators.

milieu, e.g., the presence or absence of Stat transcription family proteins, at the two mentioned cytokine gene loci, defines the ability of CREMα to mediate either an "opening" or "closure" of chromatin [26,41−47].

Regulatory CD4⁺ T cells, or *Tregs*, can modulate or suppress inflammatory responses of other immune cells. The main function of Tregs is the termination of immune responses after the successful elimination of pathogens, thus preventing or limiting tissue-damage and autoimmunity. Tregs can limit inflammatory responses through the production of the immune-regulatory cytokines IL-10 and TGF-β. Furthermore, Tregs negatively modulate effector T-cell responses by consuming IL-2, a well-recognized "growth factor" for effector T cells, excluding Th17 cells, which are inhibited by IL-2. The transcription factor forkhead box P3 (Foxp3) has been identified as the master regulator for the generation of regulatory T cells. However, the role of Foxp3 as the single necessary and sufficient master regulator of Treg generation has recently been doubted. This occurred because Foxp3 recruits to only those regulatory regions that already exhibit an "open" chromatin conformation. Furthermore, the presence of additional factors, such as ETS (E-twenty-six), Runx (Runt-related-transcription-factor) and NFAT (Nuclear factor of activated T cells) appears to be necessary to allow Foxp3 recruitment. Thus, Foxp3 appears to solidify Treg phenotypes during late stages of T-cell differentiation, utilizing predefined and active enhancer elements rather than "generating" new ones [11,48,49].

The activity, stability, and function of Foxp3 is regulated by post-translational acetylation. Interestingly, Foxp3 acetylation is mediated by enzymes that also act as epigenetic modifiers: so-called HATs. In addition to their profound effects on chromatin structure, HATs can regulate a number of transcription factors, including Foxp3. Of note, those interactions between Foxp3 and HAT enzymes, which result in its activation, also determine the suppressive effects of Foxp3 by epigenetic remodeling of certain genes. Acetylation of Foxp3 proteins, however, depends not only on the interaction with HATs (including TIP60 and p300), but also on deacetlytlation through HDACs (including HDAC1, HDAC7, HDAC9, and sirtuin-1). Complex direct and indirect interactions between Foxp3, HATs, and HDACs regulate the activity, recruitment, function, and stability of Foxp3. Acetylation of Foxp3 by the HAT TIP60, for instance, enhances its binding to the murine *IL2* gene. The presence of HDACs in the Foxp3 complex, however, may in turn exert inhibitory functions on IL-2 expression. Thus, the master transcription factor Foxp3 blocks *trans*-activation and epigenetic "opening" at distinct genomic regions through its binding to other transcriptional regulators or their replacement [29,32,49,50].

However, Foxp3 does not only mediate chromatin remodeling, its expression also is controlled on the epigenomic level. Furthermore, a number of additional Treg-specific transcription factors and surface receptor proteins are regulated through epigenetic mechanisms. In Tregs, regions with hypomethylated DNA have been detected in a number of genes including *Foxp3*, *Ctla4*, and *Ikzf4* encoding for the zinc finger protein Eos. Thus, it appears likely that also Treg differentiation, stability, and function is under the close control of epigenetic mechanisms [29,32,49,50].

Since the classification of CD4$^+$ T cells into T helper subsets is over-simplified and does not completely reflect the highly complex situation *in vivo*, investigators have distinguished between *naïve, regulatory,* and *effector T-cell subsets* [11–13,17,18,26,40]. Though our understanding concerning lineage commitment is limited and somewhat oversimplified, we know that during T-cell priming, naïve CD4$^+$ T cells are exposed to antigens. In response to antigen contact and TCR activation, T cells proliferate and differentiate into central or effector memory T cells. Central memory CD4$^+$ T cells migrate to secondary lymphatic tissues, where they wait for a secondary challenge in order to exert enhanced immune responses. Effector memory T cells, however, migrate into inflamed tissues or remain in the peripheral circulation to exert their immunological functions [17,18]. Based on surface receptor expression profiles, we, and others, investigated cytokine expression patterns of these T-cell subsets. Unprimed naïve CD4$^+$ T cells express various cytokines, including IL-2 and IL-17A. Conversely, effector memory CD4$^+$ T cells fail to express IL-2, but express the effector cytokine IL-17A [17,18,26]. We demonstrated that cytokine expression in naïve and memory T cells is controlled through epigenetic mechanisms. In naïve CD4$^+$ T cells, low levels of DNA methylation allow for IL-2 expression, while in effector memory CD4$^+$ T cells increased DNA methylation of the *IL2* promoter forbids gene expression. Conversely, effector memory T cells exhibit low levels of *IL17A* promoter methylation, allowing for IL-17A expression. We demonstrated that, at least partially, epigenetic patterns in naïve and effector T cells are controlled through the absence or presence of the transcription factor CREMα. In naïve and central memory CD4$^+$ T cells low CREMα expression allows for IL-2 production, while high CREMα expression in effector memory CD4$^+$ T cells inhibits IL-2 production and promotes the expression of the effector cytokine IL-17A. In this context, CREMα has different effects on the *IL2* and *IL17A* promoter (Figure 6.2A and B). At the *IL2* promoter, CREMα mediates *trans*-repression and increased DNA methylation through the recruitment of DNMT3a (Figure 6.2A), while, at the *IL17A* promoter, CREMα causes *trans*-activation and a reduction of DNA methylation in a yet to be determined manner (Figure 6.2B) [26,47]. It is likely that the distinct transcription factor microenvironments at the *IL2* or *IL17A* promoter influence the contrary effects of CREMα on the two genes. This concept, however, is currently in the focus of research.

6.2.2 Generation of CD3$^+$CD4$^-$CD8$^-$ DN T Cells

TCR-αβ$^+$CD3$^+$CD4$^-$CD8$^-$ DN T cells are effector T cells that are expanded in the periphery of patients with SLE and other autoimmune disorders [21−23,51,52]. In humans and lupus-prone mice, DN T cells can derive from CD8$^+$ T cells by downregulating CD8 co-receptors on the cell surface [24,25]. In a series of studies, we demonstrated that the transcription factor CREMα contributes to the downregulation of the two CD8 isoforms CD8A and CD8B in humans and mice (Figure 6.2C). CREMα *trans*-represses CD8A and CD8B transcription through its binding to a region syntenic to the murine *CD8b* promoter, replacing the CD8-promoting transcription factor cAMP response element-binding protein (CREB). Furthermore, we demonstrated that CREMα induces epigenetic silencing of the *CD8* cluster through its interactions with DNMT3a and the histone methyltransferase G9a. As a result, the *CD8* cluster undergoes DNA and histone methylation (H3K9me3 and H3K27me3), mediating stable epigenetic silencing of CD8 expression [24,25].

This is of special interest, since specialized T phenotypes are reflected by the exclusive expression of CD4 or CD8, suggesting shared molecular mechanisms regulating both surface co-receptor and subset-defining cytokine expression [26]. Recently, Crispin et al. demonstrated that DN T cells express high levels of IL-17A [22,23]. As aforementioned, CREMα is involved in epigenetic remodeling of the *IL17* gene cluster in effector CD4$^+$ T cells, which is enhanced in T cells from patients with SLE [3,26,41,42,45,46,53]. Thus, CREMα may, in analogy to Stat transcription factors, have "pioneer" effects, shaping the epigenome during T-cell differentiation. Thus, next to its role during the remodeling of the *CD8* cluster in DN T cells, CREMα-mediated epigenetic mechanisms may be central in establishing the effector T-cell phenotype in CD3$^+$ T-cell lineages: CD4$^+$, CD8$^+$, and DN T cells [1,3,26,41,42,45−47,53]. Opposing effects on lineage-specific genes ("Closing" of the *CD8* locus or *IL2*, while "opening" the *IL17* cluster) may depend on the transcription factor milieu, e.g., Stat transcription factors [5,26,41−43,46,47]. However, at this point this explanation remains hypothetical and is in the focus of ongoing research.

6.3 EPIGENETIC REGULATION OF THE IMMUNE-MODULATORY CYTOKINE IL-10

IL-10 is an immune-modulatory cytokine that plays a central role during both innate and adaptive immune responses. A large number of cells and tissues, including almost all immune cells, express IL-10. Main sources within the immune system, however, are T cells (mostly Th2 cells and Tregs), B cells, natural killer (NK) cells, mast cells, eosinophils, dendritic cells (DCs), monocytes, and macrophages. A tightly controlled, precisely timed, tissue- and receptor-specific expression of IL-10 is central to immune homeostasis. IL-10 not only has inhibiting effects on cytokine expression from T cells and antigen presenting cells (APCs), but also negatively affects major histocompatibility complex (MHC) class II expression and co-stimulation through CD80/CD86. Of note, IL-10 also promotes B-cell proliferation, differentiation, and antibody production [2,4,54−56]. Thus, increased or reduced IL-10 expression can potentially contribute to autoimmune/inflammatory phenomena. To maintain immune-homeostasis, IL-10 expression is tightly controlled on the transcriptional and post-transcriptional levels through tissue-specific mechanisms. Cell-specific transcription factor networks and the epigenetic environment represent central transcriptional control mechanisms [2,4,56–58].

In T cells, Stat transcription factors (Stat1, Stat3, Stat4, Stat5) but also zinc-finger transcription factors of the Ikaros family, GATA3, and activator protein (AP-)1, have been documented to induce IL-10 expression. As aforementioned, epigenetic patterns not only allow the recruitment of transcription factors but also can be influenced by them [2,4]. Tsuji-Takayama and colleagues reported IL-2-dependent, Stat5-mediated histone H3 acetylation of an enhancer region within the 4th intron of *IL10* in human T cells [59,60]. We at the same time reported IL-12-dependent, Stat4-mediated acetylation of the same region in murine NK cells [61]. The Stat-binding element within the enhancer was referred to as I-SRE [60]. The exact mechanisms by which epigenetic remodeling was mediated remained elusive. Recently, we demonstrated that, in human T cells, DNA methylation controls gene expression by Stat3 binding to the proximal *IL10* promoter and that Stat3 and Stat5 govern IL-10 expression through their recruitment to the intronic enhancer element (I-SRE) previously reported in B cells and NK cells. In human T cells, Stat transcription factors (Stat3 and/or Stat5) *trans*-activate *IL10* and induce epigenetic remodeling through their interaction with the transcriptional co-activator p300. In turn, p300 mediates epigenetic remodeling, resulting in a further "opening" of *IL10* through histone acetylation (H3K18ac). Interestingly, the effects of Stat3 on epigenetic remodeling were stronger when compared to Stat5. Since Stat3, which is increased in SLE T cells, replaces Stat5 at the I-SRE within the 4th intron, this at least partially explains increased IL-10 expression by T cells from SLE patients (PNAS, in revision process). Of note, human and murine T cells exhibit cell-type-specific DNA methylation patterns that are distinct from those of myeloid cells. As compared to monocytes and macrophages, the *IL10* promoter exhibits a significantly higher degree of DNA methylation in T cells, while the aforementioned intronic enhancer (I-SRE) is methylated to a lower degree [2,4,62,63]. These T-cell-specific DNA methylation patterns appear to be disrupted in such autoimmune disorders characterized by increased T-cellular IL-10 expression, such as SLE (Figure 6.3A) ([64], PNAS, in revision process).

Innate APCs, namely monocytes, macrophages, and DCs, are another important source of IL-10 [2,4,56,58]. Research indicates that IL-10 expression in macrophages is chromatin-dependent. Mosser and colleagues documented chromatin remodeling of the *IL10* promoter in murine macrophages mediated by FCγR-stimulation [65]. The same group demonstrated activation-induced extracellular signal regulated kinase (ERK)-dependent histone H3 phosphorylation at position serine 10 (H3S10p) that co-localized with the recruitment of the transcription factor specificity protein (Sp-)1 to the *IL10* promoter in murine

(A) T cells

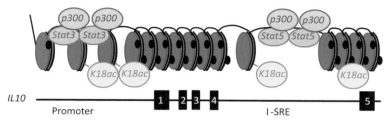

(B) Macrophages/monocytes

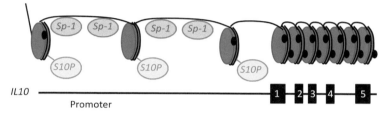

FIGURE 6.3 **Epigenetic and transcriptional regulation of IL-10 in T cells and mono-cytes/macrophages.** Cell- and signal-specific expression of the immune-modulatory cytokine IL-10 is tightly controlled on the transcriptional level. (A) In human IL-10 expressing T cells, *IL10* is controlled by Stat transcription factors. Low levels of DNA methylation (as indicated by black filled circles) allow for transcription factor recruitment to the promoter and an enhancer element in the 4th intron (I-SRE). Stat3 *trans*-activates the *IL10* promoter, while Stat5 *trans*-activates the I-SRE. Both Stat3 and Stat5 co-recruit p300 to the *IL10* gene, mediating epigenetic "opening" through histone H3 acetylation (at Lysine residue 18: K18ac). (B) In human monocytes/macrophages, MAP kinase-dependent signaling pathways mediate epigenetic remodeling of the *IL10* promoter through H3 phosphorylation (at position serine 10: S10P), allowing the recruitment of the transcription factor Sp-1, which in turn *trans*-activates the *IL10* promoter. Of note, the I-SRE within the 4th intron is highly methylated in monocytes/macrophages (as indicated by the filled black circles), thus seeming less important for myeloid-derived IL-10 expression.

macrophages [58,66]. We recently demonstrated that failure to activate the mitogen-activated protein (MAP) kinases ERK1 and ERK2 in monocytes from patients with the autoinflammatory bone disorder chronic recurrent multifocal osteomyelitis (CRMO) results in reduced IL-10 expression. Failure to activate ERK1/2 results in impaired histone H3S10p and reduced Sp-1 recruitment to the *IL10* promoter. Thus, ERK-mediated transcription factor recruitment and epigenetic remodeling of the *IL10* promoter are central steps for IL-10 expression from monocytes and macrophages (Figure 6.3B) [67,68].

The presence of tissue-specific intergenic transcripts has been indicated in the regulation of gene expression in the murine and human *IL10* cluster. Two enhancer regions in conserved noncoding (CNS) elements between the murine *IL10* and *IL19* genes have been demonstrated to express transcripts in murine IL-10-expressing Th2 cells and macrophages [2,69]. Applying a transgenic mouse model, carrying the murine and human *IL10* cluster, we documented the presence of species- and tissue-specific transcripts in the same region, reflecting DNA methylation patterns [2,62,70]. At this point, we only incompletely understand the contribution of these intergenic transcripts to *IL10* regulation. Their presence, however, suggests an involvement in solidifying epigenetic patterns across the *IL10* cluster and the potential presence of additional enhancer elements or alternative promoters within the locus [2,4].

6.4 DNA CONTRACTION DURING ANTIGEN RECEPTOR ARRANGEMENT IN T AND B CELLS

Antigen receptors require great diversity in order to recognize a large number of antigens. Diversity of antigen receptors is granted by enzymatic recombination activating gene (RAG)1- or RAG2-mediated recombination of variable (V), diversity (D), and joining (J) gene segments at antigen receptor loci [8,71]. All B- and T-cell receptor complexes are formed through this process, referred to as VDJ recombination. A constantly growing body of literature suggests that transcription-factor-mediated locus contraction is a central step for the recombination of distal V with proximal D/J regions [8].

The *TCR complex* consists of two units, either TCRα and TCRβ, or TCRδ and TCR-γ. The main challenge during the process of TCR generation is to achieve sufficient variability through rearrangement of the subunit encoding genes. It has been demonstrated that TCR genes undergo stage specific three-dimensional structural changes, so-called locus contraction. Contraction and extension of the TCRα/δ locus has been documented as cell- and stage-specific. While double positive T cells in the thymus undergo re-extension of the TCRα/δ locus, DN T cells require locus contraction for the rearrangement of V genes [72–76].

The *B-cell receptor (BCR) complex* consists of two immunoglobulin heavy chains (Igh), and two light chains (Igκ and Igγ). Heavy chains are encoded by the *Igh* locus; light chains are encoded by the *Igκ* or *Igγ* locus. In B cells, three-dimensional structural changes at the *Igh* locus contribute to the creation of a diverse antigen BCR repertoire. Thus, the *Igh* and *Igκ* loci can be found in the nuclear periphery, which is considered transcriptionally silent, in nonrecombining cell types. In recombining B cells, however, they are located in the center of the nucleus, where the *Igh* locus has also been demonstrated to be more compact, contributing to combinatory transcription. Locus compaction is regulated by a number of transcription factors and epigenetic modifiers. Animals, deficient in the transcription factors Yin Yang (YY)1, paired box protein (Pax)5, or the histone—lysine *N*-methyltransferase Ezh2, exhibit reduced *Igh* locus contraction. Interestingly, deficiency of the recombinase enzymes Rag1/2 does not interfere with locus contraction, but abrogates rearrangement [75,77—82].

The *mechanisms of DNA contraction* during antigen receptor recombination in T and B cells have not finally been determined. A series of elegant studies in mice have provided insight into several mechanisms involved in gene recombination. CTCF, also known as 11-zinc finger protein or CCCTC-binding factor, is a transcription factor primarily binding to insulator regions. Insulators are genomic regions that prevent transcriptionally inactive heterochromatin spreading from one side of the insulator to the other. In some instances, insulators exhibit enhancer-blocking activity, whenever insulators separate these regulatory regions. During VDJ recombination, CTCF binds other CTCF-occupied regions, contributing to locus compaction and DNA looping. Through these loops, active or inactive genes can be insulated from the activity of adjacent genes or domains. Indeed, CTCF has been documented to play a role during higher order organization of genes, chromatin, and chromosomes. B cells deficient in CTCF exhibited reduced DNA compaction at the *Igh* locus [75,83—85]. Cohesin regulates the separation of sister chromatids during cell division [86]. Recently, cohesin has been linked with DNA loop formation through CTCF binding. Indeed, while CTCF has been demonstrated to rather homogeneously bind to *Igh* antigen receptor loci in pro- and pre-B cells, and thymocytes in a nonlineage specific manner, cohesin only binds to the *Igh* locus in pre- and pro-B cells, suggesting a central role during DNA compaction. The role of loop-promoting transcription factors remains unclear. As aforementioned, mice deficient in YY1, Pax5, or the histone methyltransferase Ezh2 exhibit reduced *Igh* locus contraction. The presence of those factors may be responsible for long-range interactions while CTCF loops reflect local interactions, forming rosette-like structures within the *Igh* locus. Interestingly, CTCF-mediated loops differ between different stages during B-cell development. This allows the creation of a diverse antigen receptor repertoire through variable rearrangement of all V genes. The molecular mechanisms by which the rearrangement is orchestrated remain elusive [75,83—89]. Similar and highly complex mechanisms are responsible for DNA contraction resulting in TCR rearrangement and have been reviewed elsewhere (Figure 6.4A and B) [73,74,76].

Transcription of noncoding RNAs has been reported for all of the J and D genes of antigen receptors during rearrangement. Also, but to a lesser extent, V genes undergo noncoding transcription. Three major

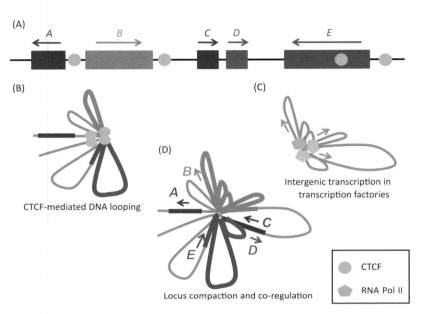

FIGURE 6.4 **Mechanisms of co-regulation or rearrangement of distant genes.** Transcription occurs in subnuclear compartments referred to as "transcription factories." Co-regulated genes can be found within the same transcription factory, regardless of the genomic distance between them. DNA contraction is a very recently appreciated mechanism involved in transcription factory formation and the co-regulation (Th2 cytokine genes) or rearrangement (e.g., BCR or TCR) of originally distant genes. (A) A number of genes are co-regulated regardless of their orientation and/or physical distance from each other. (B) The 11-zinc finger protein CTCF recruits to binding sites throughout the genome. CTCF-rich regions bind to other CTCF occupied sites, allowing the formation of DNA loops and DNA contraction. (C) Noncoding RNAs may facilitate CTCF-mediated locus compaction through transcription of distant noncoding regions within the same transcription factory. (D) Together, these mechanisms result in the formation and/or stabilization of DNA loops that allow co-regulation or rearrangement of originally distant genes.

and two minor regions with antisense RNA transcription have been identified [73]. Interestingly, transcription at the three major regions started at so-called PAIR, or Pax5 intergenic repeat elements that contain binding sites for the transcription factors Pax5a, E2A, and CTCF. As expected, and in analogy to the presence of intergenic transcripts in the previously discussed *IL10* cluster, these regions exhibited high degrees of the activating epigenetic marks (H3K4me3 and H3ac) [71,75,90].

Transcription of genes happens in so-called transcription factories, subnuclear compartments with DNA and clusters of RNA polymerases [91–93]. Co-regulated genes (such as the aforementioned Th1 cytokine genes *IFN* and *IL12*; the Th2 cytokine genes *IL4*, *IL5*, and *IL13*; or the

IL-10 family cytokine genes, particularly *IL10* and *IL19*) frequently can be found within the same compartment regardless of their genomic distance [2,69,94]. Even genes that are located on different chromosomes (such as the Th1 cytokine genes *IFN* and *IL12*) can be found within the same transcription factory. For BCR rearrangement, Choi and colleagues claimed that long-range intergenic transcription of generally distant genes within the same transcription factory, contribute to DNA looping and bringing V and D/J regions into close physical proximity, ultimately favoring rearrangement of genes (Figure 6.4C) [8,71].

The aforementioned CTCF proteins together with their partner cohesin may provide looped domains within the entire genome. Since the density of CTCF/chohesin sites is enriched in the two antigen receptor loci (BCR and TCR), they may contribute to a looped, rosette-like DNA formation. This allows compaction of antigen receptor loci during rearrangement. Noncoding RNAs furthermore facilitate locus compaction through transcription of originally distant genes within the same compact transcription factory. Choi et al. suggested that transcription of noncoding RNAs contributes to moving different parts of the V gene locus into the proximity of D/J genes, resulting in D/J rearrangement. Through such mechanisms, co-regulation of genes can be ensured and the production of a highly diverse repertoire of antibodies and TCR complexes is assured [73–76,82,88].

6.5 POSTTRANSCRIPTIONAL REGULATION OF INFLAMMATION THROUGH miRNAs

Transcription of noncoding intronic and intergenic DNA into RNA comprises gene-regulatory functions. As aforementioned, the presence of intergenic transcripts allows interactions between physically distant regulatory elements, such as core promoters and distal enhancers [2,3,7]. These interactions may happen within a gene cluster (e.g., Th2 cluster) or even between different chromosomes (e.g., the *IL12* and *Ifn* genes in Th1 cells). However, such preliminary noncoding RNAs can be further processed into more mature RNAs. This happens through two enzymes: the nuclear ribonuclease Drosha, and the cytoplasmatic enzyme Dicer. The end-result of these posttranscriptional modifications is so-called miRNAs [2,3,7].

miRNAs function as translational regulators of target genes. They can bind to messenger RNAs, usually within the 3′ untranslated region (3′ UTR), which results in duplex and translational arrest, or mRNA degradation. Of note, to date up to 1,000 miRNAs have been documented. They have been discussed in terms of their controlling one third of the human transcriptome. miRNAs have been claimed to be involved in cell differentiation, cell cycle programming, apoptosis, and immune regulation. miRNAs identified so far, have been implicated with the regulation of the immune system through three main mechanisms:

1. the regulation of innate immune responses through abnormal activation of type I IFN pathways;
2. inflammatory responses by suppressing KLF13 and RANTES; and
3. interference with other epigenetic mechanisms, including DNA methylation through DNMT3a and DNMT3b [2,3,7].

Applying animal models, several miRNAs have been implicated in T- and B-cell development. miRNA-181 for instance has been documented to play a key role during T-cell differentiation by modulating TCR signaling-related phosphatases [95]. miRNA-150 is involved in B-cell differentiation through negative regulation of the transcriptional activator c-Myb [96,97]. Mice, deficient in miRNA-150 exhibit impaired B-cell development and steady serum immunoglobulin levels. T-cellular immune responses, however, are increased in miRNA-150-deficient animals [3,7].

6.6 THE X CHROMOSOME AND IMMUNE REGULATION

The role of the X chromosome in immune regulation becomes apparent considering the female predominance in many autoimmune disorders, including SLE, primary Sjogren's syndrome, and multiple sclerosis. On the contrary, women with Turner syndrome, who lack one X chromosome (45, X0), exhibit disease incidences that are comparable to those in men (46, XY) [5,98,99]. Sex chromosomes derive from so-called ancestor autosomes, therefore still comprising genes in so-called pseudo-autosomal regions. Most X-linked genes are not gender-specific and exhibit comparable gene expression levels in women and men. Suppressing copy number-associated differences in gene expression is achieved by a complex epigenetic event, referred to as X-chromosome inactivation. All of the previously discussed epigenetic mechanisms (DNA methylation, histone modifications, and miRNAs) are involved in this process [3,5,9,10,98,99].

Studies in patients with different autoimmune disorders documented that improper inactivation of the X chromosome results in an increased expression of single genes. Reduced methylation of the *CD40L* gene, for instance, contributes to female predominance in SLE and rheumatoid arthritis. CD40 is expressed on the surface of B cells or innate APCs, while CD40L is expressed on T cells. Increased CD40L expression promotes CD40: CD40L interactions, contributing to immunoglobulin class-switch, memory B-cell differentiation, and germinal center formation, thus promoting autoimmune phenomena [3,5,6,9,10,98−101].

6.7 ENVIRONMENTALLY INDUCED EPIGENETIC DISTURBANCES IMPAIR IMMUNE REGULATION

The observation that autoimmune disorders only co-segregate to a certain extent in genetically identical (monozygotic) twins has suggested that additional, e.g., environmental factors, contribute to autoimmunity in genetically predisposed individuals [3,5,102].

A well-established example for the influence of environmental factors on immune responses is the development of lupus-like symptoms in predisposed individuals in response to certain medication. In such cases, the application of medication may result in disturbed epigenetic homeostasis [3,5,6,103,104]. Resulting biochemical modifications of chromatin structures may affect gene expression and/or act as self antigens. It has been established that procainamide, an antiarrhythmic agent, and hydralazine, an antihypertensive drug, induce the activity of poly-ADP-ribose polymerase (PADPRP) in T and B cells. PADPRP is a chromatin-bound enzyme that catalyzes the transfer of ADP-ribose to histone proteins. Histone ribosylation may increase the immunogenicity of histone proteins, thus inducing autoantibody production [103,104]. Evidence supports a strong connection between drug-induced DNA hypomethylation and autoimmune phenomena. DNA methylation patterns require replication during mitosis. Certain drugs, including the nucleoside-analog 5-azacytidine, interfere with DNMT activity. 5-Azacytidine is an irreversible inhibitor of DNMT1, which plays a central role during cell division. During mitosis, DNMT1 recognizes hemi-methylated DNA in the newly generated parent:daughter complex. DNMT1 then catalyzes the transfer of one methyl group from S-adenosyl methionine (SAM) to the cytosine residue of the primarily unmethylated daughter DNA strand. Furthermore, 5-azacytidine has been demonstrated to affect the activity of HDACs, suggesting another mechanism by which drugs may alter the epigenome and gene expression patterns. In addition to (over-) expression of previously silenced genes from immune cells and other tissues, DNA demethylation through 5-azacytidine mediates further immune activation: demethylated DNA exhibits higher immunogenicity when compared to physiologically methylated DNA [1,5,6].

Disrupted epigenetic patterns also physiologically accumulate over time, thus contributing to the increased onset of autoimmune disorders with age. The activity of DNMTs, particularly DNMT1, has been reported decreased in the elderly. Methionine, an essential amino acid, serves as substrate for DNMT1 during DNA methylation. Thus, reduced DNMT1 activity can be potentiated in individuals with reduced methionine intake. The net result is impaired DNA methylation, especially in genes that play a role during cell differentiation. Indeed, the degree of DNA methylation can even be used as an indirect measure of life expectancy. Generally, this process affects all DNMT expressing cells. To date, however, most data have been generated in T lymphocytes. The accumulation of so-called senescent T cells within the T lymphocyte pool is the result of the gradual age-related loss of DNMT activity. Senescent T cells are characterized by reduced CD28 co-receptor expression, reduced telomere length, and expression of genes that are usually silenced, including killer cell immunoglobulin-like receptors (KIR), perforin, and the signaling molecule CD70. The result of disturbed pro-inflammatory gene expression is immune stimulation that may result in autoimmune disorders [1,5,6].

6.8 EPIGENETIC MODIFICATIONS FUNCTION AS AUTOANTIGENS IN AUTOIMMUNE DISEASE

Many of the previously discussed chromatin components (histones, nucleosomes, and DNA fragments) are potential antigen sources. In patients with SLE and other autoimmune disorders, nucleosomes, histones, and both single- and double-stranded DNA can be detected in the peripheral circulation. The presence of these, usually intracellular, components in the blood has been explained by aberrant apoptosis and/or defective clearance of apoptotic material and cell debris. Furthermore, DNA from lymphocytes of SLE patients is globally hypomethylated when compared to healthy controls. Thus, it has been thought that demethylated DNA fragments in the serum of SLE patients might mimic microbial DNA (which is also generally demethylated), resulting in the generation of anti-DNA antibodies. Furthermore, histones undergo modifications during apoptosis. These modifications (among others) include acetylation, transglutamination, and phosphorylation. Of note, antibodies derived from lupus patients react with such apoptosis-related histone modifications. Based on these findings and given that peptides carrying such apoptosis-induced acetylation motifs accelerate disease onset and aggravate disease severity in lupus-prone mice, they have been considered as possible contributors to the disruption of self-tolerance in SLE and other autoimmune disorders. Most likely as a result of these anomalies, SLE patients exhibit antinuclear antibodies and antibodies against nuclear antigens (including anti-DNA antibodies) years before the onset of clinical symptoms. Given the disease-specificity of some of the aforementioned modifications to histones and other nuclear proteins, they may even prove useful as disease biomarkers or therapeutic targets [3,5,6,105].

6.9 CONCLUSION

Epigenetic mechanisms are key components in the tight regulation of pro- and anti-inflammatory responses. Impaired epigenetic control has been documented as contributing to inflammation and autoimmune disorders. This is based on the broad involvement of epigenetic mechanisms in the differentiation of cells and tissues. During immune cell differentiation, a close interplay between preexisting epigenetic marks, transcription factors, and further epigenetic mediators is responsible for signature gene expression. Some master transcriptional regulators, such as T-bet (in Th1 cells), GATA-3 (in Th2 cells), ROR-γt (in Th17 cells), and Foxp3 (in Tregs), occupy preexisting "open" chromatin and mainly promote *trans*-activation and locally limited epigenetic remodeling of promoters and enhancers, creating or stabilizing active elements. Thus, the so-called master transcription factors may exert distinct functions on enhancers or promoters of target genes without globally affecting chromatin structure. Conversely, adjunct factors for the appropriate expression of signature genes, such as Stat transcription factors potentially together with CREMα, mediate locus, gene cluster, or even more widespread epigenetic remodeling that affects regions well beyond the aforementioned enhancer elements. The ability of select transcription factors to mediate all the aforementioned mechanisms, thus shaping the epigenetic landscape in differentiating immune cells, and providing a permissive epigenetic environment for so-called master regulators of transcription, has resulted in such transcription factors more recently being referred to as "pioneer factors."

Additional factors, including CTCFF, or the presence of intergenic transcripts, can mediate interactions between distant genomic regions, contributing to co-regulation of distant genes or gene rearrangement. Such co-regulation through DNA contraction has been implicated with the co-regulation of Th2 cytokines, and the rearrangement of the highly variable TCR and BCR complexes.

Loss of epigenetic control results in impaired control of transcription factor, cytokine and chemokine, and surface receptor expression. However, epigenetic alterations can themselves be caused by the impaired expression of pioneer transcription factors; preexisting inflammation; environmental triggers, including medication; and physiological processes, including aging. Thus, epigenetic mechanisms promise large potential as biomarkers and therapeutic targets in inflammatory conditions, including infections, autoimmune disorders, allergies, and cancer.

References

[1] Hedrich CM, Tsokos GC. Bridging the gap between autoinflammation and autoimmunity. Clin Immunol 2013;147(3):151—4.
[2] Hedrich CM, Bream JH. Cell type-specific regulation of IL-10 expression in inflammation and disease. Immunol Res 2010;47(1—3):185—206.
[3] Hedrich CM, Tsokos GC. Epigenetic mechanisms in systemic lupus erythematosus and other autoimmune diseases. Trends Mol Med 2011;17(12):714—24.
[4] Hofmann SR, Rosen-Wolff A, Tsokos GC, Hedrich CM. Biological properties and regulation of IL-10 related cytokines and their contribution to autoimmune disease and tissue injury. Clin Immunol 2012;143(2):116—27.
[5] Hedrich CM, Crispin JC, Tsokos GC. Epigenetic regulation of cytokine expression in systemic lupus erythematosus with special focus on T cells. Autoimmunity 2014;47 (4):234—41.
[6] Zhang Y, Zhao M, Sawalha AH, Richardson B, Lu Q. Impaired DNA methylation and its mechanisms in CD4(+)T cells of systemic lupus erythematosus. J Autoimmun 2013;41:92—9.
[7] Thai TH, Christiansen PA, Tsokos GC. Is there a link between dysregulated miRNA expression and disease? Discov Med 2010;10(52):184—94.
[8] Choi NM, Feeney AJ. CTCF and ncRNA regulate the three-dimensional structure of antigen receptor Loci to Facilitate V(D)J recombination. Front Immunol 2014;5:49.
[9] Sawalha AH, Wang L, Nadig A, et al. Sex-specific differences in the relationship between genetic susceptibility, T cell DNA demethylation and lupus flare severity. J Autoimmun 2012;38(2—3):J216—22.
[10] Strickland FM, Hewagama A, Lu Q, et al. Environmental exposure, estrogen and two X chromosomes are required for disease development in an epigenetic model of lupus. J Autoimmun 2012;38(2—3):J135—43.
[11] Kanno Y, Vahedi G, Hirahara K, Singleton K, O'Shea JJ. Transcriptional and epigenetic control of T helper cell specification: molecular mechanisms underlying commitment and plasticity. Annu Rev Immunol 2012;30:707—31.
[12] Vahedi G, C Poholek A, Hand TW, et al. Helper T-cell identity and evolution of differential transcriptomes and epigenomes. Immunol Rev 2013;252(1):24—40.
[13] Vahedi G, Kanno Y, Sartorelli V, O'Shea JJ. Transcription factors and CD4 T cells seeking identity: masters, minions, setters and spikers. Immunology 2013;139(3):294—8.
[14] Ghoreschi K, Laurence A, Yang XP, Hirahara K, O'Shea JJ. T helper 17 cell heterogeneity and pathogenicity in autoimmune disease. Trends Immunol 2011;32(9):395—401.
[15] Hirahara K, Ghoreschi K, Laurence A, Yang XP, Kanno Y, O'Shea JJ. Signal transduction pathways and transcriptional regulation in Th17 cell differentiation. Cytokine Growth Factor Rev 2010;21(6):425—34.
[16] Hirahara K, Vahedi G, Ghoreschi K, et al. Helper T-cell differentiation and plasticity: insights from epigenetics. Immunology 2011;134(3):235—45.
[17] Geginat J, Sallusto F, Lanzavecchia A. Cytokine-driven proliferation and differentiation of human naive, central memory, and effector memory CD4(+) T cells. J Exp Med 2001;194(12):1711—19.
[18] Geginat J, Sallusto F, Lanzavecchia A. Cytokine-driven proliferation and differentiation of human naive, central memory and effector memory CD4+ T cells. Pathologie-biologie 2003;51(2):64—6.
[19] Arens R, Schoenberger SP. Plasticity in programming of effector and memory CD8 T-cell formation. Immunol Rev 2010;235(1):190—205.
[20] Nakayamada S, Takahashi H, Kanno Y, O'Shea JJ. Helper T cell diversity and plasticity. Curr Opin Immunol 2012;24(3):297—302.

[21] Apostolidis SA, Crispin JC, Tsokos GC. IL-17-producing T cells in lupus nephritis. Lupus 2011;20(2):120−4.

[22] Crispin JC, Oukka M, Bayliss G, et al. Expanded double negative T cells in patients with systemic lupus erythematosus produce IL-17 and infiltrate the kidneys. J Immunol 2008;181(12):8761−6.

[23] Crispin JC, Tsokos GC. Human TCR-alpha beta + CD4- CD8- T cells can derive from CD8+ T cells and display an inflammatory effector phenotype. J Immunol 2009;183 (7):4675−81.

[24] Hedrich CM, Crispin JC, Rauen T, et al. cAMP responsive element modulator (CREM) alpha mediates chromatin remodeling of CD8 during the generation of CD3 + CD4- CD8- T cells. J Biol Chem 2014;289(4):2361−70.

[25] Hedrich CM, Rauen T, Crispin JC, et al. cAMP-responsive element modulator alpha (CREMalpha) trans-represses the transmembrane glycoprotein CD8 and contributes to the generation of CD3 + CD4- CD8- T cells in health and disease. J Biol Chem 2013;288(44):31880−7.

[26] Hedrich CM, Crispin JC, Rauen T, et al. cAMP response element modulator alpha controls IL2 and IL17A expression during CD4 lineage commitment and subset distribution in lupus. Proc Natl Acad Sci USA. 2012;109(41):16606−11.

[27] O'Shea JJ, Lahesmaa R, Vahedi G, Laurence A, Kanno Y. Genomic views of STAT function in CD4+ T helper cell differentiation. Nat Rev Immunol 2011;11(4):239−50.

[28] Bonelli M, Shih HY, Hirahara K, et al. Helper T Cell plasticity: impact of extrinsic and intrinsic signals on transcriptomes and epigenomes. Curr Top Microbiol Immunol 2014;381:279−326.

[29] Hirahara K, Poholek A, Vahedi G, et al. Mechanisms underlying helper T-cell plasticity: implications for immune-mediated disease. J Allergy Clin Immunol 2013;131(5):1276−87.

[30] Rothenberg EV. The chromatin landscape and transcription factors in T cell programming. Trends Immunol 2014;35(5):195−204.

[31] Lee GR, Kim ST, Spilianakis CG, Fields PE, Flavell RA. T helper cell differentiation: regulation by cis elements and epigenetics. Immunity 2006;24(4):369−79.

[32] Josefowicz SZ. Regulators of chromatin state and transcription in CD4 T-cell polarization. Immunology 2013;139(3):299−308.

[33] Baguet A, Bix M. Chromatin landscape dynamics of the Il4-Il13 locus during T helper 1 and 2 development. Proc Natl Acad Sci USA 2004;101(31):11410−15.

[34] Fields PE, Lee GR, Kim ST, Bartsevich VV, Flavell RA. Th2-specific chromatin remodeling and enhancer activity in the Th2 cytokine locus control region. Immunity 2004;21(6):865−76.

[35] Lee GR, Spilianakis CG, Flavell RA. Hypersensitive site 7 of the TH2 locus control region is essential for expressing TH2 cytokine genes and for long-range intrachromosomal interactions. Nat Immunol 2005;6(1):42−8.

[36] Spilianakis CG, Lalioti MD, Town T, Lee GR, Flavell RA. Interchromosomal associations between alternatively expressed loci. Nature 2005;435(7042):637−45.

[37] Williams A, Lee GR, Spilianakis CG, Hwang SS, Eisenbarth SC, Flavell RA. Hypersensitive site 6 of the Th2 locus control region is essential for Th2 cytokine expression. Proc Natl Acad Sci USA 2013;110(17):6955−60.

[38] Crispin JC, Tsokos GC. Interleukin-17-producing T cells in lupus. Curr Opin Rheumatol 2010;22(5):499−503.

[39] Nalbandian A, Crispin JC, Tsokos GC. Interleukin-17 and systemic lupus erythematosus: current concepts. Clin Exp Immunol 2009;157(2):209−15.

[40] Weng NP, Araki Y, Subedi K. The molecular basis of the memory T cell response: differential gene expression and its epigenetic regulation. Nat Rev Immunol 2012;12 (4):306−15.

[41] Hedrich CM, Rauen T, Kis-Toth K, Kyttaris VC, Tsokos GC. cAMP-responsive element modulator alpha (CREMalpha) suppresses IL-17F protein expression in T lymphocytes from patients with systemic lupus erythematosus (SLE). J Biol Chem 2012;287(7):4715−25.

[42] Hedrich CM, Rauen T, Tsokos GC. cAMP-responsive element modulator (CREM) alpha protein signaling mediates epigenetic remodeling of the human interleukin-2 gene: implications in systemic lupus erythematosus. J Biol Chem 2011;286 (50):43429−36.

[43] Koga T, Hedrich CM, Mizui M, et al. CaMK4-dependent activation of AKT/mTOR and CREM-alpha underlies autoimmunity-associated Th17 imbalance. J Clin Invest 2014;124(5):2234−45.

[44] Lippe R, Ohl K, Varga G, et al. CREMalpha overexpression decreases IL-2 production, induces a T(H)17 phenotype and accelerates autoimmunity. J Mol Cell Biol 2012;4(2):121−3.

[45] Rauen T, Grammatikos AP, Hedrich CM, et al. cAMP-responsive element modulator alpha (CREMalpha) contributes to decreased Notch-1 expression in T cells from patients with active systemic lupus erythematosus (SLE). J Biol Chem 2012;287 (51):42525−32.

[46] Rauen T, Hedrich CM, Juang YT, Tenbrock K, Tsokos GC. cAMP-responsive element modulator (CREM)alpha protein induces interleukin 17A expression and mediates epigenetic alterations at the interleukin-17A gene locus in patients with systemic lupus erythematosus. J Biol Chem 2011;286(50):43437−46.

[47] Rauen T, Hedrich CM, Tenbrock K, Tsokos GC. cAMP responsive element modulator: a critical regulator of cytokine production. Trends Mol Med 2013;19(4):262−9.

[48] Arvey A, van der Veeken J, Samstein RM, Feng Y, Stamatoyannopoulos JA, Rudensky AY. Inflammation-induced repression of chromatin bound by the transcription factor Foxp3 in regulatory T cells. Nat Immunol 2014;15(6):580−7.

[49] Zhou Z, Song X, Li B, Greene MI. FOXP3 and its partners: structural and biochemical insights into the regulation of FOXP3 activity. Immunol Res 2008;42 (1−3):19−28.

[50] Lal G, Bromberg JS. Epigenetic mechanisms of regulation of Foxp3 expression. Blood 2009;114(18):3727−35.

[51] Apostolidis SA, Lieberman LA, Kis-Toth K, Crispin JC, Tsokos GC. The dysregulation of cytokine networks in systemic lupus erythematosus. J Interferon Cytokine Res 2011;31(10):769−79.

[52] Crispin JC, Tsokos GC. IL-17 in systemic lupus erythematosus. J Biomed Biotechnol 2010;2010:943254.

[53] Rauen T, Juang YT, Hedrich CM, Kis-Toth K, Tsokos GC. A novel isoform of the orphan receptor RORgammat suppresses IL-17 production in human T cells. Genes Immun 2012;13(4):346−50.

[54] Peng H, Wang W, Zhou M, Li R, Pan HF, Ye DQ. Role of interleukin-10 and interleukin-10 receptor in systemic lupus erythematosus. Clin Rheumatol 2013;32 (9):1255−66.

[55] Su DL, Lu ZM, Shen MN, Li X, Sun LY. Roles of pro- and anti-inflammatory cytokines in the pathogenesis of SLE. J Biomed Biotechnol 2012;2012:347141.

[56] Mosser DM, Zhang X. Interleukin-10: new perspectives on an old cytokine. Immunol Rev 2008;226:205−18.

[57] Mosser DM, Edwards JP. Exploring the full spectrum of macrophage activation. Nat Rev Immunol 2008;8(12):958−69.

[58] Zhang X, Edwards JP, Mosser DM. Dynamic and transient remodeling of the macrophage IL-10 promoter during transcription. J Immunol 2006;177(2):1282−8.

[59] Tsuji-Takayama K, Suzuki M, Yamamoto M, et al. IL-2 activation of STAT5 enhances production of IL-10 from human cytotoxic regulatory T cells, HOZOT. Exp Hematol 2008;36(2):181−92.

[60] Tsuji-Takayama K, Suzuki M, Yamamoto M, et al. The production of IL-10 by human regulatory T cells is enhanced by IL-2 through a STAT5-responsive intronic enhancer in the IL-10 locus. J Immunol 2008;181(6):3897−905.

[61] Grant LR, Yao ZJ, Hedrich CM, et al. Stat4-dependent, T-bet-independent regulation of IL-10 in NK cells. Genes Immun 2008;9(4):316−27.

[62] Hedrich CM, Ramakrishnan A, Dabitao D, Wang F, Ranatunga D, Bream JH. Dynamic DNA methylation patterns across the mouse and human IL10 genes during CD4+ T cell activation; influence of IL-27. Mol Immunol 2010;48(1−3): 73−81.

[63] Hofmann SR, Moller J, Rauen T, et al. Dynamic CpG-DNA methylation of Il10 and Il19 in CD4+ T lymphocytes and macrophages: effects on tissue-specific gene expression. Klin Padiatr 2012;224(2):53−60.

[64] Zhao M, Tang J, Gao F, et al. Hypomethylation of IL10 and IL13 promoters in CD4+ T cells of patients with systemic lupus erythematosus. J Biomed Biotechnol 2010;2010:931018.

[65] Gerber JS, Mosser DM. Reversing lipopolysaccharide toxicity by ligating the macrophage Fc gamma receptors. J Immunol 2001;166(11):6861−8.

[66] Lucas M, Zhang X, Prasanna V, Mosser DM. ERK activation following macrophage FcgammaR ligation leads to chromatin modifications at the IL-10 locus. J Immunol 2005;175(1):469−77.

[67] Hofmann SR, Morbach H, Schwarz T, Rosen-Wolff A, Girschick HJ, Hedrich CM. Attenuated TLR4/MAPK signaling in monocytes from patients with CRMO results in impaired IL-10 expression. Clin Immunol 2012;145(1):69−76.

[68] Hofmann SR, Schwarz T, Moller JC, et al. Chronic non-bacterial osteomyelitis is associated with impaired Sp1 signaling, reduced IL10 promoter phosphorylation, and reduced myeloid IL-10 expression. Clin Immunol 2011;141(3):317−27.

[69] Jones EA, Flavell RA. Distal enhancer elements transcribe intergenic RNA in the IL-10 family gene cluster. J Immunol 2005;175(11):7437−46.

[70] Ranatunga D, Hedrich CM, Wang F, et al. A human IL10 BAC transgene reveals tissue-specific control of IL-10 expression and alters disease outcome. Proc Natl Acad Sci USA 2009;106(40):17123−8.

[71] Choi NM, Loguercio S, Verma-Gaur J, et al. Deep sequencing of the murine IgH repertoire reveals complex regulation of nonrandom V gene rearrangement frequencies. J Immunol 2013;191(5):2393−402.

[72] Kondilis-Mangum HD, Shih HY, Mahowald G, Sleckman BP, Krangel MS. Regulation of TCRbeta allelic exclusion by gene segment proximity and accessibility. J Immunol 2011;187(12):6374−81.

[73] Shih HY, Hao B, Krangel MS. Orchestrating T-cell receptor alpha gene assembly through changes in chromatin structure and organization. Immunol Res 2011;49 (1−3):192−201.

[74] Shih HY, Krangel MS. Distinct contracted conformations of the Tcra/Tcrd locus during Tcra and Tcrd recombination. J Exp Med 2010;207(9):1835−41.

[75] Shih HY, Krangel MS. Chromatin architecture, CCCTC-binding factor, and V(D)J recombination: managing long-distance relationships at antigen receptor loci. J Immunol 2013;190(10):4915−21.

[76] Shih HY, Verma-Gaur J, Torkamani A, Feeney AJ, Galjart N, Krangel MS. Tcra gene recombination is supported by a Tcra enhancer- and CTCF-dependent chromatin hub. Proc Natl Acad Sci USA 2012;109(50):E3493−502.

[77] Jhunjhunwala S, van Zelm MC, Peak MM, et al. The 3D structure of the immunoglobulin heavy-chain locus: implications for long-range genomic interactions. Cell 2008;133(2):265−79.

[78] Jhunjhunwala S, van Zelm MC, Peak MM, Murre C. Chromatin architecture and the generation of antigen receptor diversity. Cell 2009;138(3):435−48.

[79] Kind J, van Steensel B. Genome-nuclear lamina interactions and gene regulation. Curr Opin Cell Biol 2010;22(3):320−5.

[80] Kind J, van Steensel B. Stochastic genome-nuclear lamina interactions: modulating roles of Lamin A and BAF. Nucleus 2014;5(2):124−30.

[81] Kosak ST, Skok JA, Medina KL, et al. Subnuclear compartmentalization of immunoglobulin loci during lymphocyte development. Science 2002;296(5565):158−62.

[82] Roldan E, Fuxa M, Chong W, et al. Locus 'decontraction' and centromeric recruitment contribute to allelic exclusion of the immunoglobulin heavy-chain gene. Nat Immunol 2005;6(1):31−41.

[83] Botta M, Haider S, Leung IX, Lio P, Mozziconacci J. Intra- and inter-chromosomal interactions correlate with CTCF binding genome wide. Mol Syst Biol 2010;6:426.

[84] Dixon JR, Selvaraj S, Yue F, et al. Topological domains in mammalian genomes identified by analysis of chromatin interactions. Nature 2012;485(7398):376−80.

[85] Hou C, Li L, Qin ZS, Corces VG. Gene density, transcription, and insulators contribute to the partition of the *Drosophila* genome into physical domains. Mol Cell 2012;48 (3):471−84.

[86] Nasmyth K, Haering CH. Cohesin: its roles and mechanisms. Annu Rev Genet 2009;43:525−58.

[87] Parelho V, Hadjur S, Spivakov M, et al. Cohesins functionally associate with CTCF on mammalian chromosome arms. Cell 2008;132(3):422−33.

[88] Rubio ED, Reiss DJ, Welcsh PL, et al. CTCF physically links cohesin to chromatin. Proc Natl Acad Sci USA 2008;105(24):8309−14.

[89] Wendt KS, Yoshida K, Itoh T, et al. Cohesin mediates transcriptional insulation by CCCTC-binding factor. Nature 2008;451(7180):796−801.

[90] Gopalakrishnan S, Majumder K, Predeus A, et al. Unifying model for molecular determinants of the preselection Vbeta repertoire. Proc Natl Acad Sci USA 2013;110 (34):E3206−15.

[91] Mitchell JA, Fraser P. Transcription factories are nuclear subcompartments that remain in the absence of transcription. Genes Dev 2008;22(1):20−5.

[92] Osborne CS, Chakalova L, Brown KE, et al. Active genes dynamically colocalize to shared sites of ongoing transcription. Nat Genet 2004;36(10):1065−71.

[93] Schoenfelder S, Sexton T, Chakalova L, et al. Preferential associations between co-regulated genes reveal a transcriptional interactome in erythroid cells. Nat Genet 2010;42(1):53−61.

[94] Yao X, Zha W, Song W, et al. Coordinated regulation of IL-4 and IL-13 expression in human T cells: 3C analysis for DNA looping. Biochem Biophys Res Commun 2012;417(3):996−1001.

[95] Lashine YA, Seoudi AM, Salah S, Abdelaziz AI. Expression signature of microRNA-181-a reveals its crucial role in the pathogenesis of paediatric systemic lupus erythematosus. Clin Exp Rheumatol 2011;29(2):351−7.

[96] Alevizos I, Alexander S, Turner RJ, Illei GG. MicroRNA expression profiles as biomarkers of minor salivary gland inflammation and dysfunction in Sjogren's syndrome. Arthritis Rheum 2011;63(2):535−44.

[97] Alevizos I, Illei GG. MicroRNAs in Sjogren's syndrome as a prototypic autoimmune disease. Autoimmun Rev 2010;9(9):618−21.

[98] Pennell LM, Galligan CL, Fish EN. Sex affects immunity. J Autoimmun 2012;38(2−3): J282−91.

[99] Tiniakou E, Costenbader KH, Kriegel MA. Sex-specific environmental influences on the development of autoimmune diseases. Clin Immunol 2013;149(2):182−91.

[100] Lu Q, Wu A, Tesmer L, Ray D, Yousif N, Richardson B. Demethylation of CD40LG on the inactive X in T cells from women with lupus. J Immunol 2007;179(9):6352−8.

[101] Zhou Y, Yuan J, Pan Y, et al. T cell CD40LG gene expression and the production of IgG by autologous B cells in systemic lupus erythematosus. Clin Immunol 2009;132 (3):362−70.

[102] Crispin JC, Hedrich CM, Tsokos GC. Gene-function studies in systemic lupus erythematosus. Nat Rev Rheumatol 2013;9(8):476−84.

[103] Hobbs RN, Clayton AL, Bernstein RM. Antibodies to the five histones and poly (adenosine diphosphate-ribose) in drug induced lupus: implications for pathogenesis. Ann Rheum Dis 1987;46(5):408−16.

[104] Kanai Y, Kawaminami Y, Miwa M, Matsushima T, Sugimura T. Naturally-occurring antibodies to poly(ADP-ribose) in patients with systemic lupus erythematosus. Nature 1977;265(5590):175−7.

[105] Jonsen A, Bengtsson AA, Nived O, Truedsson L, Sturfelt G. Gene-environment interactions in the aetiology of systemic lupus erythematosus. Autoimmunity 2007;40(8):613−17.

7

Malignant Transformation and Epigenetics

Yixing Han[1], Jianke Ren[1], Weishi Yu[1], Minoru Terashima[1], and Kathrin Muegge[1,2]

[1]Mouse Cancer Genetics Program, Center for Cancer Research, National Cancer Institute, Frederick, MD [2]Basic Science Program, Leidos Biomedical Research, Inc., Mouse Cancer Genetics Program, Frederick National Laboratory for Cancer Research, Frederick, MD

7.1 INTRODUCTION

The transition from normal cells to transformed cells during tumorigenesis involves an extensive reconfiguration of the genome's expression program. Epigenetics is defined as a molecular mechanism that changes gene expression patterns without alteration of the primary genetic sequence. Epigenetic mechanisms alter chromatin structure and function.

Genomic DNA is packaged as chromatin in the nucleus by highly conserved proteins termed histones. DNA and histones are assembled into nucleosomes to form the fundamental building blocks of chromatin. Two copies each of H2A, H2B, H3, and H4 constitute the four core histone proteins of the octamer; these are wrapped around 147 base pair DNA to form the nucleosome [1]. The spacing of nucleosomes, the internucleosomal interactions, and the interactions of nucleosomes with chromatin factors all influence the folding and compaction of chromatin. The nucleosome and the higher order chromatin structure form a hindrance for DNA-binding factors and thus impose an accessibility problem for DNA-based processes such as transcription, DNA repair, or recombination. Epigenetic modifications can alter chromatin structure and thus facilitate the access of DNA-binding factors to DNA sequences, and ultimately control gene transcription.

© 2015 Elsevier Inc. All rights reserved.

Epigenetic changes are heritable and reversible, and determine cell-type and tissue specificity during cell growth and the development of an organism. Epigenetic changes are also at the heart of cellular transformation, since aberrant epigenetic modifications induce changes in gene expression during the process of transformation. The abnormal gene expression profile alters the cellular identity and contributes to the switch into a malignant phenotype.

Recent advances in epigenetics provide an improved molecular understanding about the process of carcinogenesis, disease progression, and metastasis. Studying epigenetics leads to the discovery of potential tumor biomarkers for early diagnosis, disease monitoring, and disease prognosis. Inhibition of specific epigenetic marks has shown promise in preclinical trials, and the first epigenetic-based therapies for cancer treatment have been approved. Basic knowledge about chromatin function and how it contributes to tumorigenesis can lead to the discovery of novel epigenetic drugs. Furthermore, it may lead to improved molecular targeting of disease processes with less cellular toxicity.

Abnormal expression of epigenetic modifiers and aberrant epigenetic changes have been documented in many cancer types. High-throughput sequencing techniques have detected genetic mutations in enzymatic activities that regulate epigenetics, thus providing a novel link between epigenetics and tumorigenesis. The focus of this review is on two aspects in the field of epigenetics and cancer: we report on chromatin modifiers for which genetic mutations have been identified in cancer. It is hypothesized that these mutations act as drivers of tumorigenesis and that chromatin aberrations caused by the mutant proteins will alter the phenotype of the cell [2]. In addition, we describe murine studies in which genetic deletion or overexpression of a chromatin modifier influences cancer in mice, thus providing causal evidence for the role of epigenetics in tumorigenesis.

Epigenetic modifications include at least three basic mechanisms, which are as follows:

1. DNA methylation
2. Posttranslational modification of histone
3. SNF chromatin-remodeling complexes

We present a short overview for each chromatin modification, describe specific members, and then discuss the occurrence of genetic mutations that have been identified for some chromatin-modifying enzymes in human malignancies.

Recently, a cross talk between noncoding RNA and chromatin has been reported, and miRNAs, forming a subclass of noncoding RNA, are aberrantly expressed and play a role in cancer [3–5]. However, miRNAs are typically known as posttranscriptional gene silencers, either through translational repression or by destabilizing mRNA. Since epigenetics modulates the structure of the chromatin, thereby controlling transcription, we have not included the role of miRNA in cancer in this review. The reader interested in noncanonical pathways of miRNA function, the role of epigenetics in deregulation of miRNA expression, and newly discovered genetic mutations in dicer is referred to recent reviews [6–9]. A summary of epigenetic changes and alteration of the expression of epigenetic modifiers in cancer can be found elsewhere [10–14]. The use of epigenetic marks as biomarkers to detect cancer and link the disease to a certain prognosis, or the use of epigenetic enzymes as drug targets for cancer therapy, is covered elsewhere [15,16].

7.2 REGULATION OF DNA METHYLATION

DNA methylation is a covalent modification of cytosine that can confer stable silencing on imprinted genes, on genes located on the inactive X chromosomes, and on germ-cell-specific genes [17]. DNA methylation is mediated by a group of highly conserved DNA methyltransferases, including Dnmt1, the major maintenance DNA methyltransferase residing at the replication fork and copying DNA methylation patterns at hemi-methylated DNA onto the newly synthesized strand [18]. The *de novo* DNA methyltransferases Dnmt3a and 3b are associated with nucleosomes and can methylate newly replicated DNA, but may also assist in maintaining methylation patterning at highly methylated genomic regions [19]. The Tet proteins, comprising a family of at last three members, can oxidize methylated cytosine into 5-hydroxyl-methyl cytosine, carboxyl-methyl-cytosine, and formyl-methyl-cytosine [20–23]. The generation of oxidized derivatives is not recognized by Dnmt1 thus leading to passive demethylation during replication. On the other hand, oxidized methyl-cytosine may be actively removed via DNA repair pathways resulting ultimately in unmethylated cytosine. The methyl-CpG-binding domain (MBD) protein family comprises several members that can be regarded as "readers" of DNA methylation. MBD proteins may recruit transcriptional repressors to methylated sites. Binding of MecP2, a member of the MBD family, is also influenced by hydroxyl-methyl cytosine [24]. The mutated DNA methylation regulators and their associated cancers are summarized in Table 7.1.

7.2.1 Animal Studies Involving Modulation of DNA Methylation

Several mouse models that exhibit depletion or overexpression of chromatin remodelers have been used to study the effect on tumorigenesis. The modulation of DNA methyltransferase activity has led to tumor

TABLE 7.1   DNA Methylation Regulators in Cancer

Methyltransferase	Mutations	Tumors
DNMT3A	M, F, N, S	AML, MDS, MPD
Hydroxymethylation	Mutations	Tumors
TET1	T	AML
TET2	M, N, F	AML, MPD, MDS, CMML

promotion as well as suppression. For example, in a conditional mouse lung cancer model, it was demonstrated that *Dnmt3a* deletion promotes lung cancer progression, and thus Dnmt3a protects against malignant growth [25]. *Dnmt3b* heterozygotes (expressing only one *Dnmt3b* allele) in a B-cell lymphoma mouse model (expressing the *Eμ-myc* transgene) show accelerated lymphoma development suggesting Dnmt3b has a tumor-suppressing effect [26]. Dnmt3b deletion in T cells in a lymphoma mouse model also accelerated tumor development and demethylated a number of promoter regions [27]. Overexpression of Dnmt3b in a mouse model of intestinal tumors (*Apc(Min/+)* mice) increased the number and size of tumors [28]. Dnmt3b overexpression caused loss of imprinting, increased expression of Igf2, and led to repression of several tumor-suppressor genes [28]. These studies are consistent with a model in which *de novo* DNA methyltransferase activity inhibits tumor formation (e.g., suppressing oncogene expression) and inactivation of *Dnmt3a/b* promotes cancer.

On the other hand, several studies suggest that impaired Dnmt1 activity may inhibit the process of cancer development. For example, deletion of *Dnmt1* completely inhibited leukemia development in the MLL-AF9-induced acute myeloid leukemia (AML) mouse model [29]. Mice with a hypomorphic *Dnmt1* allele show reduced tumor incidence in a mouse intestinal tumor model (*Apc(Min/+)* mice) compared to control mice [30]. The opposing effects of Dnmt *deletion* on tumorigenesis may be in part explained by different target genes that are affected (e.g., tumor-suppressor genes versus oncogenes). It should be noted that reduction of DNA methyltransferase activity can have opposing effects dependent on the stages of tumor development [31,32] or on the tumor type. For example, a hypomorphic *Dnmt1* allele reduces tumor incidence of intestinal tumors in *Mlh1* −/− mice (with a DNA mismatch repair deficiency), but leads also to a higher frequency of lymphoma [33].

Mice with Tet2 deficiency showed an increased pool of hematopoietic stem cells and an increase in myeloproliferation and leukemia development, suggesting a tumor-suppressor function of Tet2 [34−36].

Genetic depletion of *Mbd2* impairs tumor formation and increases the life span in mice that develop intestinal tumors (*Apc(Min/+)* mice). These results suggest that the failure to recruit corepressor complexes to methylated DNA may prevent intestinal tumorigenesis [37].

7.2.2 Human Cancer and DNA Methylation

Abnormal methylation profiles that are detected in cancer include hypermethylation of CpG islands located at transcriptional start sites leading to transcriptional repression [14,15,17,38,39]. Gene silencing may be in part mediated by elevated occupancy of methyl-DNA-binding proteins [40,41] that recruit in turn transcriptional repressors and render histones hypoacetylated at promoter regions. DNA hypermethylation occurs preferentially at genes marked by H3K27me3, a histone modification mediated by polycomb proteins. Thus, DNA methylation may promote a switch from a reversible (polycomb-mediated repression) to a more stable form of gene silencing. Examples of gene silencing by DNA hypermethylation include the repression of the tumor-suppressor gene *VHL* (von Hippel–Lindau) in renal cancer and the DNA repair gene *MLH1* (MutL Homolog 1) in colon cancer; interestingly, for both genes frequent genetic mutations have been documented [15]. Other examples of repression mediated by hypermethylation include silencing of miRNA which in turn results in upregulation of proteins regulated by miRNA or deregulation of noncoding RNA [42,43].

Cancer tissue shows not only locus-specific hypermethylation, but also reduced DNA methylation. The regions of hypomethylation comprise large chromosomal domains located in lamin-associated regions [44]. DNA hypomethylation may influence transcription of genes located within these domains or DNA or hypomethylation may contribute to genomic instability [45]. A third example of aberrant methylation profiles in cancer occurs at imprinted genes. Methylation of an imprinted region upstream of the *H19* gene can lead to aberrant placement of the chromatin insulator CTCF which in turn leads to biallelic expression of *IGF2* [46]. The overexpression of the growth factor Igf2 has been implicated in Wilm's tumor formation [47,48].

Though abnormal DNA methylation (hyper and hypo) has been known for decades, and is frequently accompanied by abnormal expression of enzymes involved in DNA methylation, it has only recently been reported that human cancer shows genetic mutations of enzymes involved in DNA methylation. Thus in some cancers, genetic changes may contribute directly to epigenetic alterations.

7.2.3 DNA Methyltransferase DNMT3A

Genetic mutations of a DNA methyltransferase are detected in diverse blood-cell-related disorders. For example, genetic mutations of *DNMT3A* have been described in subsets of acute myelogenous leukemia patients [49–51]. However, it should be pointed out that no specific changes in DNA methylation or gene expression could be linked to the *DNMT3A* mutation [49]. The frequency of *DNMT3A* mutation reached up to 22%

and the majority involved mutations in the catalytic domain, thus resulting in impaired DNA methyltransferase activity [49,52,53]. The *DNMT3A* mutants are heterozygotic and can act as negative-dominant mutants on the wild type allele [53]. Expression of a mutant DNMT3A form in bone marrow cells leads to the development of chronic myelomonocytic leukemia in mice after a period of 12 months [54] indicating a transforming capacity of mutant DNMT3A. *DNMT3A* mutations can be also detected at lower frequency than in AML in other blood-related disorders, such as the myelodysplastic syndrome (MDS) and myelodysplastic neoplasms [55–57]. The occurrence of *DNMT3A* mutations correlates with reduced survival and poor prognosis of patients [51,57,58].

7.2.4 TET Proteins and Demethylation of DNA

The *TET1* (ten–eleven translocation 1) gene had been found to be a fusion partner of the *MLL* gene (a chromatin modifier, see below) in a subset of patients with AML carrying a translocation t [10,11] (q22; q23) [59,60]. Although it remains unclear how the TET1 fusion protein promotes leukemia, a recent study reported that rearranged MLL fusion proteins directly target the *TET1* gene [61]. This results in an overexpression of the TET1 protein, a common feature of MLL-rearranged leukemias. TET1 protein can convert 5-methyl-cytosine to 5-hydroxyl-methyl-cytosine at specific genes (such as *HOXA9* or *MEIS1*), which are known target genes of MLL fusion proteins. In this way, TET1 may support the oncogenic role of MLL fusion proteins [61].

TET2 mutations occur in certain subsets of leukemias, including myeloproliferative neoplasms, MDS, and subsets of acute and chronic myeloid leukemias with frequencies of the mutation varying between 3% and 42% [62–65]. A recent screening of 23 patients with chronic myelomonocytic leukemia revealed a frequency of *TET2* mutations as high as 65% of patient samples [66]. The prognostic value of *TET2* mutations may vary by disease, since reduced survival in AML [62,65], as well as enhanced survival in MDS has been reported in association with *TET2* mutations [67]. On the other hand, some studies did not detect any prognostic value for survival [68]. *TET2* mutations can lead to truncated proteins or introduce missense mutations that impair TET2 enzymatic activity [64].

Although hydroxy-methyl-cytosine levels are reduced in patient samples with *TET2* mutations, a correlation with gene expression was not evident [64,69]. It has been hypothesized that *TET2* mutations may reduce hydroxyl-methyl-cytosine which in turn could lead to CG hypermethylation. However, CG hypermethylation was only detected at selected CpG island genes, whereas most CpG sites showed unexpectedly CG hypomethylation [64,70]. For this reason, it remains unclear if *TET2* mutations mediate their transforming activity by altering DNA methylation. Alternatively, the hydroxy-methyl-cytosine may provide a signal in its own right, and reduced 5hmC may alter recruitment of other chromatin-interacting proteins. For example, Mecp2 occupancy can specifically be modulated by hydroxyl-methyl-cytosines [24]. In this way, the mutation of TET proteins may influence the recruitment of other epigenetic modifiers and consequently alter chromatin function.

A recent study reported *TET1*, *TET2*, and *TET3* mutations occur in primary human colon tumors, albeit at very low frequency [71]. Whether *TET* mutations play any role in colon cancer requires further investigation.

7.3 HISTONE MODIFICATIONS

Histone tails are subject to a number of covalent modifications, and it is thought that the faithful transfer of these posttranslational modifications between cell generations is essential for epigenetic inheritance. Until now at least 16 types of posttranslational histone modifications (PTMs) have been found at nearly 70 different amino acid residues on histones [72] (summarized in Table 7.2). A prominent histone modification is the transfer of one, two, or three methyl groups to lysine (K) or arginine (L) residues of histone 3. Numerous histone-modifying enzymes that catalyze histone methylation (HMT) share a conserved sequence motif known as the SET domain. Histone modifications at specific residues characterize genomic regulatory elements; for instance, active promoters are enriched in trimethylated H3 at lysine 4 (H3K4me3), active enhancers are enriched in monomethylated H3 at lysine 4 (H3K4me1) and acetylated H3 at lysine 27 (H3K27ac), while repressed promoters are enriched in trimethylated H3 at lysine 27 (H3K27me3) and lysine 9 (H3K9me3). Many enzymes have been identified as being "writer" (enzymes that set up PTMs), "eraser" (enzymes that remove PTMs), or "reader" (factors that interpret PTMs patterns) [73–75] (Table 7.3).

Correlative evidence supports the idea that aberrant patterning of histone modifications contributes to tumorigenesis, and histone modifications are altered on a genome-wide level and at discrete gene loci. In addition, the expression level of enzymes that alter histone modifications is perturbed in cancer and probably adds to malignant transformation and the disease process [76,77]. Recent genome-wide sequencing approaches have revealed genetic mutations in several histone methyltransferases, as well as demethylases.

TABLE 7.2   Histone Modifications, Readers, and Their Function

Histone modification	Nomenclature	Reader motif	Attributed function
Acetylation	K-ac	Bromodomain Tandem, PHD fingers	Transcription, repair, replication, and condensation
Methylation (lysine)	K-me1, K-me2, K-me3	Chromodomain, Tudor domain, MBT domain, PWWP domain, PHD fingers, WD40/b propeller	Transcription and repair
Methylation (arginine)	R-me1, R-me2s, R-me2a	Tudor domain	Transcription
Phosphorylation (serine and threonine)	S-ph, T-ph	14-3-3, BRCT	Transcription, repair, and condensation
Phosphorylation (tyrosine)	Y-ph	SH2	Transcription and repair
Ubiquitination	K-ub	UIM, IUIM	Transcription and repair
Sumoylation	K-su	SIM	Transcription and repair
ADP ribosylation	E-ar	Macro domain, PBZ domain	Transcription and repair
Deimination	R > Cit	Unknown	Transcription and decondensation
Proline isomerization	P-cis > P-trans	Unknown	Transcription
Crotonylation	K-cr	Unknown	Transcription
Propionylation	K-pr	Unknown	Unknown
Butyrylation	K-bu	Unknown	Unknown
Formylation	K-fo	Unknown	Unknown
Hyroxylation	Y-oh	Unknown	Unknown
O-GlcNAcylation (serine and threonine)	S-GlcNAc; T-GlcNAc	Unknown	Transcription

Modifications: me1, monomethylation; me2, dimethylation; me3, trimethylation; me2s, symmetrical dimethylation; me2a, asymmetrical dimethylation; and Cit, citrulline. Reader domains: MBD, methyl-CpG-binding domain; PHD, plant homeodomain; MBT, malignant brain tumor domain; PWWP, proline-tryptophan-tryptophan-proline domain; BRCT, BRCA1 C terminus domain; UIM, ubiquitin interaction motif; IUIM, inverted ubiquitin interaction motif; SIM, sumo interaction motif; and PBZ, poly ADP-ribose binding zinc finger.

TABLE 7.3   Histone-Modifying Enzymes in Cancer

Lysine methyltransferase	Residues modified	Mutations	Tumors
KMT2A (MLL1)	H3K4	T, PTD	AML, ALL, TCC, hematological malignancies
KMT2B (MLL2)	H3K4	N, F, M	Medulloblastoma, renal, DLBCL, FL glioma, pancreatic cancer
KMT2C (MLL3)	H3K4	N	Medulloblastoma, TCC, breast hematological malignancies, colon cancer
KMT3A (SETD2)		N, F, S, M	Renal, breast
KMT3B (NSD1)	H3K36 H4K20	T, M	AML, neuroblastoma, Wilms' tumors, and hematological malignancies
G9a	H3K9	Aberrant expression	Colon, gastric, breast cancer, HCC, cervical, uterine, ovarian
KMT6 (EZH2)	H3K27	M, aberrant expression	DLBCL, MPD, MDS, prostate, breast ovarian cancer, bladder, colon, pancreas, liver, gastric, uterine tumors, melanoma, lymphoma, myeloma, and Ewing's sarcoma
SUZ12	H3K27		Prostate, breast ovarian cancer
Lysine demethylase	**Residues modified**	**Mutations**	**Tumors**
KDM1A/LSD1/ BHC110/AOF2	H3K4 H3K9	M	Breast, prostate cancer, bladder cancer, neuroblastoma, lung carcinoma, and colorectal carcinoma
KDM1B/LSD2	H3K4	Amplified, overexpressed	Urothelial carcinoma
KDM6A/UTX	H3K27	M, D, N, F, S	AML, TCC, renal, esophageal, multiple myeloma, squamous cell carcinoma, renal cell carcinomas, bladder, breast, kidney, lung, pancreas, esophagus, colon, uterus, brain cancer

Abbreviations for the cancers are as follows: AML, acute myeloid leukemia; ALL, acute lymphoid leukemia; B-NHL, B-cell non-Hodgkin's lymphoma; DLBCL, diffuse large B-cell lymphoma; HNSCC, head and neck squamous cell carcinoma; FL, follicular lymphoma; MDS, myelodysplastic syndromes; MPD, myeloproliferative diseases; and TCC, transitional cell carcinoma of the urinary bladder. Mutation types are as follows: M, missense; F, frameshift; N, nonsense; S, splice site mutation; T, translocation; D, deletion; and PTD, partial tandem duplication.

7.3.1 MLL, Histone 3 Lysine 4 Methyltransferases

MLL (mixed lineage leukemia) gene is an example in which a histone methyltransferase plays a causative role in cancer [78]. A translocation involving the *MLL* gene is found in subsets of leukemia. MLL mediates H3K4me3 methylation, a chromatin mark associated with gene transcription. The chromosomal rearrangement of *MLL* leads to fusion of the 5' end of the *MLL* genes with a translocation partner of which about 70 different partners are known. The two most frequent partners of *MLL* fusion are AF9 and AF4. The *MLL* fusion protein leads to transcriptional upregulation of target genes, many of which play a role in cancer development and eventually to leukemic transformation. As an alternative pathway, it has been reported that some of the *MLL* fusion products can interact with other epigenetic modifiers. For example, interaction with DOT1L, a histone methyltransferase, results in H3 lysine 79 methylation at target genes and contributes to leukemia initiation and maintenance [79]. In addition, widespread DNA hypomethylation has been detected in AML patients with translocations of the *MLL* gene, suggesting yet another link with an epigenetic mechanism [80].

Genetically engineered MLL-Af9 fusion proteins cause leukemia in mice thus giving evidence that MLL fusion proteins can be a direct cause of tumorigenesis [80a].

Most recently, somatic mutations of the *MLL2* gene have been detected in a subset of lymphoma [81].

7.3.2 EZH2, Histone 3 Lysine 27 Methyltransferases

Mutations have been identified in *EZH2*, a component of the polycomb repressor complex (PRC). Genetic aberrations are frequent in MDS and neoplasm, follicular lymphoma, and subsets of B-cell lymphoma, with frequencies ranging from 6% to 22% of patient samples [82–87]. A subtype of acute lymphoblastic leukemia shows mutations in *EZH2*, *EED*, or *SUZ1,2*, which are all components of the PRC with a combined frequency of 42% [87]. Another study reported a frequency of 8% for *EZH2* and 3% for either *EED* or *SUZ1,2* mutations in myeloid neoplasm. The *EZH2* mutations included truncations and deletions. These types of genetic changes lead to impaired PRC activity. Indeed, consistent with reduced PRC activity, a decrease in H3K27me3 level was detected at some *Hox* genes, which are classical polycomb target genes [88]. Likewise, subsets of lymphoma showed specific *EZH2* mutations and *in vitro* assays confirmed reduced H3K27me3 activity [83]. On the other hand, gene amplification and overexpression of *EZH2* can be detected in some breast and prostate cancer samples [89]. Recent analysis of point mutations suggests, rather, hyperactivity of the mutated EZH2 proteins resulting in increases of H3K27me3 level in tumor samples [90–92].

Thus, depending on cofactors of the PRC or depending on the cancer type, an oncogenic as well as a tumor-suppressor function has been suggested for genetic mutants of PRC components.

Interestingly, somatic mutations have been also detected in genes encoding H3K27me3 demethylases. H3K27me3 demethylases are expected to counteract polycomb-mediated repression and to increase H3K27me3 levels upon mutation. For example, *UTX* (*KDM6*) mutations are found in diverse cancers, being most prevalent with a frequency of 10% in multiple myeloma, followed by esophageal cancer (8%) and subsets of kidney cancer (1.4% of renal clear cell carcinomas) [93]. One study found that up to 20% of patient samples with a subtype of bladder cancer carried *UTX* mutations [94]. However, global H3K27me3 levels were not altered in patients with *UTX* mutations [93].

7.3.3 Other Histone Methyltransferases and Demethylases

SETDB2 mutations are found in renal cell carcinoma, and in some types of glioma ranging in frequency from 8% to 15% of patient samples. *SETDB2* confers H3K36me3 modifications, and mutations involve loss of function of gene activity leading to reduced H3K36m3 levels. A subtype of kidney cancer shows genetic mutations in *SETDB2*, *JARID1C* (mediating demethylation of H3K4me3), and *UTX* (demethylase of H3K27me3) at a combined frequency of less than 15% [95] suggesting an imbalance of several histone modifications in this cancer type.

There is growing evidence that genetic disruption of SET-domain-containing proteins, such as NSD1, plays a key role in cellular transformation [96]. NSD1 has histone methyltransferase activity, mediating methylation of H3K36 and H4K20 [97]. Patients with germ line genetic disruption of one *NSD1* allele have an increased risk of developing malignant tumors before adulthood, including neuroblastoma, Wilms' tumor, and hematological malignancies. Therefore, it was proposed that *NSD1* has a tumor-suppressor function. Human neuroblastoma and glioma cells frequently have DNA hypermethylation at the *NSD1* genes leading to gene silencing [98]. This type of epigenetic inactivation is associated with global diminished levels of trimethylated histone lysine residues H4K20 and H3K36. In addition, NSD1 occupancy was reduced at promoter regions of NSD1 target genes, such as *MEIS1*. Furthermore, chromosomal translocations in leukemia can involve *NSD1*. The fusion protein NUP98-NSD1 shows aberrant H3K36 methylation at *HOXA* target genes that are involved in leukemia development [99]. Chromosomal rearrangements that involve the *NSD1* locus have been reported in breast cancer cell lines [100].

LSD1/KDM1 as an HMT eraser can convert dimethyl or monomethyl H3K4 and H3K9 to the unmodified state [101,102]. LSD1 is a component of the NURD—Mi-2 repressive complexes which suppresses the invasiveness and metastasis of breast cancer cells. Knocking down LSD1 increases the invasive and metastatic potential of breast cancer cells. On the other hand, overexpression of LSD1 reduces the invasiveness of breast cancer cells [103]. In addition to the tumor-suppressive function of LSD1, oncogenic properties have been reported for LSD1. The latter activity may be mediated by targeting and suppressing transforming growth factor-β signaling and the p53-dependent pathway.

G9A is a histone methyltransferase that mediates H3K9 methylation and contributes to the epigenetic silencing of several tumor-suppressor genes. G9A is overexpressed in aggressive lung cancer cells, and its elevated expression correlates with poor prognosis. Knockdown of G9A reduces H3K9 dimethylation and decreases the recruitment of several transcriptional repressors (HP1, DNMT1, and HDAC1) to the *Ep-CAM* promoter. The silencing of *Ep-CAM* leads to reduced migration and invasion of highly invasive lung cancer cells [104].

Other cancers that carry multiple mutations in more than one histone-modifying enzymes are subtypes of acute lymphoblastic leukemia carrying mutations in PRC proteins (EZH2; SUZ1,2; and EED) mediating H3K27me3 modification, SETD2 histone methyltransferase mediating H3K9 methylation, and in EP300, an enzyme altering histone acetylation level [87] or bladder cancer with genetic mutations in histone methyltransferase MLL3, the histone demethylase UTX, histone acetylase EP300, and the chromatin-remodeling enzymes ARID1A and CHD6 [94].

7.4 SNF FACTORS OF CHROMATIN REMODELING

The SWI/SNF (SWItch/sucrose nonfermentable) complex is a multi-subunit chromatin-remodeling complex, which was described as a critical regulator of gene regulation, cell specification, and cancer development [105–111]. The SWI/SNF subunit binds directly to nucleosome cores. It uses the energy derived from ATP hydrolysis to disrupt the interaction between DNA and histones. As a consequence nucleosomal sliding, histone exchange, or nucleosomal eviction may occur. Ultimately, the chromatin-remodeling activity of SNF factors regulates access of DNA-binding factors to DNA sequences [112]. Genetic alterations that inactivate SWI/SNF subunits are observed in around 20% of human cancers [113], suggesting that the SWI/SNF complex may have a functional role in preventing tumor development in diverse tissues. It is thought that SNF components regulate the expression of cell cycle regulators and tumor-suppressor proteins [114,115]. Genetic mutations of three subunits of nucleosome-remodeling factors are most prominent, including *SNF5*, *BRM* (Brahma), and *BRG1* (Brahma-related gene 1). These enzymes are highly conserved across species [116,117].

7.4.1 SNF Factor SNF5

The *SNF5/INI1* gene, also called *SMARCB1*, encodes a widely expressed subunit of the SWI/SNF chromatin-remodeling complex playing a role in chromatin remodeling and transcription [118–120]. *SNF5* genetic mutations were first reported in malignant rhabdoid tumors [121]. The biallellic frameshift or nonsense mutations of *SNF5/INI1* in rhabdoid tumors lead to functional inactivity, suggesting a tumor-suppressor role of *SNF5* [121]. Mice with a deletion of one *SNF5* gene copy (heterozygotic allele) develop spontaneous tumors (such as sarcoma) indicating a tumor-suppressor function [122]. *SNF5* is critical for cellular differentiation of embryonic stem cells, in part by modulating nucleosomal positioning at pluripotency genes and silencing targeted genes, such as *Oct4* [123]. A patient with early childhood tumors, including brain and renal rhabdoid tumors, had a germ line mutation of *SNF5*. In addition, it was reported that SNF5 expression was undetectable in both types of tumors [124].

Systematic screening of a variety of tumors revealed that *SNF5* genetic defects are most prevalent in rhabdoid tumors (90%), in choroid plexus carcinomas (63%), and in a subset of neuroectodermal tumors (13%) [125]. The high frequency of *SNF5* mutations suggests a causative role ("driver mutation") of *SNF5* inactivation in tumorigenesis. Patients with chronic myeloid leukemia show deletions involving *SNF5* in 36% of samples during blast crisis [126]. Low frequencies of *SNF5* mutations have been also observed in a variety of hematopoietic neoplasms, including non-Hodgkin lymphoma, acute lymphoblastic leukemia, and multiple myeloma [127]. However, the relevance of *SNF5* mutations in hematologic neoplasms has been disputed [128].

7.4.2 SNF Factor BRG1 (SMARCA4)

BRG1 (SMARCA4) and BRM (SMARCA2) are two crucial components of the SWI/SNF complex. They share approximately 75% identity at the protein level [129]. Both proteins belong to the SNF2 family of chromatin-remodeling proteins. They contain an ATPase domain that utilizes ATP to generate energy which is critical for their nucleosomal remodeling function [130]. The first somatic genetic mutation of *BRM* was reported in human non-melanoma skin cancers [131]. *BRG1* mutations are present in lung cancer cell lines, and the mutant cell lines show a loss of BRG1 expression [132,133]. Moreover, primary lung adenocarcinomas show a frequency of *BRG1* mutations of about 12% and *BRM* mutations of about 17% [133]. The loss of BRG1 and BRM is associated with poor prognosis [133]. In another study, *BRG1* mutations in lung adenocarcinomas were higher in smokers [134,135]. A subset of primary lung tumors, lymphoma, and medulloblastomas show somatic point mutations in the ATPase domain, which may affect the ATPase function of the protein [136−139].

Recently, evidence of germ line mutations in BRG1 has been detected that can lead to a subtype of ovarian cancer. The germ line mutation leads to functional loss of the affected allele, and all patients showed loss of heterozygosity (due to somatic deletion or mutation of the wild type allele). Altogether 38/40 tumor samples of this subtype of ovarian cancer (small-cell carcinoma of the ovary, hypercalcemic type (SCCOHT)) showed BRG1 mutations, suggesting that BRG1 mutations are the major cause of this cancer type [140]. A point mutation in BRG1 (leading to a single amino acid substitution) can lead to loss of BRG1 expression and release of RB-mediated cell-cycle arrest. [141]. On the other hand, reexpression of BRG1 in BRG1-deficient cells can induce growth arrest [142].

Genetic evidence suggests that also other components of SNF complexes can show mutations in diverse human cancers, including ARID1A (BAF250A), ARID1B (BAF250B) or PBRM1 (BAF180), and BAP1 [143−147]. Notably, the cancers with the highest SWI/SNF mutation rates were 75% in ovarian clear cell carcinoma, 57% in clear cell renal cell carcinoma, 40% in hepatocellular carcinoma, 34% in melanoma, 36% in gastric cancer, 23% in pancreatic cancer, 16% in diffuse large B-cell lymphoma and multiple myeloma, 14% in glioblastoma, and 11% in head and neck cancers from 7 to 106 cases [4,7,9,143,148−154].

7.5 EPIGENETIC THERAPY FOR CANCER

Since epigenetic modifications appear to play a role in the development and proliferation of cancer, and chromatin-modifying enzymes may act as drivers of cancer, the molecular targeting of chromatin modifiers has been suggested as "epigenetic" therapy for cancer. Up to now only a few drugs are approved by the FDA for cancer treatment [16].

Among FDA approved drugs are the DNA methyltransferase inhibitors azacitidine and decitabine. Both drugs are applied for use in clinical treatment of various MDS and have been shown to delay the onset of leukemia [150,155]. The efficacy of the drugs may be due to demethylation and reactivation of tumor-suppressor genes [156], but also in part by eliciting an alteration of the antitumor response [157], inducing apoptosis and causing DNA damage [145,158]. The potential use of demethylation drugs in solid tumors (not yet FDA approved) is discussed elsewhere [154].

Histone acetylation is closely associated with gene activation and histone deacetylation with gene silencing. Although genes encoding histone deacetylases (HDACs) are not listed among driver mutations in human cancer [2], HDAC inhibitors were used as anticancer agents after empirical screens had established *in vitro* antitumor efficacy [148,149]. Among the FDA approved drugs are HDAC inhibitors vorinostat and romidepsin [16]. They are currently applied for treatment of certain types of T-cell lymphoma with cutaneous manifestations [16]. Their cytostatic effect may be in part explained by inhibition of abnormal HDAC expression in tumor cells and by interfering with aberrant HDAC recruitment to promoter regions (e.g., by oncogenic fusion proteins) [159]. On the other hand, histone acteylation and deacetylation play a widespread role in cell survival and proliferation and it remains unclear how tumor specific the effect of HDAC inhibitor is. In addition, mouse studies that inhibit HDAC function have revealed tumor-suppressor function of HDACs indicating specific function of HDAC in different tumors [151–153,160]. Finally, it should be noted that HDACs may also affect other targets, since hundreds of proteins, other than histones, are acetylated and possibly substrates of HDACs, including transcription factors, DNA repair proteins, and cell cycle regulators [161]. Other recently developed drugs that may interfere with the function of chromatin-modifying enzymes and may have potential use for future epigenetic therapy in cancer are discussed elsewhere [162,163].

7.6 CONCLUSION

In conclusion, tumorigenesis is a process depending on genetic mutations and epigenetic abnormalities. Numerous studies have detected genetic mutations of epigenetic modifiers in human cancer, with hematopoietic neoplasm being prevalent. The current data suggests that mutations of chromatin modifiers may cause an epigenetic instability that alters the phenotype of cells and thus contributes to the process of tumorigenesis.

Many questions remain with respect to epigenetics, including how gene expression is normally regulated in development and how an epigenetic memory is established during cellular differentiation. With respect to cancer it remains unresolved if and how environmentally induced stimuli alter the epigenome and modulate the epigenetic memory. How are epigenetic signatures influenced by specific diets, exercise, hormonal changes, environmental toxins, and inflammation and how are risks factors involved in epigenetic changes? Do epigenetic reprogramming events occur in tumorigenesis, leading to an "induced cancer stem cell?" Finally, how are aberrant epigenetic patterns preserved through cellular replication, which abnormal epigenetic events are reversible, and what are potential novel drugs that can interfere with a deviant epigenome?

In the future, the field of epigenetics will not only provide further insights into cancer biology, into initiation, progression, and metastasis, but will also provide novel tools for improved diagnosis, prognosis, cancer classification, and likely will contribute to novel therapeutic strategies with individually tailored cancer treatment plans.

Acknowledgments

This project has been funded in whole or in part with federal funds from the Frederick National Laboratory for Cancer Research, National Institutes of Health, under contract HHSN261200800001E. The content of this publication does not necessarily reflect the views or policies of the Department of Health and Human Services, nor does mention of trade names, commercial products, or organizations imply endorsement by the US Government.

This research was supported in part by the Intramural Research Program of NIH, Frederick National Lab, Center for Cancer Research.

References

[1] Margueron R, Trojer P, et al. The key to development: interpreting the histone code? Curr Opin Genet Dev 2005;15(2):163—76.
[2] Vogelstein B, Papadopoulos N, et al. Cancer genome landscapes. Science 2013;339 (6127):1546—58.
[3] Costa FF. Non-coding RNAs: meet thy masters. Bioessays 2010;32(7):599—608.
[4] Fabbri M, Calin GA. Epigenetics and miRNAs in human cancer. Adv Genet 2010;70:87—99.
[5] Sato F, Tsuchiya S, et al. MicroRNAs and epigenetics. FEBS J 2011;278(10):1598—609.
[6] Fabbri M, Calore F, et al. Epigenetic regulation of miRNAs in cancer. Adv Exp Med Biol 2013;754:137—48.
[7] Kala R, Peek GW, et al. MicroRNAs: an emerging science in cancer epigenetics. J Clin Bioinforma 2013;3(1):6.
[8] Adams BD, Kasinski AL, et al. Aberrant regulation and function of microRNAs in cancer. Curr Biol 2014;24(16):R762—76.
[9] Foulkes WD, Priest JR, et al. DICER1: mutations, microRNAs and mechanisms. Nat Rev Cancer 2014;14:662—72.
[10] Robertson KD. DNA methylation, methyltransferases, and cancer. Oncogene 2001;20 (24):3139—55.
[11] Feinberg AP. The epigenetics of cancer etiology. Semin Cancer Biol 2004;14(6):427—32.
[12] Laird PW. Cancer epigenetics. Hum Mol Genet 2005;14(Spec No 1):R65—76.
[13] Taby R, Issa JP. Cancer epigenetics. CA Cancer J Clin 2010;60(6):376—92.
[14] Sandoval J, Esteller M. Cancer epigenomics: beyond genomics. Curr Opin Genet Dev 2012;22(1):50—5.
[15] Baylin SB, Jones PA. A decade of exploring the cancer epigenome—biological and translational implications. Nat Rev Cancer 2011;11(10):726—34.
[16] Mummaneni P, Shord SS. Epigenetics and oncology. Pharmacotherapy 2014;34 (5):495—505.
[17] Jones PA. Functions of DNA methylation: islands, start sites, gene bodies and beyond. Nat Rev Genet 2012;13(7):484—92.

[18] Ooi SK, O'Donnell AH, et al. Mammalian cytosine methylation at a glance. J Cell Sci 2009;122(Pt 16):2787–91.

[19] Jones PA, Liang G. Rethinking how DNA methylation patterns are maintained. Nat Rev Genet 2009;10(11):805–11.

[20] Kriaucionis S, Heintz N. The nuclear DNA base 5-hydroxymethylcytosine is present in Purkinje neurons and the brain. Science 2009;324(5929):929–30.

[21] Tahiliani M, Koh KP, et al. Conversion of 5-methylcytosine to 5-hydroxymethylcytosine in mammalian DNA by MLL partner TET1. Science 2009;324(5929):930–5.

[22] He YF, Li BZ, et al. Tet-mediated formation of 5-carboxylcytosine and its excision by TDG in mammalian DNA. Science 2011;333(6047):1303–7.

[23] Ito S, Shen L, et al. Tet proteins can convert 5-methylcytosine to 5-formylcytosine and 5-carboxylcytosine. Science 2011;333(6047):1300–3.

[24] Mellen M, Ayata P, et al. MeCP2 binds to 5hmC enriched within active genes and accessible chromatin in the nervous system. Cell 2012;151(7):1417–30.

[25] Gao Q, Steine EJ, et al. Deletion of the *de novo* DNA methyltransferase Dnmt3a promotes lung tumor progression. Proc Natl Acad Sci USA 2011;108(44):18061–6.

[26] Vasanthakumar A, Lepore JB, et al. Dnmt3b is a haploinsufficient tumor suppressor gene in Myc-induced lymphomagenesis. Blood 2013;121(11):2059–63.

[27] Hlady RA, Novakova S, et al. Loss of Dnmt3b function upregulates the tumor modifier Ment and accelerates mouse lymphomagenesis. J Clin Invest 2012;122(1):163–77.

[28] Linhart HG, Lin H, et al. Dnmt3b promotes tumorigenesis *in vivo* by gene-specific *de novo* methylation and transcriptional silencing. Genes Dev 2007;21(23): 3110–22.

[29] Trowbridge JJ, Sinha AU, et al. Haploinsufficiency of Dnmt1 impairs leukemia stem cell function through derepression of bivalent chromatin domains. Genes Dev 2012;26(4):344–9.

[30] Trasler J, Deng L, et al. Impact of Dnmt1 deficiency, with and without low folate diets, on tumor numbers and DNA methylation in Min mice. Carcinogenesis 2003;24 (1):39–45.

[31] Lin H, Yamada Y, et al. Suppression of intestinal neoplasia by deletion of Dnmt3b. Mol Cell Biol 2006;26(8):2976–83.

[32] Kinney SR, Moser MT, et al. Opposing roles of Dnmt1 in early- and late-stage murine prostate cancer. Mol Cell Biol 2010;30(17):4159–74.

[33] Trinh BN, Long TI, et al. DNA methyltransferase deficiency modifies cancer susceptibility in mice lacking DNA mismatch repair. Mol Cell Biol 2002;22(9):2906–17.

[34] Li Z, Cai X, et al. Deletion of Tet2 in mice leads to dysregulated hematopoietic stem cells and subsequent development of myeloid malignancies. Blood 2011;118(17):4509–18.

[35] Moran-Crusio K, Reavie L, et al. Tet2 loss leads to increased hematopoietic stem cell self-renewal and myeloid transformation. Cancer Cell 2011;20(1):11–24.

[36] Quivoron C, Couronne L, et al. TET2 inactivation results in pleiotropic hematopoietic abnormalities in mouse and is a recurrent event during human lymphomagenesis. Cancer Cell 2011;20(1):25–38.

[37] Sansom OJ, Berger J, et al. Deficiency of Mbd2 suppresses intestinal tumorigenesis. Nat Genet 2003;34(2):145–7.

[38] Issa JP. CpG-island methylation in aging and cancer. Curr Top Microbiol Immunol 2000;249:101–18.

[39] Esteller M. Epigenetics in cancer. N Engl J Med 2008;358(11):1148–59.

[40] Lopez-Serra L, Ballestar E, et al. A profile of methyl-CpG binding domain protein occupancy of hypermethylated promoter CpG islands of tumor suppressor genes in human cancer. Cancer Res 2006;66(17):8342–6.

[41] Esteller M. Epigenetic gene silencing in cancer: the DNA hypermethylome. Hum Mol Genet 2007;16(Spec No 1):R50–9.

[42] Saito Y, Liang G, et al. Specific activation of microRNA-127 with downregulation of the proto-oncogene BCL6 by chromatin-modifying drugs in human cancer cells. Cancer Cell 2006;9(6):435−43.

[43] Lujambio A, Portela A, et al. CpG island hypermethylation-associated silencing of non-coding RNAs transcribed from ultraconserved regions in human cancer. Oncogene 2010;29(48):6390−401.

[44] Berman BP, Weisenberger DJ, et al. Regions of focal DNA hypermethylation and long-range hypomethylation in colorectal cancer coincide with nuclear lamina-associated domains. Nat Genet 2012;44(1):40−6.

[45] Gaudet F, Hodgson JG, et al. Induction of tumors in mice by genomic hypomethylation. Science 2003;300(5618):489−92.

[46] Bell AC, Felsenfeld G. Methylation of a CTCF-dependent boundary controls imprinted expression of the Igf2 gene. Nature 2000;405(6785):482−5.

[47] Moulton T, Crenshaw T, et al. Epigenetic lesions at the H19 locus in Wilms' tumour patients. Nat Genet 1994;7(3):440−7.

[48] Steenman MJ, Rainier S, et al. Loss of imprinting of IGF2 is linked to reduced expression and abnormal methylation of H19 in Wilms' tumour. Nat Genet 1994;7(3):433−9.

[49] Ley TJ, Ding L, et al. DNMT3A mutations in acute myeloid leukemia. N Engl J Med 2010;363(25):2424−33.

[50] Yamashita Y, Yuan J, et al. Array-based genomic resequencing of human leukemia. Oncogene 2010;29(25):3723−31.

[51] Thol F, Damm F, et al. Incidence and prognostic influence of DNMT3A mutations in acute myeloid leukemia. J Clin Oncol 2011;29(21):2889−96.

[52] Yan XJ, Xu J, et al. Exome sequencing identifies somatic mutations of DNA methyltransferase gene DNMT3A in acute monocytic leukemia. Nat Genet 2011;43(4):309−15.

[53] Russler-Germain DA, Spencer DH, et al. The R882H DNMT3A mutation associated with AML dominantly inhibits wild-type DNMT3A by blocking its ability to form active tetramers. Cancer Cell 2014;25(4):442−54.

[54] Xu J, Wang YY, et al. DNMT3A Arg882 mutation drives chronic myelomonocytic leukemia through disturbing gene expression/DNA methylation in hematopoietic cells. Proc Natl Acad Sci USA 2014;111(7):2620−5.

[55] Brecqueville M, Cervera N, et al. Rare mutations in DNMT3A in myeloproliferative neoplasms and myelodysplastic syndromes. Blood Cancer J 2011;1(5):e18.

[56] Ewalt M, Galili NG, et al. DNMT3a mutations in high-risk myelodysplastic syndrome parallel those found in acute myeloid leukemia. Blood Cancer J 2011;1(3):e9.

[57] Walter MJ, Ding L, et al. Recurrent DNMT3A mutations in patients with myelodysplastic syndromes. Leukemia 2011;25(7):1153−8.

[58] Renneville A, Boissel N, et al. Prognostic significance of DNA methyltransferase 3A mutations in cytogenetically normal acute myeloid leukemia: a study by the Acute Leukemia French Association. Leukemia 2012;26(6):1247−54.

[59] Ono R, Taki T, et al. LCX, leukemia-associated protein with a CXXC domain, is fused to MLL in acute myeloid leukemia with trilineage dysplasia having t(10;11)(q22;q23). Cancer Res 2002;62(14):4075−80.

[60] Lorsbach RB, Moore J, et al. TET1, a member of a novel protein family, is fused to MLL in acute myeloid leukemia containing the t(10;11)(q22;q23). Leukemia 2003;17(3):637−41.

[61] Huang H, Jiang X, et al. TET1 plays an essential oncogenic role in MLL-rearranged leukemia. Proc Natl Acad Sci USA 2013;110(29):11994−9.

[62] Abdel-Wahab O, Mullally A, et al. Genetic characterization of TET1, TET2, and TET3 alterations in myeloid malignancies. Blood 2009;114(1):144−7.

[63] Delhommeau F, Dupont S, et al. Mutation in TET2 in myeloid cancers. N Engl J Med 2009;360(22):2289−301.

[64] Ko M, Huang Y, et al. Impaired hydroxylation of 5-methylcytosine in myeloid cancers with mutant TET2. Nature 2010;468(7325):839–43.

[65] Weissmann S, Alpermann T, et al. Landscape of TET2 mutations in acute myeloid leukemia. Leukemia 2012;26(5):934–42.

[66] Perez C, Martinez-Calle N, et al. TET2 mutations are associated with specific 5-methylcytosine and 5-hydroxymethylcytosine profiles in patients with chronic myelomonocytic leukemia. PLOS ONE 2012;7(2):e31605.

[67] Kosmider O, Gelsi-Boyer V, et al. TET2 mutation is an independent favorable prognostic factor in myelodysplastic syndromes (MDSs). Blood 2009;114(15):3285–91.

[68] Nibourel O, Kosmider O, et al. Incidence and prognostic value of TET2 alterations in *de novo* acute myeloid leukemia achieving complete remission. Blood 2010;116(7):1132–5.

[69] Konstandin N, Bultmann S, et al. Genomic 5-hydroxymethylcytosine levels correlate with TET2 mutations and a distinct global gene expression pattern in secondary acute myeloid leukemia. Leukemia 2011;25(10):1649–52.

[70] Yamazaki J, Taby R, et al. Effects of TET2 mutations on DNA methylation in chronic myelomonocytic leukemia. Epigenetics 2012;7(2):201–7.

[71] Seshagiri S, Stawiski EW, et al. Recurrent R-spondin fusions in colon cancer. Nature 2012;488(7413):660–4.

[72] Tan M, Luo H, et al. Identification of 67 histone marks and histone lysine crotonylation as a new type of histone modification. Cell 2011;146(6):1016–28.

[73] Dawson MA, Kouzarides T. Cancer epigenetics: from mechanism to therapy. Cell 2012;150(1):12–27.

[74] You JS, Jones PA. Cancer genetics and epigenetics: two sides of the same coin? Cancer Cell 2012;22(1):9–20.

[75] Hojfeldt JW, Agger K, et al. Histone lysine demethylases as targets for anticancer therapy. Nat Rev Drug Discov 2013;12(12):917–30.

[76] Berdasco M, Ropero S, et al. Epigenetic inactivation of the Sotos overgrowth syndrome gene histone methyltransferase NSD1 in human neuroblastoma and glioma. Proc Natl Acad Sci USA 2009;106(51):21830–5.

[77] Fullgrabe J, Kavanagh E, et al. Histone onco-modifications. Oncogene 2011;30 (31):3391–403.

[78] Neff T, Armstrong SA. Recent progress toward epigenetic therapies: the example of mixed lineage leukemia. Blood 2013;121(24):4847–53.

[79] Okada Y, Feng Q, et al. hDOT1L links histone methylation to leukemogenesis. Cell 2005;121(2):167–78.

[80] Akalin A, Garrett-Bakelman FE, et al. Base-pair resolution DNA methylation sequencing reveals profoundly divergent epigenetic landscapes in acute myeloid leukemia. PLOS Genet 2012;8(6):e1002781.

[80a] Corral J, Lavenir I, et al. An Mll-AF9 fusion gene made by homologous recombination causes acute leukemia in chimeric mice: a method to create fusion oncogenes. Cell 1996;85(6):853–61.

[81] Morin RD, Mendez-Lago M, et al. Frequent mutation of histone-modifying genes in non-Hodgkin lymphoma. Nature 2011;476(7360):298–303.

[82] Ernst T, Chase AJ, et al. Inactivating mutations of the histone methyltransferase gene EZH2 in myeloid disorders. Nat Genet 2010;42(8):722–6.

[83] Morin RD, Johnson NA, et al. Somatic mutations altering EZH2 (Tyr641) in follicular and diffuse large B-cell lymphomas of germinal-center origin. Nat Genet 2010;42(2):181–5.

[84] Nikoloski G, Langemeijer SM, et al. Somatic mutations of the histone methyltransferase gene EZH2 in myelodysplastic syndromes. Nat Genet 2010;42(8):665–7.

[85] Abdel-Wahab O, Pardanani A, et al. Concomitant analysis of EZH2 and ASXL1 mutations in myelofibrosis, chronic myelomonocytic leukemia and blast-phase myeloproliferative neoplasms. Leukemia 2011;25(7):1200–2.

[86] Bodor C, O'Riain C, et al. EZH2 Y641 mutations in follicular lymphoma. Leukemia 2011;25(4):726−9.

[87] Zhang J, Ding L, et al. The genetic basis of early T-cell precursor acute lymphoblastic leukaemia. Nature 2012;481(7380):157−63.

[88] Khan SN, Jankowska AM, et al. Multiple mechanisms deregulate EZH2 and histone H3 lysine 27 epigenetic changes in myeloid malignancies. Leukemia 2013;27(6):1301−9.

[89] Garnis C, Buys TP, et al. Genetic alteration and gene expression modulation during cancer progression. Mol Cancer 2004;3:9.

[90] Sneeringer CJ, Scott MP, et al. Coordinated activities of wild-type plus mutant EZH2 drive tumor-associated hypertrimethylation of lysine 27 on histone H3 (H3K27) in human B-cell lymphomas. Proc Natl Acad Sci USA 2010;107(49):20980−5.

[91] Yap DB, Chu J, et al. Somatic mutations at EZH2 Y641 act dominantly through a mechanism of selectively altered PRC2 catalytic activity, to increase H3K27 trimethylation. Blood 2011;117(8):2451−9.

[92] Majer CR, Jin L, et al. A687V EZH2 is a gain-of-function mutation found in lymphoma patients. FEBS Lett 2012;586(19):3448−51.

[93] van Haaften G, Dalgliesh GL, et al. Somatic mutations of the histone H3K27 demethylase gene UTX in human cancer. Nat Genet 2009;41(5):521−3.

[94] Gui Y, Guo G, et al. Frequent mutations of chromatin remodeling genes in transitional cell carcinoma of the bladder. Nat Genet 2011;43(9):875−8.

[95] Dalgliesh GL, Furge K, et al. Systematic sequencing of renal carcinoma reveals inactivation of histone modifying genes. Nature 2010;463(7279):360−3.

[96] Schneider R, Bannister AJ, et al. Unsafe SETs: histone lysine methyltransferases and cancer. Trends Biochem Sci 2002;27(8):396−402.

[97] Rayasam GV, Wendling O, et al. NSD1 is essential for early post-implantation development and has a catalytically active SET domain. EMBO J 2003;22(12):3153−63.

[98] Berdasco M, Esteller M. Aberrant epigenetic landscape in cancer: how cellular identity goes awry. Dev Cell 2010;19(5):698−711.

[99] Wang GG, Cai L, et al. NUP98-NSD1 links H3K36 methylation to Hox-A gene activation and leukaemogenesis. Nat Cell Biol 2007;9(7):804−12.

[100] Zhao Q, Caballero OL, et al. Transcriptome-guided characterization of genomic rearrangements in a breast cancer cell line. Proc Natl Acad Sci USA 2009;106(6):1886−91.

[101] Shi Y, Lan F, et al. Histone demethylation mediated by the nuclear amine oxidase homolog LSD1. Cell 2004;119(7):941−53.

[102] Metzger E, Wissmann M, et al. LSD1 demethylates repressive histone marks to promote androgen-receptor-dependent transcription. Nature 2005;437(7057):436−9.

[103] Wang Y, Zhang H, et al. LSD1 is a subunit of the NuRD complex and targets the metastasis programs in breast cancer. Cell 2009;138(4):660−72.

[104] Chen MW, Hua KT, et al. H3K9 histone methyltransferase G9a promotes lung cancer invasion and metastasis by silencing the cell adhesion molecule Ep-CAM. Cancer Res 2010;70(20):7830−40.

[105] Kadam S, Emerson BM. Transcriptional specificity of human SWI/SNF BRG1 and BRM chromatin remodeling complexes. Mol Cell 2003;11(2):377−89.

[106] Das AV, James J, et al. SWI/SNF chromatin remodeling ATPase Brm regulates the differentiation of early retinal stem cells/progenitors by influencing Brn3b expression and Notch signaling. J Biol Chem 2007;282(48):35187−201.

[107] Faralli H, Martin E, et al. Teashirt-3, a novel regulator of muscle differentiation, associates with BRG1-associated factor 57 (BAF57) to inhibit myogenin gene expression. J Biol Chem 2011;286(26):23498−510.

[108] Shema-Yaacoby E, Nikolov M, et al. Systematic identification of proteins binding to chromatin-embedded ubiquitylated H2B reveals recruitment of SWI/SNF to regulate transcription. Cell Rep 2013;4(3):601–8.

[109] Tolstorukov MY, Sansam CG, et al. Swi/Snf chromatin remodeling/tumor suppressor complex establishes nucleosome occupancy at target promoters. Proc Natl Acad Sci USA 2013;110(25):10165–70.

[110] Eroglu E, Burkard TR, et al. SWI/SNF complex prevents lineage reversion and induces temporal patterning in neural stem cells. Cell 2014;156(6):1259–73.

[111] Jegu T, Latrasse D, et al. The BAF60 subunit of the SWI/SNF chromatin-remodeling complex directly controls the formation of a gene loop at FLOWERING LOCUS C in *Arabidopsis*. Plant Cell 2014;26(2):538–51.

[112] Cote J, Peterson CL, et al. Perturbation of nucleosome core structure by the SWI/SNF complex persists after its detachment, enhancing subsequent transcription factor binding. Proc Natl Acad Sci USA 1998;95(9):4947–52.

[113] Kadoch C, Hargreaves DC, et al. Proteomic and bioinformatic analysis of mammalian SWI/SNF complexes identifies extensive roles in human malignancy. Nat Genet 2013;45(6):592–601.

[114] Medina PP, Sanchez-Cespedes M. Involvement of the chromatin-remodeling factor BRG1/SMARCA4 in human cancer. Epigenetics 2008;3(2):64–8.

[115] Cancer Genome Atlas Research Network. Comprehensive molecular characterization of clear cell renal cell carcinoma. Nature 2013;499(7456):43–9.

[116] Muchardt C, Yaniv M. A human homolog of *Saccharomyces cerevisiae* Snf2/Swi2 and *Drosophila*-Brm genes potentiates transcriptional activation by the glucocorticoid receptor. EMBO J 1993;12(11):4279–90.

[117] Phelan ML, Sif S, et al. Reconstitution of a core chromatin remodeling complex from SWI/SNF subunits. Mol Cell 1999;3(2):247–53.

[118] Kalpana GV, Marmon S, et al. Binding and stimulation of HIV-1 integrase by a human homolog of yeast transcription factor SNF5. Science 1994;266(5193):2002–6.

[119] Muchardt C, Sardet C, et al. A human protein with homology to *Saccharomyces cerevisiae* SNF5 interacts with the potential helicase hbrm. Nucleic Acids Res 1995;23 (7):1127–32.

[120] Wang W, Xue Y, et al. Diversity and specialization of mammalian SWI/SNF complexes. Genes Dev 1996;10(17):2117–30.

[121] Versteege I, Sevenet N, et al. Truncating mutations of hSNF5/INI1 in aggressive paediatric cancer. Nature 1998;394(6689):203–6.

[122] Guidi CJ, Sands AT, et al. Disruption of Ini1 leads to peri-implantation lethality and tumorigenesis in mice. Mol Cell Biol 2001;21(10):3598–603.

[123] You JS, De Carvalho DD, et al. SNF5 is an essential executor of epigenetic regulation during differentiation. PLOS Genet 2013;9(4):e1003459.

[124] Kusafuka T, Miao J, et al. Novel germ-line deletion of SNF5/INI1/SMARCB1 gene in neonate presenting with congenital malignant rhabdoid tumor of kidney and brain primitive neuroectodermal tumor. Genes Chromosomes Cancer 2004;40 (2):133–9.

[125] Sevenet N, Lellouch-Tubiana A, et al. Spectrum of hSNF5/INI1 somatic mutations in human cancer and genotype–phenotype correlations. Hum Mol Genet 1999;8 (13):2359–68.

[126] Grand F, Kulkarni S, et al. Frequent deletion of hSNF5/INI1, a component of the SWI/SNF complex, in chronic myeloid leukemia. Cancer Res 1999;59(16):3870–4.

[127] Yuge M, Nagai H, et al. HSNF5/INI1 gene mutations in lymphoid malignancy. Cancer Genet Cytogenet 2000;122(1):37–42.

[128] Mori N, Inoue K, et al. Absence of mutations on the SNF5 gene in hematological neoplasms with chromosome 22 abnormalities. Acta Haematol 2011;126(2):69–75.

[129] Khavari PA, Peterson CL, et al. BRG1 contains a conserved domain of the SWI2/SNF2 family necessary for normal mitotic growth and transcription. Nature 1993;366(6451):170−4.

[130] Muchardt C, Yaniv M. ATP-dependent chromatin remodelling: SWI/SNF and Co. are on the job. J Mol Biol 1999;293(2):187−98.

[131] Moloney FJ, Lyons JG, et al. Hotspot mutation of Brahma in non-melanoma skin cancer. J Invest Dermatol 2009;129(4):1012−15.

[132] Medina PP, Romero OA, et al. Frequent BRG1/SMARCA4-inactivating mutations in human lung cancer cell lines. Hum Mutat 2008;29(5):617−22.

[133] Matsubara D, Kishaba Y, et al. Lung cancer with loss of BRG1/BRM, shows epithelial mesenchymal transition phenotype and distinct histologic and genetic features. Cancer Sci 2013;104(2):266−73.

[134] Liu J, Lee W, et al. Genome and transcriptome sequencing of lung cancers reveal diverse mutational and splicing events. Genome Res 2012;22(12):2315−27.

[135] Seo JS, Ju YS, et al. The transcriptional landscape and mutational profile of lung adenocarcinoma. Genome Res 2012;22(11):2109−19.

[136] Medina PP, Carretero J, et al. Genetic and epigenetic screening for gene alterations of the chromatin-remodeling factor, SMARCA4/BRG1, in lung tumors. Genes Chromosomes Cancer 2004;41(2):170−7.

[137] Love C, Sun Z, et al. The genetic landscape of mutations in Burkitt lymphoma. Nat Genet 2012;44(12):1321−5.

[138] Pugh TJ, Weeraratne SD, et al. Medulloblastoma exome sequencing uncovers subtype-specific somatic mutations. Nature 2012;488(7409):106−10.

[139] Robinson G, Parker M, et al. Novel mutations target distinct subgroups of medulloblastoma. Nature 2012;488(7409):43−8.

[140] Witkowski L, Carrot-Zhang J, et al. Germline and somatic SMARCA4 mutations characterize small cell carcinoma of the ovary, hypercalcemic type. Nat Genet 2014;46(5):438−43.

[141] Bartlett C, Orvis TJ, et al. BRG1 mutations found in human cancer cell lines inactivate Rb-mediated cell-cycle arrest. J Cell Physiol 2011;226(8):1989−97.

[142] Wong AK, Shanahan F, et al. BRG1, a component of the SWI−SNF complex, is mutated in multiple human tumor cell lines. Cancer Res 2000;60(21):6171−7.

[143] Varela I, Tarpey P, et al. Exome sequencing identifies frequent mutation of the SWI/SNF complex gene PBRM1 in renal carcinoma. Nature 2011;469(7331):539−42.

[144] Fujimoto A, Totoki Y, et al. Whole-genome sequencing of liver cancers identifies etiological influences on mutation patterns and recurrent mutations in chromatin regulators. Nat Genet 2012;44(7):760−4.

[145] Ruiz-Magana MJ, Rodriguez-Vargas JM, et al. The DNA methyltransferase inhibitors zebularine and decitabine induce mitochondria-mediated apoptosis and DNA damage in p53 mutant leukemic T cells. Int J Cancer 2012;130(5):1195−207.

[146] Shain AH, Giacomini CP, et al. Convergent structural alterations define SWItch/Sucrose NonFermentable (SWI/SNF) chromatin remodeler as a central tumor suppressive complex in pancreatic cancer. Proc Natl Acad Sci USA 2012;109(5):E252−9.

[147] Helming KC, Wang X, et al. ARID1B is a specific vulnerability in ARID1A-mutant cancers. Nat Med 2014;20(3):251−4.

[148] Leder A, Leder P. Butyric acid, a potent inducer of erythroid differentiation in cultured erythroleukemic cells. Cell 1975;5(3):319−22.

[149] Riggs MG, Whittaker RG, et al. n-Butyrate causes histone modification in HeLa and Friend erythroleukaemia cells. Nature 1977;268(5619):462−4.

[150] Heyn H, Esteller M. DNA methylation profiling in the clinic: applications and challenges. Nat Rev Genet 2012;13(10):679−92.

[151] Dovey OM, Foster CT, et al. Histone deacetylase 1 and 2 are essential for normal T-cell development and genomic stability in mice. Blood 2013;121(8):1335–44.

[152] Heideman MR, Wilting RH, et al. Dosage-dependent tumor suppression by histone deacetylases 1 and 2 through regulation of c-Myc collaborating genes and p53 function. Blood 2013;121(11):2038–50.

[153] Santoro F, Botrugno OA, et al. A dual role for Hdac1: oncosuppressor in tumorigenesis, oncogene in tumor maintenance. Blood 2013;121(17):3459–68.

[154] Nie J, Liu L, et al. Decitabine, a new star in epigenetic therapy: the clinical application and biological mechanism in solid tumors. Cancer Lett 2014;354(1):12–20.

[155] Yamazaki J, Issa JP. Epigenetic aspects of MDS and its molecular targeted therapy. Int J Hematol 2013;97(2):175–82.

[156] Estecio MR, Issa JP. Dissecting DNA hypermethylation in cancer. FEBS Lett 2011;585(13):2078–86.

[157] Russ BE, Prier JE, et al. T cell immunity as a tool for studying epigenetic regulation of cellular differentiation. Front Genet 2013;4:218.

[158] Yang D, Torres CM, et al. Decitabine and vorinostat cooperate to sensitize colon carcinoma cells to Fas ligand-induced apoptosis *in vitro* and tumor suppression *in vivo*. J Immunol 2012;188(9):4441–9.

[159] Falkenberg KJ, Johnstone RW. Histone deacetylases and their inhibitors in cancer, neurological diseases and immune disorders. Nat Rev Drug Discov 2014;13(9):673–91.

[160] Bhaskara S, Knutson SK, et al. Hdac3 is essential for the maintenance of chromatin structure and genome stability. Cancer Cell 2010;18(5):436–47.

[161] Choudhary C, Kumar C, et al. Lysine acetylation targets protein complexes and co-regulates major cellular functions. Science 2009;325(5942):834–40.

[162] Dhanak D, Jackson P. Development and classes of epigenetic drugs for cancer. Biochem Biophys Res Commun 2014; pii: S0006-291X(14)01220-0. PMID: 25016182.

[163] Simo-Riudalbas L, Esteller M. Targeting the histone orthography of cancer: drugs for writers, erasers and readers. Br J Pharmacol 2014. PMID: 25039449.

CHAPTER

8

Epigenetic Mechanisms of Sirtuins in Dermatology

Alexander Lo[1], Melissa Serravallo[2], and Jared Jagdeo[2,3,4]

[1]SUNY Downstate College of Medicine, Brooklyn, NY [2]Department of Dermatology, SUNY Downstate Medical Center, Brooklyn, NY [3]Department of Dermatology, University of California at Davis, Sacramento, CA [4]Dermatology Service, Sacramento VA Medical Center, Mather, CA

8.1 INTRODUCTION

Sirtuins are a family of proteins that play an important role as key epigenetic regulators in health and disease. In dermatology, sirtuin activity is implicated in skin aging, inflammatory skin diseases, hyperproliferative disease, scarring, skin cancer, and inherited diseases. Therefore, sirtuins are potential novel targets for new treatment therapies and epigenetic research in a variety of disease modalities. In this chapter, we describe the function of sirtuins in epigenetic mechanisms related to dermatology and discuss the potential role of sirtuin modulation in dermatological therapy.

This begins with an understanding of the multistep process by which sirtuin function can impact the pathogenesis of various pathologies. First, sirtuins have enzymatic activity that allows them to modify the structure and activity of various histone and nonhistone substrates. These epigenetic modifications then result in altered activity of those modified substrates that are involved in a variety of cellular processes including apoptotic signaling, DNA repair, telomere maintenance, cell cycle regulation, and cell proliferation. Finally, these changes result in promotion or inhibition of relevant disease processes. The chapter is formatted as

© 2015 Elsevier Inc. All rights reserved.

follows: the first few sections examine the epigenetic mechanisms that sirtuins use to modulate cellular processes and the following sections describe dermatological processes and diseases that sirtuins are linked to, and examine the specific associated epigenetic mechanisms.

8.2 BACKGROUND

8.2.1 A Brief History

Mammalian sirtuins are a family of seven proteins (SIRT1−7) that are defined by their homology to the Sir2 protein found in the budding yeast, *Saccharomyces cerevisiae*. This silent information regulator 2 (Sir2) was first identified in 1979 in yeast as the *mar1* (mating type regulator) gene for its effect on yeast mating ability, and characterized in 1984 as functioning through gene silencing [1]. Over the next few decades, Sir2 became well associated with longevity, beginning with the discovery of Sir2's role in extending lifespan in yeast in 1999 and then in higher eukaryotes, including the nematode worm and fruit fly, in the years 2001 and 2004, respectively [2−4]. Since then, a growing body of research has further investigated the role that sirtuins play in human cells, revealing that they play a role in a number of cellular processes.

8.2.2 Sirtuins Interact with Substrates via Enzymatic Activity

Human sirtuins SIRT1−7 were found to act in two major enzymatic modalities that contribute to their epigenetic function in human cells. The first, nicotinamide adenine dinucleotide (NAD^+)-dependent histone deacetylase (HDAC) activity, results from a highly conserved catalytic domain across all sirtuins that include an NAD^+-binding domain and small zinc ribbon that provides structural stability [5]. This enzymatic activity was first characterized in Sir2 and is subsequently used to classify all sirtuins as class III HDACs [6]. This class designation distinguishes sirtuins from other classes of HDACs, which do not share their unique NAD^+ dependency. NAD^+ is a cofactor required for energy metabolism that, together with NADH, alters sirtuin activity in response to different metabolic states. Among the seven proteins within the mammalian family, HDAC activity has been demonstrated in SIRT1, SIRT2, SIRT3, SIRT5, SIRT6, and SIRT7 [7−11].

The second enzymatic reaction that sirtuins participate in is monoadenosine diphosphate ribosyl transferase (ADPRT) activity, which has been demonstrated in SIRT1, SIRT4, and SIRT6 [7−10]. Together these enzymatic mechanisms allow sirtuins to reversibly modify many histone and nonhistone targets that play a role in the various cellular signaling machinery within a cell (Table 8.1). However, these two reactions may not yet encompass all of the enzymatic activity through which sirtuins function. There may be other enzymatic reactions that sirtuins play a role in as well. For example, recently SIRT5 has also been shown to exhibit greater NAD^+-dependent protein lysine demalonylase and desuccinylase activity than deacetylase activity [12]. However, the significance of this enzymatic activity has not yet been determined.

The location in the cell and enzymatic activity of each of the seven mammalian sirtuins correlate with their cellular effect (Table 8.2). SIRT1, SIRT6, and SIRT7 function primarily in the nucleus, SIRT2 predominantly

TABLE 8.1   Overview of Sirtuin Targets in Dermatology

Sirtuin targets	Specific examples
Histones	Please refer to Table 8.3
Transcription regulators	NF-κB, FOX family (FOXO1, FOXO3a, FOXO4), PPARγ, p300 HAT, MOF, TIP60, E2F1, c-Jun, dnmt1, SUV39H1, STAT3
Tumor suppressors	p53, p73, Rb
Enzymes	ICDH2, PARP1, MMP, Pol I, eNOS
Structural proteins	α-Tubulin
DNA repair proteins	Ku70, SNF2H, MRE11-RAD50-NBS1, XPA
Cell signaling proteins	SMAD7, 14-3-3β/γ

TABLE 8.2   Location and Demonstrated Enzymatic Activity of Each Sirtuin [7–12]

Location	Sirtuin	Demonstrated enzymatic activity
Nucleus	SIRT1	HDAC and ADPRT
Nucleus	SIRT6	HDAC and ADPRT
Nucleolus	SIRT7	HDAC
Cytoplasm	SIRT2	HDAC
Mitochondria	SIRT3	HDAC
Mitochondria	SIRT4	ADPRT
Mitochondria	SIRT5	HDAC, lysine demalonylase, and desuccinylase

in the cytoplasm, and SIRT3, SIRT4, and SIRT5 mainly work in the mitochondria [7–11]. Sirtuins play a dynamic role within the cell, however, and can influence processes in several subcellular locations in response to a variety of factors including cell type, cell stress, and molecular signaling. SIRT1 and SIRT2 can both localize to either the nucleus and/or cytoplasm in response to certain cell states and subsequently interact in signaling pathways involving both nuclear and cytosolic proteins [13–15]. SIRT3 has also been shown to localize to the nucleus or mitochondria under conditions of stress [16,17].

8.2.3 Sirtuin Enzyme Reactions

In the sirtuin deacetylation reaction, NAD^+ and an acetyl lysine substrate bind on opposite sides of the sirtuin conserved catalytic core [18]. An acetyl group is then transferred from the substrate to NAD^+, producing nicotinamide (NAM), the deacetylated substrate, and O-acetyl-ADP-ribose (OAADPR) [19]. NAM is notable as a noncompetitive inhibitor of sirtuins by reacting with a catalytic intermediate to re-form β-NAD^+ through a base exchange process [20,21]. This interaction therefore provides a negative feedback loop that plays a role in regulation of sirtuin activity. OAADPR may also play a role in certain signaling pathways, but its function is still in the early stages of investigation [22].

Sirtuin ADPRT activity provides another mechanism for modifying different substrates, and either functions in addition to, or instead of, HDAC activity [23]. ADP ribosylation is a method of posttranslational protein modification in which one or more ADP ribose groups are added to a specific site. As such, ADP ribosylation is important in many of the processes that sirtuins are implicated in, including inter- and intracellular signaling, transcriptional regulation, DNA repair, telomere maintenance, cell differentiation, proliferation, and apoptosis [5,11]. SIRT4 possesses NAD^+-dependent mono-ADP-ribosyl transferase activity [24], while SIRT1 and SIRT6 have been shown to have both substrate-specific deacetylase activity and auto-ADP-ribosyl transferase activity that is hypothesized to function as a self-regulating mechanism [25,26].

8.3 EPIGENETIC MECHANISMS OF SIRTUIN FUNCTION

The primary epigenetic mechanisms by which sirtuins exert an effect on various cellular pathways are chromatin regulation, including histone modification and DNA methylation, and direct interaction with nonhistone substrates [5,11,27,28]. Each of these mechanisms involves sirtuin binding to a substrate and enzymatically modifying it such as via deacetylation or ADP ribosylation. These epigenetic changes then result in alterations of gene expression and changes in the activity of nonhistone substrates that persist after sirtuin modification (until reversed by another enzyme).

8.3.1 Heritability of Epigenetic Changes

Heritability of epigenetic changes is not necessary for epigenetic mechanisms to result in enduring changes in cells as the genetic state of the cell (stage of mitosis) may be indirectly affected by sirtuin activity [29]. However, it is an interesting topic to consider with respect of the possible long-term implications this might have on therapies involving sirtuin modulation. Notably, while the heritability of DNA methylation between generations is relatively well understood (see Section 8.4.2), it is less clear how histone modifications are propagated through generations. Meanwhile, sirtuin interaction with nonhistone substrates that are unrelated to chromatin may induce heritable transcription states via other mechanisms. One possibility is that sirtuin activity causes the transcription of a product that is able to maintain the activity of that gene. An example of such a system would be a bistable promoter response model wherein small changes in transcription factor activity past a given threshold can lead to one of two stable gene expression states among identical cells [30]. Such a mechanism may include the recruitment of various histone modification enzymes such as sirtuins or histone acetyltransferases (HATs) that can then result in changes to the epigenetic state of nearby nucleosomes and affect their transcriptional activity [30]. Other research demonstrates that transient transcriptional errors present in the mRNA encoding a transcription factor can alone promote heritable epimutations by reprogramming a transcriptional network, all without altering the DNA sequence [31]. These mechanisms may be associated with the formation of an RNA transcript that encodes a product that maintains its own activity [31].

8.4 SIRTUIN MODIFICATION OF CHROMATIN STRUCTURE AND FUNCTION IMPACTS GENE EXPRESSION

8.4.1 Histone Modification

Chromatin regulation is a highly conserved feature of sirtuin function that is epigenetic by nature. Chromatin consists of repeating segments of DNA coiled around a histone octamer that contains two copies of each core histone—H2A, H2B, H3, and H4 [32]. The exposed

N-terminal domains or histone tails of each core are frequently subject to posttranslational modifications including, but not limited to, acetylation, methylation, phosphorylation, ubiquitination, ADP ribosylation (transfers ADP-ribose from NAD^+ to arginine, glutamic acid, aspartic acid), glycosylation, sumoylation, and citrullination (modification of arginine to citrulline) [5]. These alterations regulate the overall epigenetic state of chromatin structure and function [27,33].

One epigenetic mechanism by which sirtuins regulate various cellular processes is via deacetylation of core histones at specific residues, a reversible modification that results in transcriptional silencing of various genes [11]. Histone deacetylation increases the positive charges on the already positively charged amine-containing histones and enhances their binding with the negatively charged phosphate groups of DNA, resulting in the formation of heterochromatin structures [27]. Chromatin compaction then results in decreased gene expression, because DNA is less accessible to cellular transcriptional machinery. This allows sirtuins to participate in epigenetic silencing of a wide variety of genes under various circumstances. Conversely, histone acetylation is generally associated with regions of active transcription, and many transcriptional coactivators contain HAT activity [27,34,35].

Together with these HATs, which add acetyl groups, sirtuins regulate the level of acetylation and, subsequently, the transcriptional activity at a given location by directly deacetylating specific histone targets [11,36]. Epigenetic research has focused primarily on therapies targeting the various histone modification enzymes: HATs, HDACs, histone methyltransferases (HMTs), histone demethylases (HDMs), kinases, and ADP-ribosyl transferases (ADPRTs) [6]. Sirtuins are unique, however, because of their key roles as mediators of cell survival under stress and their ability to respond to metabolic fluctuations induced by environmental stimuli via their NAD^+ dependency.

Sirtuin activity is consistently involved with loci associated with nucleolar rDNA transcription and at subtelomeric regions, and the majority of sirtuins display direct deacetylase activity of histones 3 and 4 (Table 8.3) [11]. Two of these specific histone targets, lysine 16 of histone H4 (H4K16) and lysine 9 of histone H3 (H3K9), have been well investigated and play a role in gene silencing as well as signaling of DNA damage and its related cellular processes [11,35]. Acetylation/deacetylation of H4K16 is associated with heterochromatin formation in yeast and mammals and is linked to DNA repair at sites of double strand breakage (DSB) [5,35]. Sirtuin activity with H3K9 functions differently. Deacetylation is required for methylation at the same site,

TABLE 8.3 Histone Targets—Sirtuins' Role in Chromatin Structure and Function

Sirtuin	Histone substrate	Effect
SIRT1	Preferentially deacetylates H4K16Ac and H3K9Ac. Also recruits H1 to establish repressive chromatin and catalyzes the deacetylation of H1K26	Deacetylation of both core histones (H3 and H4) increases compact chromatin and coordinates formation of heterochromatin (both constitutive and facultative), repressing gene expression [5,11]
		H4K16 deacetylation is linked with methylation at the same site, a modification that also leads to chromatin compaction [11,35]
	Deacetylates H3K56Ac upon DNA damage	Implicated in the signaling of DNA damage during the cell cycle [37]
	Binds to rDNA via interaction with DNA methyltransferase (Dnmt1)	Suggests an epigenetic caretaker role for Dnmt1 via maintaining nucleolar structure
		Dnmt1 deficient cells have disorganized nucleoli [38]
SIRT2	Deacetylates H4K16Ac during mitosis and the G_2/M transition	Leads to global decrease in acetylated H4K16 levels in the G_2/M transition and may be important in the formation of condensed chromatin/compaction of chromosome during mitosis. Lack of SIRT2 affects proper entry and progression through the S phase [39]
	Deacetylates H3K56Ac upon DNA damage	Implicated in the signaling of DNA damage during the cell cycle [11,40]
SIRT3	Deacetylates H4K16Ac and H3K9Ac	Also results in chromatin compaction and repression of gene expression [35]
	Deacetylates H3K56Ac upon DNA damage	Suspected to be involved in signaling of DNA damage [40]

(Continued)

TABLE 8.3   (Continued)

Sirtuin	Histone substrate	Effect
SIRT4	Unknown role in chromatin structure and function	Unknown role in chromatin structure and function
SIRT5	Unknown role in chromatin structure and function	Unknown role in chromatin structure and function
SIRT6	H3K9Ac-specific deacetylase activity at telomeric chromatin	Required for proper telomere function and maintenance
		SIRT6 is responsible for the low levels of acetylated H3K9 at telomeric chromatin observed in the S phase and regulates WRN association with telomeres during the S phase that prevents aberrant loss of telomeres [25]
	Deacetylation of H3K9 in response to DNA double stranded breaks (DSBs)	SIRT6 forms a complex with DNA-dependent protein kinase (DNA-PK) and mobilizes the DNA-PK subunit to chromatin in order to facilitate DSB repair [41]
	Deacetylation of H3K9 at NF-κB target gene promoters and is recruited to the gene loci via interaction with the RelA subunit	Modifies chromatin at NF-κB target genes, leading to decreased age-associated NF-κB-dependent expression of genes such as IAP2, MnSOD2, ICAM1, NFKBIA, and NFKB1 [42]
SIRT7	Part of RNA Polymerase I (Pol I) machinery that interacts with the promoter, ribosomal DNA (rDNA) coding sequence, H2A, and H2B	SIRT7 activity is linked to RNA Pol I transcription and expression of ribosomal RNA genes
Highly specific deacetylase activity with H3K18Ac	Linked to oncogenic maintenance	SIRT7 depletion stops cell proliferation and triggers apoptosis; therefore SIRT7-mediated Pol I transcription is believed to be required for cell growth and viability [43]
		Implicates SIRT7 in a pro-proliferative capacity [43,44]

resulting in H3K9me2/3, which is a hallmark of chromatin compaction and of heterochromatin [11]. SIRT1 and SIRT6 are important deacetylases of H3K9Ac, while deacetylation of H4K16Ac is mainly restricted to SIRT1—SIRT3 [11].

Other histone targets of note include H3K56, which is implicated in signaling of DNA damage during the cell cycle, and H3K18, which has been linked to oncogenic maintenance [11,44]. SIRT1–SIRT3 are associated with deacetylase activity of H3K56Ac when DNA damage occurs [37,40]. SIRT7, meanwhile, exhibits highly specific deacetylase activity with H3K18Ac, but it is unclear how this interaction impacts tumorigenesis as research using models of SIRT7 knockdown or overexpression have provided conflicting evidence of SIRT7 in proliferative and antiproliferative capacities [43,44]. SIRT1 activity is associated with remaining sites of histone deacetylation, such as H1K26 [45]. A summary of these interactions can be found in Table 8.3.

8.4.2 DNA Methylation

Sirtuins, particularly SIRT1, also play a role in chromatin regulation via DNA methylation, a process wherein DNA methyltransferases reversibly modify the DNA molecule via the addition of methyl groups to cytosine bases in CG pairs. Sirtuins are linked to DNA methylation by either directly acetylating methyltransferases to impact their level of activity or by deacetylating histones that promote DNA methylation at the same site. Highly methylated DNA then results in increased chromatin compaction and transcriptional repression [46]. SIRT1 has been associated with two human methyltransferases, including DNA methyltransferase1 (Dnmt1) and a HMT named SUV39H1, both of which play a role in gene silencing by the formation of heterochromatin-associated histone marks [38,47].

SUV39H1 specifically targets H3K9. Meanwhile, heritability of methylation states is established by certain enzymes, like Dnmt1, that have a preferred affinity for hemimethylated CpG di-nucleotides in the DNA [48]. When cell division occurs, Dnmt1 methylates the other half of a hemimethylated portion of DNA, thereby contributing to the persistence of a cell's methylation state [48].

8.4.3 Sirtuin Interaction with Nonhistone Substrates

Sirtuins target other enzymes as well, some of which interact in chromatin regulation, and others that play a direct role in other cellular processes. HATs and DNA repair-associated protein complexes are nonhistone substrates that affect chromatin regulation and whose activity is also regulated by sirtuin deacetylation [11]. An important HAT that sirtuins interact with is the p300 HAT, which acetylates a number of core histone targets and acts as a required transcriptional co-activator in a number of cellular processes that sirtuins also play a role in. SIRT1 and SIRT2 have been shown to deacetylate p300 HAT and are associated with repression of p300 HAT enzymatic activity [49,50]. In a similar fashion, SIRT1 deacetylates and represses MOF and Tip60, two members of the MYST family of HATs that are involved in DNA damage signaling [11,51].

Sirtuin nonhistone substrates that are unrelated to chromatin include transcription factors, other enzymes, structural proteins, DNA repair complexes, cell signaling proteins, and transport proteins (Table 8.1). These nonhistone targets expand the repertoire of cellular processes that sirtuins can affect, although many targets are involved in the transcriptional machinery of the cell [27]. By directly binding to these substrates and modifying them via the same enzymatic reactions used in chromatin regulation, such as deacetylation or ADP ribosylation, sirtuins can epigenetically modulate the activity of the target substrate.

8.5 SIRTUIN FUNCTION IN CELLULAR PROCESSES COMMON TO SKIN DISEASES

8.5.1 Sirtuins in Apoptosis

SIRT1, SIRT2, SIRT3, and SIRT7 have been identified to play a role in the regulation of apoptosis in several epigenetic capacities. First, SIRT1 and SIRT2 activities have been found to have an inhibitory effect on tumor suppressor and cell cycle regulator p53 [52,53]. SIRT1 regulation of p53 is discussed in section 8.6.4, "Sirtuin Modulation of Apoptosis in Photoaging." SIRT2 downregulates p53 activity by interacting with the 14-3-3 protein family of regulatory molecules that modulate p53-mediated apoptosis via Akt (also known as Protein Kinase B (PKB)) signaling [52]. Sirtuin modulation of p53 activity may be useful in cancer cells that have developed apoptotic resistance via a mutation of p53, since inhibition of SIRT1 expression with a variety of sirtuin modulators was demonstrated to restore apoptotic sensitivity in p53-mutated human keratinocytes [54].

Other targets of sirtuin activity that play a role in the regulation of apoptosis include the Forkhead box (FOX) family of transcription factors, tumor suppressor p73, and DNA repair protein Ku70 [55–57]. These interactions implicate SIRT1 as a regulator of cell survival and SIRT2 as a regulator of cell cycle progression in response to DNA damage.

SIRT1 directly deacetylates and suppresses p73, a protein that normally activates genes such as p21 and Bax to induce apoptosis and suppress cell growth and differentiation in a manner similar to p53 [55]. In addition, SIRT1 deacetylation represses FOXO1, FOXO3a, and FOXO4, resulting in the inhibition of apoptotic mechanisms within the cell [56]. SIRT2, on the other hand, also directly deacetylates FOXO3a in response to oxidative stress, and is instead implicated in enhancing apoptosis [57]. The proposed mechanism is that SIRT2 deacetylation increases FOX binding to DNA and elevates the expression of target genes that include p27 (Kip1), manganese superoxide dismutase (MnSOD), and Bim to mitigate the ROS-induced damage in response to oxidative stress [57]. Bim, however, is a proapoptotic factor and likely promotes cell death in cells that are under severe stress.

Ku70 is notably involved in multiple mechanisms of DNA repair including DNA DSB and nonhomologous end joining (NHEJ) repair, which has been associated with telomere maintenance and silencing [58]. SIRT1 and SIRT3 have both been shown to bind with and deacetylate Ku70, leading to a decrease in Ku70 interaction with proapoptotic protein Bax that promotes cell survival [59,60]. SIRT1 inhibition leads to increased Ku70 acetylation and an increase in TNF-related apoptosis-inducing ligand (TRAIL) induced apoptosis in human hepatoma HepG2 cells [61].

SIRT7's role in cell proliferation is complicated by studies that implicate SIRT7 in both pro-proliferative and antiproliferative roles, indicating a need for further research to elucidate the relevant mechanisms [44,62,63]. However, SIRT7 is known to be essential for cell survival and is linked to activation of Pol I transcription and expression of ribosomal RNA genes [43]. Depletion of SIRT7 stops cell proliferation and triggers apoptosis [43]. SIRT7 is suggested to be critical in the resumption of ribosomal transcription in telophase, a process that may contribute to the observed apoptosis that occurs in the absence of SIRT7 [64]. SIRT1's role in tumorigenesis is similarly complicated by studies that demonstrate SIRT1 can function as a tumor promoter or suppressor [65].

8.5.2 Cell Cycle Regulation in Cell Proliferation

Cell cycle arrest is often used by cells to facilitate DNA repair before cell proliferation. If, however, the DNA damage is too severe, other signaling mechanisms work to induce cell senescence or apoptosis to prevent malignancies. SIRT1 interacts with Rb and E2F1 to control cell cycle progression before the S phase [66]. These interactions are described later in Section 8.8.

SIRT2 is meanwhile associated with the normal progression of mitosis and may function as a tumor suppressor in the setting of uncontrolled cell proliferation, as overexpression of SIRT2 was found to induce cell cycle arrest [67,68]. SIRT2 expression is markedly increased during mitosis, during the G2/M phase, and SIRT2 deacetylation of H4K16 is suggested to work as a mitotic checkpoint in response to DNA damage or stress, similarly to that of SIRT1 deacetylation of p53. SIRT2 activity is implicated in the formation of condensed chromatin associated with mitotic exit [39,69]. This is supported by a study that showed SIRT2 deficient animals exhibit a phenotype with various signs of genomic instability, including chromosomal aberrations and tumorigenesis [69,70]. SIRT2 is also known to deacetylate α-tubulin, which likely has a role in mitotic exit by regulating chromosomal segregation [68,71].

Another role that SIRT2 may play in the cell cycle is via regulating the deposition of H4K20me1, a marker crucial to chromatin compaction during S-phase progression and genome stability. SIRT2 activity affects H4K20 by increasing the activity of PR-Set7, an HMT that catalyzes the monomethylation of H4K20 to H4K20me1 [70]. The epigenetic mechanism that contributes to this effect is twofold: (i) SIRT2 deacetylates H4K16Ac, a modification that supports localization of PR-Set7 to histone sites and increases its activity, and (ii) SIRT2 also directly deacetylates PR-Set7 resulting in a promotion of PR-Set7 activity [70].

8.6 SIRTUINS IN SKIN AGING

Skin aging, like cellular aging, is a complex process that consists of both intrinsic changes, such as chronological aging, and extrinsic environmentally induced factors, such as photoaging, that result in a progressive decline in function within a cell [72]. General characteristics that are associated with skin aging include decreased cellularity, decreased elastic fibers, fragile capillaries, and slowed collagen metabolism, which result in skin displaying slight atrophy of the epidermis, straightening of the rete ridges, and a decrease in the number of Langerhans cells [73]. Lower intrinsic epidermal fat and increased transepidermal water loss (TEWL) also occur in skin aging, leading to dryness of the skin [73].

Sirtuins function in skin aging primarily by mitigating the damage from reactive oxygen species (ROS), generated by UV radiation, to various cell components including DNA, lipids, proteins, and other macromolecules that are associated with photoaging [74]. By epigenetically regulating cellular processes that include DNA repair, antioxidant regeneration, cell cycle regulation, telomere maintenance, apoptosis, and inflammation, sirtuins then reduce factors such as accumulation of DNA damage and cell senescence that contribute to skin aging. It is possible then that modulation of SIRT1, SIRT2, SIRT3, and SIRT6, which function in these various capacities, may prove to be an effective strategy to protect against and/or to treat the progression of skin aging [74]. One such clinical study of an active skin care product containing yeast biopeptides that are SIRT1 upregulators yielded promising results, with topical application demonstrating a decrease in the cell senescence marker beta-galactosidase and DNA fragmentation induced by UVB stress in culture and on healthy skin samples *ex vivo* [75].

8.6.1 Telomere Maintenance in Skin Aging

Although important in other diseases as well, progressive telomere shortening is a mechanism that relates cellular senescence to aging. SIRT6 plays a potential role in treating photoaging and other diseases by maintaining proper telomere function across cell generations [25]. The epigenetic mechanism here involves SIRT6 deacetylation of H3K9 during DNA replication, an interaction that is required for the stable association of WRN, also known as Werner syndrome ATP-dependent helicase, at telomeric chromatin [25]. Werner syndrome (WS) is an inherited autosomal recessive disease in which the WRN gene is mutated, resulting in the characteristic appearance of premature aging. Symptoms include graying hair, skin atrophy, loss of subcutaneous fat, osteoporosis, and malignancies. The stabilization of WRN by SIRT6 is thought to contribute to the propagation of a specialized epigenetic state of telomeric chromatin that is required for proper function and maintenance of telomere length [25].

These findings are supported by studies of SIRT6 knockdown cells (S6KD) that displayed premature cellular senescence not due to defective base excision repair (BER) or oxidative stress, but rather telomere dysfunction involving end-to-end chromosomal fusions that resemble defects seen in WS [25]. SIRT6 may therefore have a role in the pathogenesis of WS, aging, and possibly other diseases.

8.6.2 Sirtuins in DNA Repair Signaling

Sirtuins have been found to play a multifaceted role in mitigating DNA damage from various sources, including, but not limited to, oxidative stress. Subsequently, sirtuin function in DNA repair has a pivotal role in mediating several other cellular processes including cell cycle regulation and cellular mechanisms promoting apoptosis or cellular senescence, all of which play key roles in numerous dermatological pathologies in addition to skin aging. SIRT1, 2, 3, and 6 are all associated with DNA repair in some capacity.

SIRT6 mediates DNA repair by more than one mechanism in response to DNA damage. First, SIRT6 deacetylation of chromatin at both H3K9 and H3K56 is required for mobilization of separate DNA repair proteins to sites of DNA damage, particularly DSBs [41,76]. Deacetylation at H3K9 was found to recruit the catalytic subunit of DNA-dependent protein kinase (DNA-PK) to sites of DSBs [41]. Meanwhile, SNF2H, a SWI family member with helicase activity, binds preferentially to the C terminus of SIRT6, which then directly recruits SNF2H to sites of DNA damage where it opens chromatin at DSB sites to promote DNA repair at those sites [77]. This mobilization of SNF2H is proposed to be associated with H3K56 deacetylation upon DNA damage [77].

SIRT6 also promotes DNA repair via epigenetic modification of nonhistone substrates including C-terminal interacting protein (CtIP) and poly-(ADP-ribose) polymerase 1 (PARP1), both of which play a role in DSB repair. DNA end resection, which is an important step of DSB repair, is promoted by SIRT6-mediated deacetylation of CtIP [78,79]. Meanwhile, SIRT6 associates with and then mono-ADP-ribosylates PARP1 at lysine residue 521, resulting in the stimulation of PARP1 activity and enhancement of DSB repair [79]. This interaction has been demonstrated in cells under oxidative stress [79].

SIRT2 and SIRT3 have also been shown to deacetylate H3K56Ac *in vivo*, which is a marker that is increased following DNA damage and is colocalized to sites of DNA repair along with ATM, CHK2, and p53 [40]. However, no research that we reviewed has yet demonstrated a direct connection of SIRT2 or SIRT3 expression on H3K56 with DNA repair.

Lastly, SIRT1 mediates several forms of DNA repair by directly deacetylating several DNA repair-associated substrates. These substrates include: (i) the DNA damage sensor MRE11−RAD50−NBS1 nuclease complex, which responds to DSBs by delaying cell cycle progression, (ii) repair factor Ku70, which is involved in NHEJ and is discussed in Section 8.5.1, and (iii) DNA repair factor XPA, which is crucial for nucleotide excision repair (NER) caused by UV damage and is mutated in various manifestations of xeroderma pigmentosum (XP) [11,80,81]. SIRT1 deacetylation of XPA, which is enhanced in cells after UV irradiation, likely deacetylates XPA at lysines 63 and 67 after acetylation occurs there [80]. Research implicates this deacetylase activity in the downregulation of Akt (PKB) phosphorylation, which leads to suppression of the XPC gene [82]. XPC is also linked to XP. Lastly, SIRT1 activity plays a significant role in the efficiency of DNA repair factor NBS1 phosphorylation at Ser343 by ataxia telangiectasia mutated (ATM) protein kinase in response to DSBs [81].

8.6.3 Sirtuins in Oxidative Stress

Oxidative stress occurs when cellular defenses, such as natural antioxidants, are no longer able to compensate for the level of ROS found within a cell and often results from extrinsic factors that generate additional ROS on top of the basal level of ROS formed during normal cellular metabolism. These oxygen-derived free radicals, when unchecked, can then oxidize DNA bases and cause DNA damage. UV radiation is a common and ubiquitous example, both in research and in life, of an environmental effect that often induces oxidative stress.

Sirtuins function in managing oxidative stress primarily through two epigenetic mechanisms: by mitigating oxidative stress related DNA damage with upregulation of DNA repair mechanisms or by regeneration of antioxidant molecules within the cell. While sirtuins have been demonstrated in many DNA repair mechanisms as described in detail in Section 8.6.2, there have been significantly less pathways identified in which sirtuin activity helps via regeneration of antioxidants. One of these pathways involves direct deacetylation of FOX transcription factors by SIRT2 in response to oxidative stress, during which the expression of the antioxidant manganese superoxide dismutase MnSOD, along with p27 (Kip1) and Bim, is increased [57]. The other pathway involves SIRT3 deacetylation of isocitrate dehydrogenase 2 (ICDH2) in the mitochondria, a modification that stimulates ICDH2 activity and results in increased production of nicotinamide adenine dinucleotide phosphate (NADPH) [83]. The specific sites deacetylated by SIRT3 were found to be at K75 and K241 [83]. NADPH is then used in the regeneration of endogenous antioxidants.

8.6.4 Sirtuin Modulation of Apoptosis in Photoaging

There is research to suggest that sirtuin regulation of signaling pathways in apoptosis and inflammation may also play a role in the progression of skin aging. SIRT1, 3, and 7 were found to have decreased expression in adult donor skin samples and cultured skin when exposed to UVB irradiation versus normal skin [84–86]. No difference in sirtuin expression was identified in normal, chronologically aged skin, however, meaning that sirtuin modulation therapy may only be useful in preventing the effects of photoaging on skin aging [86].

Different studies have suggested that the impact of decreased SIRT1 expression in photoaging may result in decreased cell survival, based on its regulation of p53-mediated apoptosis [84,85]. There are two epigenetic mechanisms by which SIRT1 suppresses p53 activity: (i) by directly binding to and deacetylating p53, resulting in the inhibition of p53 activity, and also (ii) by deacetylating and negatively regulating TIP60, a HAT that typically acetylates p53 following UV-induced DNA damage. SIRT1 deacetylation of TIP60 suppresses p53 acetylation [51,53]. Studies involving sirtuin modulators support the effect of SIRT1 on skin aging induced by oxidative stress via a p53-related mechanism. Resveratrol, a natural polyphenol and proven antioxidant in skin [87], is a known activator of SIRT1 activity that has been shown to decrease p53-mediated apoptosis in cultured skin keratinocytes exposed to UVB irradiation and hydrogen peroxide, two sources of oxidative stress [84]. Meanwhile, pretreatment of the same cells with three SIRT1 inhibitors, sirtinol, NAM, and SIRT1-siRNA, resulted in the opposite effect by promoting p53-mediated apoptosis [84]. This suggests that sirtuin modulation of p53-mediated apoptosis may have potential application in the treatment of skin aging, and other modalities such as cancer, that have altered SIRT1 expression.

Other sirtuins including SIRT2, 3, and 7 also play an epigenetic role in mediating apoptosis, though none of these sirtuins is yet linked to skin aging. A more thorough discussion of their role in apoptosis can be found in Section 8.5.1, along with other mechanisms of SIRT1 regulation.

8.6.5 Inflammation in Photoaging

Photoaged skin is characterized by decreased levels of collagen expression and increased levels of matrix metalloproteinases (MMPs), implicating SIRT1 in another capacity to affect skin aging [72]. These changes have been found in human skin fibroblasts from older aged donors as well and are attributed to UV radiation induced stimulation of genes encoding MMP-1 (collagenase), MMP-3 (stromelysin-1), and MMP-9 (gelatinase) in skin cells [88,89].

An epigenetic mechanism related to inflammation in photoaging is SIRT1-mediated reduction of matrix metalloproteinase 9 (MMP-9), an interaction that has been demonstrated in skin and monocyte cell lines and plays a crucial role in the formation and destruction of dermal collagen [72,88,90]. This mechanism is proposed to be via SIRT1 targeting of the c-Jun component of the transcription factor Activator Protein-1 (AP-1), which regulates gene expression in response to inflammatory stimuli. SIRT1 hence decreases MMP-9 expression by suppressing AP-1 activity [91].

SIRT1 agonists such as resveratrol have been shown to reduce UV-stimulated MMP expression in human epithelial keratinocytes and dermal fibroblasts [88]. This interaction may play a role in other disease modalities as well because MMPs are also suggested to be modulators of inflammation and subsequently play an important role in cutaneous wound healing, photoaging, dermal fibrosis, bullous disease, and the progression of malignant melanoma [92,93].

Increased SIRT1 and/or SIRT6 activity may protect against skin aging by decreasing NF-κB expression, which is a key regulator of the immune response and the resulting inflammation [42,94]. Upregulation of the NF-κB-binding domain is strongly associated with the aged epidermis, and blockade of NF-κB has been shown to reverse some aspects of aged epidermis in mouse models, including increased epidermal cell proliferation and decreased senescence markers [95]. Therefore, downregulation of NF-κB activity may have protective effects in skin aging. However, the potential implication of these findings for dermatological therapies is limited because the exact mechanism of age-associated NF-κB activation is unknown and because the NF-κB pathway interacts with numerous cellular processes such that inhibition of the entire NF-κB pathway may be harmful [95,96]. Working via separate epigenetic mechanisms, SIRT1 directly deacetylates the RelA/p65 subunit of NF-κB at K310, while SIRT6 deacetylates H3K9 at NF-κB target gene promoters, leading to an attenuation NF-κB signaling [42,97].

This inhibition also affects the expression of TNF-α, a key inflammatory factor that is involved in the pathogenesis of psoriasis and other diseases, discussed next in section 8.7. In a few studies, activation of SIRT1 activity using resveratrol decreased NF-κB mediated transcription of TNF-α, while exogenous NAM has been linked to upregulation of TNF-α translation in the presence of adequate NAD$^+$ [97,98].

8.7 SIRTUINS IN SKIN INFLAMMATION, INFLAMMATORY DISEASES, AND AUTOIMMUNE DISEASES

Inflammation is an important component in the pathogenesis of skin aging and a variety of dermatological diseases including inflammatory and autoimmune skin diseases such as autoimmune blistering disorders, atopic dermatitis (AD), collagen vascular disorders, psoriasis, and systemic lupus erythematosus (SLE) [73,99]. Research is still limited, however, and sirtuin function in skin inflammation is still primarily attributed to sirtuin regulation of the NF-κB inflammatory pathway and interaction with MMPs, discussed earlier in section 8.6.5. Emerging research suggests that sirtuins may be linked to autoimmune and inflammatory skin diseases via additional epigenetic mechanisms.

For psoriasis, which is currently treated by TNF-α antagonists, sirtuin modulation may be a novel target for epigenetic therapies involving inflammatory skin disease [100]. Research demonstrates that TNF-α decreases expression of two epidermal barrier proteins, filaggrin (FLG) and loricrin (LOR), in lesional and nonlesional skin of psoriasis patients [101,102]. The mechanism is proposed to be via a c-Jun N-terminal kinase-dependent pathway and suggests that improving the epidermal barrier in psoriasis may be of therapeutic benefit [101]. Interestingly, FLG and LOR are also highly associated with AD. Flaky tail (Flg(ft)) mice, which characteristically have mutated FLG, are used as murine models of AD, while LOR gene and protein expression is significantly decreased in acute and nonlesional skin samples from AD subjects [102–104].

Sirtuin inhibition of TNF-α activity may subsequently play a role in the pathogenesis of AD as well as in psoriasis. AD is an inflammatory disease characterized by the relapse of itchy rashes commonly localized to the flexural regions of the arms or legs, face, and neck. A relationship between sirtuins and AD is supported by research that indicates that SIRT1 expression is decreased in Flg(ft) mice [103,105]. In addition, SIRT1 inhibitor sirtinol was shown to induce downregulation of the same proteins that are found to be downregulated in the skin of Flg(ft) mice, suggesting that disruption of SIRT1 activity may be associated with the pathogenesis of AD [105].

SIRT1 may have a TNF-α independent role in psoriasis too. IL-22, a key cytokine in the pathogenesis of psoriasis, is primarily mediated by STAT3 activity, a transcription factor whose activity relies on the acetylation status of Lys685 [106]. SIRT1 has subsequently been shown to deacetylate Lys685 and inhibit the activity of STAT3, indicating that SIRT1 activity may also inhibit the activity of IL-22 [107,108]. A recent study confirmed that SIRT1 activity opposes STAT3-induced IL-22 effects in keratinocytes via deacetylation of STAT3 [109]. SIRT1 levels were also found to be decreased in psoriatic skin lesions, suggesting that SIRT1 activation may be of potential therapeutic use in mitigating the pathogenesis of psoriasis [109].

SIRT1 epigenetic modulation of inflammation, and NF-κB in particular, is implicated in the development of autoimmune diseases such as SLE, which is characterized by symptoms of skin inflammation, particularly malar rash and discoid rash [110]. No specific mechanistic pathway has yet been determined. However, research indicates that global hypoacetylation of histone H3 and H4 in CD4$^+$ T cells of SLE patients may correlate with SLE disease activity in lupus patients [111]. Splenocytes in MRL-lpr/lpr mice, a lupus mouse model, demonstrate this aberrant hypoacetylation, which was found to be corrected by the administration of histone deacetylase inhibitors (HDI), trichostatin A (TSA), and suberoylanilide hydroxamic acid (SAHA) [112,113]. While these modulators are specific to only class I HDACs, SIRT1 was demonstrated in a later study to be overexpressed in MRL/lpr mice [114]. In addition, the administration of SIRT1-siRNA, which suppresses SIRT1, in these mice resulted in significant decrease in specific indicators of SLE, including serum anti-dsDNA antibody levels, renal IgG deposition, and renal pathological scores [114]. These data suggest that SIRT1 activity may be implicated in the pathogenesis of SLE and that further studies of sirtuin modulation in SLE are necessary to elucidate the epigenetic mechanisms involved.

Vitiligo, an autoimmune disease characterized by melanocyte death or dysfunction, is another skin pathology associated with SIRT1 activity [115]. In this instance, the mechanism is demonstrated to be via SIRT1 deacetylation and activation of Akt [115]. Akt then phosphorylates apoptosis signal-regulating kinase-1 (ASK-1) at Ser-83, resulting in an epigenetic modification that maintains ASK-1 in an inactive form [115]. ASK-1 plays a role in various apoptotic effects and is implicated in downregulation by SIRT1 activity [115]. SIRT1 activation is suggested to have a protective role in vitiligo by regulating MAPK signaling associated with this Akt-ASK-1 interaction. Deacetylation of Akt results in the downregulation of proapoptotic molecules and a reduction in oxidative stress and apoptosis in vitiligo keratinocytes [115].

8.8 SIRTUINS IN HYPERPROLIFERATIVE SKIN DISEASE

Cellular processes that impact cell proliferation have a direct relationship to the development of hyperproliferative skin diseases, such as psoriasis, as well as skin cancers. Sirtuins, which epigenetically regulate many of those signaling pathways, are implicated in the pathogenesis of hyperproliferative skin disease and scarring. SIRT1 expression itself is associated with keratinocyte proliferation. One study demonstrated that overexpression or underexpression of SIRT1 significantly affects both the proliferation as well as the expression of differentiation markers in normal human keratinocytes [116]. Modulation of sirtuin activity with sirtinol, a SIRT1 inhibitor, and resveratrol, a SIRT1 activator, yielded confirmatory results that suggest SIRT1 activity promotes keratinocyte differentiation [116]. While the exact mechanism for this proliferative activity is not clear, SIRT1 interactions with Rb and E2F1 are thought to be potential mechanisms by which SIRT1 affects cell proliferation [117].

SIRT1 plays a role in cell cycle control at the G1 to S transition via deacetylation of the tumor suppressor retinoblastoma protein (Rb) [66]. Transcription factor E2F1, which preferentially binds to Rb to mediate cell proliferation and apoptosis, has also been shown to be inhibited by SIRT1 directly [117]. The two form a negative feedback loop, whereby E2F1 induces SIRT1 expression at the transcriptional level, and SIRT1 binds to and inhibits E2F1. Knockdown of SIRT1 via small interference RNA (siRNA) subsequently shows an increase in E2F1 transcriptional and apoptotic functions [117].

8.9 SIRTUINS IN SKIN REPAIR AND SCARRING

Other studies implicate sirtuins in skin repair, particularly keratinocyte proliferation, via SIRT1 epigenetic deacetylation and activation of endothelial nitric oxide synthase (eNOS) at lysines 496 and 506 of the calmodulin-binding domain [118,119]. This activation of eNOS results in the release of nitric oxide (NO), a molecule that plays a key role in all stages of wound healing [118]. *In vivo* studies with sirtuin activators resveratrol and MC2526, a synthetic activator of SIRT1, 2, and 3, showed accelerated wound repair while sirtinol had no effect [118]. Further experiments performed with these modulators found that sirtuin activation was associated with: (i) a significant increase in cell proliferation, (ii) enhanced NO production in keratinocytes, (iii) stabilization of the SIRT−eNOS complex with phosphorylation in serine 1177, and (iv) significant H4K16 deacetylation when compared with controls [118].

The same study also proposed that HDAC2, a class I HDAC whose expression is increased in normal and keloid scar formation, could be a downstream effector of sirtuin-dependent NO signaling since *S*-nitrosylation of HDAC2 directly depends on NO [118,120]. However, further research is needed to elucidate the relationship between SIRT1 and HDAC2.

SIRT6 may also affect keloid formation via a different epigenetic mechanism. Tissue samples of hypertrophic or keloid scarring are associated with telomere shortening and increased oxidative stress [121]. A statistically significant ($P < 0.05$) decrease in telomere length in a study of 20 patients with keloids was demonstrated when a comparison was made with results for normal skin [121]. Therefore modulation of SIRT6, which plays a major role in telomeric stability and maintenance, may be a potential target for investigation or in therapeutic benefit in scarring.

8.10 SIRTUINS IN SKIN CANCER

With the rates of melanoma and nonmelanoma skin cancer (NMSC) rising in the United States, there is a growing need for new strategies to prevent and treat the development of malignancies [122,123]. Modulation of sirtuin function may prove to yield novel treatments by targeting the epigenetic changes that are associated with malignancy, rather than the genetic mutations. These epigenetic treatments have the potential not only to prevent the progression of cancer but also to reverse some aspects of cancer pathogenesis.

Alteration of sirtuin expression is clearly demonstrated in malignant melanoma and NMSCs by multiple studies [124,125]. SIRT1 is overexpressed in clinical melanoma tissue and melanoma cell lines Sk-Mel2, WM35, G361, A375, and Hs294T, and corresponds with a noticeable shift in SIRT1 localization from nuclear to cytoplasmic in melanoma tissue [126]. Meanwhile, a study that examined SIRT1 protein expression in 87 cases of NMSCs and other skin tumors using immunohistochemical analysis found that SIRT1 expression was increased in all but the eccrine poroma and keratocanthoma cases when a comparison was made with results for normal epidermis samples [124]. Sirtuin expression was similarly found to be upregulated in other skin cancer cells: (i) SIRT1 and SIRT3 in A431 epidermoid carcinoma cells, (ii) SIRT2, 3, 5, 6, and 7 in actinic keratoses, and (iii) SIRT1−7 in squamous cell carcinoma (SCC) [127]. Human papillomavirus (HPV) oncoprotein E7 has also been demonstrated to upregulate SIRT1 activity in human keratinocytes and cervical cancer cells, a mechanism that is proposed to contribute to malignant transformation in HPV [128].

SIRT2 expression has conversely been demonstrated in a recent study to be downregulated in human SCC samples, whereas SIRT2 deletion increases tumor growth [129]. These results suggest that SIRT2 may function as a tumor suppressor in skin; however, these data did not tie in with a specific epigenetic mechanism [129]. Research also demonstrates that sirtuin expression is altered in human skin in response to photodamage. The upregulation of SIRT1 and SIRT4 in keratinocytes after exposure to solar simulated light (SSL) suggests that sirtuin function may play a role in mediating the early stages of carcinogenesis because photodamage is a primary initiator of skin malignancy [127].

Epigenetic modifications of many of the cellular signaling pathways that have already been discussed, including apoptosis, cell cycle regulation, DNA repair, oxidative stress, and cell proliferation, are likely to also contribute to sirtuin function in carcinogenesis. A summarization of sirtuin function in these cellular processes along with the related epigenetic mechanisms can be found in Table 8.4.

8.11 MODULATORS OF SIRTUIN FUNCTION

8.11.1 Overview of Sirtuin Modulators and Related Limitations

Several naturally occurring products modulate sirtuins. Resveratrol is well-known natural polyphenol antioxidant and SIRT1 activator that has been used in many animal and clinical studies. However, the use of resveratrol in sirtuin research has a few limitations. First, resveratrol has numerous cellular targets other than SIRT1 that play a role in other cellular processes, so it is difficult to attribute any effects of resveratrol addition solely to SIRT1 activation [130]. Second, resveratrol is also limited by relatively low potency and stability in solution [130]. Sirtuin modulators are nevertheless important tools that can be used to further elucidate the role of sirtuin activity in various pathologies or in the

TABLE 8.4   Cellular Processes Related to Sirtuin Function

Cellular processes	Sirtuin	Function	Epigenetic mechanism
Apoptosis	SIRT1	Promotes cell survival	Directly binds to and deacetylates p53, resulting in inhibition of p53 activity
			Also deacetylates and negatively regulates TIP60, which decreases HAT acetylation of p53 [51,53]
			SIRT1 direct deacetylation represses FOXO1, FOXO3a, and FOXO4, which inhibits apoptotic mechanisms within the cell [56]
			Deacetylates Ku70, which leads to decreased interaction with proapoptotic protein Bax [60]
			Deacetylates Akt (PKB), which interacts with (ASK-1), a MAPK kinase kinase with antiapoptotic effects [115]
	SIRT2	Promotes cell survival	Downregulates p53 activity by deacetylating 14-3-3 protein family of regulatory molecules that modulate p53-mediated apoptosis [52]
		Enhances apoptosis	SIRT2 directly deacetylates FOXO3a in response to oxidative stress and is implicated in enhancing apoptosis [57]
	SIRT3	Promotes cell survival	Deacetylates Ku70, which leads to decreased interaction with proapoptotic protein Bax [59]
	SIRT7	SIRT7 depletion stops cell proliferation and triggers apoptosis	SIRT7 activity is linked to RNA Pol I transcription and expression of ribosomal RNA genes. Exact mechanism is unclear [43]
Telomere maintenance	SIRT6	Maintains telomere structure and function	SIRT6 deacetylates H3K9 during DNA replication at telomeric chromatin, allows for the stabilization of WRN DNA helicase [25]
DNA repair	SIRT1	Delays cell cycle progression	Deacetylates DNA damage sensor MRE11−RAD50−NBS1 nuclease complex [11,81]

(Continued)

TABLE 8.4   (Continued)

Cellular processes	Sirtuin	Function	Epigenetic mechanism
		Involved in NHEJ	Deacetylates repair factor Ku70 [11]
		Delays cell cycle progression	Deacetylates DNA damage sensor MRE11−RAD50−NBS1 nuclease complex [81]
		Increases repair response to DSBs	Increases the efficiency of DNA repair factor NBS1 phosphorylation in response to DSB [81]
	SIRT2 and SIRT3	May mediate DNA repair	Deacetylate H3K56Ac *in vivo* and are colocalized to sites of DNA repair along with phosphor-ATM, CHK2, and p53 [40]
	SIRT6	Promotes DNA repair at DSBs	Deacetylation of chromatin at H3K9 and H3K56 increases mobilization of separate DNA repair proteins to sites of DNA damage [41,76]
		Promotes DNA repair	Deacetylation at H3K9 recruits the catalytic subunit of DNA-PK to sites of DSBs [79]
		Opens chromatin at sites of DNA damage to promote DNA repair	SNF2H binds preferentially to the C terminus of SIRT6, which recruits SNF2H to sites of DNA damage [77]
		Promotes DNA end resection	Deacetylates CtIP [78]
		Enhances DSB repair under oxidative stress	Mono-ADP-ribosylates PARP1 at lysine residue 521 under oxidative stress [79]
Oxidative stress	SIRT2	Antioxidant regeneration of MnSOD	Deacetylation of FOX transcription factors, which increase expression of MnSOD [57]
	SIRT3	General antioxidant regeneration	Deacetylates ICDH2 in the mitochondria at K75 and K241, which stimulate the production of NADPH [83]

(Continued)

1. BIOLOGICAL AND HISTORICAL ASPECTS OF EPIGENETICS

TABLE 8.4   (Continued)

Cellular processes	Sirtuin	Function	Epigenetic mechanism
Inflammation	SIRT1	Decreases NF-κB related gene expression	SIRT1 directly deacetylates the RelA/p65 subunit of NF-κB at K310 [117]
		Decreases MMP-9 expression	SIRT1 activity targets the c-Jun component of the transcription factor Activator Protein-1 (AP-1), which decreases MMP-9 expression among others [91]
		Decreases the effects of cytokine IL-22 in psoriatic skin lesions	SIRT1 deacetylates Lys685 of STAT3 and inhibits its ability to upregulate IL-22 signaling in psoriasis lesioned skin [109]
	SIRT6	Attenuates NF-κB signaling	Deacetylates H3K9 at NF-κB target gene promoters [42,97]
Cell proliferation	SIRT1	Promotes keratinocyte differentiation	SIRT1 deacetylates tumor suppressor Rb [66]
		Decreases apoptotic functions	SIRT1 binds to and inhibits E2F1 [117]
		Increased release of NO, which promotes skin regeneration	SIRT1 activates eNOS via deacetylation of lysines 496 and 506 of the calmodulin-binding domain [118,119]
Cell cycle regulation	SIRT1	Increases cell cycle control at G1 to S phase	SIRT1 directly deacetylates the RelA/p65 subunit of NF-κB at K310 [117]
	SIRT2	Promotes cell cycle arrest	SIRT2 deacetylation of H4K16Ac is suggested to work as a mitotic checkpoint in response to DNA damage or stress [39]
		Regulates chromosomal segregation during division	SIRT2 deacetylates α-tubulin [68,71]
		Required for S-phase progression and promotes genome stability	SIRT2 increases H4K20me1 deposition which is necessary for chromatin compaction by increasing PR-Set7 activity via SIRT2 direct deacetylation of PR-Set7 and deacetylation of H4K16Ac to promote PR-Set7 activity [70]

development of epigenetic therapies. Due to the limited biomodulatory activity of natural activators, there is a significant effort to develop synthetic compounds with higher potency and greater binding capacity. Several of these have already been studied in animal models and clinical trials for various pathologies [131]. Additional sirtuin modulators, activators and inhibitors, can be found in Tables 8.5 (sirtuin activators) and 8.6 (sirtuin inhibitors).

8.11.2 Modulation of Sirtuin Activity in Skin Cancer

Of all the dermatological conditions impacted by sirtuins, sirtuin modulation in skin cancer is currently receiving the greatest attention. Sirtuin modulation in skin cancer cells has been shown to have various effects on their proliferation, giving conflicting evidence of SIRT1 as both a tumor suppressor and promoter [65]. Human malignant melanoma A375-S2 cells treated with evodiamine, a bioactive compound with demonstrated antitumor effects, were associated with a decrease in SIRT1 expression and an increase in Bax- and p53-mediated apoptosis [157]. This result suggests that SIRT1 functions as a tumor promoter in these cells.

In a different study involving UV-irradiated A375-S2 cells, SIRT1 polyphenol activator silymarin suppressed apoptosis by inducing SIRT1 deacetylation of Ku70, which results in a decrease in Bax activation [140]. This interaction also supports the hypothesis that SIRT1 functions as a tumor promoter. In addition, silymarin pretreatment reverses the S-phase arrest normally seen in UV-exposed melanoma cells and increases the G_2/M phase cells [140]. This silymarin-induced cell cycle modulation is speculated to allow more time for DNA damage repair before mitosis [140].

Meanwhile, sirtuin inhibitors tenovin-1 and -6 were found to delay growth of the highly malignant melanoma ARN8 cell line in single digit micromolar concentrations, suggesting that SIRT1 may function as a tumor promoter [156]. These inhibitors also demonstrated an associated decrease in SIRT1 and SIRT2 activity in human malignant melanoma and normal dermal fibroblast cells [156]. However, normal human dermal fibroblasts are less susceptible than ARN8 cells to the growth arresting effects of tenovin-1, suggesting that treatment with tenovin may be useful in specifically targeting melanoma cells [156]. Similar treatment of melanoma cell lines A375, Hs294T, and G361 with tenovin-1 demonstrated a significant decrease in cell growth and cell viability associated with an increase in p53 activity [126].

Another interesting study of sirtuin modulators involves human MM418 and mouse B16F1 melanoma cells, in which SIRT1 was found

TABLE 8.5 Activators of Sirtuin Activity

Activator	Sirtuin(S) affected	Origin/source	Modulation
Butein	Yeast Sir2, SIRT1	Stems of *Rhus verniciflua* (Stokes plant)	8.5-fold increase in SIRT1 deacetylase activity [132,133]
EGCg or ECg + vitamin C	SIRT1	Green tea	EGCg or ECg stabilized with vitamin C-stimulated recombinant SIRT1 deacetylase activity [133]
1,4-Dihydropyridine derivatives	SIRT1, SIRT2, SIRT3	Synthetic (Enzo Life Sciences)	Reduced number of senescent primary human mesenchymal cells by up to 40% and showed a dose-dependent increase of mitochondrial function in murine myoblasts [134]
Myricetin + vitamin C	SIRT1	Red wine, walnuts, grapes, berries	Myricetin stabilized with vitamin C-stimulated recombinant SIRT1 deacetylase activity [133]
Oxazolo[4,5-b] pyridines	SIRT1 (*in vitro*)	Synthetic (Sirtris Pharmaceuticals, Inc.)	Some compounds identified as potent SIRT1 activators, such as a benzimidazole that increased SIRT1 activity 8-fold [135]
Piceatannol	Yeast Sir2, SIRT1	Hydroxylated, naturally occurring analog of resveratrol	7.9-fold increase in SIRT1 deacetylase activity [132,133]
Quercetin	Yeast Sir2, SIRT1	Capers, lovage, apples, tea, onion, citrus fruits, green vegetables, most berries	4.6-fold increase in SIRT1 deacetylase activity [132,133]
Quinoxalines	SIRT1 (*in vitro*)	Synthetic (Cayman Chemical)	Showed antiobesity, anti-inflammatory, and antidiabetic properties [136]

Resveratrol (resVida™) (Longevinex)	Yeast Sir2, SIRT1	Grapeskins, peanuts, red wine, plums, raspberries, blueberries	The trans-isomer of resveratrol is responsible for most of the health benefits and promotes cell survival via SIRT1-mediated p53 inhibition [137] Reformulated versions with improved bioavailability have been developed [138,139]
Silymarin	SIRT1	Milk thistle plant (*Silybum marianum*)	Activates SIRT1 expression and prevents UV-induced apoptosis in human melanoma cells [140]
SRT501	SIRT1	Synthetic (Sirtris Pharmaceuticals, Inc.)	Clinical trials for treatment of: − Colon cancer (phase I) [141] (Identifier: NCT00920803) − Multiple myeloma (phase II) [141] (Identifier: NCT00920556)
SRT1720	SIRT1	Synthetic (Sirtris Pharmaceuticals, Inc.)	Reported to be 800- to 1000-fold more effective than resveratrol [142] Recent studies show it is nonspecific for SIRT1 activation and interacts with over 100 targets [143]
SRT2104	SIRT1	Synthetic (Sirtris Pharmaceuticals, Inc.)	Clinical trials for treatment of: − Moderate-to-severe plaque-type psoriasis (phase II) [141] (Identifier: NCT01154101) − Type 2 DM (phase II) [141] (Identifier: NCT01018017 and NCT00937326) − Skeletal Muscle Atrophy (phase I) [141] (Identifier: NCT01039909)
SRT2172	SIRT1	Synthetic (Sirtris Pharmaceuticals, Inc.)	More effective than resveratrol in inhibiting MMP-9 production in monocytes [90]

TABLE 8.6 Inhibitors of Sirtuin Activity

Inhibitor	Sirtuin (S) affected	Origin/source	Modulation
AC-93253	SIRT2 (selective and potent)	Synthetic (multiple companies, including: Santa Cruz Biotechnology, Inc.; Sigma-Aldrich)	Cytotoxic at low concentrations in pancreas, prostate, and lung cancer cell lines with less toxicity in endothelial and epithelial cells [144]
	SIRT1, SIRT3 (lower potency)		Potent antiproliferative activity at even lower concentrations [144]
AGK2	SIRT2	Synthetic (multiple companies, including: Tocris Bioscience; Sigma-Aldrich; Biovision; Merck4Biosciences; Enzo Life Sciences; Santa Cruz Biotechnology, Inc.; Cayman Chemical)	Selective for SIRT2 inhibition. Potential utility in the treatment of Parkinson's disease [145]
β-Phenylsplitomicins (splitomicin derivative)	SIRT2	Synthetic (Research Labs: Institute of Pharmaceutical Sciences, Freiburg, Germany, and Gladstone Institute of Virology & Immunology, San Francisco, CA)	Active in low micromolar concentrations [146]
Cambinol (splitomicin derivative)	SIRT1, SIRT2	Synthetic (multiple companies, including: Santa Cruz Biotechnology, Inc.; Sigma-Aldrich; Merck4Biosciences; Otava Chemicals)	Highly toxic to Burkitt's lymphoma cells (dependent on Bcl-6 expression) [147]
EX-527 (an indole)	SIRT1, SIRT2 (higher specificity for SIRT1)	Synthetic (multiple companies, including: Elixir Pharmaceuticals, Inc. & Siena Biotech; Biovision; Tocris Bioscience; Cayman Chemical; Selleck Chemicals)	Induces cell cycle arrest at G1 in breast cancer cell lines [148]
			Phase I clinical trial in Europe for treatment of Huntington's disease (funded by Elixir and Siena) [149]
	SIRT6 [5]		Potent inhibition (>50%) of SIRT6 deacetylation at H3K56 [150]

Compound	Target	Source	Notes
Quercetin	SIRT6	Capers, lovage, apples, tea, onion, citrus fruits, green vegetables, most berries	Potent inhibition (>50%) of SIRT6 deacetylation at H3K56 [150]
Salermide	SIRT1, SIRT2	Synthetic (multiple companies, including: Santa Cruz Biotechnology, Inc.; Cayman Chemical; Merck4Biosciences; Sigma-Aldrich)	Induces apoptosis in cancer cells not normal cells via SIRT-mediated, p53-independent mechanisms [151] Induces p53-mediated apoptosis in breast cancer cell lines via inhibition of both SIRT1 and SIRT2 [148]
Sirtinol	Yeast Sir2, SIRT1, SIRT2	Synthetic (multiple companies, including: Santa Cruz Biotechnology, Inc.; Tocris Bioscience; Enzo Life Sciences; Merck4Biosciences)	Induces p53-mediated apoptosis in breast cancer cell lines via inhibition of both SIRT1 and SIRT2 [148]
Splitomicin	Yeast Sir2, SIRT1	Synthetic (multiple companies, including: Biovision; Santa Cruz Biotechnology, Inc.; Tocris Bioscience; Merk4Biosciences)	Mediates reversal of gene silencing in Fragile X syndrome via SIRT1 inhibition [152,153]
Suramin	SIRT5, SIRT1, SIRT2	Synthetic (multiple companies, including: Santa Cruz Biotechnology, Inc.; Enzo Life Sciences; Merck4Biosciences; SGC (nonprofit))	Treatment of sleeping sickness and onchocerciasis [154] Severe neurotoxicity limits therapeutic potential [155]
Tenovin-1	SIRT1, SIRT2	Synthetic (multiple companies, including: Santa Cruz Biotechnology, Inc.; Cayman Chemical; Merck4Biosciences)	Cytotoxic to aggressive melanoma cells [156]
Tenovin-6	SIRT1, SIRT2	Synthetic (multiple companies, including: Santa Cruz Biotechnology, Inc.; Cayman Chemical; Dundee Cell Products)	More water soluble than tenovin-1. More toxic to highly aggressive melanoma cells than tenovin-1 [156]

to play a role in melanoma cell migration, whereas other sirtuins were not [158]. Sirtuin activators, NAD^+ and resveratrol, were associated with increased cell migration, wound healing, and lamellipodium formation *in vitro* while sirtuin inhibitors NAM and splitomicin were associated with the opposite effects [158]. These findings were echoed by *in vivo* experiments in which SIRT1-knockdown and treatment with NAM resulted in the inhibition of abdominal metastasis of B16F1 cells, suggesting that SIRT1 inhibition may be a potential therapeutic target to control melanoma metastasis [158]. The mechanism for this interaction is proposed to be via SIRT1 deacetylation of Akt (PKB) in the PIP_3-binding region, which promotes activation and membrane localization of Akt [158].

8.12 CONCLUSION

Sirtuin activity has demonstrated several roles in epigenetic mechanisms that impact a wide variety of cellular processes related to skin health and disease. Modulation of sirtuin activity may prove to be important to the development of epigenetic therapies in dermatology. Though research into improving sirtuin modulator via synthetic compounds is rapidly increasing, further research is necessary to elucidate the epigenetic mechanisms of sirtuin function in skin diseases. Sirtuin research is an evolving field that presents us with novel targets in the treatment of epigenetic changes associated with longevity and disease.

List of Abbreviations

AD	atopic dermatitis
ADP	adenosine diphosphate
ADPRT	ADP-ribosyl transferase
AK	actinic keratosis
AP-1	activator protein-1
ASK-1	apoptosis signal-regulating kinase-1
ATM	ataxia telangiectasia mutated
Bax	bcl-2 associated X protein
BER	base excision repair
Bim	bcl-2-interacting mediator of cell death
CtIP	C-terminal interacting protein
DNA-PK	DNA-dependent protein kinase
Dnmt	DNA methyltransferase
DSB	double strand break
ECG	epicatechin galate
EGCG	epigallocatechin gallate
eNOS	endothelial nitric oxide synthase

FLG	filaggrin
FOX	forkhead box
HAT	histone acetyltransferase
HDAC	histone deacetylase
HDI	histone deacetylase inhibitors
HDM	histone demethylase
HMT	histone methyltransferase
HPV	human papillomavirus
ICDH	isocitrate dehydrogenase
KA	keratoacanthoma
LOR	loricrin
MMP	matrix metalloproteinase
MnSOD	manganese superoxide dismutase
NAD$^+$	nicotinamide adenine dinucleotide
NADPH	nicotinamide adenine dinucleotide phosphate
NAM	nicotinamide
NER	nucleotide excision repair
NF-κB	nuclear factor kappa-light chain enhances activated B cells
NHEJ	nonhomologous end joining
NMSC	nonmelanoma skin cancer
NO	nitric oxide
OAADPR	*O*-acetyl-ADP-ribose
PARP1	poly (ADP-ribose) polymerase 1
PIP$_3$	phosphatidylinositol (3,4,5)-triphosphate
PKB	protein kinase B
PPAR-γ	peroxisome proliferator-activated receptor gamma
Rb	retinoblastoma (protein)
rDNA	ribosomal deoxyribonucleic acid
ROS	reactive oxygen species
SAHA	suberoylanilide hydroxamic acid
SCC	squamous cell carcinoma
Sir	silent information regulator
SIRT	silent information regulator homolog (sirtuin)
SLE	systemic lupus erythematosus
SSL	solar simulated light
TEWL	transepidermal water loss
TNF-α	tumor necrosis factor alpha
TRAIL	TNF-related apoptosis-inducing ligand
TSA	trichostatin A
UV	ultraviolet (radiation)
WS	Werner syndrome
XP	xeroderma pigmentosum

References

[1] Shore D, Squire M, Nasmyth KA. Characterization of two genes required for the position-effect control of yeast mating-type genes. EMBO J 1984;3(12):2817–23.

[2] Tissenbaum HA, Guarente L. Increased dosage of a Sir-2 gene extends lifespan in *Caenorhabditis elegans*. Nature 2001;410(6825):227–30.

[3] Rogina B, Helfand SL. Sir2 mediates longevity in the fly through a pathway related to calorie restriction. Proc Natl Acad Sci USA 2004;101(45):15998–6003.

[4] Kaeberlein M, McVey M, Guarente L. The SIR2/3/4 complex and SIR2 alone promote longevity in *Saccharomyces cerevisiae* by two different mechanisms. Genes Dev 1999;13(19):2570–80.

[5] Vaquero A. The conserved role of sirtuins in chromatin regulation. Int J Dev Biol 2009;53(2–3):303–22.

[6] Keppler B, Archer T. Chromatin-modifying enzymes as therapeutic targets—Part 1. Expert Opin Ther Targets 2008;12:1301–12.

[7] Haigis M, Sinclair D. Mammalian sirtuins: biological insights and disease relevance. Annu Rev Pathol 2010;5:253–95.

[8] Hallows WC, Albaugh BN, Denu JM. Where in the cell is SIRT3?—functional localization of an NAD$^+$-dependent protein deacetylase. Biochem J 2008;411(2):e11–13.

[9] Michan S, Sinclair D. Sirtuins in mammals: insights into their biological function. Biochem J 2007;404:1–13.

[10] Michishita E, Park J, Burneskis J, Barrett J, Horikawa I. Evolutionarily conserved and nonconserved cellular localizations and functions of human SIRT proteins. Mol Biol Cell 2005;16:4623–35.

[11] Martinez-Redondo P, Vaquero A. The diversity of histone versus nonhistone sirtuin substrates. Genes Cancer 2013;4(3–4):148–63.

[12] Du J, Zhou Y, Su X, Yu JJ, Khan S, Jiang H, et al. Sirt5 is a NAD-dependent protein lysine demalonylase and desuccinylase. Science 2011;334(6057):806–9.

[13] Cohen HY, Lavu S, Bitterman KJ, Hekking B, Imahiyerobo TA, Miller C, et al. Acetylation of the C terminus of Ku70 by CBP and PCAF controls Bax-mediated apoptosis. Mol Cell 2004;13(5):627–38.

[14] Tanno M, Sakamoto J, Miura T, Shimamoto K, Horio Y. Nucleocytoplasmic shuttling of the NAD$^+$-dependent histone deacetylase SIRT1. J Biol Chem 2007;282(9):6823–32.

[15] North BJ, Verdin E. Interphase nucleo-cytoplasmic shuttling and localization of SIRT2 during mitosis. PLOS ONE 2007;2(8):e784.

[16] Scher MB, Vaquero A, Reinberg D. SirT3 is a nuclear NAD$^+$-dependent histone deacetylase that translocates to the mitochondria upon cellular stress. Genes Dev 2007;21 (8):920–8.

[17] Nakamura Y, Ogura M, Tanaka D, Inagaki N. Localization of mouse mitochondrial SIRT proteins: shift of SIRT3 to nucleus by co-expression with SIRT5. Biochem Biophys Res Commun 2008;366(1):174–9.

[18] Yuan H, Marmorstein R. Structural basis for sirtuin activity and inhibition. J Biol Chem 2012;287:42428–35.

[19] Sauve AA, Youn DY. Sirtuins: NAD(+)-dependent deacetylase mechanism and regulation. Curr Opin Chem Biol 2012;16(5–6):535–43.

[20] Bitterman KJ, Anderson RM, Cohen HY, Latorre-Esteves M, Sinclair DA. Inhibition of silencing and accelerated aging by nicotinamide, a putative negative regulator of yeast Sir2 and human SIRT1. J Biol Chem 2002;277(47):45099–107.

[21] Sauve AA, Schramm VL. Sir2 regulation by nicotinamide results from switching between base exchange and deacetylation chemistry. Biochemistry 2003;42(31):9249–56.

[22] Tong L, Denu JM. Function and metabolism of sirtuin metabolite O-acetyl-ADP-ribose. Biochim Biophys Acta 2010;1804(8):1617–25.

[23] Denu J. Linking chromatin function with metabolic networks: Sir2 family of NAD (+)-dependent deacetylases. Trends Biochem Sci 2003;28:41–8.

[24] Haigis M, Mostoslavsky R, Haigis K, Fahie K, Christodoulou D, et al. SIRT4 inhibits glutamate dehydrogenase and opposes the effects of calorie restriction in pancreatic beta cells. Cell 2006;126:941–54.

[25] Michishita E, McCord R, Berber E, Kioi M, Padilla-Nash H, et al. SIRT6 is a histone H3 lysine 9 deacetylase that modulates telomeric chromatin. Nature 2008;452:492–6.

[26] Liszt G, Ford E, Kurtev M, Guarente L. Mouse Sir2 homolog SIRT6 is a nuclear ADP-ribosyl transferase. J Biol Chem 2005;280(22):21313−20.

[27] Feige JN, Auwerx J. Transcriptional targets of sirtuins in the coordination of mammalian physiology. Curr Opin Cell Biol 2008;20(3):303−9.

[28] Turner B. Histone acetylation and an epigenetic code. Bioessays 2000;22:836−45.

[29] Gruenert DC, Cozens AL. Inheritance of phenotype in mammalian cells: genetic vs. epigenetic mechanisms. Am J Physiol 1991;260(6 Pt 1):L386−94.

[30] Sneppen K, Micheelsen MA, Dodd IB. Ultrasensitive gene regulation by positive feedback loops in nucleosome modification. Mol Syst Biol 2008;4:182.

[31] Gordon AJ, Satory D, Halliday JA, Herman C. Heritable change caused by transient transcription errors. PLOS Genet 2013;9(6):e1003595.

[32] Wu J, Grunstein M. 25 years after the nucleosome model: chromatin modifications. Trends Biochem Sci 2000;25(12):619−23.

[33] Yuan H, Su L, Chen WY. The emerging and diverse roles of sirtuins in cancer: a clinical perspective. Onco Targets Ther 2013;6:1399−416.

[34] Shogren-Knaak M, Ishii H, Sun JM, Pazin MJ, Davie JR, Peterson CL. Histone H4-K16 acetylation controls chromatin structure and protein interactions. Science 2006;311(5762):844−7.

[35] Vaquero A, Sternglanz R, Reinberg D. NAD$^+$-dependent deacetylation of H4 lysine 16 by class III HDACs. Oncogene 2007;26(37):5505−20.

[36] Brownell JE, Allis CD. Special HATs for special occasions: linking histone acetylation to chromatin assembly and gene activation. Curr Opin Genet Dev 1996;6(2):176−84.

[37] Yuan J, Pu M, Zhang Z, Lou Z. Histone H3-K56 acetylation is important for genomic stability in mammals. Cell Cycle 2009;8(11):1747−53.

[38] Espada J, Ballestar E, Santoro R, Fraga M, Villar-Garea A, et al. Epigenetic disruption of ribosomal RNA genes and nucleolar architecture in DNA methyltransferase 1 (Dnmt1) deficient cells. Nucleic Acids Res 2007;35:2191−8.

[39] Vaquero A, Scher MB, Lee DH, Sutton A, Cheng HL, Alt FW, et al. SirT2 is a histone deacetylase with preference for histone H4 Lys 16 during mitosis. Genes Dev 2006;20(10):1256−61.

[40] Vempati RK, Jayani RS, Notani D, Sengupta A, Galande S, Haldar D. p300-mediated acetylation of histone H3 lysine 56 functions in DNA damage response in mammals. J Biol Chem 2010;285(37):28553−64.

[41] McCord RA, Michishita E, Hong T, Berber E, Boxer LD, Kusumoto R, et al. SIRT6 stabilizes DNA-dependent protein kinase at chromatin for DNA double-strand break repair. Aging (Albany, NY) 2009;1(1):109−21.

[42] Kawahara TL, Michishita E, Adler AS, Damian M, Berber E, Lin M, et al. SIRT6 links histone H3 lysine 9 deacetylation to NF-kappaB-dependent gene expression and organismal life span. Cell 2009;136(1):62−74.

[43] Ford E, Voit R, Liszt G, Magin C, Grummt I, Guarente L. Mammalian Sir2 homolog SIRT7 is an activator of RNA polymerase I transcription. Genes Dev 2006;20:1075−80.

[44] Barber MF, Michishita-Kioi E, Xi Y, Tasselli L, Kioi M, Moqtaderi Z, et al. SIRT7 links H3K18 deacetylation to maintenance of oncogenic transformation. Nature 2012;487(7405):114−18.

[45] Vaquero A, Scher M, Lee D, Erdjument-Bromage H, Tempst P, Reinberg D. Human SirT1 interacts with histone H1 and promotes formation of facultative heterochromatin. Mol Cell 2004;16(1):93−105.

[46] Espada J, Esteller M. DNA methylation and the functional organization of the nuclear compartment. Semin Cell Dev Biol 2010;21(2):238−46.

[47] Vaquero A, Scher M, Erdjument-Bromage H, Tempst P, Serrano L, Reinberg D. SIRT1 regulates the histone methyl-transferase SUV39H1 during heterochromatin formation. Nature 2007;450(7168):440−4.

[48] Robertson KD, Wolffe AP. DNA methylation in health and disease. Nat Rev Genet 2000;1(1):11−19.

[49] Bouras T, Fu M, Sauve A, Wang F, Quong A, et al. SIRT1 deacetylation and repression of p300 involves lysine residues 1020/1024 within the cell cycle regulatory domain 1. J Biol Chem 2005;280:10264−76.

[50] Black JC, Mosley A, Kitada T, Washburn M, Carey M. The SIRT2 deacetylase regulates autoacetylation of p300. Mol Cell 2008;32(3):449−55.

[51] Wang J, Chen J. SIRT1 regulates autoacetylation and histone acetyltransferase activity of TIP60. J Biol Chem 2010;285:11458−64.

[52] Jin Y, Kim Y, Kim D, Baek K, Kang B, et al. Sirt2 interacts with 14-3-3 beta/gamma and down-regulates the activity of p53. Biochem Biophys Res Commun 2008;368:690−5.

[53] Vaziri H, Dessain S, Ng E, Imai S, Frye R, et al. hSIR2(SIRT1) functions as an NAD-dependent p53 deacetylase. Cell 2001;107:149−59.

[54] Herbert KJ, Cook AL, Snow ET. SIRT1 inhibition restores apoptotic sensitivity in p53-mutated human keratinocytes. Toxicol Appl Pharmacol 2014;277(3):288−97.

[55] Dai J, Wang Z, Sun D, Lin R, Wang S. SIRT1 interacts with p73 and suppresses p73-dependent transcriptional activity. J Cell Physiol 2007;210:161−6.

[56] Motta M, Divecha N, Lemieux M, Kamel C, Chen D, Gu W, et al. Mammalian SIRT1 represses forkhead transcription factors. Cell 2004;116:551−63.

[57] Wang F, Nguyen M, Qin FX, Tong Q. SIRT2 deacetylates FOXO3a in response to oxidative stress and caloric restriction. Aging Cell 2007;6:505−14.

[58] Boulton SJ, Jackson SP. Components of the Ku-dependent non-homologous end-joining pathway are involved in telomeric length maintenance and telomeric silencing. EMBO J 1998;17(6):1819−28.

[59] Sundaresan N, Samant S, Pillai V, Rajamohan S, Gupta M. SIRT3 is a stress-responsive deacetylase in cardiomyocytes that protects cells from stress-mediated cell death by deacetylation of Ku70. Mol Cell Biol 2008;28:6384−401.

[60] Jeong J, Juhn K, Lee H, Kim SH, Min BH, Lee KM, et al. SIRT1 promotes DNA repair activity and deacetylation of Ku70. Exp Mol Med 2007;39(1):8−13.

[61] Kim MJ, Hong KS, Kim HB, Lee SH, Bae JH, Kim DW, et al. Ku70 acetylation and modulation of c-Myc/ATF4/CHOP signaling axis by SIRT1 inhibition lead to sensitization of HepG2 cells to TRAIL through induction of Dr5 and down-regulation of c-FLIP. Int J Biochem Cell Biol 2013;45(3):711−23.

[62] Vakhrusheva O, Braeuer D, Liu Z, Braun T, Bober E. Sirt7-dependent inhibition of cell growth and proliferation might be instrumental to mediate tissue integrity during aging. J Physiol Pharmacol 2008;59(Suppl. 9):201−12.

[63] Ashraf N, Zino S, Macintyre A, Kingsmore D, Payne AP, George WD, et al. Altered sirtuin expression is associated with node-positive breast cancer. Br J Cancer 2006;95(8):1056−61.

[64] Grob A, Roussel P, Wright JE, McStay B, Hernandez-Verdun D, Sirri V. Involvement of SIRT7 in resumption of rDNA transcription at the exit from mitosis. J Cell Sci 2009;122(Pt 4):489−98.

[65] Fang Y, Nicholl MB. A dual role for sirtuin 1 in tumorigenesis. Curr Pharm Des 2014;20(15):2634−6.

[66] Wong S, Weber J. Deacetylation of the retinoblastoma tumour suppressor protein by SIRT1. Biochem J 2007;407:451−60.

[67] Inoue T, Hiratsuka M, Osaki M, Oshimura M. The molecular biology of mammalian SIRT proteins: SIRT2 in cell cycle regulation. Cell Cycle 2007;6 (9):1011–18.

[68] Inoue T, Hiratsuka M, Osaki M, Yamada H, Hishimoto I, et al. SIRT2, a tubulin deacetylase, acts to block the entry to chromosome condensation in response to mitotic stress. Oncogene 2007;26:945–57.

[69] Dryden S, Nahhas F, Nowak J, Goustin A, Tainsky M. Role for human SIRT2 NAD-dependent deacetylase activity in control of mitotic exit in the cell cycle. Mol Cell Biol 2003;23:3173–85.

[70] Serrano L, Martinez-Redondo P, Marazuela-Duque A, Vazquez BN, Dooley SJ, Voigt P, et al. The tumor suppressor SirT2 regulates cell cycle progression and genome stability by modulating the mitotic deposition of H4K20 methylation. Genes Dev 2013;27 (6):639–53.

[71] North BJ, Marshall BL, Borra MT, Denu JM, Verdin E. The human Sir2 ortholog, SIRT2, is an NAD$^+$-dependent tubulin deacetylase. Mol Cell 2003;11 (2):437–44.

[72] Rabe JH, Mamelak AJ, McElgunn PJ, Morison WL, Sauder DN. Photoaging: mechanisms and repair. J Am Acad Dermatol 2006;55(1):1–19.

[73] Goihman-Yahr M. Skin aging and photoaging: an outlook. Clin Dermatol 1996;14:153–60.

[74] Riley PA. Free radicals in biology: oxidative stress and the effects of ionizing radiation. Int J Radiat Biol 1994;65(1):27–33.

[75] Moreau M, Neveu M, Stephan S, Noblesse E, Nizard C, et al. Enhancing cell longevity for cosmetic application: a complementary approach. J Drugs Dermatol 2007;6:s14–19.

[76] Michishita E, McCord RA, Boxer LD, Barber MF, Hong T, Gozani O, et al. Cell cycle-dependent deacetylation of telomeric histone H3 lysine K56 by human SIRT6. Cell Cycle 2009;8(16):2664–6.

[77] Toiber D, Erdel F, Bouazoune K, Silberman DM, Zhong L, Mulligan P, et al. SIRT6 recruits SNF2H to DNA break sites, preventing genomic instability through chromatin remodeling. Mol Cell 2013;51(4):454–68.

[78] Kaidi A, Weinert BT, Choudhary C, Jackson SP. Human SIRT6 promotes DNA end resection through CtIP deacetylation. Science 2010;329(5997):1348–53.

[79] Mao Z, Hine C, Tian X, Van Meter M, Au M, Vaidya A, et al. SIRT6 promotes DNA repair under stress by activating PARP1. Science 2011;332(6036):1443–6.

[80] Fan W, Luo J. SIRT1 regulates UV-induced DNA repair through deacetylating XPA. Mol Cell 2010;39(2):247–58.

[81] Yuan Z, Seto E. A functional link between SIRT1 deacetylase and NBS1 in DNA damage response. Cell Cycle 2007;6(23):2869–71.

[82] Ming M, Shea C, Guo X, Li X, Soltani K, et al. Regulation of global genome nucleotide excision repair by SIRT1 through xeroderma pigmentosum C. Proc Natl Acad Sci USA 2010;107:22623–8.

[83] Schlicker C, Gertz M, Papatheodorou P, Kachholz B, Becker CF, Steegborn C. Substrates and regulation mechanisms for the human mitochondrial sirtuins Sirt3 and Sirt5. J Mol Biol 2008;382(3):790–801.

[84] Cao C, Lu S, Kivlin R, Wallin B, Card E, et al. SIRT1 confers protection against UVB-and H$_2$O$_2$-induced cell death via modulation of p53 and JNK in cultured skin keratinocytes. J Cell Mol Med 2009;13:3632–43.

[85] Chen M, Li J, Xiao W, Sun L, Tang H, et al. Protective effect of resveratrol against oxidative damage of UVA irradiated HaCaT cells. J Cent South Univ (Engl Ed) 2006;31:635–9.

[86] Lasserre C, D'Arcangelis A, Mildner M, Bhatt P, Tschachler E. 2007 ESDR abstracts: the effect of ultraviolet irradiation on sirtuin expression in human skin. J Invest Dermatol 2007;127:S57.

[87] Jagdeo J, Adams L, Lev-Tov H, Siemiska J, Michl J, Brody N. Dose-dependent antioxidant function of resveratrol demonstrated via modulation of reactive oxygen species in normal human skin fibroblasts *in vitro*. J Drugs Dermatol 2010;9 (12):1523—6.

[88] Lee J, Park K, Min H, Lee S, Kim J, et al. Negative regulation of stress-induced matrix metalloproteinase-9 by Sirt1 in skin tissue. Exp Dermatol 2010;19:1060—6.

[89] Afaq F, Mukhtar H. Botanical antioxidants in the prevention of photocarcinogenesis and photoaging. Exp Dermatol 2006;15(9):678—84.

[90] Nakamaru Y, Vuppusetty C, Wada H, Milne J, Ito M, et al. A protein deacetylase SIRT1 is a negative regulator of metalloproteinase-9. FASEB J 2009;23:2810—19.

[91] Gao Z, Ye J. Inhibition of transcriptional activity of c-JUN by SIRT1. Biochem Biophys Res Commun 2008;376:793—6.

[92] Parks WC, Wilson CL, Lopez-Boado YS. Matrix metalloproteinases as modulators of inflammation and innate immunity. Nat Rev Immunol 2004;4(8):617—29.

[93] Serravallo M, Jagdeo J, Glick SA, Siegel DM, Brody NI. Sirtuins in dermatology: applications for future research and therapeutics. Arch Dermatol Res 2013;305 (4):269—82.

[94] Yeung F, Hoberg J, Ramsey C, Keller M, Jones D, et al. Modulation of NF-kappaB-dependent transcription and cell survival by the SIRT1 deacetylase. EMBO J 2004;23:2369—80.

[95] Adler AS, Sinha S, Kawahara TL, Zhang JY, Segal E, Chang HY. Motif module map reveals enforcement of aging by continual NF-kappaB activity. Genes Dev 2007;21 (24):3244—57.

[96] Salimen A, Kaarniranta K. NF-kappaB signaling in the aging process. J Clin Immunol 2009;29:397—405.

[97] Zhu X, Liu Q, Wang M, Liang M, Yang X, et al. Activation of Sirt1 by resveratrol inhibits TNF-α induced inflammation in fibroblasts. PLOS ONE 2011;6:e27081.

[98] VanGool F, Galli M, Gueydan C, Kruys V, Bedalov A, et al. Intracellular NAD levels regulate tumor necrosis factor protein synthesis in a sirtuin-dependent manner. Nat Med 2009;15:206—10.

[99] Nagel A, Hert M, Eming R. B-cell-directed therapy for inflammatory skin diseases. J Invest Dermatol 2009;129:289—301.

[100] Menter A, Gottlieb A, Feldman S, Voorhees AV, Leonardi C, et al. Guidelines of care for the management of psoriasis and psoriatic arthritis: Section 1. Overview of psoriasis and guidelines of care for the treatment of psoriasis with biologics. J Am Acad Dermatol 2008;58:826—50.

[101] Kim BE, Howell MD, Guttman-Yassky E, Gilleaudeau PM, Cardinale IR, Boguniewicz M, et al. TNF-alpha downregulates filaggrin and loricrin through c-Jun N-terminal kinase: role for TNF-alpha antagonists to improve skin barrier. J Invest Dermatol 2011;131(6):1272—9.

[102] Kim BE, Leung DY, Boguniewicz M, Howell MD. Loricrin and involucrin expression is down-regulated by Th2 cytokines through STAT-6. Clin Immunol 2008;126(3):332—7.

[103] Moniaga CS, Kabashima K. Filaggrin in atopic dermatitis: flaky tail mice as a novel model for developing drug targets in atopic dermatitis. Inflamm Allergy Drug Targets 2011;10(6):477—85.

[104] Palmer CN, Irvine AD, Terron-Kwiatkowski A, Zhao Y, Liao H, Lee SP, et al. Common loss-of-function variants of the epidermal barrier protein filaggrin are a major predisposing factor for atopic dermatitis. Nat Genet 2006;38(4):441—6.

[105] Nakai K, Yoneda K, Hosokawa Y, Moriue T, Presland RB, Fallon PG, et al. Reduced expression of epidermal growth factor receptor, E-cadherin, and occludin in the skin of flaky tail mice is due to filaggrin and loricrin deficiencies. Am J Pathol 2012;181(3):969−77.

[106] Hao JQ. Targeting interleukin-22 in psoriasis. Inflammation 2014;37(1):94−9.

[107] Yuan ZL, Guan YJ, Chatterjee D, Chin YE. Stat3 dimerization regulated by reversible acetylation of a single lysine residue. Science 2005;307(5707):269−73.

[108] Nie Y, Erion DM, Yuan Z, Dietrich M, Shulman GI, Horvath TL, et al. STAT3 inhibition of gluconeogenesis is downregulated by SirT1. Nat Cell Biol 2009;11(4):492−500.

[109] Sestito R, Madonna S, Scarponi C, Cianfarani F, Failla CM, Cavani A, et al. STAT3-dependent effects of IL-22 in human keratinocytes are counterregulated by sirtuin 1 through a direct inhibition of STAT3 acetylation. FASEB J 2011;25(3):916−27.

[110] Kalergis A, Iruretagoyena M, Barrientos M, Gonzalez P, Herrada A, et al. Modulation of nuclear factor-kappaB activity can influence the susceptibility to systemic lupus erythematosus. Immunology 2008;128(1 Suppl.):e306−14.

[111] Hu N, Qiu X, Luo Y, Yuan J, Li Y, Lei W, et al. Abnormal histone modification patterns in lupus CD4$^+$ T cells. J Rheumatol 2008;35(5):804−10.

[112] Mishra N, Reilly CM, Brown DR, Ruiz P, Gilkeson GS. Histone deacetylase inhibitors modulate renal disease in the MRL-lpr/lpr mouse. J Clin Invest 2003;111(4):539−52.

[113] Garcia BA, Busby SA, Shabanowitz J, Hunt DF, Mishra N. Resetting the epigenetic histone code in the MRL-lpr/lpr mouse model of lupus by histone deacetylase inhibition. J Proteome Res 2005;4(6):2032−42.

[114] Hu N, Long H, Zhao M, Yin H, Lu Q. Aberrant expression pattern of histone acetylation modifiers and mitigation of lupus by SIRT1-siRNA in MRL/lpr mice. Scand J Rheumatol 2009;38:464−71.

[115] Becatti M, Fiorillo C, Barygina V, Cecchi C, Lotti T, Prignano F, et al. SIRT1 regulates MAPK pathways in vitiligo skin: insight into the molecular pathways of cell survival. J Cell Mol Med 2014;18(3):514−29.

[116] Blander G, Bhimavarapu A, Mammone T, Maes D, Elliston K, et al. SIRT1 promotes differentiation of normal human keratinocytes. J Invest Dermatol 2009;129:41−9.

[117] Wang C, Chen L, Hou X, Li Z, Kabra N, et al. Interactions between E2F1 and SirT1 regulate apoptotic response to DNA damage. Nat Cell Biol 2006;8:1025−31.

[118] Spallotta F, Cencioni C, Straino S, Nanni S, Rosati J, Artuso S, et al. A nitric oxide-dependent cross-talk between class I and III histone deacetylases accelerates skin repair. J Biol Chem 2013;288(16):11004−12.

[119] Mattagajasingh I, Kim CS, Naqvi A, Yamamori T, Hoffman TA, Jung SB, et al. SIRT1 promotes endothelium-dependent vascular relaxation by activating endothelial nitric oxide synthase. Proc Natl Acad Sci USA 2007;104(37):14855−60.

[120] Fitzgerald O'Connor EJ, Badshah II, Addae LY, Kundasamy P, Thanabalasingam S, Abioye D, et al. Histone deacetylase 2 is upregulated in normal and keloid scars. J Invest Dermatol 2012;132(4):1293−6.

[121] DeFelice B, Wilson R, Nacca M. Telomere shortening may be associated with human keloids. BMC Med Genet 2009;10:110.

[122] Rogers H, Weinstock M, Harris A, Hinckley M, Feldman S, et al. Incidence estimate of nonmelanoma skin cancer in the United States, 2006. Arch Dermatol 2010;146:283−7.

[123] Markovic S, Erickson L, Rao R, Weenig R, Pockaj B, et al. Malignant melanoma in the 21st century, Part 1: epidemiology, risk factors, screening, prevention, and diagnosis. Mayo Clin Proc 2007;83:364−80.

[124] Hida Y, Kubo Y, Murao K, Arase S. Strong expression of a longevity-related protein, SIRT1, in Bowen's disease. Arch Dermatol Res 2007;299(2):103−6.

[125] Lennerz V, Fatho M, Gentilini C, Frye R, Lifke A, et al. The response of autologous T cells to a human melanoma is dominated by mutated neoantigens. Proc Natl Acad Sci USA 2005;102:16013−18.

[126] Wilking MJ, Singh C, Nihal M, Zhong W, Ahmad N. SIRT1 deacetylase is overexpressed in human melanoma and its small molecule inhibition imparts antiproliferative response via p53 activation. Arch Biochem Biophys 2014;563:94−100.

[127] Benavente C, Schnell S, Jacobson E. Effects of niacin restriction on sirtuin and PARP responses to photodamage in human skin. PLOS ONE 2012;7:e42276.

[128] Allison S, Jiang M, Milner J. Oncogenic viral protein HPV E7 up-regulates the SIRT1 longevity protein in human cervical cancer cells. Aging 2009;1:316−27.

[129] Ming M, Qiang L, Zhao B, He YY. Mammalian SIRT2 inhibits keratin 19 expression and is a tumor suppressor in skin. Exp Dermatol 2014;23(3):207−9.

[130] Chung S, Yao H, Caito S, Hwang JW, Arunachalam G, Rahman I. Regulation of SIRT1 in cellular functions: role of polyphenols. Arch Biochem Biophys 2010;501(1):79−90.

[131] Camins A, Sureda F, Junyent F, Verdaguer E, Folch J, et al. Sirtuin activators: designing molecules to extend life span. Biochim Biophys Acta 2010;1799:740−9.

[132] Howitz KT, Bitterman KJ, Cohen HY, Lamming DW, Lavu S, Wood JG, et al. Small molecule activators of sirtuins extend *Saccharomyces cerevisiae* lifespan. Nature 2003;425(6954):191−6.

[133] de Boer VC, de Goffau MC, Arts IC, Hollman PC, Keijer J. SIRT1 stimulation by polyphenols is affected by their stability and metabolism. Mech Ageing Dev 2006;127(7):618−27.

[134] Mai A, Valente S, Meade S, Carafa V, Tardugno M, Nebbioso A, et al. Study of 1,4-dihydropyridine structural scaffold: discovery of novel sirtuin activators and inhibitors. J Med Chem 2009;52(17):5496−504.

[135] Bemis JE, Vu CB, Xie R, Nunes JJ, Ng PY, Disch JS, et al. Discovery of oxazolo[4,5-b]pyridines and related heterocyclic analogs as novel SIRT1 activators. Bioorg Med Chem Lett 2009;19(8):2350−3.

[136] Nayagam VM, Wang X, Tan YC, Poulsen A, Goh KC, Ng T, et al. SIRT1 modulating compounds from high-throughput screening as anti-inflammatory and insulin-sensitizing agents. J Biomol Screen 2006;11(8):959−67.

[137] Borra MT, Smith BC, Denu JM. Mechanism of human SIRT1 activation by resveratrol. J Biol Chem 2005;280(17):17187−95.

[138] Timmers S, Konings E, Bilet L, Houtkooper RH, van de Weijer T, Goossens GH, et al. Calorie restriction-like effects of 30 days of resveratrol supplementation on energy metabolism and metabolic profile in obese humans. Cell Metab 2011;14(5):612−22.

[139] Fujitaka K, Otani H, Jo F, Jo H, Nomura E, Iwasaki M, et al. Modified resveratrol Longevinex improves endothelial function in adults with metabolic syndrome receiving standard treatment. Nutr Res 2011;31(11):842−7.

[140] Li L, Wu L, Tashiro S, Onodera S, Uchiumi F, Ikejima T. Activation of the SIRT1 pathway and modulation of the cell cycle were involved in silymarin's protection against UV-induced A375-S2 cell apoptosis. J Asian Nat Prod Res 2007;9:242−52.

[141] ClinicalTrials.gov [Internet]. Bethesda, MD: National Library of Medicine (US). 2000 [cited 15.07.14]. Available from: <http://clinicaltrials.gov/ct2/search> NLM Identifier.

[142] Milne JC, Lambert PD, Schenk S, Carney DP, Smith JJ, Gagne DJ, et al. Small molecule activators of SIRT1 as therapeutics for the treatment of type 2 diabetes. Nature 2007;450(7170):712−16.

[143] Pacholec M, Bleasdale JE, Chrunyk B, Cunningham D, Flynn D, Garofalo RS, et al. SRT1720, SRT2183, SRT1460, and resveratrol are not direct activators of SIRT1. J Biol Chem 2010;285(11):8340–51.

[144] Zhang Y, Au Q, Zhang M, Barber JR, Ng SC, Zhang B. Identification of a small molecule SIRT2 inhibitor with selective tumor cytotoxicity. Biochem Biophys Res Commun 2009;386(4):729–33.

[145] Outeiro TF, Kontopoulos E, Altmann SM, Kufareva I, Strathearn KE, Amore AM, et al. Sirtuin 2 inhibitors rescue alpha-synuclein-mediated toxicity in models of Parkinson's disease. Science 2007;317(5837):516–19.

[146] Neugebauer RC, Uchiechowska U, Meier R, Hruby H, Valkov V, Verdin E, et al. Structure-activity studies on splitomicin derivatives as sirtuin inhibitors and computational prediction of binding mode. J Med Chem 2008;51(5):1203–13.

[147] Heltweg B, Gatbonton T, Schuler AD, Posakony J, Li H, Goehle S, et al. Antitumor activity of a small-molecule inhibitor of human silent information regulator 2 enzymes. Cancer Res 2006;66(8):4368–77.

[148] Peck B, Chen CY, Ho KK, Di Fruscia P, Myatt SS, Coombes RC, et al. SIRT inhibitors induce cell death and p53 acetylation through targeting both SIRT1 and SIRT2. Mol Cancer Ther 2010;9(4):844–55.

[149] Elixir Pharmaceuticals Inc. Elixir Announces First Sirtuin Inhibitor Clinical Trial [Press Release]. Bloomberg.com. N.p., 07 January 2010. Web. 15 July 2014. <http://www.bloomberg.com/apps/news?pid = newsarchive&sid = a.pko1st.BrE>.

[150] Kokkonen P, Rahnasto-Rilla M, Jarho E, Lahtela-Kakkonen M, Kokkola T. Studying SIRT6 regulation using H3K56 based substrate and small molecules. Eur J Pharm Sci 2014;63:71–6.

[151] Lara E, Mai A, Calvanese V, Altucci L, Lopez-Nieva P, Martinez-Chantar ML, et al. Salermide, a Sirtuin inhibitor with a strong cancer-specific proapoptotic effect. Oncogene 2009;28(6):781–91.

[152] Bedalov A, Gatbonton T, Irvine WP, Gottschling DE, Simon JA. Identification of a small molecule inhibitor of Sir2p. Proc Natl Acad Sci USA 2001;98(26):15113–18.

[153] Biacsi R, Kumari D, Usdin K. SIRT1 inhibition alleviates gene silencing in Fragile X mental retardation syndrome. PLOS Genet 2008;4(3):e1000017.

[154] Schuetz A, Min J, Antoshenko T, Wang CL, Allali-Hassani A, Dong A, et al. Structural basis of inhibition of the human NAD$^+$-dependent deacetylase SIRT5 by suramin. Structure 2007;15(3):377–89.

[155] Alcain FJ, Villalba JM. Sirtuin inhibitors. Expert Opin Ther Pat 2009;19(3):283–94.

[156] Lain S, Hollick J, Campbell J, Staples O, Higgins M, et al. Discovery, *in vivo* activity, and mechanism of action of a small-molecule p53 activator. Cancer Cell 2008;13:454–63.

[157] Wang C, Wang M, Tashiro S, Onodera S, Ikejima T. Roles of SIRT1 and phosphoinositide 3-OH kinase/protein kinase C pathways in evodiamine-induced human melanoma A375-S2 cell death. J Pharmacol Sci 2005;97:494–500.

[158] Kunimoto R, Jimbow K, Tanimura A, Sato M, Horimoto K, Hayashi T, et al. SIRT1 regulates lamellipodium extension and migration of melanoma cells. J Invest Dermatol 2014;134(6):1693–700.

MicroRNAs in Skin Diseases

Marianne B. Løvendorf and Lone Skov

Department of Dermato-Allergology, Gentofte Hospital, University of
Copenhagen, Hellerup, Denmark

9.1 INTRODUCTION

Epigenetic mechanisms have been an intensive area of research within the last couple of years. In addition to chromatin remodeling and DNA methylation, microRNAs (miRNAs) are the most important epigenetic modulators in the regulation of protein expression. miRNAs are small noncoding RNA molecules that modulate gene expression posttranscriptionally. The understanding of miRNA-mediated gene regulation has increased significantly in recent years and miRNAs have been shown to have a significant impact on several physiological and pathophysiological cellular processes including development, differentiation, proliferation, and apoptosis. Several studies have revealed that epigenetic aberrations including miRNA deregulation have a major impact on gene regulation in various skin diseases. In this chapter, we summarize our current knowledge of the aberrant expression and biological role of miRNAs in skin diseases. In addition, we present the most recent progress within the clinical utility of miRNAs as biomarkers for various skin diseases.

9.2 MicroRNAs

9.2.1 miRNA Discovery

miRNAs comprise an abundant class of highly conserved small noncoding RNA molecules that modulate gene expression at the posttranscriptional level [1–3]. The first miRNAs, lin-4 and let-7, were discovered through genetic screens in *Caenorhabditis elegans* as genes important for developmental timing [4,5]. In 2000, the first human miRNA was discovered [6] and the term miRNA was introduced. Since the discovery of miRNAs, they have received considerable attention in basic research and much progress has been made to elucidate miRNA expression patterns and functions. Currently, 2,588 human mature miRNAs have been registered in the miRBase database (release 21.0, June 2014, www.miRBase.org) and it is estimated that more than 60% of all human protein-coding genes are under selective pressure to maintain pairing to miRNAs [7]. Hence, it has now become evident that miRNAs constitute one of the most abundant classes of gene-regulatory molecules.

9.2.2 miRNA Biogenesis

miRNAs regulate gene expression by sequence-specific base pairing, typically within the 3′ untranslated region (UTR) of the target mRNA. In mammalian cells more than 70% of miRNAs are located within an intron, with the majority residing in a protein-coding gene. The remaining miRNAs are located in noncoding RNA transcription units [8]. Most of the miRNAs are transcribed by RNA polymerase II [9]. The transcribed long primary miRNA (pri-miRNA) is processed by the microprocessor that includes Drosha, an RNase III enzyme, and its cofactor DGCR8 (DiGeorge syndrome critical region gene 8), producing the 60–70 nucleotide hairpin precursor miRNA (pre-miRNA) [8]. After nuclear processing the pre-miRNA is exported to the cytosol by Exportin 5 [10] where another RNase III enzyme, Dicer, processes the pre-miRNA into the 21–24 nucleotide duplex miRNA. The strand designated to be the mature sequence is then incorporated into the effector complex known as the RNA-induced silencing complex (RISC) [11] in which core components are members of the Argonaute protein family [2]. The mature miRNA strand serves as a guide to target mRNAs. With a few exceptions miRNA-binding sites in metazoan mRNAs are located in the 3′ UTR. miRNA target recognition requires among others stringent Watson–Crick base pairing to the 5′ end of the miRNA region centered on nucleotides 2–7, representing the seed region [12]. Depending on the level of miRNA–mRNA complementarity [2,9] two different mechanisms of RISC-mediated gene regulation occur. At sites with extensive complementarity the miRNA can direct mRNA cleavage, while at sites with lower degrees of complementarity, translational repression or mRNA destabilization might be the outcome [13], the latter occurring more commonly in mammals [14]. An overview of the miRNA biogenesis process is illustrated in Figure 9.1.

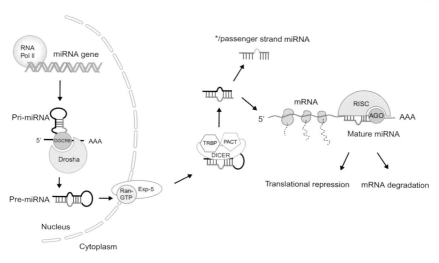

FIGURE 9.1 **miRNA biogenesis.** Similarly to protein-coding genes, most miRNAs are transcribed by RNA polymerase II in mammalian cells. Transcription by polymerase III has been shown to occur for miRNAs that are near Alu repeats. The primary miRNA transcript (pri-miRNA) is usually several kilobases long, polyadenylated at its 3′ end and 5′-capped as with mRNAs. The pri-miRNA is cleaved by the microprocessor that includes Drosha, an RNase III enzyme, and its cofactor DGCR8 (DiGeorge syndrome critical region gene 8), producing the 60−70 nucleotide hairpin precursor miRNA (pre-miRNA). The cofactor DGCR8 is essential for the Drosha activity and is thought to initiate pri-miRNA binding, as it contains two dsRNA-binding domains. A subset of miRNAs, named mirtrons, depends on splicing, instead of microprocessor activity, for release of the precursor forms. The pre-miRNA is actively transported from the nucleus to the cytoplasm by Exportin 5 (Exp-5) in the presence of its cofactor, Ran-guanine triphosphatase (Ran-GTP). In the cytoplasm the pre-miRNA is further processed by a second RNase III enzyme, Dicer. In humans, Dicer interacts with the dsRNA-binding domain proteins TAR RNA-binding protein (TRBP) and protein activator of PKR (PACT). The processing by Dicer results in the generation a small 21−24 nucleotide RNA duplex. The two strands of the duplex are separated and one of the strands is incorporated into a RISC in which the core components are members of the Argonaute protein family (AGO). Strand selection is determined by inherent features of the miRNA duplex, including thermodynamic asymmetry/stability. The other strand (the passenger strand or miRNA*) is presumed to be excluded from RISC incorporation and subsequently degraded. The mature miRNA guides RISC to its specific mRNA target and directs translation inhibition or mRNA degradation [8].

9.2.3 miRNA Function

Given that miRNAs regulate more than half of mammalian mRNAs [7], it is evident that miRNAs have a significant impact on several physiological and pathophysiological cellular processes including development, differentiation, proliferation, and apoptosis [15]. Modest repression appears to be the more common regulatory outcome of an miRNA–mRNA target interaction and, hence, miRNAs have been described as fine-tuners of gene expression [16]. Many factors influence the impact of an individual miRNA on target gene regulation. miRNA-mediated regulation is believed to have a widespread impact on protein levels; hence, a single miRNA can target >100 mRNA targets [12,16]. Simultaneously, a single mRNA can be regulated by multiple miRNAs [3]; thus identifying miRNA targets is a challenge. Currently, several computational approaches for predicting miRNA targets exist, e.g., miRWalk, TargetScan, PicTar, and RNAhybrid [17]. Other factors that likely influence the impact of an miRNA on target gene regulation include the temporal and tissue-specific expression of an miRNA and its targets, the ratio between target and miRNA concentration, competition with other miRNAs for RISC, co-expression of functionally redundant miRNAs [3] and expression of decoys [18]. Moreover, it has been demonstrated that miRNAs can activate translation [19] and target the 5′ UTR of mRNA [20]. Finally, miRNAs are suggested to be subject to as much complex regulation as they exert [8]. Indeed, epigenetic modifications, such as DNA methylation or histone acetylation, have been demonstrated to affect miRNA expression, altogether making miRNA-mediated gene regulation extraordinarily complex.

Considering the high number of human miRNAs identified so far, their diverse expression patterns, and their complex regulation, along with the significant number of putative mRNA targets, it is not surprising that miRNAs have been implicated in a broad range of diseases, for example cancers [21], immunological diseases [22], and several diseases of the skin.

9.2.4 miRNAs in Normal Skin

The first evidence that miRNAs are important for the development of the skin came from the observation that mice carrying an epidermal-specific Dicer or DGCR8 deletion, failed to produce mature miRNAs and embryonic hair germs were found to evaginate into the epidermis rather than invaginate normally toward the dermis [23,24]. Concordantly, high expression of several miRNAs in the epidermis and hair follicles was found to be necessary for normal skin development [23,24]. Currently, the best characterized miRNA in the skin is miR-203 which has been found to be highly expressed in the suprabasal layer of the epidermis [25]. By altering miR-203's spatio-temporal expression *in vivo*, Yi and colleagues showed that miR-203 promotes epidermal differentiation by restricting the proliferative potential and inducing cell-cycle exit [26]. The transcription factor ΔNp63α, which is essential for the maintenance of "stemness" in the skin, was identified as a conserved target for miR-203 [26,27]. Together, these studies suggested that miR-203 defines a molecular boundary between proliferative basal progenitors and terminally differentiating suprabasal cells [26]. Since these pioneering studies which emphasized the importance of miRNAs in controlling mammalian skin development and function [23–27], miRNAs have been subject to significant interest in dermatology. Currently, several observations indicate a role of the miRNAs in the pathogenesis of various cutaneous diseases, including psoriasis, atopic dermatitis (AD), allergic contact dermatitis (ACD), cutaneous T-cell lymphoma (CTCL), and malignant melanoma (MM). In the following section, we present some of the most important inflammatory and malignant cutaneous diseases wherein miRNAs have been studied. We apologize to all the authors whose work could not be discussed and cited due to space limitations.

9.3 miRNAs IN INFLAMMATORY SKIN DISEASES

9.3.1 Psoriasis

Psoriasis is a systemic inflammatory skin disease affecting approximately 2–3% of the world's population [28]. The cause of psoriasis still remains unknown; however, the disease is considered to have a multifactorial etiology involving environmental factors and genetic susceptibility. Histologically, psoriasis is defined by epidermal hyperproliferation with parakeratosis, elongation of the rete ridges, exaggerated vascularity, and dermal infiltration of activated immune cells [28].

The first evidence that miRNAs were implicated in psoriasis came from a study by Sonkoly et al. demonstrating that psoriasis-affected skin has a specific miRNA expression profile when compared with healthy human skin or with another chronic inflammatory skin disease, AD [25]. Accordingly, Zibert et al. identified several deregulated miRNAs when comparing lesional and nonlesional psoriatic skin with healthy skin [29] including a significant overlap with the findings by Sonkoly et al. [25]. Subsequently, we demonstrated that miRNA detection in psoriatic skin is robust independently of preservation method [30]. Two recent studies using next-generation sequencing (NGS) have added additional deregulated miRNAs and noncanonical miRNAs to lesional psoriatic skin [31,32]. Collectively, the present miRNAs reported as aberrantly expressed in psoriatic lesions have been shown to be involved in inflammation (e.g., miR-21, -31, -142-3p, -146a) and keratinocyte (KC) proliferation and differentiation (e.g., miR-99a, -100, -125b, -203) (Figure 9.2). The previously established differences in global miRNA expression in lesional psoriatic skin compared with nonlesional psoriatic skin and healthy skin were based on whole tissue extracts and thus, represent global miRNA

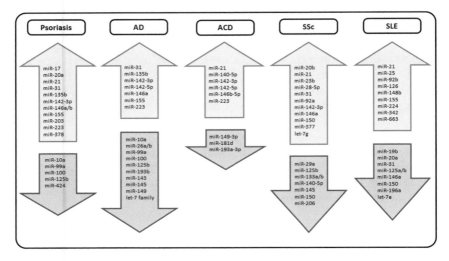

FIGURE 9.2   **miRNAs deregulated in inflammatory skin diseases.** A selected panel of miRNAs that have been identified as either increased (blue arrow) or decreased (red arrow) in the inflammatory skin diseases psoriasis, AD, ACD, scleroderma (SSc), and SLE.

changes from a mixture of cells including KCs, leukocytes, and endothelial cells. To increase the understanding of the role of miRNAs in psoriasis we recently investigated whether certain disease-related miRNAs could be specifically confined to the epidermis or to immune cells in the dermal inflammatory infiltrates [33]. By combining laser capture microdissection (LCM) and barcoded small RNA cDNA libraries coupled with NGS, we identified several highly abundant miRNAs in the psoriatic epidermis, which previously have been reported to be involved in KC proliferation and differentiation, for example, miR-203 [25–27] and miR-205 [34]. In agreement with prior studies using whole tissue extracts, miR-99a and miR-125b were found to be significantly downregulated in lesional epidermis [25,31]. miRNA-99a has been shown to be expressed in the upper part of the epidermis; most pronounced in the stratum spinosum and much less in the basal layer [35]. In the same study, miR-99a was proposed to target IGF1R (insulin-like growth factor 1 receptor), which has been shown to be upregulated in the proliferating basal layer of psoriatic skin lesions [36], altogether suggesting a role of miR-99a in the differentiation of KCs. Recently, miR-125b was reported to target FGFR2, thereby modulating KC proliferation and differentiation [37]. Moreover, using *in situ* hybridization, Xu and colleagues showed that miR-125b was decreased in all epidermal layers in psoriatic skin compared with healthy skin and they proposed that the major cell type responsible for the decreased miR-125b expression in whole tissue samples is the KCs [37]. However, from our study using LCM it seems that miR-125b is more abundant in the dermal compartment compared with the epidermal compartment based on the total read frequencies [33]. Since miR-125b has been shown to be downregulated during differentiation toward effector T cells [38] we suggest that besides having an important role in the KC proliferation, miR-125b may also play an important role in the immune cells present in the dermal inflammatory infiltrates; however, further investigations are warranted to elucidate this in more detail. Other deregulated miRNAs in psoriatic epidermis are miR-21, -142-3p, -146a, and -223, which all previously have been associated with inflammation [39]. miRNA-146a and -223 are preferentially expressed by immune cells [25,40]; hence, the observed increase in their expression in the psoriatic epidermis is likely caused by an influx of inflammatory cells into the epidermis. Using *in situ* hybridization Joyce et al. reported the expression of miR-142-3p to lumen-like structures within the psoriatic epidermis [31]. Additionally, they observed staining of miR-142-3p in dermal immune cells in psoriatic skin [31]. From the study using LCM we also identified 13 deregulated miRNAs in psoriatic epidermis which have not previously been reported as differentially expressed in psoriatic skin [33]. Among those 13 deregulated miRNAs several have also been found to be deregulated in among others MM [41], indicating a linking role for the hyperproliferation present in both pathologies.

Psoriatic skin is characterized by a significant increase of infiltrating leukocytes such as T cells, dendritic cells (DCs), and, to a lesser degree, B cells. The importance of these inflammatory aggregates in dermis is supported by the observation that they are not present in nonlesional skin and that effective treatment leads to their disappearance [42]. Several miRNAs were found to be deregulated in the cells present in these dermal inflammatory infiltrates. Among those were miR-155, -21, -142-3p, -146a, -150, -223, -99a, and -125b, which all have been reported to be expressed by various immune cell subsets including T cells, B cells, Treg cells, granulocytes, and DCs [38–40,43]. Many miRNAs seem to be active in multiple immune cells, making it challenging to define specific functions of single miRNAs. Another possible role of the deregulated miRNAs could be involvement in the altered angiogenesis seen in psoriasis. miRNA-21, -100, and -378 are among the most abundant deregulated miRNAs in lesional skin [31]. Interestingly, they have all been reported to be involved in angiogenesis [44]; thus these miRNAs may contribute to the characteristic exaggerated vascularity and increased level of vascular endothelial growth factor (VEGF) seen in psoriatic lesions.

Notably, a recent study demonstrated that inhibition of miR-21 by locked nucleic acid (LNA)-modified anti-miR-21 compounds ameliorated disease pathology in patient-derived psoriatic skin xenotransplants in mice and in a psoriasis-like mouse model [45]. The findings suggested a causal role for miR-21 in the pathogenesis of epidermal hyperplasia involving inhibition of TIMP3 (tissue inhibitor of matrix metalloproteinase 3) followed by an increase in TACE (tumor necrosis factor-α-converting enzyme)/ADAM17 (a disintegrin and metalloproteinase 17) and thereby TNF-α, which previously was proposed as a mechanism involved in psoriasis [29]. However, as miR-21 has been found to be highly abundant and increased in the dermal inflammatory infiltrates and dermal CD3$^+$ T cells [33,46], the inhibition of miR-21 may actually be beyond the epidermis. Hence, further studies are warranted to elucidate the influence of miR-21 on other cells, pathways, and disease mechanism in psoriasis in order to establish the anti-miR-21 oligonucleotide as a new treatment for psoriasis.

In psoriasis, few studies have investigated miRNA expression in the sera for service as potential biomarkers [47−49]. We recently identified several deregulated miRNAs in whole blood, plasma, and peripheral blood mononuclear cells (PBMCs) from patients with psoriasis [50]. Interestingly, among the deregulated miRNAs in PBMCs we observed a noteworthy overlap with deregulated miRNAs in the dermal inflammatory infiltrates of psoriatic skin, suggesting that alterations in the miRNA expression levels in psoriatic lesions appear to be reflected in the blood of these patients [33]. To evaluate the potential of miRNAs as blood biomarkers for psoriasis, we measured the expression of a panel of miRNAs in PBMCs obtained from patients with psoriasis at baseline, after 3−5 weeks and after 7−10 weeks of treatment with methotrexate (MTX) [50]. MTX is the standard therapy for moderate to severe psoriasis, and it appears to exert its effects by acting as both immune-modulatory agent and antimetabolite. Although not statistically significant a pairwise comparison of the miRNA expression levels at baseline and after MTX treatment identified both miR-223 and miR-143 to be downregulated in the PBMCs already after 3−5 weeks of treatment [50]. Pivarcsi et al. reported a similar decrease in the expression level of a number of miRNAs including miR-223, albeit in the sera, after 12 weeks of anti-TNF-α treatment, but not after MTX treatment [49]. This discrepancy could be caused by the fact that different compartments were investigated (PBMCs vs. serum). Altogether these findings suggest that a downregulation of miR-223 is not specifically related to a specific treatment, but, rather, associated with the immunosuppressive effect on cells in which miR-223 is abundantly expressed. In summary, in line with psoriatic skin, blood from patients with psoriasis is characterized by a specific miRNA signature and the initial findings support the idea that miRNAs indeed have the potential to serve as novel blood biomarkers for psoriasis activity [50]. Lately, the expression of hair-root miR-19a and hair-shaft miR-424 have been proposed as diagnostic biomarkers for psoriasis [51,52], exemplifying another interesting and noninvasive approach in the search for biomarkers in psoriasis.

In summary, several studies have provided evidence that psoriatic skin and blood are indeed characterized by specific miRNA expression profiles. An overview of a selected panel of deregulated miRNAs in psoriasis can be found in Figure 9.2. miRNA-21, -142-3p, and -146a have been identified as significantly upregulated in all prior studies involving global miRNA expression in psoriatic skin, emphasizing their importance in psoriasis. Even though several miRNAs have been restricted to specific cell and regions in psoriatic skin, further studies are warranted to unravel the specific functions of the miRNAs in the pathogenesis of psoriasis.

9.3.2 Atopic Dermatitis

AD is a common chronic inflammatory skin disease characterized by relapsing pruritic skin lesions. It often presents in infancy and childhood and can persist throughout adulthood. The disease is complex in nature with changes in the skin barrier function but also in the innate and adaptive immune system. The exact cause of AD is unknown, but the disease is likely to result from interactions between genetic and environmental factors.

Only a few studies have investigated the expression of miRNAs in AD. miRNAs were first associated with AD when a small study using genome-wide analysis of miRNA expression revealed a number of deregulated miRNAs in skin lesions obtained from patients with AD compared with psoriatic lesions and healthy skin samples [25]. Recently, miR-155 was found to be one of the highest-ranked upregulated miRNAs in skin from patient with AD [53]. miRNA-155 was mainly expressed in infiltrating immune cells including CD4$^+$ T cells and DCs [53]. The authors showed that miR-155 was increased during T-cell differentiation/activation. Functionally, the cytotoxic T lymphocyte-associated protein 4 (CTLA-4), which is an important negative regulator of T-cell responses, was shown to be a direct target of miR-155, altogether suggesting that miR-155 contributes to the chronic skin inflammation by increasing the proliferative response of T cells through the inhibition of CTLA-4 [53]. Recently, Rebane et al. demonstrated that miR-146a alleviated chronic skin inflammation in AD through suppression of innate immune responses in KCs [54]. The expression of miR-146a was significantly increased in KCs and chronic lesional skin of patients with AD and the authors showed that miR-146a inhibited the expression of several pro-inflammatory factors including CCL5 and CCL8 [54]. Besides miR-155 and miR-146a, another 43 miRNAs have been found to be deregulated in AD lesions [53]. Notably, ~45% of those have also been described as altered in psoriatic skin, indicating a linking role for the inflammation present in both pathologies (Figure 9.2).

9.3.3 Allergic Contact Dermatitis

ACD is a clinical term used to describe an inflammatory reaction that is caused by type IV delayed type hypersensitivity responses to allergens that come in contact with the skin. The clinical manifestations of acute ACD are often erythema, papules, and vesicles. There are two distinct phases in ACD, the sensitization (induction) phase and the challenge (effector) phase. Only miRNA expression in the challenge phase has been studied. Vennegaard et al. were the first to describe aberrant miRNA expression in ACD [55]. They analyzed the global miRNA expression in skin from subjects who were sensitized with diphenylcyclopropenone (DPCP), a strong contact allergen. miRNA-21, -223, -142-3p, and -142-5p were found to be the most upregulated miRNAs in the inflammatory lesions [55] (Figure 9.2). Notably, these four miRNAs were also found to be upregulated in the skin of a contact hypersensitivity mouse model during the challenge reaction to 2,4-dinitrofluorobenzene (DNFB), suggesting that mouse models are valuable tools for further study of the involvement of miRNAs in ACD [55]. In addition, using barcoded small RNA sequencing, Gulati et al. recently confirmed the aberrant expression of, among others, miR-21, -223, -142-3p, and -142-5p in skin lesions obtained 3 and 14 days after DPCP challenge in sensitized subjects [56]. Whether the aberrant miRNA identified after challenge with DPCP is specifically related to DPCP awaits further clarification.

Given that ACD is mainly a T-cell-mediated skin disease and that miR-21, -223, -142-3p, and -142-5p all previously have been associated with the T-cell-driven skin diseases psoriasis and AD [25,29,31,53], the upregulation of these four miRNAs in ACD is likely related to the increased infiltration of T cells into the skin and their activation upon challenge with the allergen. Interestingly, Gulati et al. also identified six miRNAs that were significantly upregulated 120 days after challenge where the lesions clinically were fully resolved [56]. This may indicate a long-lasting allergen-mediated immune reaction in the skin, which explains the persistent upregulation of some miRNAs. Across the three time points studied (3, 14, and 120 days), many but not all miRNAs were uniquely expressed at certain time points, and one miRNA in particular (miR-140-5p) progressively increased expression over time [56]. This may indicate that the exclusive expression of miRNAs at the different time points is related to changes in either adaptive immune responses or to regulatory mechanisms in the skin.

9.3.4 Connective Tissue Diseases

Scleroderma (systemic sclerosis, SSc) is a complex autoimmune disease of unknown etiology that includes the progressive fibrotic replacement of normal tissue architecture in multiple organs, including the skin. The first evidence that miRNAs were implicated in SSc came from a study by

Maurer et al., who found that miR-29a was strongly downregulated in SSc skin and fibroblasts, and the authors suggested that expression of collagen was regulated by miR-29a [57]. Following this, Li and colleagues identified 24 differentially expressed miRNAs in skin tissue from SSc compared to healthy controls including miR-206, -125b, and let-7g [58]. Since then, a number of studies have described miRNAs involved in key processes that contribute to fibrosis in SSc, including TGF-β signaling, extracellular matrix deposition, and fibroblast proliferation and differentiation [59]. Finally, the biomarker potential of several circulating miRNAs has also been investigated in patients with SSc. Analysis of the expression of serum miR-142-3p demonstrated that miR-142-3p was significantly higher in the serum from SSc patients compared to serum samples from patients with closely related diseases such as systemic lupus erythematosus (SLE) and dermatomyositis, but also healthy subjects, indicating that miR-142-3p may be a useful diagnostic marker for the presence of SSc [60].

SLE is a systemic autoimmune disease characterized by the activation of both the innate and adaptive immune system and the production of autoantibodies against multiple organs. The exact etiology is unclear but the majority of the patients display dermatological symptoms including a characteristic facial butterfly rash or thick, red scaly patches on the skin. The first study of miRNA expression in SLE was published in 2007, when the analysis of PBMCs from 23 Chinese patients with SLE identified 16 deregulated miRNAs in SLE compared to controls [61]. Following this, several groups investigated the miRNA expression in SLE. Generally, there is a lack of a consistent expression pattern on individual deregulated miRNAs in SLE; however, this is, among other considerations, caused by the fact that different biological samples (e.g., PBMCs, T-cell subsets, and circulating miRNAs) were investigated. Nevertheless, the aberrant expression of different miRNAs exhibit overlapping functional outcomes in processes considered as important players in the pathogenesis of SLE including perturbed type 1 interferon signaling cascade, DNA hypomethylation, and hyperactivation of T and B cells (Figure 9.2) [62,63]. The miRNA expression profiling in skin samples obtained from patients with SLE is very sparse and more studies are needed to unravel their specific expression and functions in skin lesions from patients with SLE.

9.4 miRNAs IN MALIGNANT SKIN DISEASES

9.4.1 Basal Cell Carcinoma

Basal cell carcinomas (BCCs) are KC tumors that histologically resemble the basal layer of the epidermis [64]. BCCs are caused by a number of environmental factors, for example ultraviolet irradiation through long-term sun exposure and inherited elements. Although BCC represents one of the most common cancers in human, it was not until recently that the role of miRNAs in its pathogenesis was investigated. Heffelfinger et al. investigated eight nodular and eight infiltrative BCCs and showed that miR-21, -143, -148a, -378, -182, and let-7 family members were the most highly expressed [65]. In addition, they described 20 aberrantly expressed miRNAs between nodular and infiltrative BCCs [65]. These initial findings were subsequently followed by those of Sonkoly et al., who compared the miRNA expression profile from BCC lesions with that of healthy skin [66]. Their work demonstrated that BCC tumors display a deregulated miRNA expression pattern involving 64 miRNAs including miR-182, -221, -99a, -100, and -29c [25]. In line with previous reports on miRNA expression in cancer [67], the majority of the miRNAs were repressed [66]. The most downregulated miRNA in BCC was the KC-related miR-203. The authors demonstrated that miR-203 suppresses KC proliferation and directly targets c-JUN and suggest that the reduced level of miR-203 expression in BCCs is likely sustaining an undifferentiated phenotype of the KCs with high proliferative capacity, thereby contributing to the pathogenic events in BCC [66]. Through analyses using miRNA microarray and quantitative real-time polymerase chain reaction (qRT-PCR), Sand et al. confirmed the deregulation of miR-182 and miR-29c in BCC albeit in lesional versus nonlesional BCC [68]. In addition, Sand and coworkers identified another 24 miRNAs with an altered expression in lesions from patients with BCC compared with nonlesional skin (Figure 9.3). Several of these have been reported to be associated with tumorigenesis pathways such as the MAPK/ERK signaling cascade [68].

9.4.2 Cutaneous Squamous Cell Carcinoma

Squamous cell carcinoma (SCC) is an epidermal KC-derived skin tumor. SCC typically manifests as a variety of progressively advanced malignancies, ranging from premalignant actinic keratosis (AK) to SCC *in situ*, invasive SCC, and metastatic SCC [69]. The most common risk factor is excessive long-term exposure to the sun and immunosuppression. Similarly to studies relating to BCC, investigations of the involvement of miRNAs in SCC have recently been commenced. Dziunycz et al. initially described a limited number of KC-related miRNAs in SCC [70]. They found an increased expression of miR-21 and miR-184 and decreased expression of miR-203 in SCC compared with normal skin. Interestingly, they showed that UVA radiation increased the expression of miR-21, -203, and -205 whereas UVB increased miR-203 but decreased miR-205 [70]. As UVB radiation also has been shown to alter the expression of miR-21 and miR-125b in psoriatic skin [71], it seems apparent that UV radiation impacts miRNA

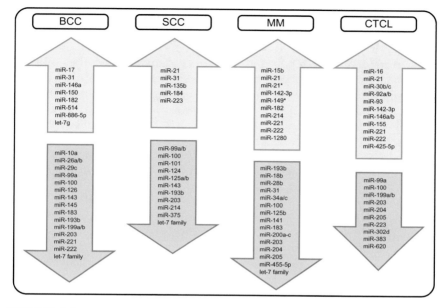

FIGURE 9.3 **miRNAs deregulated in malignant skin diseases.** A selected panel of miRNAs that have been identified as either increased (blue arrow) or decreased (red arrow) in the malignant skin diseases BCC, SCC, MM, and CTCL.

expression; however, the exact mechanism remains to be investigated. Subsequently, two studies undertook a global miRNA expression profiling approach to examine the miRNAs role in SCC [72,73]. Xu et al. identified 58 miRNAs that were differentially expressed in SCC relative to healthy skin [72]. In accordance with the miRNA signature in BCC [66], the majority of deregulated miRNAs in SCC were downregulated including miR-375, miR-125a/b family, let-7 family, miR-99a/b/100 family, miR-143, and miR-101. Only miR-21, -31, -135b, and -223 were found to be upregulated in SCC [72] (Figure 9.3). Notably, Xu et al. observed that different members of the deregulated miRNA gene families tended to be co-expressed in SCC, indicating that deregulation of miRNA expression in SCC occurs in a regulated manner [72]. miRNA-125b was found to be decreased already in AK and Xu et al. suggested that loss of miR-125b expression may be an early step during the pathogenesis of SCC. In addition, they showed that miR-125b suppressed proliferation, colony formation, migratory, and invasive capacity of SCC cells and that matrix metallopeptidase 13 (MMP13) was a direct target of miR-125b, together providing a novel molecular mechanism by which MMP13 is upregulated in SCCs and indicating that miR-125b plays a tumor-suppressive role in SCC [72]. More recently, Yamane et al. demonstrated that miR-124 and miR-214 were significantly downregulated in SCC *in vitro* and *in vivo* [74]. In addition, they showed that miR-214 is the regulator of the extracellular-signal-regulated kinase 1 (ERK1), whereas ERK2 was regulated by both miR-124 and miR-214 [74]. Recently, miR-365 was found to be increased in both lesions from SCC and in cell lines origi-nating from SCC [75]. Zhou and colleagues demonstrated that miR-365 induced tumorigenicity of HaCat cells and silencing of miR-365, using an antagomir-365 reduced tumor growth *in vivo*, suggesting that miR-365 may act as an oncomiR in SCC [75]. In contrast to the results of Zhou et al., Gastaldi et al. recently found a decreased expression of the miR-193b/365a cluster in SCC [76]. Ectopic expression of miR-193b and miR-365 in tumor cells inhibited proliferation and migration. Furthermore, the authors identi-fied the oncogene KRAS and the *MYC* partner MAX as targets suppressed by miR-193b and miR-365, as a whole suggesting that these genes might mediate the tumor-suppressive action of miR-193b and miR-365 [76].

Besides investigating the miRNA expression profiles in skin epithelial cancers, Sand et al. was one of the first to document the aberrant expres-sion of two of the most important enzymes in the miRNA machinery in epithelial skin cancers, Drosha and Dicer [77]. Accordingly, they investi-gated the expression levels of components in the miRNA maturing microprocessor complex (DGCR8) and the RISC complex (Argonaute-1, Argonaute-2, PACT, TARBP1, and TARBP2) [78] in skin lesions from patients with AK, BCC, and SCC, comparing their findings with results from skin from healthy controls. The expression of Drosha, DGCR8, Argonaute-1, Argonaute-2, PACT, and TARBP1 were significantly higher in BCC and SCC compared with healthy controls [77,78]. Collectively, these findings demonstrate that the miRNA biogenesis pathway is deregulated in epithelial skin cancers, indicating a possible role in the process of carcinogenesis [77,78]. The specific consequences of the aber-rant expressions await further clarification.

9.4.3 Malignant Melanoma

MM is derived from epidermal melanocytes and is the most aggressive type of skin cancer. MM constitutes various subtypes including the most prevalent superficial spreading melanoma (SSM) and nodular melanoma (NM) [79]. The cause of MM still remains unknown; however, the disease is considered to have a multifactorial etiology involving genetic and envi-ronmental factors such as exposure to the sun, especially that resulting in sunburn. Considerable progress has been made in elucidating the molecu-lar events in MM. Several studies have investigated miRNA expression in MM and the possible involvement of miRNAs in the development and pro-gression of the disease [80]. A number of miRNAs have been described in terms of their differential expression in MM (Figure 9.3). Lu et al. initially described miRNAs in MM when systematically investigating the global miRNA expression across 334 samples, including multiple human cancers [67]. Subsequently, a number of studies have identified both tumor-suppressive and oncogenic miRNAs in MM. For instance, Felicetti et al. identified the promyelocytic leukemia zinc finger (PLZF) transcription factor as a repressor of miR-221 and miR-222 in MM [81]. PLZF silencing in melanomas unblocked miR-221 and miR-222, which in turn controlled the progression of the neoplasia through down-modulation of p27Kip1/CDKN1B and c-KIT receptor, leading to enhanced proliferation and differentiation blockade of the melanoma cells, respectively [81]. *In vitro* and *in vivo* functional studies confirmed the key role of miR-221/-222 in regulating the progression of human melanoma [81]. Chen et al. found 31 miRNAs that were differentially expressed in metastatic melanomas relative to benign nevi [82]. Notably, cyclin D1 (CCND1), a key player involved in the control of the cell-cycle progression, was identified as a direct target of miR-193b and the authors propose that the downregulation of miR-193b in MM may play an important role in melanoma development [82]. Finally, the expression of a number of miRNAs has been reported to be subject to epigenetic silencing [80]. Asangani et al. found that downregulation of miR-31 was a common event in melanoma cell lines and was associated with genomic loss in a subset of MM samples [83]. Furthermore, these researchers showed that the decrease in miR-31 was a result of epigenetic silencing by DNA methylation and via EZH2-mediated histone methylation, together suggesting a tumor-suppressive role of miR-31 in MM [83]. Moreover, Dar et al. reported that the suppression of miR-18b in melanoma cell lines and human samples is due to DNA hypermethylation [84]. Finally, in line with studies from nonmelanoma skin cancers, Zhang et al. were the first to describe the presence of high frequency copy abnormalities in Dicer and Argonaute-2 in primary cultured melanoma cell lines [85]. Sand et al. showed that the miRNA machinery components Argonaute-1, TARBP2, and SND1 are deregulated in MM compared to benign melanocytic nevi [86]. Additionally, Völler et al. observed a strong reduction of Argonaute-2 expression in melanoma, indicating that these components may play a role in the process of malignant transformation [87].

Recently, miRNAs have emerged as new blood biomarkers for various human cancers. Currently, only a few studies have reported the potential of circulating miRNAs in MM. Leidinger et al. initially reported a specific miRNA signature in the blood cells from patients with MM compared with healthy controls [88]. They identified 51 differentially expressed miRNAs and, using a subset of 16 significantly deregulated miRNAs, they demonstrated that blood samples from patients with MM and healthy controls can be well differentiated from each other, achieving a classification accuracy of 97.4% by supervised analysis [88]. Friedman et al. identified a signature consisting of five miRNAs that classified MM patients into high- and low-recurrence risk groups [89]. More recently, Saldanha et al. described miR-21 in plasma from patients with MM as an independent prognostic factor [90]. In addition, miR-21 plasma levels decreased in postoperative samples compared with preoperative samples, suggesting that the plasma miR-21 reflects tumor burden and therefore may have a role in monitoring disease activity [90].

Collectively, the miRNAs have been studied extensively in MM compared to other malignant skin diseases. Apart from specific miRNA signatures and the miRNAs potential as biomarkers in MM, studies have also revealed that miRNAs in MM are subject to epigenetic modulation. This highlights the complexity of miRNAs and emphasizes the need for future studies to increase our understanding of epigenetic regulation in MM.

9.4.4 Cutaneous T-Cell Lymphoma

CTCLs are the most frequent primary lymphomas of the skin [91]. Mycosis fungoides (MF) is the most prevalent clinical form (70%) followed by Sézary syndrome (SzS), which is a more aggressive leukemic T-cell disorder associated with generalized erythroderma and lymphadenopathy. Van Kester et al. provided the first evidence that skin lesions from patients with MF were characterized by a specific miRNA expression profile that was different from that found in benign inflammatory skin diseases including eczema and lichen planus [92]. They identified 49 miRNAs that were differentially expressed in tumor stage MF and the majority of those (30/49) including miR-155, -92a, and -93 were increased [92] (Figure 9.3). Concurrently, the same group elucidated the complete miRNome of purified T cells from patients diagnosed with SzS [93]. Their findings revealed the presence of a specific miRNA expression profile that was distinct from that in CD4$^+$ T cells from healthy controls and B-cell lymphomas. Among the most discriminative up- and downregulated miRNAs were miR-143, -145, -574-5p, and -200c (up) and miR-223, -150, -342, and -181a (down) [93].

STAT (signal transducer and activator of transcription) signaling plays a key role in malignant inflammation in CTCL. Recently, Kopp et al. demonstrated that malignant T-cell lines obtained from patients with CTCL constitutively express high levels of miR-155 and its host gene BIC (B-cell integration cluster) [94]. This is consistent with prior studies reporting increased expression levels of miR-155 in lesions from patients with CTCL compared with lesions from benign inflammatory skin diseases and with healthy controls [92,95]. In addition, Kopp and colleagues showed that BIC is a target of the transcription factor STAT5, which is aberrantly activated in malignant T cells [94]. STAT5 induces expression of oncogenic BIC/miR-155 in cancer and the authors suggest that the STAT5/BIC/miR-155 promotes proliferation of malignant T cells [94]. Using *in situ* hybridization, miR-155 was reported to be expressed in both malignant and nonmalignant

T cells in lesions from MF patients [96]. The endothelial-related miR-126 was included as a control and was as expected confined to endothelial cells and not expressed in T cells with neoplastic morphology [96]. Recently, miR-155 expression was associated with severe disease (MF tumors compared with both early stage MF lesions and controls) and the authors suggest that miR-155 is involved in the switch from the indolent early stage to the aggressive tumor stage of the disease [97]. Given that miR-155 promotes T-cell proliferation [94], miR-155 may play a direct role in disease progression in CTCL.

Other miRNAs that likely contribute to CTCL are miR-122 and miR-223. miR-122 was detectable in MF lesional skin and its expression was confined to both epidermis and the malignant T-cell infiltrates [98]. In addition, miR-122 was present in T cells purified from patients with SzS [98]. *In situ* hybridization showed that miR-122 increased in advanced stage MF and overexpression of miR-122 decreased the sensitivity to chemotherapy-induced apoptosis [98]. The authors describe a negative regulatory loop in which miR-122 is induced by chemotherapeutic agents via p53 and increases cell resistance to apoptosis via stimulation of Akt kinase [98]. Recently, McGirt et al. investigated the role of miR-223 in CTCL and, in accordance with prior studies [93,95], the expression of miR-223 was found to be reduced in MF lesions and further decreased in advanced stage disease and PBMCs from patients with CTCL [99]. Furthermore, the authors demonstrated that miR-223 targets the thymocyte selection-associated high mobility group box (TOX) and the oncogenic proteins E2F1 and MEF2C, and they propose that miR-223 may be an inhibitor of CTCL development and/or progression [99]. Finally, a retrospective study using immunohistochemistry investigated whether the expression level of Dicer, an essential protein involved in the processing of miRNAs, is altered in lesions from patients with CTCL. The results demonstrated that the expression of Dicer was significantly associated with shorter disease-specific survival, constituting a possible molecular prognostic marker for CTCL [100].

Recently, Ralfkiaer et al. provided the first evidence that an miRNA classifier based on only three miRNAs can differentiate CTCL from benign skin disorders [95]. Given that MF, the earliest stage of CTCL, presents initially as patches and plaques resembling benign inflammatory skin lesions such as psoriasis and dermatitis, early diagnosis of MF is a challenge. The authors assessed 148 patient samples, including 63 samples from patients with CTCL and 85 samples from patients with benign lesions, using microarray miRNA profiling [95]. The most induced miRNAs in CTCL were miR-326, -663b, and -711 whereas the most repressed were miR-203 and -205. A qRT-PCR-based classifier consisting of miR-155, -203, and -205 distinguished CTCL from benign skin disorders with high specificity and sensitivity and with a classification accuracy of 95% indicating that miRNAs have a high diagnostic potential in CTCL [95]. This was further supported by validation of the three-miRNA classifier in a new, independent cohort of 78 patients using a standard TaqMan qRT-PCR assay [101]. Here, all but four patients in this large cohort were classified correctly and the minimal miRNA classifier distinguished between malignant and benign patients with a sensitivity of 90%, specificity of 97.9%, and an overall accuracy of 93.7% [101].

9.5 THE CLINICAL APPLICATIONS OF miRNAs

9.5.1 miRNAs as Biomarkers

miRNAs have been shown to be valuable as disease biomarkers. In an effort to improve the use of miRNAs as biomarkers, Renwick et al. developed multicolor miRNA FISH (fluorescence *in situ* hybridization) to enable visualization in formalin-fixed paraffin-embedded (FFPE) tissues [102]. They used BCC and Merkel cell carcinoma (MCC) as models, which share histological features but have distinct cellular origin. Both BCC and MCC have been associated with specific miRNA signatures [65,66,68,103]. Using multicolor FISH Renwick et al. identified tumor-specific miR-205 and miR-375 in BCC and MCC, respectively, and used their specific expression to classify BCC and MCC [102]. Together, the establishment of a reliable method for parallel visualization of differentially expressed miRNAs in FFPE tumor tissues reported by Renwick et al. may be important for future molecular diagnostic purposes [102]. The method must be widely applicable; however, the clinical value needs to be addressed in larger sample collections including samples from other skin diseases.

In 2008, the current concept of intracellular miRNA regulation was challenged when a number of studies reported the presence of circulating miRNAs in the blood [104,105]. Plasma and serum miRNAs were found to be remarkably stable as compared with mRNAs despite the presence of high levels of RNases [104,105]. Notably, specific serum miRNA signatures were identified for patients with prostate cancer and lung cancer, respectively, providing evidence that circulating serum miRNAs contain fingerprints for various diseases which can serve as novel diagnostic biomarkers [104,105]. Since the discovery of miRNAs in blood, miRNAs have been detected in a variety of other human body fluids including saliva, breast milk, cerebrospinal fluid, and urine [106]. Many reports have confirmed the existence of plasma and serum miRNAs and intensive research has made an attempt to explore whether miRNAs can be used as diagnostic and prognostic tools for various human diseases [107]. The substantial interest in circulating miRNAs is reflected in the exponential increase in the number of publications involving miRNAs in the blood (>4,000 since 2008 (www.PubMed.gov)). Still, the origin and biological significance of circulating miRNAs remain unclear; however, most circulating miRNAs may actually derive from blood cells [108], exosomes [109], apoptotic bodies [110], high/low-density lipoprotein [111] and/or RNA-binding proteins [112]. As cells in the blood respond rapidly to changes in their environment [113], miRNA expression levels may show changes immediately after sample collection. Thus, factors such as blood phlebotomy procedure (e.g., duration of stasis) and time before processing of the samples need to be addressed thoroughly before implementation in a clinical setting.

An ideal biomarker is, among other considerations, characterized by a high degree of specificity and sensitivity; easy accessibility, preferably through a minimal or noninvasive method; the capability to distinguish pathologies, allowing early detection; stability and consistency within the sample; and the ability for rapid and accurate detection. The fact that miRNAs offer many features relevant for an ideal biomarker makes them an attractive class of biomarkers. It has been shown that miRNA expression patterns can be used to correctly classify cancers of unknown origin to their primary origin, and that this classification can be more accurate than the classification achieved by using mRNA gene expression patterns [67,114]. miRNAs are also distinct from other biomarkers as their expression is often tissue-specific and the expression patterns are related to disease pathology, suggesting that the miRNAs have a pathogenic role in the disease process and are not only by-products of the disease state. Finally, the ease by which miRNAs can be isolated from tissue, blood, plasma, and serum and detected by methods such as qRT-PCR makes them promising candidates as novel clinical biomarkers.

9.5.2 Importance of miRNAs in Skin Diseases

Since the discovery of miRNAs, intensive research has provided us with new insights into our understanding of gene regulation. Within the field of cutaneous biology, several studies have shown that lesions from different skin diseases are characterized by specific miRNAs signatures. The specific role and importance of these aberrantly expressed miRNAs still remain unclear. Future functional studies including parallel mRNA and protein expression profiling will provide us with an increased understanding of the pathophysiological impact of miRNAs in cellular processes involved in various skin diseases. As skin is easily accessible this may indicate new directions for therapeutic application. The first miRNA-based therapeutic approach has reached a phase IIa clinical trial. Lanford et al. demonstrated that miravirsen (SPC3649,

Santaris Pharma), an LNA-modified oligonucleotide targeting miR-122 which is crucial for the functional infection of hepatitis C virus (HCV), has long-lasting suppressive effect on HCV viremia in chronically HCV-infected chimpanzees without any obvious side effects [115]. Since the report of their nonhuman primate study, miravirsen has gone through two phase I clinical trials (NCT00688012, NCT00979927), successfully proving that the drug is safe even in humans, and now a phase IIa clinical trial (NCT01200420) is underway for patients with treatment-naïve chronic HCV infection to monitor tolerability, safety, pharmacokinetics, and efficacy on HCV viral titer [116]. The results so far are encouraging and provide a landmark breakthrough for miRNA-based therapeutics.

Finally, the vast majority of studies on miRNA expression in skin diseases are confined to skin tissue samples. However, since the discovery of circulating miRNAs many reports have proposed miRNAs as diagnostic biomarkers for various diseases. For a number of skin disorders, e.g., psoriasis, several attempts have been made to identify soluble biomarkers [117]; however, there are still no specific biomarkers that can accurately predict disease progression and therapeutic response [117]. Initial studies indicate that miRNAs indeed have the potential to serve as biomarkers for skin diseases including psoriasis. To establish miRNAs as biomarkers for diagnosis, prognosis, response to treatment, and even biomarkers related to comorbidities in skin diseases, larger prospective studies are warranted. Preferentially, combinations of deregulated miRNAs or known biomarkers may increase specificity and sensitivity of the biomarker, together striving to achieve improved patient management and outcome.

9.6 CONCLUSION

In recent years, intensive research has provided us with an improved understanding of how miRNAs epigenetically modulate gene expression. Even though the role of miRNAs in skin biology is still incomplete, it is now evident that miRNAs indeed are involved in various cellular processes in both normal and diseased skin. Still, only a few skin-specific miRNAs have been described. Until now the majority of studies have characterized specific patterns of miRNA expression levels in diseased skin relative to healthy skin and, thus, both benign inflammatory skin diseases such as psoriasis and AD and malignant skin diseases including BCC and MM have been associated with specific miRNA signatures. Some miRNAs appear to be consistently deregulated across several skin diseases indicating a common role in fundamental cellular processes in these pathologies. For instance, aberrant expression of miRNAs involved in the innate and adaptive immune system have been reported for several inflammatory skin diseases, including miR-21, -142-3p, -146a, -150, -155, -181, -210, and -223, representing both the innate and the adaptive immune system. Mounting evidence has demonstrated the importance of miRNAs in normal immune cell functions, and the aberrant expressions seen in many inflammatory skin diseases may be related to infiltration of immune cells and disruption of normal immune cell development, homeostasis, and function. On the other hand, the reported aberrantly expressed miRNAs within the same skin diseases is often only partially consistent. One explanation for this inconsistency could be the use of whole tissue extracts, which, for example, contain various degrees of infiltrating immune cells in dermis and epidermis and do not allow the study of miRNA expression in specific cells and regions. Indeed, most previous studies concerning global miRNA expression in different skin diseases were based on whole tissue extracts including both epidermis and dermis and, thus, represent global miRNA changes from a mixture of cells. Using LCM to isolate epidermal and dermal cells from psoriatic skin, respectively, we found that the enrichment of specific cells and, thereby, the additional information about aberrantly expressed miRNAs which can be achieved, is favorable and we thereby suggest that LCM should be considered as a supplement to whole tissue skin samples aiming to investigate miRNA expression in skin diseases [33]. Hence, the current descriptive studies provide a basis for further investigations which should cover analysis of specific subsets of cells and more functional and mechanistic studies involving antagomirs and miRNA mimics. In addition, to improve our understanding of the role of miRNAs in skin diseases, future studies should address whether the aberrant miRNA expression observed in skin diseases is causative or is simply a result of altered gene expression in specific cells. Given that the majority of miRNAs are transcribed by RNA polymerase II, they are themselves subject to regulation, including epigenetic modifications by chromatin remodeling and DNA methylation. Aberrant expression of components involved in miRNA biogenesis may also influence the miRNA signatures and, together with the fact that each miRNA may regulate numerous mRNAs, the complexity of miRNAs is indeed evident and the current understanding of the exact role of miRNAs in skin physiology and pathology still remain unclear.

Recently, the miRNAs have proven extremely valuable as biomarkers in various diseases. This was supported by Ralfkiaer et al., who provided the first evidence that an miRNA classifier can distinguish CTCL from benign inflammatory skin diseases with very high accuracy and a high level of robustness [95]. The clinical applications of miRNAs are indeed appealing. miRNA detection in the skin is robust irrespective of preservation method [30,118,119], enabling retrospective studies using, for instance, archived FFPE samples. Furthermore, the miRNAs can be quantitatively measured using standard procedures such as qRT-PCR and microarray and their tissue distribution can be detected using *in situ* hybridization [102]. Lately, intensive research has focused on the potential of circulating miRNAs as diagnostic and prognostic biomarkers. In relation to skin diseases, initial findings indicate that, for example, blood from patients with psoriasis is characterized by a specific miRNA signature and that aberrant expression of specific miRNAs may be used as a biomarker for psoriasis activity [50]. Given that circulating miRNAs were initially described in 2008 [104,105], many questions and challenges still remain. As miRNAs are considered to be "fine-tuners" of gene regulation resulting in often relatively smallfold changes, large-scale studies are warranted to increase power and establish miRNAs as reliable blood biomarkers for skin diseases in future clinical settings. Finally, although promising studies have emerged concerning miRNA-based therapeutics, only very few miRNA-based therapies have reached the clinical trial stage [116]. A recent study demonstrated that antagomiR-21 ameliorated disease pathology in patient-derived psoriatic skin xenotransplants in mice [45]. The results are encouraging and future studies will reveal the potential of miRNAs as novel therapeutic targets in skin diseases.

List of Abbreviations

AD	atopic dermatitis
ACD	allergic contact dermatitis
AGO	Argonaute
AK	actinic keratosis
BCC	basal cell carcinoma
BIC	B-cell integration cluster
CTCL	cutaneous T-cell lymphoma
DCs	dendritic cells
DGCR8	DiGeorge syndrome critical region gene 8
DPCP	diphenylcyclopropenone
dsRNA	double-stranded RNA
Exp-5	Exportin 5
FFPE	formalin-fixed, paraffin-embedded
HCV	hepatitis C virus
KCs	keratinocytes
LCM	laser capture microdissection
LNA	locked nucleic acid
MCC	Merkel cell carcinoma
MF	mycosis fungoides
miR-	microRNA
miRNA	microRNA
MM	malignant melanoma
MTX	methotrexate
NGS	next-generation sequencing
NM	nodular melanoma

PACT	protein activator of PKR
PBMCs	peripheral blood mononuclear cells
qRT-PCR	quantitative real-time polymerase chain reaction
RISC	RNA-induced silencing complex
SCC	squamous cell carcinoma
SLE	systemic lupus erythematosus
SSc	scleroderma (systemic sclerosis)
SSM	superficial spreading melanoma
STAT	signal transducers and activators of transcription
SzS	Sézary syndrome
TRBP	TAR RNA-binding protein
UTR	untranslated region
VEGF	vascular endothelial growth factor

References

[1] Ambros V, Lee RC, Lavanway A, Williams PT, Jewell D. MicroRNAs and other tiny endogenous RNAs in *C. elegans*. Curr Biol 2003;13:807–18.
[2] Bartel DP. MicroRNAs: genomics, biogenesis, mechanism, and function. Cell 2004;116(2):281–97.
[3] Bartel DP. MicroRNAs: target recognition and regulatory functions. Cell 2009;136 (2):215–33.
[4] Lee RC, Feinbaum RL, Ambros V. The *C. elegans* heterochronic gene lin-4 encodes small RNAs with antisense complementarity to lin-14. Cell 1993;75:843–54.
[5] Reinhart BJ, Slack FJ, Basson M, Pasquinelli AE, Bettinger JC, Rougvie AE, et al. The 21-nucleotide let-7 RNA regulates developmental timing in *Caenorhabditis elegans*. Nature 2000;403(6772):901–6.
[6] Pasquinelli AE, Reinhart BJ, Slack F, Martindale MQ, Kuroda MI, Maller B, et al. Conservation of the sequence and temporal expression of let-7 heterochronic regulatory RNA. Nature 2000;408(6808):86–9.
[7] Friedman RC, Farh KK, Burge CB, Bartel DP. Most mammalian mRNAs are conserved targets of miRNAs. Genome Res 2009;19:92–105.
[8] Finnegan EF, Pasquinelli AE. MicroRNA biogenesis: regulating the regulators. Crit Rev Biochem Mol Biol 2013;48(1):51–68.
[9] Lee Y, Kim M, Han J, Yeom KH, Lee S, Baek SH, et al. MicroRNA genes are transcribed by RNA polymerase II. EMBO J 2004;23:4051–60.
[10] Lund E, Güttinger S, Calado A, Dahlberg JE, Kutay U. Nuclear export of microRNA precursors. Science 2004;303:95–8.
[11] Hutvanger G, Zamore PD. A microRNA in a multiple turnover RNAi enzyme complex. Science 2002;297:2056–60.
[12] Lewis BP, Burge CB, Bartel DP. Conserved seed pairing, often flanked by adenosines, indicates that thousands of human genes are microRNA targets. Cell 2005;120:15–20.
[13] Huntzinger E, Izaurralde E. Gene silencing by microRNAs: contributions of translational repression and mRNA decay. Nat Rev Genet 2011;12:99–110.
[14] Filipowicz W, Bhattacharyy SN, Sonenberg N. Mechanisms of post-transcriptional regulation by microRNAs: are the answers in sight? Nat Rev Genet 2008;9:102–14.
[15] Kloosterman WP, Plasterk RH. The diverse functions of microRNAs in animal development and disease. Dev Cell 2006;11(4):441–50.
[16] Baek D, Villén J, Shin C, Camargo FD, Gygi SP, Bartel DP. The impact of microRNAs on protein output. Nature 2008;455:64–71.
[17] Vlachos IS, Hatzigeorgiou AG. Online resources for miRNA analysis. Clin Biochem 2013;46:879–900.

[18] Poliseno L, Salmena L, Zhang J, Carver B, Haveman WJ, Pandolfi PP. A coding-independent function of gene and pseudogene mRNAs regulates tumour biology. Nature 2010;465(7301):1033–8.

[19] Vasudevan S, Tong Y, Steitz JA. Switching from repression to activation: microRNAs can up-regulate translation. Science 2007;318:1931–4.

[20] Lytle JR, Yario TA, Steitz JA. Target mRNAs are repressed as efficiently by microRNA-binding sites in the 5′ UTR as in the 3′ UTR. Proc Natl Acad Sci USA 2007;104(23):9667–72.

[21] Kong YW, Ferland-McCollough D, Jackson TJ, Bushell M. MicroRNAs in cancer management. Lancet Oncol 2012;13(6):e249–58. http://dx.doi.org/10.1016/S1470-2045(12)70073-6.

[22] O'Connell RM, Rao DS, Chaudhuri AA, Baltimore D. Physiological and pathological roles for microRNA in the immune system. Nat Rev Immunol 2010;10:111–22.

[23] Andl T, Murchison EP, Liu F, Zhang Y, Yunta-Gonzalez M, Tobias JW, et al. The miRNA-processing enzyme dicer is essential for the morphogenesis and maintenance of hair follicles. Curr Biol 2006;16:1041–9.

[24] Yi R, O'Carroll D, Pasolli HA, Zhang Z, Dietrich FS, Tarakhovsky A, et al. Morphogenesis in skin is governed by discrete sets of differentially expressed miRNAs. Nat Genet 2006;38:356–62.

[25] Sonkoly E, Wei T, Janson PC, Sääf A, Lundeberg L, Tengvall-Linder M, et al. MicroRNAs: novel regulators involved in the pathogenesis of psoriasis? PLOS ONE 2007;2(7):e610.

[26] Yi R, Poy MN, Stoffel M, Fuchs E. A skin microRNA promotes differentiation by repressing "stemness." Nature 2008;452(7184):225–9.

[27] Lena AM, Shalom-Feuerstein R, Rivetti di Val Cervo P, Aberdam D, Knight RA, Melino G, et al. miR-203 represses 'stemness' by repressing DeltaNp63. Cell Death Differ 2008;15(7):1187–95.

[28] Lowes MA, Suárez-Fariñas M, Krueger JG. Immunology of psoriasis. Annu Rev Immunol 2014;32:227–55.

[29] Zibert JR, Løvendorf MB, Litman T, Olsen J, Kaczkowski B, Skov L. MicroRNAs and potential target interactions in psoriasis. J Dermatol Sci 2010;58(3):177–85.

[30] Løvendorf MB, Zibert JR, Hagedorn PH, Glue C, Ødum N, Røpke MA, et al. Comparison of microRNA expression using different preservation methods of matched psoriatic skin samples. Exp Dermatol 2012;21:299–319.

[31] Joyce CE, Zhou X, Xia J, Ryan C, Thrash B, Menter A, et al. Deep sequencing of small RNAs from human skin reveals major alterations in the psoriasis miRNAome. Hum Mol Genet 2011;20(20):4025–40.

[32] Xia J, Joyce CE, Bowcock AM, Zhang W. Noncanonical microRNAs and endogenous siRNAs in normal and psoriatic human skin. Hum Mol Genet 2013;22(4):737–48.

[33] Løvendorf MB, Mitsui H, Zibert JR, Røpke MA, Hafner M, Dyring-Andersen B, et al. Laser capture microdissection followed by next-generation sequencing identifies disease-related microRNAs in psoriatic lesions that reflect systemic microRNA changes in psoriasis. Exp Dermatol 2014 Nov 28. http://dx.doi.org/10.1111/exd.12604. [Epub ahead of print].

[34] Gregory PA, Bracken CP, Bert AG, Goodall GJ. MicroRNAs as regulators of epithelial-mesenchymal transition. Cell Cycle 2008;7:3112–18.

[35] Lerman G, Avivi C, Mardoukh C, Barzilai A, Tessone A, Gradus B, et al. MiRNA expression in psoriatic skin: reciprocal regulation of hsa-miR-99a and IGF-1R. PLOS ONE 2011;6(6):e20916.

[36] Wraight CJ, White PJ, McKean SC, Fogarty RD, Venables DJ, Liepe IJ, et al. Reversal of epidermal hyperproliferation in psoriasis by insulin-like growth factor 1 receptor antisense oligonucleotides. Nat Biotechnol 2000;18:521–6.

[37] Xu N, Brodin P, Wei T, Meisgen F, Eidsmo L, Nagy N, et al. MiR-125b, a MicroRNA downregulated in psoriasis, modulates keratinocyte proliferation by targeting FGFR2. J Invest Dermatol 2011;131:1521–9.

[38] Rossi RL, Rossetti G, Wenandy L, Curti S, Ripamonti A, Bonnal RJ, et al. Distinct microRNA signatures in human lymphocyte subsets and enforcement of the naive state in CD4+ T cells by the microRNA miR-125b. Nat Immunol 2011;12:796–803.

[39] Curtale G, Citarella F. Dynamic nature of noncoding RNA regulation of adaptive immune responses. Int J Mol Sci 2013;14:17347–77.

[40] Allantaz F, Cheng DT, Bergauer T, Ravindran P, Rossier MF, Ebeling M, et al. Expression profiling of human immune cell subsets identifies miRNA-mRNA regulatory relationships correlated with cell type specific expression. PLOS ONE 2012;7:e29979.

[41] Kozubek J, Ma Z, Fleming E, Duggan T, Wu R, Shin DG, et al. In-depth characterization of microRNA transcriptome in melanoma. PLOS ONE 2013;8(9):e72699. http://dx.doi.org/10.1371/journal.pone.0072699.

[42] Zaba LC, Cardinale I, Gilleaudeau P, Sullivan-Whalen M, Suárez-Fariñas M, Fuentes-Duculan J, et al. Amelioration of epidermal hyperplasia by TNF inhibition is associated with reduced Th17 responses. J Exp Med 2007;204:3183–94.

[43] Turner ML, Schnorfeil FM, Brocker T. MicroRNAs regulate dendritic cell differentiation and function. J Immunol 2011;187:3911–17.

[44] Wang S, Olson EN. AngiomiRs—key regulators of angiogenesis. Curr Opin Genet Dev 2009;19:205–11.

[45] Guinea-Viniegra J, Jiménez M, Schonthaler HB, Navarro R, Delgado Y, Concha-Garzón MJ, et al. Targeting miR-21 to treat psoriasis. Sci Transl Med 2014;6 (225):225re1. http://dx.doi.org/10.1126/scitranslmed.3008089.

[46] Meisgen F, Xu N, Wei T, Janson PC, Obad S, Broom O, et al. Mir-21 is upregulated in psoriasis and suppresses T cell apoptosis. Exp Dermatol 2012;21:312–14.

[47] Oyama R, Jinnin M, Kakimoto A, Kanemaru H, Ichihara A, Fujisawa A, et al. Circulating miRNA associated with TNF-a signaling pathway in patients with plaque psoriasis. J Dermatol Sci 2011;61:206–17.

[48] Ichihara A, Jinnin M, Oyama R, Yamane K, Fujisawa A, Sakai K, et al. Increased serum levels of miR-1266 in patients with psoriasis vulgaris. Eur J Dermatol 2012;22:68–71.

[49] Pivarcsi A, Meisgen F, Xu N, Ståhle M, Sonkoly E. Changes in the level of serum miRNAs in patients with psoriasis after antitumour necrosis factor-α therapy. Br J Dermatol 2013;169:563–70.

[50] Løvendorf MB, Zibert JR, Gyldenløve M, Røpke MA, Skov L. MicroRNA-223 and miR-143 are important systemic biomarkers for disease activity in psoriasis. J Dermatol Sci 2014;75(2):133–9.

[51] Hirao H, Jinnin M, Ichihara A, Fujisawa A, Makino K, Kajihara I, et al. Detection of hair root miR-19a as a novel diagnostic marker for psoriasis. Eur J Dermatol 2013;23 (6):807–11.

[52] Tsuru Y, Jinnin M, Ichihara A, Fujisawa A, Moriya C, Sakai K, et al. miR-424 levels in hair shaft are increased in psoriatic patients. J Dermatol 2014;41(5):382–5. http://dx.doi.org//10.1111/1346-8138.12460.

[53] Sonkoly E, Janson P, Majuri ML, Savinko T, Fyhrquist N, Eidsmo L, et al. MiR-155 is overexpressed in patients with atopic dermatitis and modulates T-cell proliferative responses by targeting cytotoxic T lymphocyte-associated antigen 4. J Allergy Clin Immunol 2010;126(3):581–9.

[54] Rebane A, Runnel T, Aab A, Maslovskaja J, Rückert B, Zimmermann M, et al. MicroRNA-146a alleviates chronic skin inflammation in atopic dermatitis through suppression of innate immune responses in keratinocytes. J Allergy Clin Immunol 2014;134(4):836–47. http://dx.doi.org/10.1016/j.jaci.2014.05.022.

[55] Vennegaard MT, Bonefeld CM, Hagedorn PH, Bangsgaard N, Løvendorf MB, Odum N, et al. Allergic contact dermatitis induces upregulation of identical microRNAs in humans and mice. Contact Dermatitis 2012;67(5):298–305.

[56] Gulati N, Løvendorf MB, Zibert JR, Levis WR, Renwick N, Krueger JG. Unique miRNAs appear at different times during the course of a delayed-type hypersensitivity (DTH) reaction in human skin. Abstr. #050, Annual meeting May 2014, Soc Invest Dermatol (SID), Albuquerque NM, USA.

[57] Maurer B, Stanczyk J, Jüngel A, Akhmetshina A, Trenkmann M, Brock M, et al. MicroRNA-29a, a key regulator of collagen expression in systemic sclerosis. Arthritis Rheum 2010;62(6):1733–43.

[58] Li H, Yang R, Fan X, Gu T, Zhao Z, Chang D, et al. MicroRNA array analysis of microRNAs related to systemic scleroderma. Rheumatol Int 2012;32:307–13.

[59] Babalola O, Mamalis A, Lev-Tov H, Jagdeo J. The role of microRNAs in skin fibrosis. Arch Dermatol Res 2013;305:763–76.

[60] Makino K, Jinnin M, Kajihara I, Honda N, Sakai K, Masuguchi S, et al. Circulating miR-142-3p levels in patients with systemic sclerosis. Clin Exp Dermatol 2012;37(1):34–9.

[61] Dai Y, Huang YS, Tang M, Lv TY, Hu CX, Tan YH, et al. Microarray analysis of microRNA expression in peripheral blood mononuclear cells of systemic lupus erythematosus patients. Lupus 2007;16(12):939–46.

[62] Yan S, Yim LY, Lu L, Lau CS, Chan VSF. MicroRNA regulation in systemic lupus erythematosus pathogenesis. Immune Netw 2014;14(3):138–48.

[63] Miao CG, Yang YY, He X, Huang C, Huang Y, Zhang L, et al. The emerging role of microRNAs in the pathogenesis of systemic lupus erythematosus. Cell Signal 2013;25 (9):1828–36.

[64] Epstein EH. Basal cell carcinomas: attack of the hedgehog. Nat Rev Cancer 2008;8:743–54.

[65] Heffelfinger C, Ouyang Z, Engberg A, Leffell DJ, Hanlon AM, Gordon PB, et al. Correlation of global MicroRNA expression with basal cell carcinoma subtype. G3 (Bethesda) 2012;2(2):279–86.

[66] Sonkoly E, Lovén J, Xu N, Meisgen F, Wei T, Brodin P, et al. MicroRNA-203 functions as a tumor suppressor in basal cell carcinoma. Oncogenesis 2012;1:e3. http://dx.doi.org/10.1038/oncsis.2012.3.

[67] Lu J, Getz G, Miska EA, Alvarez-Saavedra E, Lamb J, Peck D, et al. MicroRNA expression profiles classify human cancers. Nature 2005;435(7043):834–8.

[68] Sand M, Skrygan M, Sand D, Georgas D, Hahn SA, Gambichler T, et al. Expression of microRNAs in basal cell carcinoma. Br J Dermatol 2012;167(4):847–55.

[69] Ratushny V, Gober MD, Hick R, Ridky TW, Seykora JT. From keratinocyte to cancer: the pathogenesis and modeling of cutaneous squamous cell carcinoma. J Clin Invest 2012;122(2):464–72.

[70] Dziunycz P, Iotzova-Weiss G, Eloranta JJ, Läuchli S, Hafner J, French LE, et al. Squamous cell carcinoma of the skin shows a distinct microRNA profile modulated by UV radiation. J Invest Dermatol 2010;130(11):2686–9.

[71] Gu X, Nylander E, Coates PJ, Nylander K. Effect of narrow-band ultraviolet B phototherapy on p63 and microRNA (miR-21 and miR-125b) expression in psoriatic epidermis. Acta Derm Venereol 2011;91(4):392–7.

[72] Xu N, Zhang L, Meisgen F, Harada M, Heilborn J, Homey B, et al. MicroRNA-125b down-regulates matrix metallopeptidase 13 and inhibits cutaneous squamous cell carcinoma cell proliferation, migration, and invasion. J Biol Chem 2012;287 (35):29899–908.

[73] Sand M, Skrygan M, Georgas D, Sand D, Hahn SA, Gambichler T, et al. Microarray analysis of microRNA expression in cutaneous squamous cell carcinoma. J Dermatol Sci 2012;68(3):119–26.

[74] Yamane K, Jinnin M, Etoh T, Kobayashi Y, Shimozono N, Fukushima S, et al. Downregulation of miR-124/-214 in cutaneous squamous cell carcinoma mediates abnormal cell proliferation via the induction of ERK. J Mol Med (Berl) 2013;91(1):69–81.

[75] Zhou M, Liu W, Ma S, Cao H, Peng X, Guo L, et al. A novel onco-miR-365 induces cutaneous squamous cell carcinoma. Carcinogenesis 2013;34(7):1653—9.

[76] Gastaldi C, Bertero T, Xu N, Bourget-Ponzio I, Lebrigand K, Fourre S, et al. miR-193b/365a cluster controls progression of epidermal squamous cell carcinoma. Carcinogenesis 2014;35(5):1110—20.

[77] Sand M, Gambichler T, Skrygan M, Sand D, Scola N, Altmeyer P, et al. Expression levels of the microRNA processing enzymes Drosha and dicer in epithelial skin cancer. Cancer Invest 2010;28(6):649—53.

[78] Sand M, Skrygan M, Georgas D, Arenz C, Gambichler T, Sand D, et al. Expression levels of the microRNA maturing microprocessor complex component DGCR8 and the RNA-induced silencing complex (RISC) components argonaute-1, argonaute-2, PACT, TARBP1, and TARBP2 in epithelial skin cancer. Mol Carcinog 2012;51 (11):916—22.

[79] Duncan LM. The classification of cutaneous melanoma. Hematol Oncol Clin North Am 2009;23(3):501—13.

[80] Sun V, Zhou WB, Majid S, Kashani-Sabet M, Dar AA. MicroRNA-mediated regulation of melanoma. Br J Dermatol 2014;171(2):234—41. http://dx.doi.org/10.1111/bjd.12989.

[81] Felicetti F, Errico MC, Bottero L, Segnalini P, Stoppacciaro A, Biffoni M, et al. The promyelocytic leukemia zinc finger-microRNA-221/-222 pathway controls melanoma progression through multiple oncogenic mechanisms. Cancer Res 2008;68(8):2745—54.

[82] Chen J, Feilotter HE, Paré GC, Zhang X, Pemberton JG, Garady C, et al. MicroRNA-193b represses cell proliferation and regulates cyclin D1 in melanoma. Am J Pathol 2010;176(5):2520—9.

[83] Asangani IA, Harms PW, Dodson L, Pandhi M, Kunju LP, Maher CA, et al. Genetic and epigenetic loss of microRNA-31 leads to feed-forward expression of EZH2 in melanoma. Oncotarget 2012;3(9):1011—25.

[84] Dar AA, Majid S, Rittsteuer C, de Semir D, Bezrookove V, Tong S, et al. The role of miR-18b in MDM2-p53 pathway signaling and melanoma progression. J Natl Cancer Inst 2013;105(6):433—42.

[85] Zhang L, Huang J, Yang N, Greshock J, Megraw MS, Giannakakis A, et al. MicroRNAs exhibit high frequency genomic alterations in human cancer. Proc Natl Acad Sci USA 2006;103(24):9136—41.

[86] Sand M, Skrygan M, Georgas D, Sand D, Gambichler T, Altmeyer P, et al. The miRNA machinery in primary cutaneous malignant melanoma, cutaneous malignant melanoma metastases and benign melanocytic nevi. Cell Tissue Res 2012;350 (1):119—26.

[87] Völler D, Reinders J, Meister G, Bosserhoff AK. Strong reduction of AGO2 expression in melanoma and cellular consequences. Br J Cancer 2013;109(12):3116—24.

[88] Leidinger P, Keller A, Borries A, Reichrath J, Rass K, Jager SU, et al. High-throughput miRNA profiling of human melanoma blood samples. BMC Cancer 2010;10:262. http://dx.doi.org/10.1186/1471-2407-10-262.

[89] Friedman EB, Shang S, de Miera EV, Fog JU, Teilum MW, Ma MW, et al. Serum microRNAs as biomarkers for recurrence in melanoma. J Transl Med 2012;10:155. http://dx.doi.org/10.1186/1479-5876-10-155.

[90] Saldanha G, Potter L, Shendge P, Osborne J, Nicholson S, Yii N, et al. Plasma microRNA-21 is associated with tumor burden in cutaneous melanoma. J Invest Dermatol 2013;133(5):1381—4.

[91] Bradford PT, Devesa SS, Anderson WF, Toro JR. Cutaneous lymphoma incidence patterns in the United States: a population-based study of 3884 cases. Blood 2009;113 (21):5064—73.

[92] van Kester MS, Ballabio E, Benner MF, Chen XH, Saunders NJ, van der Fits L, et al. miRNA expression profiling of mycosis fungoides. Mol Oncol 2011;5(3):273—80.

[93] Ballabio E, Mitchell T, van Kester MS, Taylor S, Dunlop HM, Chi J, et al. MicroRNA expression in Sezary syndrome: identification, function, and diagnostic potential. Blood 2010;116(7):1105–13.

[94] Kopp KL, Ralfkiaer U, Gjerdrum LM, Helvad R, Pedersen IH, Litman T, et al. STAT5-mediated expression of oncogenic miR-155 in cutaneous T-cell lymphoma. Cell Cycle 2013;12(12):1939–47.

[95] Ralfkiaer U, Hagedorn PH, Bangsgaard N, Løvendorf MB, Ahler CB, Svensson L, et al. Diagnostic microRNA profiling in cutaneous T-cell lymphoma (CTCL). Blood 2011;118(22):5891–900.

[96] Kopp KL, Ralfkiaer U, Nielsen BS, Gniadecki R, Woetmann A, Ødum N, et al. Expression of miR-155 and miR-126 in situ in cutaneous T-cell lymphoma. APMIS 2013;121(11):1020–4.

[97] Moyal L, Barzilai A, Gorovitz B, Hirshberg A, Amariglio N, Jacob-Hirsch J, et al. miR-155 is involved in tumor progression of mycosis fungoides. Exp Dermatol 2013;22(6):431–3.

[98] Manfè V, Biskup E, Rosbjerg A, Kamstrup M, Skov AG, Lerche CM, et al. miR-122 regulates p53/Akt signalling and the chemotherapy-induced apoptosis in cutaneous T-cell lymphoma. PLOS ONE 2012;7(1):e29541. http://dx.doi.org/10.1371/journal.pone.0029541.

[99] McGirt LY, Adams CM, Baerenwald DA, Zwerner JP, Zic JA, Eischen CM. miR-223 regulates cell growth and targets proto-oncogenes in mycosis fungoides/cutaneous T-cell lymphoma. J Invest Dermatol 2014;134(4):1101–7.

[100] Valencak J, Schmid K, Trautinger F, Wallnöfer W, Muellauer L, Soleiman A, et al. High expression of Dicer reveals a negative prognostic influence in certain subtypes of primary cutaneous T cell lymphomas. J Dermatol Sci 2011;64(3):185–90.

[101] Marstrand T, Ahler CB, Ralfkiaer U, Clemmensen A, Kopp KL, Sibbesen NA, et al. Validation of a diagnostic microRNA classifier in cutaneous T-cell lymphomas. Leuk Lymphoma 2014;55(4):957–8.

[102] Renwick N, Cekan P, Masry PA, McGeary SE, Miller JB, Hafner M, et al. Multicolor microRNA FISH effectively differentiates tumor types. J Clin Invest 2013;123(6):2694–702.

[103] Ning MS, Kim AS, Prasad N, Levy SE, Zhang H, Andl T. Characterization of the Merkel cell carcinoma miRNome. J Skin Cancer 2014;289548. http://dx.doi.org/10.1155/2014/289548.

[104] Chen X, Ba Y, Ma L, Cai X, Yin Y, Wang K, et al. Characterization of microRNAs in serum: a novel class of biomarkers for diagnosis of cancer and other diseases. Cell Res 2008;18:997–1006.

[105] Mitchell PS, Parkin RK, Kroh EM, Fritz BR, Wyman SK, Pogosova-Agadjanyan EL, et al. Circulating microRNAs as stable blood-based markers for cancer detection. Proc Natl Acad Sci USA 2008;105:10513–18.

[106] Kosaka N, Iguchi H, Ochiya T. Circulating microRNA in body fluid: a new potential biomarker for cancer diagnosis and prognosis. Cancer Sci 2010;101:2087–92.

[107] Reid G, Kirschner MB, Zandwijk NV. Circulating microRNAs: association with disease and potential use as biomarkers. Crit Rev Oncol Hematol 2011;80:193–208.

[108] Pritchard CC, Kroh E, Wood B, Arroyo JD, Dougherty KJ, Miyaji MM, et al. Blood cell origin of circulating microRNAs: a cautionary note for cancer biomarker studies. Cancer Prev Res 2012;5(3):492–7.

[109] Valadi H, Ekström K, Bossios A, Sjöstrand M, Lee JJ, Lötvall JO. Exosome-mediated transfer of mRNAs and microRNAs is a novel mechanism of genetic exchange between cells. Nat Cell Biol 2007;9(6):654–9.

[110] Zernecke A, Bidzhekov K, Noels H, Shagdarsuren E, Gan L, Denecke B, et al. Delivery of microRNA-126 by apoptotic bodies induces CXCL12-dependent vascular protection. Sci Signal 2009;2(100):ra81. http://dx.doi.org/10.1126/scisignal.2000610.

[111] Vickers KC, Palmisano BT, Shoucri BM, Shamburek RD, Remaley AT. MicroRNAs are transported in plasma and delivered to recipient cells by high-density lipoproteins. Nat Cell Biol 2011;13(4):423−33.

[112] Arroyo JD, Chevillet JR, Kroh EM, Ruf IK, Pritchard CC, Gibson DF, et al. Argonaute2 complexes carry a population of circulating microRNAs independent of vesicles in human plasma. Proc Natl Acad Sci USA 2011;108(12):5003−8.

[113] Liew CC, Ma J, Tang HC, Zheng R, Dempsey AA. The peripheral blood transcriptome dynamically reflects system wide biology: a potential diagnostic tool. J Lab Clin Med 2006;147(3):126−32.

[114] Rosenfeld N, Aharonov R, Meiri E, Rosenwald S, Spector Y, Zepeniuk M, et al. MicroRNAs accurately identify cancer tissue origin. Nat Biotechnol 2008;26 (4):462−9.

[115] Lanford RE, Hildebrandt-Eriksen ES, Petri A, Persson R, Lindow M, Munk ME, et al. Therapeutic silencing of microRNA-122 in primates with chronic hepatitis C virus infection. Science 2010;327:198−201.

[116] Van Rooij E, Kauppinen S. Development of microRNA therapeutics is coming of age. EMBO Mol Med 2014;6(7):851−64.

[117] Villanova F, Meglio PD, Nestlé FO. Biomarkers in psoriasis and psoriatic arthritis. Ann Rheum Dis 2013;72(Suppl. 2):ii104−10. http://dx.doi.org/10.1136/annrheumdis-2012-203037.

[118] Liu A, Tetzlaff MT, Vanbelle P, Elder D, Feldman M, Tobias JW, et al. MicroRNA expression profiling outperforms mRNA expression profiling in formalin-fixed paraffin-embedded tissues. Int J Clin Exp Pathol 2009;2:519−27.

[119] Glud M, Klausen M, Gniadecki R, Rossing M, Hastrup N, Nielsen FC, et al. MicroRNA expression in melanocytic nevi: the usefulness of formalin-fixed, paraffin-embedded material for miRNA microarray profiling. J Invest Dermatol 2009;129:1219−24.

IMMUNOLOGIC SKIN DISEASES

10

Systemic Lupus Erythematosus

Bruce C. Richardson

Division of Rheumatology, Department of Internal Medicine, University of
Michigan, Ann Arbor, MI

10.1 INTRODUCTION

Systemic lupus erythematosus (lupus, SLE) is a chronic relapsing autoimmune disease primarily affecting women. Lupus is characterized by the production of autoantibodies to nuclear molecules and other cellular antigens, resulting in immune complex formation and deposition in the skin, kidneys, and other organs, causing inflammation and organ damage. Lupus develops and flares when genetically predisposed people encounter environmental agents such as ultraviolet light, silica, and infectious agents, all of which cause oxidative stress [1–3]. Certain drugs can also cause a lupus-like disease in genetically predisposed people [4], and more recently dietary deficiencies have been implicated in lupus flares [5]. How these and other environmental agents initiate lupus flares in genetically predisposed people has been unclear. However, a substantial body of evidence indicates that environmental agents that inhibit DNA methylation and perhaps histone modifications and microRNAs (miRNAs) can initiate lupus flares by epigenetically altering gene expression in immune cells. How environmental agents alter immune cells to trigger lupus flares in genetically predisposed people is reviewed below.

10.2 BACKGROUND

The cutaneous manifestations of lupus were first described by Cazenave in 1851, the systemic manifestations by Kaposi in 1872, and the systemic and cutaneous manifestations by Sir William Osler in the late 1890s [6]. Lupus affects women approximately 9 times more often

© 2015 Elsevier Inc. All rights reserved.

than men, and it usually strikes women of childbearing age, although the disease can present in children as well as in adults [5]. The clinical manifestations vary widely, but common symptoms include oral ulcers, Raynaud's phenomenon, alopecia, pleurisy, pericarditis, an inflammatory symmetric polyarthritis involving the hands, wrists, and feet, an immune complex-mediated glomerulonephritis, a variety of neurologic disorders, and a variety of autoantibody-mediated hematologic abnormalities such as thrombocytopenia, leukopenia, and hemolytic anemia [6]. Lupus patients may also develop a variety of skin rashes, including malar erythema, a diffuse maculopapular rash, scarring discoid lesions, and a panniculitis [7]. Because of this variability, at least 4 of 13 distinct lupus manifestations are required to classify a patient as having lupus for study purposes [8].

Despite this variability in presentation and disease course, all lupus patients have autoantibodies to components of the nucleus including histones and DNA, generally referred to as antinuclear antibodies (ANAs) [6]. ANAs are not specific for lupus however, and can be found in patients with other autoimmune diseases as well [9]. ANAs also develop with aging in otherwise healthy people [10]. These false positives make the ANA a sensitive but relatively nonspecific test [11]. Reports that some drugs cause a lupus-like disease in genetically predisposed people but only ANAs in people lacking the necessary genes [4], and that infections and sun exposure trigger flares in lupus patients, support the contention that both an environmental contribution and a genetic predisposition are necessary for lupus to develop and flare. How the environment alters the immune system to cause lupus though has been unclear. However, a large body of evidence acquired over the past 20 years or so indicates that environmental agents may contribute to lupus development and flares in genetically predisposed people by inhibiting T-cell DNA methylation, and that oxidative stress plays an important role in initiating lupus flares. More recently, other epigenetic mechanisms including histone modifications and miRNAs have been studied in human lupus and may also contribute to disease onset and flares. These studies are reviewed in this chapter.

10.3 DNA METHYLATION, GENE EXPRESSION, AND T-CELL FUNCTION

DNA methylation refers to the postsynthetic methylation of deoxycytosine (dC) bases, primarily in CG pairs, to form deoxymethylcytosine (d^mC) and is a transcriptionally repressive modification. Gene expression

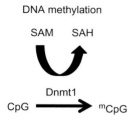

DNA methylation

SAM   SAH

Dnmt1

CpG ⟶ ^{m}CpG

FIGURE 10.1 **ERK pathway signaling and Dnmt1.** Signals transmitted through the ERK pathway upregulate Dnmt1 and T cells enter mitosis.

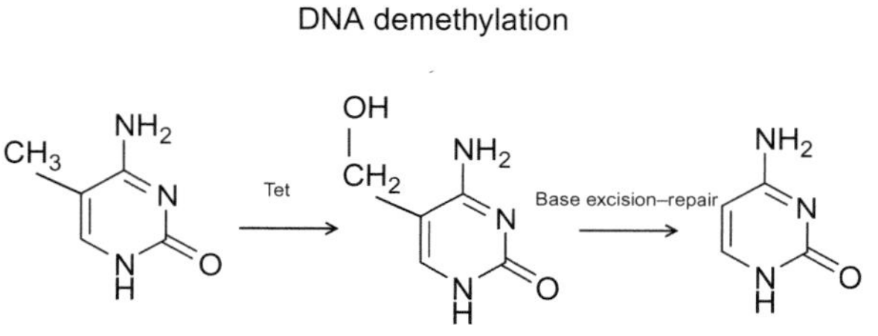

DNA demethylation

FIGURE 10.2 **DNA methylation reaction.** Dnmt1 catalyzes transfer of the methyl group from SAM to dC bases in CpG pairs, producing d^{m}C and SAH.

patterns change during cellular differentiation, and those genes unnecessary or detrimental to the function of a mature cell are silenced in part by DNA methylation. One X chromosome is also silenced in part by DNA methylation in women [12]. DNA methylation patterns are established during development by the *de novo* DNA methyltransferases Dnmt3a and Dnmt3b. Methylcytosine-binding domain proteins (MBD) such as MBD1, MBD2, MBD4, and MeCP2 then bind d^{m}C and tether chromatin inactivation complexes which locally alter the chromatin structure into a transcriptionally repressive conformation inaccessible to the transcription initiating machinery [4].

Following differentiation, the methylation patterns are replicated each time a cell divides by the maintenance DNA methyltransferase Dnmt1. Dnmt1 is upregulated during mitosis by signals transmitted through the ERK pathway (Figure 10.1). Dnmt1 then binds the replication fork and "reads" CG pairs. Where the parent strand is methylated, Dnmt1 catalyzes transfer of the methyl group from S-adenosylmethionine (SAM) to carbon 5 in the cytosine ring, producing d^{m}C and S-adenosylhomocysteine (SAH), thereby replicating the methylation patterns and maintaining gene silencing (Figure 10.2) [4]. This may be written as Dnmt1 + SAM + − CpG − → Dnmt1 + SAH + − ^{m}CpG − . Importantly, SAH is a competitive inhibitor of transmethylation reactions [13]. Cytosine demethylation can also be mediated by Tet proteins, which oxidize the methyl group on d^{m}C, and the altered base is then removed by excision−repair mechanisms and replaced by an

ERK pathway signaling regulates Dnmt1

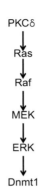

FIGURE 10.3 **Tet proteins and DNA demethylation.** Tet proteins hydroxylate the methyl group on d^mC bases, leading to excision of the modified base and replacement with unmodified dC.

unmethylated dC (Figure 10.3) [14]. However, this mechanism has not been reported to play a role in lupus T-cell DNA demethylation.

The importance of DNA methylation is evidenced by reports that null mutations in any of the DNA methyltransferase enzymes are lethal to developing embryos [4]. Interestingly though, nonlethal DNA methyltransferase mutations primarily affect the brain and immune system [4], two of the most functionally complex organs, suggesting that neurons and lymphocytes rely on DNA methylation to regulate gene expression to a greater degree than other cell types.

10.3.1 DNA Methylation and T-Cell Gene Expression

DNA methylation plays a particularly important role in regulating $CD4^+$ T-cell gene expression and function. $CD4^+$ T cells differentiate throughout life into multiple subsets including Th1, Th2, Th9, T17, Treg, TFh, and Th9, each with a distinct repertoire of effector functions [15]. However, T cells have a finite number of transcription factors and rely on DNA methylation to prevent the binding of transcription factors to the regulatory elements of genes inappropriate for the mature cell's effector functions. This is evidenced by reports that inhibiting the replication of DNA methylation patterns in dividing $CD4^+$ T cells with Dnmt1 inhibitors like 5-azacytidine (5-azaC) induces IFN-γ expression in Th2 cells [16], the cytotoxic effector molecule perforin in $CD4^+$ T cells [17], CCR6 in naïve T cells [18], and expression of the killer-cell immunoglobulin-like receptor gene family, normally expressed clonally on NK cells [19].

DNA methylation also plays a critical role in regulating $CD4^+$ T-cell antigen responses. $CD4^+$ T cells mature in the thymus, where thymocytes with high affinity receptors for self MHC molecules and thymocytes with receptors that do not recognize class II MHC determinants are eliminated. The surviving cells develop into mature CD4$^+$ T lymphocytes. Mature CD4$^+$ T cells recognize and respond to antigenic peptides bound in the antigen-binding cleft of "self" class II MHC molecules on the surface of antigen presenting cells (APCs) such as macrophages and dendritic cells. However, the interaction between the CD4$^+$ T-cell antigen/MHC receptor (TCR) and class II MHC molecules is of relatively low affinity and is stabilized by T-cell lymphocyte function-associated antigen 1 (LFA-1, CD11a/CD18) binding intracellular adhesion molecules (ICAM) on the APC. These complexes surround the TCR/MHC complex to form the "immunologic synapse," allowing the TCR as well as the LFA-1 signaling subunits to transmit signals that activate the T cell [20].

Inhibiting the replication of DNA methylation patterns during mitosis in human or mouse CD4$^+$ T cells, as well as in cloned, antigen-specific CD4$^+$ T cells, increases LFA-1 expression through demethylation and overexpression of ITGAL (CD11a). This causes T cells to lose the requirement for a specific antigenic peptide in the binding cleft of the class II molecule and respond to class II MHC molecules independent of the peptide in the binding cleft [21]. A similar autoreactivity develops when CD18 transfection is used to increase LFA-1 expression. The autoreactivity may be due to overstabilization of the normally low affinity interaction between the TCR and self class II MHC molecules without the appropriate peptide in the antigen-binding cleft, to increased costimulatory signaling by the overexpressed LFA-1 molecules, or both [21].

10.3.2 Interactions of Experimentally Demethylated T Cells with Macrophages and B Cells

Macrophages present antigen to CD4$^+$ T cells to initiate an immune response. Following CD4$^+$ T-cell activation, the macrophages undergo apoptosis. This response may serve to prevent the macrophage from continuously activating antigen-specific T cells after the source of the antigen has been eliminated [22]. In contrast to normal T cells, CD4$^+$ T cells made autoreactive with DNA methylation inhibitors respond to and cause apoptosis of autologous or syngeneic macrophages without added antigen through mechanisms including T-cell TWEAK, TRAIL, and FasL [23,24]. This autoreactive killing causes release of apoptotic chromatin from the dying macrophage. Further, since macrophages clear apoptotic material, the macrophage killing may promote accumulation of apoptotic nuclei [25]. Importantly, injecting apoptotic cells into mice is sufficient to cause anti-DNA antibodies [26]. Thus, macrophage killing by the demethylated, autoreactive T cells may provide a source of antigenic apoptotic chromatin to stimulate antichromatin antibodies.

In contrast to macrophages, culturing experimentally demethylated, autoreactive T cells with autologous or syngeneic B cells stimulates immunoglobulin production [27]. These effects are mediated in part by increased expression of B-cell stimulating cytokines such as IFN-γ, IL-4, and IL-6 [28] as well as overexpression of the B-cell costimulatory molecules CD70 and CD40L [29,30]. 5-azaC also activates expression of the TNF/lymphotoxin (TNF/LT) locus, encoding factors that are key components of the immediate early innate immune response, as well as expression of the pro-inflammatory cytokines IL-17A and IL-17F, markers of the Th17 subset [31]. Together these cytokines may augment antichromatin antibody production and promote inflammation.

10.4 DNA METHYLATION AND LUPUS

The response of experimentally demethylated CD4$^+$ T cells to self class II MHC molecules without added antigen resembles the response of CD4$^+$ T cells to host class II MHC molecules in the murine chronic graft-versus-host disease model. In this model, mice receiving semi-allogeneic CD4$^+$ T cells, capable of responding to host class II MHC molecules without being rejected, develop an autoimmune disease that resembles human lupus [32] as well as other forms of autoimmunity depending on the genetic makeup of the recipient [33]. Similarly, patients receiving allogeneic bone marrow transplants can also develop a chronic graft-versus-host disease that resembles autoimmune diseases such as lupus, Sjogren's syndrome, and scleroderma [34,35]. These reports indicate that T cells responding to host MHC determinants *in vivo* may contribute to the development of a spectrum of systemic autoimmune diseases. The functional significance of 5-azaC-induced T-cell autoreactivity to self class II MHC determinants was tested by treating CD4$^+$ murine T cells with 5-azaC and injecting the epigenetically modified cells into syngeneic recipients. Mice receiving treated, but not untreated, T cells developed a lupus-like disease with anti-DNA antibodies and an immune complex glomerulonephritis [28].

10.4.1 T-Cell DNA Methylation and Drug-Induced Lupus

Hydralazine and procainamide also give rise to ANAs in a majority of patients receiving these drugs, and ANAs together with a lupus-like disease in people metabolize these drugs slowly due to genetically determined slow acetylation [36]. The similarity of hydralazine- and procainamide-induced lupus to lupus caused by 5-azaC-treated T cells in murine models raised the possibility that procainamide and hydralazine might also inhibit T-cell DNA methylation and cause autoreactivity. This hypothesis was tested in experiments where human $CD4^+$ T cells were treated with procainamide or hydralazine. Both drugs caused T-cell DNA demethylation and autoreactivity closely resembling that caused by 5-azaC [37]. Further, murine $CD4^+$ T cells treated with these drugs also caused a lupus-like disease when transferred into syngeneic recipients [38]. Procainamide was subsequently found to be a competitive Dnmt1 inhibitor [39]. In contrast, hydralazine was found to inhibit ERK pathway signaling to prevent Dnmt1 upregulation during mitosis, and murine $CD4^+$ T cells treated with the MEK inhibitor U0126 to block ERK pathway signaling also caused a lupus-like disease in mice [40]. Together, these early studies indicate that T-cell DNA demethylation could contribute to the development of lupus-like autoimmunity.

10.4.2 T-Cell DNA Methylation, Gene Expression, and Idiopathic Lupus

Idiopathic human lupus also requires both a genetic contribution and environmental exposures to develop and flare. The drug-induced murine lupus studies reviewed above suggest that impaired $CD4^+$ T-cell DNA demethylation might also contribute to idiopathic human lupus in genetically predisposed people. The first study of T-cell DNA methylation in human lupus used high-pressure liquid chromatography (HPLC) to measure total genomic dC and d^mC in T cells from patients with inactive and active lupus as well as other diseases including rheumatoid arthritis, multiple sclerosis, gout, osteoarthritis, and chronic osteomyelitis. Patients with active, but not inactive, lupus had lower total d^mC levels relative to patients with inactive lupus and the controls, as did patients with rheumatoid arthritis [41]. Subsequent studies revealed that Dnmt1 levels are also decreased in T cells from patients with active lupus, and that Dnmt1 levels correlate inversely with lupus flare severity [42], suggesting that low Dnmt1 levels may contribute to the decreased T-cell DNA methylation in patients with active lupus.

Other reports identified a number of methylation-sensitive genes that were demethylated and overexpressed in lupus including cytokines IL-4, IL-6, IL-10, and IL-13 [43,44], and the B-cell costimulatory molecules CD70 [29] and CD40L [30], as well as PP2a, a serine—threonine phosphatase [45]. These studies were extended by Javierre et al. [46], who used bead arrays to compare more than 1500 CG sites in 807 gene promoters in leukocytes from 30 monozygotic twins discordant for SLE, rheumatoid arthritis, and dermatomyositis. This study identified 49 differentially methylated genes only in the twins with lupus. These genes were implicated in defense response, cell activation, immune response, cell proliferation, and cytokine production using gene ontology analysis. They also had significant methylation differences as well as functional relevance to SLE, and most showed diminished methylation levels. Demethylation of eight of these genes (*IFNGR2, MMP-14, LCN2, CSF3R, PECAM1, CD9, AIM2,* and *PDX1*) was confirmed using bisulfite sequencing. Global loss of d^mC was also found in the affected twins relative to their healthy siblings [46].

Coit et al. [47] used Illumina Infinium HumanMethylation450 BeadChip arrays to interrogate 27,578 CG sites in promoter regions of 14,495 genes in CD4$^+$ T cells from 12 matched SLE patients and controls. This study identified 341 CG loci with significant methylation differences. Hypomethylated genes were more prevalent than hypermethylated genes in all differentially methylated loci, having a twofold preponderance. Among the hypomethylated genes, 11 were overrepresented in the category of "development of connective tissue." The hypomethylated genes included matrix metalloproteinase 9 (*MMP-9*), platelet-derived growth factor receptor α (*PDGFRα*), *BST2* (tetherin), and *CD9*. Other studies have reported higher levels of MMP-9 in lupus patients [48,49], which has been associated with neuropsychiatric lupus [50,51] and lupus nephritis [52]. However, it should be noted that decreases in lupus T-cell MMP-9 have also been reported [53,54]. PDGFRα autoantibodies have been reported in up to 46% of lupus patients and may contribute to the development of autoimmune hemolytic anemia [55]. Interestingly, *CD9* is a T-cell coactivator and is also found to be demethylated in twins discordant for lupus [56]. Genes found to be hypermethylated included *FOLH1* and *GGH*, involved in "folate biosynthesis," as well as the transcription factor *RUNX3*.

10.4.3 Mechanisms of T-Cell DNA Demethylation in Lupus

Unstimulated T cells have relatively low Dnmt1 levels, these levels being upregulated as T cells enter S-phase by signals transmitted through the ERK and JNK signaling pathways [40,57]. Others have reported that T cells from lupus patients have multiple signaling abnormalities [58]. The cause of low Dnmt1 levels in human lupus T cells was traced to decreased ERK pathway signaling (Figure 10.3), and treating CD4$^+$ T cells with ERK pathway inhibitors decreased Dnmt1 levels and total d^mC content, similarly to lupus T cells [59]. Interestingly, as noted above, the lupus-inducing drug hydralazine also inhibits ERK pathway signaling and decreases Dnmt1 in T cells [40], suggesting that decreased T-cell ERK pathway signaling might contribute to T-cell DNA demethylation in human lupus through effects on Dnmt1 levels.

The clinical relevance of decreased T-cell ERK pathway signaling to lupus was first tested by treating CD4$^+$ murine T cells with MEK inhibitors and injecting the cells into syngeneic mice. Mice receiving the treated T cells developed lupus-like autoimmunity [40]. Subsequent studies used transgenic C57BL/6 mice in which expression of a dominant negative MEK (dnMEK) could be induced in T cells by adding doxycycline to their drinking water. Administering doxycycline activated the dnMEK, decreased Dnmt1 levels, increased expression of methylation-sensitive genes, and caused anti-DNA antibodies but not kidney disease in this strain [60]. However, C57BL6 mice are not genetically predisposed to lupus. The transgenic C57BL6 mice were therefore crossed with SJL mice, a lupus-prone strain [61], and then the BL6xSJL mice were given doxycycline. These mice developed higher anti-DNA antibody levels and an immune complex glomerulonephritis [62]. The different disease manifestations between the BL6 and BL6xSJL mice, caused by the same environmental signal, is similar to human drug-induced lupus where procainamide and hydralazine only induce an ANA in patients without a genetic predisposition to drug-induced lupus and a symptomatic lupus-like disease in those with predisposing lupus genes [63].

The defect in the lupus T-cell ERK pathway was identified by stimulating CD4$^+$ T cells from patients with active lupus and then testing phosphorylation of each of the signaling molecules in the ERK pathway signaling cascade. The defect localized to protein kinase C delta (PKCδ), which failed to phosphorylate following phorbol myristate acetate stimulation. Pharmacologic inhibition of PKCδ with the pharmacologic inhibitor rottlerin, or transfection with a dominant negative PKCδ mutant, caused demethylation of the *TNFSF7* (CD70) promoter and CD70 overexpression similarly to human lupus T cells and hydralazine-treated T cells [64]. A role for PKCδ inactivation in lupus pathogenesis is also supported by reports that PKCδ "knockout" mice also develop a lupus-like disease [56]. Further, hydralazine was also found to inhibit PKCδ activation [64]. Together, these studies indicate that impaired PKCδ signaling may be fundamental to the pathogenesis of both idiopathic and hydralazine-induced lupus.

10.5 ENVIRONMENTAL CAUSES OF T-CELL DNA DEMETHYLATION

10.5.1 Oxidative Stress

Lupus develops and flares when genetically predisposed people encounter environmental agents that cause oxidative stress, such as ultraviolet light, silica exposure, and infections [5]. Further, serum proteins from patients with active lupus are nitrated, caused by superoxide (O_2^-)

reacting with nitric oxide (NO) to form peroxynitrite (ONOO$^-$), which reacts with tyrosine to form 3-nitrotyrosine [47,65], and the level of nitration is directly related to disease activity [65]. These observations further implicate oxidative stress in lupus pathogenesis.

Gorelik et al. [66] found that PKCδ is similarly nitrated in T cells from patients with active lupus, and that nitrated PKCδ levels correlate directly with disease activity. Immunoprecipitations with antibodies to 3-nitrotyrosine demonstrated that nitrated PKCδ from lupus T cells, as well as nitrated PKCδ from ONOO$^-$-treated T cells, is refractory to PMA stimulation. Confirming studies demonstrated that both H$_2$O$_2$ and ONOO$^-$ inhibit PKCδ activation and ERK pathway signaling, decreasing Dnmt1 levels, and causing demethylation and overexpression of genes previously shown to be suppressed by DNA methylation in T cells from patients with active lupus [48]. Together these studies link oxidative stress to T-cell DNA demethylation in causing lupus flares.

10.5.2 Diet

As noted above, the replication of DNA methylation patterns during mitosis depends not only on Dnmt1 levels but also on levels of the methyl donor SAM, and is inhibited by SAH [13]. The forward velocity of this reaction (V) is directly related to Dnmt1 and SAM levels and inversely to SAH. This is shown in the following equation, where k is a constant to correct for units.

$$V = \frac{k[\text{Dnmt1}][\text{SAM}]}{[\text{SAH}]}$$

This equation implies that when Dnmt1 levels decrease, SAM levels must increase, or SAH levels decrease, in order to maintain the efficiency of this reaction. Importantly, SAM levels depend on serum levels of methionine, an essential amino acid derived from the diet, as well as other dietary transmethylation micronutrients [49], and SAH levels depend on homocysteine, which is elevated in the serum of lupus patients [50]. Thus, decreases in Dnmt1 together with decreases in SAM and increases in SAH, such as might be found in lupus patients with inadequate diets and elevated Hcy levels that are undergoing oxidative stress caused by infections or other environmental stressors, would synergize to demethylate DNA and promote lupus flares. This relationship thus suggests that diet may also be an "environmental" agent that contributes to T-cell DNA methylation abnormalities in patients with active lupus.

The effect of diet on lupus disease activity was tested using the tet-on dnMEK mouse lupus model described above [62]. Varying dietary Met levels had no effect on T-cell DNA methylation or disease manifestations in mice not receiving doxycycline and therefore normal Dnmt1 levels. However, when T-cell Dnmt1 levels were decreased with doxycycline, a diet high in methyl donors decreased anti-DNA antibody levels and kidney disease severity, while a diet poor in methyl donors increased anti-DNA antibody levels and increased kidney disease severity as measured by histologic changes [51]. This indicates that diet may also be an "environmental" contributor to lupus flares.

10.6 GENETIC/EPIGENETIC INTERACTIONS IN LUPUS

10.6.1 DNA Methylation, Lupus, and Gender

Lupus afflicts women 9–10 times more often than men. While estrogen contributes to this female predisposition [52], prepubertal girls and postmenopausal women also develop lupus with a higher incidence than in age matched boys and men [53], indicating that other factors besides estrogen are also involved. As noted above, one X chromosome is silenced in women by mechanisms including DNA methylation [12], and X chromosome DNA methylation patterns are also replicated each time a cell divides. Inhibiting the replication of DNA methylation patterns in $CD4^+$ T cells from women and men causes demethylation and overexpression of the X-linked genes *CD40LG, CXCR3*, and *OGT*, and 18 miRNAs including miR-98, let-7f-2*, miR-188-3p, miR-421, and miR-503 only in $CD4^+$ T cells from women. Expression of these genes was not altered by 5-azaC in T cells from men [12]. Demethylation and reactivation of the same X-linked genes were also found in $CD4^+$ T cells from women with active lupus, suggesting that overexpression of X-linked genes may play a role in the female predisposition to this disease [12]. This hypothesis is supported by reports that men with Klinefelter's syndrome (XXY) develop lupus at the same rate as do women [54], and that women with Turner's syndrome (XO) are resistant to lupus [67]. Together these observations suggest that demethylation of the inactive X chromosome may further increase the risk of lupus in women, in addition to the effects of estrogen.

Nonetheless, some men with only one X chromosome and normal hormone levels still develop lupus. As noted above, lupus develops and flares when genetically predisposed women encounter environmental agents that trigger the flares, and the epigenetic studies reviewed above indicate that the environment contributes to lupus flares at least in part through T-cell DNA demethylation. Since genes and DNA demethylation combine to cause lupus flares in women, Sawalha et al. hypothesized that men would likely need either a greater degree of T-cell DNA demethylation, a greater total lupus genetic risk, or both to develop a lupus flare equal in severity to a flare in women. This was tested in a study where methylation of the gene encoding perforin (*PRF1*) and *KIR2DL4* was measured in T cells in men and women with active and inactive lupus [55]. The same subjects were also genotyped for 32 lupus polymorphisms, and total genetic risk determined by multiplying the odds ratio for each gene by the number of alleles (0, 1, or 2) and adding the results. Comparing the relationship between the SLEDAI and DNA methylation levels showed no significant differences between men and women. However, comparing the relationship of the SLEDAI to the ratio of the genetic risk to either *PRF1* or *KIR2DL4* methylation demonstrated that men require a greater total genetic lupus risk and/or a greater degree of T-cell DNA demethylation to develop a flare equal in severity to that in women [55]. Together these studies support genetic/epigenetic interactions in determining lupus flares. They also support the concept that lupus genetic risk is not an all or nothing predisposition, but is variable, depending on the number and relative risk of the inherited genes.

10.6.2 Age, Genetic Risk, DNA Methylation, and Lupus

The age of lupus onset may also vary depending on total SLE genetic risk and the degree of T-cell DNA demethylation. As noted above, women are predisposed to lupus in part because of estrogen and in part because of their second X chromosome. Since T-cell DNA demethylates with age [68], Sawalha et al. hypothesized that women might develop lupus earlier in life than men due at least in part to demethylation of their second X chromosome. This hypothesis was confirmed in a study comparing gender, total lupus genetic risk, and age of lupus onset between men and women with lupus. This study demonstrated that women with early lupus onset have a greater number of lupus alleles compared to women developing lupus later in life [69]. Similarly, men tend to develop lupus later in life than do women, also consistent with an age-dependent decline in DNA methylation together with a decreased risk for lupus due to the lack of a second, methylated X chromosome [53].

10.7 HISTONE MODIFICATIONS AND LUPUS

As described elsewhere in this text, histone modifications are another epigenetic mechanism regulating gene expression. DNA is packaged into the nucleus as chromatin, consisting of two turns of DNA wrapped around a histone core to form the nucleosome. Each nucleosome contains two subunits each of H2A, H2B, H3, and H4. The histone proteins tails protrude from the nucleosome and are modified by a number of covalent modifications, including acetylation, methylation, phosphorylation, and ubiquitination [70]. Of these, acetylation is the best studied, and global increases in histone acetylation have been reported in T cells from lupus patients [71]. Acetylation of some lupus T-cell immune genes correlate with active gene transcription and suppression of others [72]. Tsokos and colleagues reported that the transcription factor CREMα may be involved in altering histone acetylation in T cells from patients with active lupus. They found that CREMα is overexpressed in lupus T cells and contributes to silencing of IL-2 expression. CREMα recruits histone deacetylases to Cre sites in the IL-2 promoter, causing histone de-acetylation and contributing to the silencing of IL-2 in lupus T cells [72]. This is an actively growing field, and a wide range of histone alterations have been reported in lupus, although the functional consequences are not completely understood at this time [73].

10.8 miRNAs AND LUPUS

miRNAs are small (21−25 bp) RNA molecules that bind mRNAs leading to their degradation and represent another epigenetic mechanism regulating gene expression and possibly contributing to lupus. Reports of miRNAs in lupus are starting to appear in the literature: Zhao et al. reported that miR-126 is upregulated in CD4$^+$ T cells from patients with active lupus, and that the degree of overexpression is inversely correlated to Dnmt1 protein levels. They also demonstrated that miR-126 directly inhibits Dnmt1 translation via interaction with its 3′ untranslated region, and that overexpression of miR-126 in CD4$^+$ T cells significantly reduces Dnmt1 protein levels. Further, the overexpression of miR-126 in CD4$^+$ T cells from healthy donors caused the demethylation and upregulation of genes encoding CD11a and CD70, causing T-cell and B-cell hyperactivity [74].

Similarly, Ding et al. reported that miR-142-3p and miR-142-5p are significantly downregulated in SLE CD4$^+$ T cells compared with healthy controls and that miR-142-3p/5p levels correlated inversely with the putative SLE-related protein "signaling lymphocytic activation molecule-associated protein" (SAP), CD84, and IL-10. These investigators also found that miR-142-3p and miR-142-5p directly inhibit SAP, CD84, and IL-10 translation, and that reduced miR-142-3p/5p expression in CD4$^+$ T cells significantly increases protein levels of these target genes [75].

10.9 SUMMARY/CONCLUSIONS

Together, these studies provide compelling evidence that inhibiting the replication of DNA methylation patterns in dividing CD4$^+$ T cells is sufficient to give rise to ANAs in people and mice lacking lupus genes, and a lupus-like disease in people and mice with a sufficient number of lupus genes. Histone modifications and miRNAs may contribute as well. These studies also indicate that common environmental exposures such as oxidative stress and a poor diet likely contribute to the onset of lupus flares in genetically predisposed people. Future therapeutic approaches will target the epigenetically altered T cells, and preventative measures may include dietary antioxidants and transmethylation micronutrients.

List of Abbreviations

5-azaC	5-azacytidine
APC	antigen-presenting cell
dmC	deoxymethylcytosine
HPLC	high pressure liquid chromatography
LFA-1	lymphocyte function antigen 1
LT	lymphotoxin
MBD	methyl binding domain
MHC	major histocompatibility complex
miRNA	microRNA
MMP-9	matrix metalloproteinase
PDGFRα	platelet derived growth factor receptor α
PMA	phorbol myristate acetate
SAH	S-adenosylhomocysteine
SAM	S-adenosylmethionine
SAP	signaling lymphocytic activation molecule-associated protein
SLE	systemic lupus erythematosus
TCR	T-cell receptor
TNF	tumor necrosis factor

References

[1] Farhat SC, Silva CA, Orione MA, Campos LM, Sallum AM, Braga AL. Air pollution in autoimmune rheumatic diseases: a review. Autoimmun Rev 2011;11(1):14−21.

[2] Leitinger N. The role of phospholipid oxidation products in inflammatory and autoimmune diseases: evidence from animal models and in humans. Subcell Biochem 2008;49:325−50.

[3] Zandman-Goddard G, Solomon M, Rosman Z, Peeva E, Shoenfeld Y. Environment and lupus-related diseases. Lupus 2012;21(3):241−50.

[4] Richardson B. Primer: epigenetics of autoimmunity. Nat Clin Pract Rheumatol 2007;3(9):521−7.

[5] Somers E, Richardson B. Environmental exposures, epigenetic changes and the risk of lupus. Lupus 2014;23(6):568−76.

[6] Hess E. Lupus: the clinical entity. In: Kammer GM, Tsokos GC, editors. Lupus: molecular and cellular pathogenesis. Totowa, NJ: Humana Press; 1999. pp. 1−12.

[7] Clarke JT, Werth VP. Rheumatic manifestations of skin disease. Curr Opin Rheumatol 2010;22(1):78−84.

[8] Petri M. Review of classification criteria for systemic lupus erythematosus. Rheum Dis Clin North Am 2005;31(2):245−54.

[9] Pisetsky DS. Antinuclear antibodies in rheumatic disease: a proposal for a function-based classification. Scand J Immunol 2012;76(3):223−8.

[10] Nilsson BO, Skogh T, Ernerudh J, Johansson B, Löfgren S, Wikby A, et al. Antinuclear antibodies in the oldest-old women and men. J Autoimmun 2006;27 (4):281−8.

[11] Richardson B, Epstein WV. Utility of the fluorescent antinuclear antibody test in a single patient. Ann Intern Med 1981;95(3):333−8.

[12] Hewagama A, Gorelik G, Patel D, Liyanarachchi P, McCune WJ, Somers E, et al. Overexpression of X-linked genes in T cells from women with lupus. J Autoimmun 2013;41:60−71.

[13] Li Y, Liu Y, Strickland FM, Richardson B. Age-dependent decreases in DNA methyl-transferase levels and low transmethylation micronutrient levels synergize to promote overexpression of genes implicated in autoimmunity and acute coronary syndromes. Exp Gerontol 2010;45(4):312−22.

[14] Hill PW, Amouroux R, Hajkova P. DNA demethylation, Tet proteins and 5-hydroxymethylcytosine in epigenetic reprogramming: an emerging complex story. Genomics 2014;104:324−33.

[15] Christie D, Zhu J. Transcriptional regulatory networks for CD4 T cell differentiation. Curr Top Microbiol Immunol 2014;381:125−72.

[16] Young HA, Ghosh P, Ye J, Lederer J, Lichtman A, Gerard JR, et al. Differentiation of the T helper phenotypes by analysis of the methylation state of the IFN-gamma gene. J Immunol 1994;153(8):3603−10.

[17] Lu Q, Wu A, Ray D, Deng C, Attwood J, Hanash S, et al. DNA methylation and chromatin structure regulate T cell perforin gene expression. J Immunol 2003;170(10):5124−32.

[18] Suarez-Alvarez B, Baragaño Raneros A, Ortega F, López-Larrea C. Epigenetic modulation of the immune function: a potential target for tolerance. Epigenetics 2013;8(7):694−702.

[19] Liu Y, Chen Y, Richardson B. Decreased DNA methyltransferase levels contribute to abnormal gene expression in "senescent" CD4(+)CD28(−) T cells. Clin Immunol 2009;132(2):257−65.

[20] Chen W, Zhu C. Mechanical regulation of T-cell functions. Immunol Rev 2013;256 (1):160−76.

[21] Richardson B, Powers D, Hooper F, Yung RL, O'Rourke K. Lymphocyte function-associated antigen 1 overexpression and T cell autoreactivity. Arthritis Rheum 1994;37(9):1363−72.

[22] Richardson BC, Buckmaster T, Keren DF, Johnson KJ. Evidence that macrophages are programmed to die after activating autologous, cloned, antigen-specific, CD4+ T cells. Eur J Immunol 1993;23(7):1450−5.

[23] Kaplan MJ, Ray D, Mo RR, Yung RL, Richardson BC. TRAIL (Apo2 ligand) and TWEAK (Apo3 ligand) mediate CD4+ T cell killing of antigen-presenting macrophages. J Immunol 2000;164(6):2897−904.

[24] Richardson BC, Lalwani ND, Johnson KJ, Marks RM. Fas ligation triggers apoptosis in macrophages but not endothelial cells. Eur J Immunol 1994;24(11):2640−5.

[25] Denny MF, Chandaroy P, Killen PD, Caricchio R, Lewis EE, Richardson BC, et al. Accelerated macrophage apoptosis induces autoantibody formation and organ damage in systemic lupus erythematosus. J Immunol 2006;176(4):2095−104.

[26] Van Bruggen MC, Kramers C, Berden JH. Autoimmunity against nucleosomes and lupus nephritis. Ann Med Interne (Paris) 1996;147(7):485−9.

[27] Richardson BC, Liebling MR, Hudson JL. CD4+ cells treated with DNA methylation inhibitors induce autologous B cell differentiation. Clin Immunol Immunopathol 1990;55(3):368−81.

[28] Quddus J, Johnson KJ, Gavalchin J, Amento EP, Chrisp CE, Yung RL, et al. Treating activated CD4+ T cells with either of two distinct DNA methyltransferase inhibitors,

5-azacytidine or procainamide, is sufficient to cause a lupus-like disease in syngeneic mice. J Clin Invest 1993;92(1):38−53.

[29] Oelke K, Lu Q, Richardson D, Wu A, Deng C, Hanash S, et al. Overexpression of CD70 and overstimulation of IgG synthesis by lupus T cells and T cells treated with DNA methylation inhibitors. Arthritis Rheum 2004;50(6):1850−60.

[30] Lu Q, Wu A, Tesmer L, Ray D, Yousif N, Richardson B. Demethylation of CD40LG on the inactive X in T cells from women with lupus. J Immunol 2007;179(9):6352−8.

[31] Falvo JV, Jasenosky LD, Kruidenier L, Goldfeld AE. Epigenetic control of cytokine gene expression: regulation of the TNF/LT locus and T helper cell differentiation. Adv Immunol 2013;118:37−128.

[32] Eisenberg RA, Via CS. T cells, murine chronic graft-versus-host disease and autoimmunity. J Autoimmun 2012;39(3):240−7.

[33] Tyndall A, Dazzi F. Chronic GVHD as an autoimmune disease. Best Pract Res Clin Haematol 2008;21(2):281−9.

[34] Filipovich AH. Diagnosis and manifestations of chronic graft-versus-host disease. Best Pract Res Clin Haematol 2008;21(2):251−7.

[35] Gilman AL, Serody J. Diagnosis and treatment of chronic graft-versus-host disease. Semin Hematol 2006;43(1):70−80.

[36] Drayer DE, Reidenberg MM. Clinical consequences of polymorphic acetylation of basic drugs. Clin Pharmacol Ther 1977;22(3):251−8.

[37] Cornacchia E, Golbus J, Maybaum J, Strahler J, Hanash S, Richardson B. Hydralazine and procainamide inhibit T cell DNA methylation and induce autoreactivity. J Immunol 1988;140(7):2197−200.

[38] Yung R, Chang S, Hemati N, Johnson K, Richardson B. Mechanisms of drug-induced lupus. IV. Comparison of procainamide and hydralazine with analogs *in vitro* and *in vivo*. Arthritis Rheum 1997;40(8):1436−43.

[39] Scheinbart LS, Johnson MA, Gross LA, Edelstein SR, Richardson BC. Procainamide inhibits DNA methyltransferase in a human T cell line. J Rheumatol 1991;18 (4):530−4.

[40] Deng C, Lu Q, Zhang Z, Rao T, Attwood J, Yung R, et al. Hydralazine may induce autoimmunity by inhibiting extracellular signal-regulated kinase pathway signaling. Arthritis Rheum 2003;48(3):746−56.

[41] Richardson B, Scheinbart L, Strahler J, Gross L, Hanash S, Johnson M. Evidence for impaired T cell DNA methylation in systemic lupus erythematosus and rheumatoid arthritis. Arthritis Rheum 1990;33(11):1665−73.

[42] Balada E, Ordi-Ros J, Serrano-Acedo S, Martinez-Lostao L, Rosa-Leyva M, Vilardell-Tarrés M. Transcript levels of DNA methyltransferases DNMT1, DNMT3A and DNMT3B in CD4+ T cells from patients with systemic lupus erythematosus. Immunology 2008;124(3):339−47.

[43] Mi XB, Zeng FQ. Hypomethylation of interleukin-4 and -6 promoters in T cells from systemic lupus erythematosus patients. Acta Pharmacol Sin 2008;29(1):105−12.

[44] Zhao M, Ordi-Ros J, Serrano-Acedo S, Martinez-Lostao L, Rosa-Leyva M, Vilardell-Tarrés M. Hypomethylation of IL10 and IL13 promoters in CD4+ T cells of patients with systemic lupus erythematosus. J Biomed Biotechnol 2010;2010:931018.

[45] Sunahori K, Nagpal K, Hedrich CM, Mizui M, Fitzgerald LM, Tsokos GC. The catalytic subunit of protein phosphatase 2A (PP2Ac) promotes DNA hypomethylation by suppressing the phosphorylated mitogen-activated protein kinase/extracellular signal-regulated kinase (ERK) kinase (MEK)/phosphorylated ERK/DNMT1 protein pathway in T-cells from controls and systemic lupus erythematosus patients. J Biol Chem 2013;288(30):21936−44.

[46] Javierre BM, Fernandez AF, Richter J, Al-Shahrour F, Martin-Subero JI, Rodriguez-Ubreva J. Changes in the pattern of DNA methylation associate with twin discordance in systemic lupus erythematosus. Genome Res 2010;20(2):170−9.

[47] Coit P, Jeffries M, Altorok N, Dozmorov MG, Koelsch KA, Wren JD, et al. Genome-wide DNA methylation study suggests epigenetic accessibility and transcriptional poising of interferon-regulated genes in naïve CD4+ T cells from lupus patients. J Autoimmun 2013;43:78−84. PubMed PMID: 23623029.

[48] Li Y, Gorelik G, Strickland FM, Richardson BC. Oxidative stress, T cell DNA methylation and lupus. Arthritis Rheumatol 2014;66(6):1574−82.

[49] Bertolo RF, McBreairty LE. The nutritional burden of methylation reactions. Curr Opin Clin Nutr Metab Care 2013;16(1):102−8.

[50] Lazzerini PE, Capecchi PL, Selvi E, Lorenzini S, Bisogno S, Galeazzi M, et al. Hyperhomocysteinemia: a cardiovascular risk factor in autoimmune diseases? Lupus 2007;16(11):852−62.

[51] Strickland FM, Hewagama A, Wu A, Sawalha AH, Delaney C, Hoeltzel MF, et al. Diet influences expression of autoimmune-associated genes and disease severity by epigenetic mechanisms in a transgenic mouse model of lupus. Arthritis Rheum 2013;65(7):1872−81.

[52] Lateef A, Petri M. Hormone replacement and contraceptive therapy in autoimmune diseases. J Autoimmun 2012;38(2−3):J170−6.

[53] Somers EC, Thomas SL, Smeeth L, Schoonen WM, Hall AJ. Incidence of systemic lupus erythematosus in the United Kingdom, 1990−1999. Arthritis Rheum 2007;57 (4):612−18.

[54] Sawalha AH, Harley JB, Scofield RH. Autoimmunity and Klinefelter's syndrome: when men have two X chromosomes. J Autoimmun 2009;33(1):31−4.

[55] Sawalha AH, Wang L, Nadig A, Somers EC, McCune WJ, Hughes T, et al. Sex-specific differences in the relationship between genetic susceptibility, T cell DNA demethylation and lupus flare severity. J Autoimmun 2012;38(2−3):J216−22.

[56] Miyamoto A, Nakayama K, Imaki H, Hirose S, Jiang Y, Abe M, et al. Increased proliferation of B cells and auto-immunity in mice lacking protein kinase Cdelta. Nature 2002;416(6883):865−9.

[57] Chen Y, Gorelik GJ, Strickland FM, Richardson BC. Decreased ERK and JNK signaling contribute to gene overexpression in "senescent" CD4+CD28− T cells through epigenetic mechanisms. J Leukoc Biol 2010;87(1):137−45.

[58] Moulton VR, Tsokos GC. Abnormalities of T cell signaling in systemic lupus erythematosus. Arthritis Res Ther 2011;13(2):207.

[59] Deng C, Kaplan MJ, Yang J, Ray D, Zhang Z, McCune WJ, et al. Decreased Ras-mitogen-activated protein kinase signaling may cause DNA hypomethylation in T lymphocytes from lupus patients. Arthritis Rheum 2001;44(2):397−407.

[60] Sawalha AH, Jeffries M, Webb R, Lu Q, Gorelik G, Ray D, et al. Defective T-cell ERK signaling induces interferon-regulated gene expression and overexpression of methylation-sensitive genes similar to lupus patients. Genes Immun 2008;9 (4):368−78.

[61] Smith DL, Dong X, Du S, Oh M, Singh RR, Voskuhl RR. A female preponderance for chemically induced lupus in SJL/J mice. Clin Immunol 2007;122(1):101−7.

[62] Strickland FM, Hewagama A, Lu Q, Wu A, Hinderer R, Webb R, et al. Environmental exposure, estrogen and two X chromosomes are required for disease development in an epigenetic model of lupus. J Autoimmun 2012;38(2-3):J135−43.

[63] Strandberg I, Boman G, Hassler L, Sjöqvist F. Acetylator phenotype in patients with hydralazine-induced lupoid syndrome. Acta Med Scand 1976;200(5):367−71.

[64] Gorelik G, Fang YJ, Wu A, Sawalha AH, Richardson B. Impaired T cell protein kinase C delta activation decreases ERK pathway signaling in idiopathic and hydralazine-induced lupus. J Immunol 2007;179(8):5553—63.

[65] Oates JC, Gilkeson GS. The biology of nitric oxide and other reactive intermediates in systemic lupus erythematosus. Clin Immunol 2006;121(3):243—50.

[66] Gorelik GJ, Yarlagadda S, Patel DR, Richardson BC. Protein kinase Cdelta oxidation contributes to ERK inactivation in lupus T cells. Arthritis Rheum 2012;64(9):2964—74.

[67] Cooney CM, Bruner GR, Aberle T, Namjou-Khales B, Myers LK, Feo L, et al. 46,X,del (X)(q13) Turner's syndrome women with systemic lupus erythematosus in a pedigree multiplex for SLE. Genes Immun 2009;10(5):478—81.

[68] Golbus J, Palella TD, Richardson BC. Quantitative changes in T cell DNA methylation occur during differentiation and ageing. Eur J Immunol 1990;20(8):1869—72.

[69] Webb R, Kelly JA, Somers EC, Hughes T, Kaufman KM, Sanchez E, et al. Early disease onset is predicted by a higher genetic risk for lupus and is associated with a more severe phenotype in lupus patients. Ann Rheum Dis 2011;70(1):151—6.

[70] Rothbart SB, Strahl BD. Interpreting the language of histone and DNA modifications. Biochim Biophys Acta 2014;1839(8):627—43.

[71] Hu N, Qiu X, Luo Y, Yuan J, Li Y, Lei W, et al. Abnormal histone modification patterns in lupus CD4+ T cells. J Rheumatol 2008;35(5):804—10.

[72] Oaks Z, Perl A. Metabolic control of the epigenome in systemic lupus erythematosus. Autoimmunity 2014;47(4):256—64.

[73] Hedrich CM, Tsokos GC. Epigenetic mechanisms in systemic lupus erythematosus and other autoimmune diseases. Trends Mol Med 2011;17(12):714—24.

[74] Zhao S, Wang Y, Liang Y, Zhao M, Long H, Ding S, et al. MicroRNA-126 regulates DNA methylation in CD4+ T cells and contributes to systemic lupus erythematosus by targeting DNA methyltransferase 1. Arthritis Rheum 2011;63(5):1376—86.

[75] Ding S, Liang Y, Zhao M, Liang G, Long H, Zhao S, et al. Decreased microRNA-142-3p/5p expression causes CD4+ T cell activation and B cell hyperstimulation in systemic lupus erythematosus. Arthritis Rheum 2012;64(9):2953—63.

Epigenetics in Psoriasis

Kuan-Yen Tung[1,2], Fu-Tong Liu[1], Yi-Ju Lai[1],
Chih-Hung Lee[3], Yu-Ping Hsiao[4,5],
and Yungling Leo Lee[1,2]

[1]Institute of Biomedical Sciences, Academia Sinica, Taipei, Taiwan
[2]Institute of Epidemiology and Preventive Medicine, National Taiwan
University, Taipei, Taiwan [3]Department of Dermatology, Kaohsiung Chang
Gung Memorial Hospital, Kaohsiung, Taiwan [4]Department of Medical
Education, Taichung Veterans General Hospital, Taichung, Taiwan
[5]Institute of Medicine, Chung Shan Medical University, Taichung, Taiwan

11.1 EPIDEMIOLOGY OF PSORIASIS

Psoriasis is a chronic inflammatory skin disease affected by complex interactions between genes and environmental factors. The prevalence of psoriasis has been reported to be 1.0−11.8% in different parts of the world, and is dependent on various factors such as the geographical location, ethnicity of the subjects, and the method of assessment [1]. The severity of psoriasis in a patient may worsen with age and may depend upon the patient's genetic components, gender, race, exposure to environmental factors, and the geographic location [2]. Parisi et al. [3] reported that the incidence of psoriasis is more common in adults than in young people, and that its incidence increases with age up to 39 years. The incidence rate of psoriasis is high in people between the ages of 60 and 70 years; the incidence rate tends to decline toward the end-of-life period. However, no evidence supports the influence of gender difference on the prevalence of psoriasis. Increased prevalence of metabolic syndromes and cardiovascular diseases among psoriasis patients has been reported in several countries including Italy, Israel, India, Japan, China, Tunisia, and the United States [3−10].

      © 2015 Elsevier Inc. All rights reserved.

11.2 ENVIRONMENTAL FACTORS OF PSORIASIS

The effect of exposure to environmental factors on the development of psoriasis or its contribution to the exacerbation of psoriasis remains poorly understood [11]. The initiation, maintenance, and subsequent exacerbation of psoriatic lesions are affected by the complex interplay among genes, environmental risk factors, and comorbidities of psoriasis. In addition, increasing evidence suggests that the epigenetic modification is dynamically responsive depending on the exposure to environmental factors and gender and age of the individual [12]. The epigenetic phenomena could possibly be modified by exposure to environmental risk factors and lifestyle of the psoriasis patients and may directly exacerbate psoriasis. There has been no validation of the causality relationship between environmental risk factors and the pathogenesis of psoriasis. Moreover, epigenetic modifications play an important role in controlling gene expression during cell cycle, development, biological modifications, and modifications due to exposure to environmental factors.

11.2.1 UV Irradiation

Ultraviolet (UV) irradiation has been conventionally performed in clinics for psoriasis treatment. In psoriasis, the primary initiating factor is the hyperstimulation of antigen-presenting cells and auto-reactive dermal T cells. In the past, exposure to the natural sunlight has also been seen to help in psoriasis treatment. UV irradiation has been reported to decrease T-cell responses via reduced antigen presentation and alloactivation in the skin [13]. Therefore, the efficacy of phototherapy for psoriasis can be attributed to the immunoregulatory mechanisms of T cells in the skin. On the other hand, some studies suggest that silencing of tumor suppressor genes in the epidermis occurs in photocarcinogenesis after exposure to UV radiation, and that this process is correlated with epigenetic modifications such as alterations in DNA methylation, modification of DNA methyltransferases (DNMTs), and histone acetylation. Exposure to UV radiation induces inflammation, suppression of immune reactions, oxidative stress, and DNA damage in the skin [14]. Chronic and sustained inflammation, oxidative stress, and DNA damage can significantly modify the epigenetic patterns of keratinocytes. Chen et al. [15] determined that $P16^{INK4a}$, a tumor suppressor gene, tended to be epigenetically silenced in UVA-irradiated human keratinocytes. Another study reported that hypoacetylation of H3 and H4 histones silences $p16^{INK4a}$ after UVB exposure at the transcription level, which indicates increased promoter methylation of $p16^{INK4a}$. In addition, the expression of microRNA (miRNA) also changes after exposure to UV radiation. It has also been shown that UV radiation upregulates miR-125b expression via ataxia telangiectasia mutated (ATM)-dependent NF-κB activation [16]. However, the mechanisms underlying the epigenetic modifications in psoriatic skin after exposure to UV radiation remain unclear.

11.2.2 Smoking Habits

Smoking has been reported to prolong the course of psoriasis and make the patient less responsive to treatment [17,18]. Although it is known that psoriasis is related to smoking, it remains to be determined whether smoking is an important risk factor in the pathogenesis of psoriasis [17,19]. Some evidence supports the association between the epigenetic modifications and the smoking habits of the patients. Kim et al. [20] reported that methylation of the $p16^{INK4a}$ gene is positively associated with the numbers of cigarette (in pack-years) smoked and the duration of smoking, and it is negatively associated with the time elapsed after the patient quit smoking. The miR-21 and miR-$146a$ gene expression levels are also downregulated in the placenta of mothers who smoked during pregnancy as compared with those of healthy controls [21]. In addition, Torii et al. [22] reported that smokers have a higher level of circulating Th17 cells compared to non-smokers; moreover, they reported that tobacco smoke extract can induce Th17 cell generation from the central memory T cells and alter the expression levels of IL-17 and IL-22 in a cell line model. In summary, whether smoking-associated epigenetic modifications and the increase in the circulating Th17 cells are associated with psoriasis remain to be determined.

11.2.3 Psoriasis-Related Comorbidities

Psoriasis is associated with higher prevalence of anxiety, depression, and suicidal ideation, and reduced health-related quality of life as compared with other chronic diseases such as cancer and diabetes [23–26]. Emerging data suggest that the association of psoriasis with metabolic syndrome occurs early in the course of the disease, because psoriasis is associated with obesity and elevated lipid levels even during childhood [27]. Several studies suggest that obesity is a risk factor for the future development of psoriasis as 30% of recent psoriasis cases have been attributed to obesity [28,29]. Previous studies have shown that patients with severe psoriasis had an increased risk of mortality due to cardiovascular diseases [30,31]. A meta-analysis of 22 hospital-based studies indicated that psoriasis is associated with an increased risk of cardiovascular diseases, ischemic heart disease, peripheral vascular disease, atherosclerosis, diabetes, hypertension, dyslipidemia, obesity, and metabolic syndrome [32]. However, the association between epigenetic modifications caused by comorbidities and the pathogenesis of psoriasis remains largely unknown.

11.3 EPIGENETICS AND GENETICS IN PSORIASIS

The pathogenesis of psoriasis is complex and involves both the genetic components and environmental factors. Evidence from family studies suggests the role of genetic predisposition in the pathogenesis of psoriasis. However, the exact inheritance pattern remained unknown as of 1985 [33]. In 2007, a study showed that the heritability of psoriasis is approximately 66% [34]. A strong association between the *HLA-Cw6* allele and psoriasis was determined in various races, but the concordance in monozygotic twins was found to be only 67%. In fact, the presence of the *HLA-Cw6* allele only partially explains the overall heritability of psoriasis. Therefore, we hypothesize that epigenetic modifications are involved in the pathogenesis of psoriasis. In addition, some methylation-sensitive genes implicated in the pathogenesis of lupus are also aberrantly expressed in psoriasis patients. For example, the expression of perforins increased in the epidermis of lesional skin of patients with psoriasis [35]. Another methylation-sensitive gene is *LFA-1* (*CD11a*/*CD18*); LFA-1 protein is a key regulator in the pathogenesis of psoriasis that is overexpressed in both lupus and psoriasis patients [36]. Although the detailed pathological role of the aforementioned methylation-sensitive genes remains to be investigated, the epigenetic mechanisms are considered to be the new putative links between genetic components and environmental factors in the pathogenesis of psoriasis [37].

11.3.1 Epigenetics and Pathogenesis of Psoriasis

Immune cell activation, keratinocyte differentiation, and proliferation and infiltration of inflammatory cells are the key pathogenic events in psoriasis. These events are correlated with changes in gene expression [38,39]. In these events, maintenance of cellular differentiation and immune cell activation may be considered as epigenetic phenomena. These phenomena can also be modified by exogenous factors resulting from environmental alterations [40]. Therefore, epigenetic changes are important in the pathogenesis of psoriasis, because one of the distinctive features of psoriatic lesions is the prominent thickening of the epidermis, reflecting altered differentiation, and increased proliferation of keratinocytes [40]. However, the detailed mechanisms of epigenetic modification in keratinocyte differentiation, proliferation, immune cell activation, and inflammatory cell infiltration remain unclear.

Epigenetics is defined as the study of heritable changes in gene expression caused by DNA-binding molecules as opposed to changes in the underlying DNA sequence [41]. Epigenetic modifications involve both intracellular and intercellular interactions, which lead to alterations in reversible phenomena such as cell signaling and DNA modification [42].

Recent studies have reported that the epigenetic patterns are different between psoriasis patients and healthy controls, suggesting that epigenetic modifications may also be involved in the development of this disease. There are three main epigenetic markers: DNA methylation, histone modifications, and noncoding RNAs. In the following sections, we have reviewed new epigenetic markers of psoriasis.

11.4 DNA METHYLATION

DNA methylation plays an important role in the epigenetic silencing mechanism [43]. Methylation occurring in the cytosine residues in CpG dinucleotides (5-methylcytosine) within the CpG islands is the simplest form of epigenetic modification in humans. Hypermethylation of CpG islands in gene promoters leads to gene silencing, and hypomethylation leads to active transcription. Recent studies have reported that methylation occurs in less CpG-dense regions near CpG islands (also called the CpG shores), which regulates gene expression in a specific tissue. In addition, methylation in CpG shores also leads to aberrant gene expression patterns in carcinogenesis and lineage-specific cell differentiation. These observations suggest that DNA methylation of CpG shores is an important mechanism that controls gene transcription in specific tissues [44,45]. Recent evidence supports DNA methylation being more prevalent within gene bodies such as introns or 3'-untranslated regions (UTRs) than in promoter regions [46]. The function of intragenic DNA methylation being partially regulated by transcription from alternative promoters [47]. However, other mechanisms are also involved in regulating the function of intragenic DNA methylation. DNA methylation plays an important role in the regulation of several cellular processes such as genetic transcription, genomic imprinting, and inactivation of the X chromosome [48]. In mammalian cells, DNA methylation occurs at the fifth carbon of the cytosine residues within the CpG sites, and it is catalyzed by DNMTs with S-adenosyl methionine (SAM) as the methyl donor. DNMTs are encoded by *DNMT1*, *DNMT3a*, and *DNMT3b* genes, and the enzymes establish and maintain DNA methylation patterns in the human genome [49].

11.4.1 Epigenomic Profiling in Psoriasis

Epigenetic markers can be studied by using genome-wide approaches or by focusing on candidate genes [50]. Generally, such studies first employ genome-wide approaches to search for novel targets and then employ highly specific experimental approaches to confirm the pattern of novel targets in the same cohort (internally) or in independent cohorts (externally). The microarray platforms are usually selected for profiling epigenetic markers (DNA methylation and miRNA) on a genomic scale. Several platforms and protocols are available for determining the patterning of DNA methylation and miRNA, such as the Illumina BeadChip Microarray, comprehensive analysis of relative DNA methylation (CHARM) platform, and the methylated DNA immunoprecipitation (MeDIP) array [51,52]. Microarray platforms have also been used to investigate histone modifications as well as miRNAs by chromatin immunoprecipitation (ChIP), followed by hybridization on microarrays (ChIP-chip) [52,53]. The research tools available to study epigenetic profiles and patterns are discussed in more detail in Chapter 19.

Although the complete range of epigenetic modifications is currently unknown, it can be assumed to be enormous. The human epigenome is believed to involve more than 10^8 cytosine nucleotides (of which more than 10^7 are CpGs) and more than 10^8 are histone tails; all of these regions may be different depending on the exposure to environmental factors and development of a disease [54]. Approximately 70% of human genes are linked to CpG islands on promoters, whose methylation status may influence the transcription activity. The recent development in DNA methylation profiling technologies will facilitate more feasible and accessible investigation of the status of DNA methylation in the relevant diseases on a genome-wide scale. Although other epigenetic modifications, such as histone modification, also play important roles during disease progression, the current techniques limit the high-throughput exploration of such epigenetic modifications [54]. In addition, this issue is further hampered by the lack of epigenomic data in the HapMap project, which is a reference database for the normal level of epigenetic variation in humans.

The data from high-throughput DNA methylation microarray can raise several questions. As mentioned above, cellular differentiation may be considered an epigenetic phenomenon, but the epigenetic profile may be different in different tissues or cells. In the pathogenesis of psoriasis, the infiltration of granulocytes, inflammatory cells, and the increased growth of blood vessels occur frequently in the skin. Therefore, the differences in the epigenetic modification signals between psoriatic skin and normal skin may simply reflect the differences in composition between different types of cell rather than the differences between the healthy and psoriatic skin.

11.4.2 DNA Methylation Profiling in Psoriasis

Whole-genome microarray platforms are considered appropriate for exploring the unknown epigenetic variations in tissues and cells from different diseases. Studies on methylation in disease tissues, such as those in autoimmune diseases, are limited by the availability of such tissues.

However, some studies have managed to identify methylated genes in different clinical specimens of psoriasis (Table 11.1). Global DNA methylation in the skin and peripheral blood mononuclear cells (PBMCs) from psoriasis patients is significantly higher than in healthy controls. The psoriasis area and severity index (PASI) score of psoriasis patients is correlated with the DNA methylation levels in the skin cells. In addition, the mRNA levels of DNMT1, a methyltransferase enzyme required for maintaining the DNA methylation pattern, are significantly higher in psoriasis PBMCs than in those of healthy controls, whereas two methyl-DNA binding domain (MBD) genes MBD2 and MeCP2 were found to be downregulated in psoriatic PBMCs. These results indicate that DNA methylation plays important roles in the pathogenesis of psoriasis [60]. Gervin et al. [55] analyzed the DNA methylation profiling from CD4$^+$ and CD8$^+$ cells in PBMCs between monozygotic twins by using the Illumina BeadChip microarray. In comparison with the unaffected siblings, 16 hypermethylated genes and 34 hypomethylated genes were identified in the affected ones. Furthermore, these researchers used gene ontology (GO) analysis and noted that these genes were associated with biological pathways involved with several cytokines and chemokines [55]. Another study reported that the promoter regions of 121 genes on the X chromosome of psoriasis patients had significantly increased methylation levels in T cells as compared to those from healthy controls (>4-fold) by using the ChIP-seq method [56]. In addition, Han et al. [56] also found that the immune-related genes on the X chromosome showed greater hypermethylation than did nonimmune genes on the same chromosome in naïve CD4$^+$ T cells from psoriasis patients. Mesenchymal stem cells (MSCs) have immunomodulatory functions and promote angiogenesis, and the MSCs in the dermis of psoriatic lesions are associated with the pathogenesis and development of psoriasis through a complex mechanism. Hou et al. [57] used the MeDIP-array platform to identify some methylated genes involved in cell communication, surface-receptor signaling pathways, cellular responses to stimulus, and cell migration in MSCs from psoriatic patients. Roberson et al. [58] identified KYNU, OAS2, S100A12, and SERPINB3 gene expression as being regulated by the methylation level on their promoters in psoriatic skin. GO analysis of MeDIP-Seq data showed that the aberrantly methylated genes have been implicated in several systems such as the immune system, cell cycle mechanism, and apoptotic mechanism in psoriasis [59].

11.4.3 DNA Methylation of Specific Genes in Psoriasis

Psoriasis is generally considered to result from T-cell activation initiated by environmental antigens that are processed and presented to

TABLE 11.1 Identified Methylated Genes from Different Clinical Specimens of Psoriasis Patients from High-Throughput Data Analysis

Author [Reference]	Specimen source	Hypermethylated genes	Hypomethylated genes	Gene ontology
Gervin et al. [55]	CD4$^+$ cells from PBMCs	IL13, CSF2, TNFSF9	CCL1, TNFSF11	Immune response, inflammatory response
Han et al. [56]	CD4$^+$ cells from PBMCs	SLITRK4, EMD, ZIC3, CXorf40A, HDAC6, IKBKG	—	Autoimmunity, T-cell polarization, immune system
Hou et al. [57]	Skin mesenchymal stem cells	CACNA2D3, CBX4, SRF	NRP2, S100A10, TCL1B, VASH1	Cell communication, surface–receptor signaling pathway, regulation of transmission of nerve impulse, cell migration
Roberson et al. [58]	Skin biopsy	KYNU, OAS2, S100A12, SERPINB3	C10orf99, IFI27, SERPINB4	Apoptosis, chronic inflammation, immune response, cell–cell signaling, cell membrane organization, endosome transport
Zhang et al. [59]	Skin biopsy	PDCD5	TIMP2	Antiapoptosis, regulation of cell migration, cell cycle, cell communication, signal transduction

T cells by antigen-presenting cells [61]. Cytokine production by T-cell subsets and aberrant proliferation of keratinocytes promotes formation of plaque, which is a characteristic of psoriasis [62−64]. In the studies of DNA methylated genes in the pathogenesis of psoriasis, the methylation level of some antiapoptotic genes was significantly altered in the psoriatic skin. Chen et al. [65] showed that the $p16^{INK4a}$ gene promoter was methylated in 30% of psoriasis patients. The PASI score was higher in patients who did not exhibit $p16^{INK4a}$ methylation. Hypermethylation of $p16^{INK4a}$ gene promoter leads to reduced p53 levels in keratinocytes as well as to excessive keratinocyte proliferation [65].

Another example of aberrant DNA methylation in psoriasis is of *SHP-1*, which serves as a regulator in growth and proliferation processes. The *SHP-1* promoter 2 region was found to be significantly demethylated in keratinocytes from psoriatic lesional skin [66]. A previous study on lymphoma proved that downregulation of *STAT3* mRNA expression results in *SHP-1* demethylation [67]. A study reported by Sano et al. [68] determined that upregulation of STAT3 is an important mechanism in the pathogenesis of psoriasis. Hence, STAT3 may act differently on the *SHP-1* promoter in epithelial cells. Whether the additional tissue-specific factor that triggers the opposite outcome is hypomethylation or hypermethylation remains to be determined.

Two studies employed high-proliferative-potential colony-forming cell (HPP-CFC) assay to investigate the relationship between antiapoptotic genes and proliferation/differentiation in hematopoietic stem cells (HSCs) [69,70]. The promoter regions of *p15* and *p21* are hypomethylated in psoriasis, and colony counts of HPP-CFCs in the bone marrow of psoriatic patients are significantly lower than those in healthy controls [69]. In addition, the lower colony-formation capability of HPP-CFCs from bone marrow hematopoietic progenitor cells in psoriasis patients is closely associated with the methylation levels of *p16* in HPP-CFCs cells [70].

11.5 HISTONE MODIFICATION

In eukaryotes, DNA is packaged into nucleosomes that consist of two copies each of the histone proteins H2A, H2B, H3, and H4. There are three conditions in histone modification, which include the permissive (active promoter) status, repressive (inactive promoter) status, and poised (accessible promoter) status. In addition, methylation, acetylation, phosphorylation, and ubiquitination of histone tails occur at specific sites and residues, and control of gene expression occurs by regulating DNA accessibility to RNA polymerase II and other transcription factors. Histone modifications influence the transcription process and reflect the transcriptional states of the underlying genes [71,72]. For example, tri-methylation of histone H3 tails with lysine 4 (H3K4) is strongly associated with transcriptional activation, and tri-methylation of H3K27 is frequently associated with gene silencing [73]. Methylation of H3K9 and H3K27 has also been reported to be the key regulator in red eye pigmentation of *Drosophila* (position-effect variegation) [74–76], long-term gene silencing (polycomb silencing) [76–78], and X-chromosome inactivation [79]. In addition, these two modified histones possess certain characteristics of a prototype-guided information duplication system. However, whether methylation of histone tails is inherited during mitotic divisions remains unclear. In addition to methylation, acetylation of histone tails is also considered to result in gene activation, whereas histone deacetylation is believed to result in gene silencing and is, therefore, considered a repressive marker. Acetylation of histone tails is catalyzed by histone acetyltransferases (HATs), and the acetyl groups are removed from the histone tails by deacetylases.

An emerging paradigm for epigenetic modification indicates association between DNA methylation and histone modification. DNMT3L is a regulatory factor related (in sequence) to the mammalian *de novo* DNMT3A and DNMT3B. DNMT3L was demonstrated to catalyze methylation of H3K4 via recruitment or activation of DNMT3A2 [80]. Cedar and Bergman showed a bidirectional relationship between histone and DNA methylation. They believed that methylation of the histone tail may influence gene silencing. Methylation usually occurs on closed chromatin structure during early development; this state is maintained through regulation of DNA methylation and chromatin structure following cell division [81].

11.5.1 Histone Modification in Psoriasis

The balance of acetylation and deacetylation of histone tails with lysine residues is tightly regulated by HATs and histone deacetylases (HDACs). Aberrant expression of these two enzymes has been observed in psoriasis. Tovar-Castillo et al. [82] reported that HDAC-1 mRNA is overexpressed in psoriatic skin as compared with that in normal skin of healthy controls. In recent clinical practices, HDAC inhibitors (HDAC-Is) are used to treat chronic immune and inflammatory disorders such as systemic lupus erythematosus (SLE) and psoriasis [83,84]. Next, the silent mating type information regulation 2 homologue 1 (HDAC-SIRT1) is a nicotinamide adenine dinucleotide (NAD$^+$)-dependent deacetylase that is involved in cellular metabolisms and cellular stress responses [85]. Zhang et al. [86] also found that *SIRT1* mRNA expression was significantly downregulated in PBMCs from psoriasis patients compared to those from healthy controls. In addition, it has been proposed that HDAC-SIRT1 inhibits E2F1 activity to prevent normal keratinocyte proliferation [87]. E2F1 is a member of the E2F family of transcription factors, and activation of E2F signaling is often involved in aberrant cell proliferation or apoptosis [88].

11.6 NONCODING RNAs

miRNAs are the most studied class of noncoding RNAs that regulate gene expression by binding to the 3′ UTRs of target mRNA, which ultimately leads to mRNA degradation or inhibition of protein translation [89,90]. miRNAs are encoded in the genome and are derived from the intergenic regions (encoded as a single gene or gene clusters) or the intron regions. pri-miRNA is a 70-nucleotide structure that is transcribed from the nucleus. The RNase III enzymes such as Drosha and DGCR8 (double-stranded RNA-binding protein) process pri-miRNA to form pre-miRNA with a stem-loop structure. The resulting pre-miRNA is imported in the cytoplasm by the transporter protein Exportin 5. The pre-miRNA is bound and cleaved by Dicer (RNase III enzyme) in the cytoplasm to generate mature miRNA molecules.

To date, almost 2000 mature miRNAs have been found in the human genome, the information concerning each miRNA being available on the website: http://www.mirbase.org/. miRNAs are considered both oncogenes and tumor suppressor genes, and aberrant miRNA expression has been reported to contribute to the pathogenesis of most malignancies [91]. In addition, the role of miRNAs in cardiovascular diseases and liver injury is also well-established and therefore they are regarded as therapeutic targets [92,93]. Recently, the miRNA expression microarray platform was comprehensively applied for screening novel miRNAs from different psoriasis specimens. In recent years, next-generation sequencing technologies have matured and become capable of assessing epigenetic markers on a genome-wide scale [94]. Compared with the array platform, the next-generation sequencing technologies can provide superior data quality; these technologies have been applied to the study of miRNAs (miRNA-seq) and histone markers (ChIP-seq). Some studies have shown that miRNAs play important roles in the regulation of processes during early development, cell proliferation, cell differentiation, cell apoptosis, signal transduction, and organ development [95−99]. Aberrant miRNA expression is also involved in the regulation of immune system development and in the pathogenesis of chronic inflammatory diseases [100−102]. The first evidence of the involvement of miRNAs in inflammatory diseases was obtained from the miRNA expression microarray data. Novel miRNA signature was found to be associated with psoriasis [102]. miRNAs are expressed in a nonrandom manner in healthy and psoriatic skin. However, in the mRNA expression array data and the pathway signaling of psoriasis, a set of miRNAs was found to be downregulated in psoriasis. These findings imply that miRNAs play general roles in the pathogenesis of psoriasis and that some miRNAs are specifically downregulated in psoriatic skin. In addition, some genetic regions of interest were found to overlap with the loci identified as being involved in the development of psoriasis and atopic dermatitis [103]. In recent years, increasing evidence suggests that miRNAs are also important in the pathogenesis of skin disorders such as psoriasis and atopic eczema.

11.6.1 miR-203 in Psoriasis

miR-203 is the first miRNA discovered in skin; it is overexpressed in psoriatic skin as compared with that in both normal skin from healthy controls and lesional skin from atopic eczema patients [102]. miR-203 is highly expressed in keratinocytes and plays a role in skin morphogenesis by suppressing the transcription factor p63 [104]. Sonkoly et al. [102] also reported specific expression of miR-203 in keratinocytes. Their results suggest that miR-203 plays a role in the regulation of keratinocyte functions, especially in psoriatic skin. One of the target genes of miR-203 is the suppressor of cytokine signaling-3 (*SOCS-3*), which is a negative regulator in the STAT3 pathway. In the STAT3 pathway, SOCS-3 can be activated by inflammatory cytokines and it has important functions in cell growth, survival, and differentiation [105]. A cell-based study showed that overexpression of miR-203 repressed skin inflammation by downregulating the cytokine levels of TNF-α and IL-24 in primary keratinocytes from psoriasis patients. Sonkoly et al. suggested that the inflammatory cytokines TNF-α and IL-24 are direct targets of miR-203 in keratinocytes in the pathway of epidermal remodeling and skin homeostasis in the pathogenesis of psoriasis (Figure 11.1).

11.6.2 miR-146a in Psoriasis

In addition to miR-203, Sonkoly et al. [102] also used the miRNA expression array technique to screen out three other miRNAs (miR-146a, miR-21, and miR-125b), which were differentially expressed in psoriatic skin. For miR-146a, Taganov et al. [106] showed that miR-146a inhibits the expression of TNF receptor-associated factor 6 (*TRAF-6*) and IL-1 receptor-associated kinase 1 (*IRAK-1*). These two proteins are the key adapter molecules downstream of Toll-like receptors and cytokine receptors. Moreover, promoter analysis of the *miR-146a* gene also revealed that

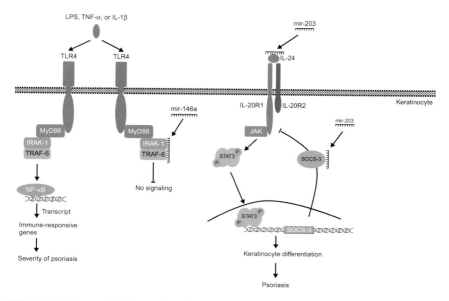

FIGURE 11.1   miR-146a and miR-203 act as key regulators in the pathogenesis of psoriasis.

NF-κB plays a critical role in the induction of miR-146a transcription after treatment with lipopolysaccharide (LPS), TNF-α, and IL-1β. Similarly, Meisgen et al. [107] reported the downregulation of IRAK-1 and TRAF-6 and suppression of NF-κB promoter-binding activity by miR-146a. Furthermore, overexpression of miR-146a was reported to significantly suppress the production of IL-8, CCL20, and TNF-α; these factors in turn suppress the chemotactic attraction of neutrophils by keratinocytes. Xia et al. [108] found that overexpression of miR-146a could inhibit *IRAK-1* expression and was positively correlated to the PASI score in psoriatic patients. Thus, these results suggest that miR-146a may be a regulator in the innate immunity of keratinocytes, which prevents the production of inflammatory mediators under homeostatic conditions.

11.6.3 miR-125b in Psoriasis

The expression of miR-125b is also downregulated in psoriasis. Unlike miR-146a, miR-125b targets TNF-α and directly inhibits its post-transcriptional modification. Therefore, downregulation of miR-125b was considered to be correlated with excess production of TNF-α in skin inflammation [109]. Xu et al. [110] found that overexpression of miR-125b suppressed cell proliferation and induced the expression of several differentiation markers, such as keratin 10 (K10) and filaggrin, in primary human keratinocytes. They also demonstrated that miR-125b modulates keratinocyte proliferation and differentiation through direct repression of fibroblast growth factor receptor 2 (FGFR2). Therefore, miR-125b may be a potential therapeutic target for psoriasis that acts by modulating proliferation and differentiation in keratinocytes.

11.6.4 miR-21 in Psoriasis

Two studies have reported that miR-21 is significantly upregulated in psoriatic skin, as determined by miRNA expression array [102,111]. miR-21 is a well-known oncogene, and some papers have reported that miR-21 plays an important role in cardiovascular diseases and inflammation [112,113]. In addition, miR-21 is also considered as an antiapoptotic and antiproliferative molecule in a variety of cell types [114,115]. Meisgen et al. [107] found that downregulation of miR-21 increases the apoptosis rate of activated T cells and that upregulation of miR-21 may contribute to T-cell-derived inflammation in psoriatic skin. Therefore, they suggested that upregulation of miR-21 in psoriatic skin was caused by infiltration of activated CD4$^+$ T cells. In addition, upregulation of miR-21 may contribute to skin inflammation by suppressing CD4$^+$ T-cell apoptosis. Therefore, regulation of miR-21 in T cells is considered to modify activated T cells in the microenvironment of psoriatic lesions.

11.6.5 Other miRNAs in Psoriasis

In recent microarray-based studies, some miRNAs were identified as being associated with the pathogenesis of psoriasis, but the detailed mechanism of their association remains unclear. Zibert et al. [116] investigated the interaction of miRNA and mRNA expression in lesional/nonlesional skin in psoriasis patients. They found 42 upregulated and 5 downregulated miRNAs expressed in lesional skin as compared with normal skin. Furthermore, they also validated that overexpression of miR-221 and miR-222 leads to degradation of TIMP3, resulting in decreased TIMP3 protein levels in a cell line model. Another study indicated that 38 circulating miRNAs were suppressed by the TNF-inhibitor etanercept in psoriasis patients, by using the TaqMan miRNA low-density array (TLDA). After validation by quantitative polymerase chain reaction (PCR), the serum levels of miR-106b, miR-26b, miR-142-3p, miR-223, and miR-126 were found to be significantly downregulated by etanercept in responders (PASI score >50%) [117]. Lerman et al. [118] found

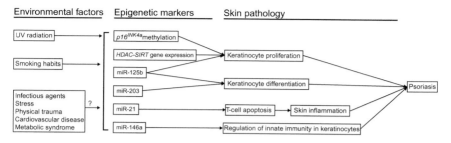

FIGURE 11.2 **The overview of published epigenetic markers between exposure to environmental factors and the pathogenesis of psoriasis.** The question mark indicates lack of evidence to prove the association between risk factors of psoriasis and epigenetic markers.

14 differentially expressed miRNAs in psoriatic skin and determined that IGF-1R and miR-99a are reciprocally expressed in the epidermis. They also found that overexpression of miR-99a in primary keratinocytes leads to downregulation of the endogenous IGF-1R protein level and inhibition of keratinocyte proliferation, and that it leads to the upregulation of keratin 10. Ichihara et al. [119] determined the role of miR-424 in psoriatic skin and serum samples. They found that downregulation of miR-424 could lead to overexpression of MEK1 and cyclin E1 in keratinocytes and reduction of serum miR-424 concentration in psoriasis patients. miR-31, a miRNA overexpressed in psoriasis keratinocytes, may contribute to skin inflammation by enhancing leukocyte migration into the skin [120].

11.7 CONCLUSION

In this review, we have summarized the epigenetic modifications of some genes that are involved in the pathogenesis of psoriasis and discussed epigenetic aberrations acting as triggers in psoriasis. We have also reviewed some candidate epigenetic markers involved in the pathogenesis of psoriasis based on the reports of the published studies summarized in this review (see Figure 11.2). However, the detailed mechanisms involved between epigenetic modifications and the pathogenesis of psoriasis need to be further examined to elucidate the mystery behind this relatively common disease. In addition, some previous studies have reported that epigenetic changes are also influenced by environmental factors such as UV radiation and smoking habits. Therefore, future studies need to determine exposure to environmental factors as a risk factor in the regulation of epigenetic changes that contribute to the pathogenesis of psoriasis.

List of Abbreviations

CHARM	comprehensive analysis of relative DNA methylation
ChIP	chromatin immunoprecipitation
DNMT	DNA methyltransferase
FGFR2	fibroblast growth factor receptor 2
GO	gene ontology
HAT	histone acetyltransferase
HDAC	histone deacetylase
HPP-CFC	high-proliferative potential colony-forming cell
HSC	hematopoietic stem cell
IRAK-1	IL-1 receptor-associated kinase 1
K10	keratin 10
LPS	lipopolysaccharide
MBD	methyl-DNA binding domain
MeDIP	methylated DNA immunoprecipitation
miRNA	microRNA
MSC	mesenchymal stem cell
PASI	psoriasis area and severity index
SAM	S-adenosyl methionine
SIRT1	silent mating type information regulation 2 homologue 1
SOCS-3	suppressor of cytokine signaling-3
TRAF-6	TNF receptor-associated factor 6
UV	ultraviolet

References

[1] Chandran V, Raychaudhuri SP. Geoepidemiology and environmental factors of psoriasis and psoriatic arthritis. J Autoimmun 2010;34(3):J314−21.

[2] Lebwohl M. Psoriasis. Lancet 2003;361(9364):1197−204.

[3] Parisi R, Symmons DP, Griffiths CE, Ashcroft DM. Global epidemiology of psoriasis: a systematic review of incidence and prevalence. J Invest Dermatol 2013;133(2):377−85.

[4] Cohen AD, Gilutz H, Henkin Y, Zahger D, Shapiro J, Bonneh DY, et al. Psoriasis and the metabolic syndrome. Acta Derm Venereol 2007;87(6):506−9.

[5] Cohen AD, Sherf M, Vidavsky L, Vardy DA, Shapiro J, Meyerovitch J. Association between psoriasis and the metabolic syndrome. A cross-sectional study. Dermatology 2008;216(2):152−5.

[6] Gisondi P, Tessari G, Conti A, Piaserico S, Schianchi S, Peserico A, et al. Prevalence of metabolic syndrome in patients with psoriasis: a hospital-based case-control study. Br J Dermatol 2007;157(1):68−73.

[7] Li F, Jin HZ, Wang BX. Prevalence of metabolic syndrome in psoriasis inpatients in Peking Union Medical College Hospital. Zhongguo Yi Xue Ke Xue Yuan Xue Bao 2010; 32(5):583−5.

[8] Love TJ, Qureshi AA, Karlson EW, Gelfand JM, Choi HK. Prevalence of the metabolic syndrome in psoriasis: results from the National Health and Nutrition Examination Survey, 2003−2006. Arch Dermatol 2011;147(4):419−24.

[9] Mebazaa A, El Asmi M, Zidi W, Zayani Y, Cheikh Rouhou R, El Ounifi S, et al. Metabolic syndrome in Tunisian psoriatic patients: prevalence and determinants. J Eur Acad Dermatol Venereol 2011;25(6):705−9.

[10] Nisa N, Qazi MA. Prevalence of metabolic syndrome in patients with psoriasis. Indian J Dermatol Venereol Leprol 2010;76(6):662–5.

[11] Christensen TE, Callis KP, Papenfuss J, Hoffman MS, Hansen CB, Wong B, et al. Observations of psoriasis in the absence of therapeutic intervention identifies two unappreciated morphologic variants, thin-plaque and thick-plaque psoriasis, and their associated phenotypes. J Invest Dermatol 2006;126(11):2397–403.

[12] Feil R, Fraga MF. Epigenetics and the environment: emerging patterns and implications. Nat Rev Genet 2011;13(2):97–109.

[13] Okada S, Weatherhead E, Targoff IN, Wesley R, Miller FW. Global surface ultraviolet radiation intensity may modulate the clinical and immunologic expression of autoimmune muscle disease. Arthritis Rheum 2003;48(8):2285–93.

[14] Katiyar SK. UV-induced immune suppression and photocarcinogenesis: chemoprevention by dietary botanical agents. Cancer Lett 2007;255(1):1–11.

[15] Chen IP, Henning S, Faust A, Boukamp P, Volkmer B, Greinert R. UVA-induced epigenetic regulation of P16(INK4a) in human epidermal keratinocytes and skin tumor derived cells. Photochem Photobiol Sci 2012;11(1):180–90.

[16] Tan G, Niu J, Shi Y, Ouyang H, Wu ZH. NF-kappaB-dependent microRNA-125b up-regulation promotes cell survival by targeting p38alpha upon ultraviolet radiation. J Biol Chem 2012;287(39):33036–47.

[17] Behnam SM, Behnam SE, Koo JY. Smoking and psoriasis. Skinmed 2005;4(3):174–6.

[18] Fortes C, Mastroeni S, Leffondre K, Sampogna F, Melchi F, Mazzotti E, et al. Relationship between smoking and the clinical severity of psoriasis. Arch Dermatol 2005;141(12):1580–4.

[19] Poikolainen K, Karvonen J, Pukkala E. Excess mortality related to alcohol and smoking among hospital-treated patients with psoriasis. Arch Dermatol 1999;135(12):1490–3.

[20] Kim DH, Nelson HH, Wiencke JK, Zheng S, Christiani DC, Wain JC, et al. p16 (INK4a) and histology-specific methylation of CpG islands by exposure to tobacco smoke in non-small cell lung cancer. Cancer Res 2001;61(8):3419–24.

[21] Maccani MA, Avissar-Whiting M, Banister CE, McGonnigal B, Padbury JF, Marsit CJ. Maternal cigarette smoking during pregnancy is associated with downregulation of miR-16, miR-21, and miR-146a in the placenta. Epigenetics 2010;5(7):583–9.

[22] Torii K, Saito C, Furuhashi T, Nishioka A, Shintani Y, Kawashima K, et al. Tobacco smoke is related to Th17 generation with clinical implications for psoriasis patients. Exp Dermatol 2011;20(4):371–3.

[23] Rapp SR, Feldman SR, Exum ML, Fleischer Jr. AB, Reboussin DM. Psoriasis causes as much disability as other major medical diseases. J Am Acad Dermatol 1999;41(3 Pt 1):401–7.

[24] Schmitt JM, Ford DE. Work limitations and productivity loss are associated with health-related quality of life but not with clinical severity in patients with psoriasis. Dermatology 2006;213(2):102–10.

[25] Schmitt JM, Ford DE. Role of depression in quality of life for patients with psoriasis. Dermatology 2007;215(1):17–27.

[26] Schmitt J, Ford DE. Understanding the relationship between objective disease severity, psoriatic symptoms, illness-related stress, health-related quality of life and depressive symptoms in patients with psoriasis—a structural equations modeling approach. Gen Hosp Psychiatry 2007;29(2):134–40.

[27] Koebnick C, Black MH, Smith N, Der-Sarkissian JK, Porter AH, Jacobsen SJ, et al. The association of psoriasis and elevated blood lipids in overweight and obese children. J Pediatr 2011;159(4):577–83.

[28] Naldi L, Chatenoud L, Linder D, Belloni Fortina A, Peserico A, Virgili AR, et al. Cigarette smoking, body mass index, and stressful life events as risk factors for psoriasis: results from an Italian case–control study. J Invest Dermatol 2005;125(1):61–7.

[29] Setty AR, Curhan G, Choi HK. Obesity, waist circumference, weight change, and the risk of psoriasis in women: Nurses' Health Study II. Arch Intern Med 2007;167 (15):1670–5.

[30] Gisondi P, Girolomoni G. Psoriasis and atherothrombotic diseases: disease-specific and non-disease-specific risk factors. Semin Thromb Hemost 2009;35(3):313–24.

[31] Mehta NN, Azfar RS, Shin DB, Neimann AL, Troxel AB, Gelfand JM. Patients with severe psoriasis are at increased risk of cardiovascular mortality: cohort study using the general practice research database. Eur Heart J 2010;31(8):1000–6.

[32] Miller IM, Ellervik C, Yazdanyar S, Jemec GB. Meta-analysis of psoriasis, cardiovascular disease, and associated risk factors. J Am Acad Dermatol 2013;69(6):1014–24.

[33] Henseler T, Christophers E. Psoriasis of early and late onset: characterization of two types of psoriasis vulgaris. J Am Acad Dermatol 1985;13(3):450–6.

[34] Grjibovski AM, Olsen AO, Magnus P, Harris JR. Psoriasis in Norwegian twins: contribution of genetic and environmental effects. J Eur Acad Dermatol Venereol 2007;21 (10):1337–43.

[35] Kastelan M, Prpic Massari L, Gruber F, Zamolo G, Zauhar G, Coklo M, et al. Perforin expression is upregulated in the epidermis of psoriatic lesions. Br J Dermatol 2004;151(4):831–6.

[36] Giblin PA, Lemieux RM. LFA-1 as a key regulator of immune function: approaches toward the development of LFA-1-based therapeutics. Curr Pharm Des 2006;12 (22):2771–95.

[37] Brooks WH, Le Dantec C, Pers JO, Youinou P, Renaudineau Y. Epigenetics and autoimmunity. J Autoimmun 2010;34(3):J207–19.

[38] Mee JB, Johnson CM, Morar N, Burslem F, Groves RW. The psoriatic transcriptome closely resembles that induced by interleukin-1 in cultured keratinocytes: dominance of innate immune responses in psoriasis. Am J Pathol 2007;171(1):32–42.

[39] Suarez-Farinas M, Lowes MA, Zaba LC, Krueger JG. Evaluation of the psoriasis transcriptome across different studies by gene set enrichment analysis (GSEA). PLOS ONE 2010;5(4):e10247.

[40] Handy DE, Castro R, Loscalzo J. Epigenetic modifications: basic mechanisms and role in cardiovascular disease. Circulation 2011;123(19):2145–56.

[41] Espada J, Esteller M. DNA methylation and the functional organization of the nuclear compartment. Seminars Cell Dev Biol 2010;21(2):238–46.

[42] Postovit LM, Costa FF, Bischof JM, Seftor EA, Wen B, Seftor RE, et al. The commonality of plasticity underlying multipotent tumor cells and embryonic stem cells. J Cell Biochem 2007;101(4):908–17.

[43] Holliday R, Pugh JE. DNA modification mechanisms and gene activity during development. Science 1975;187(4173):226–32.

[44] Doi A, Park IH, Wen B, Murakami P, Aryee MJ, Irizarry R, et al. Differential methylation of tissue- and cancer-specific CpG island shores distinguishes human induced pluripotent stem cells, embryonic stem cells and fibroblasts. Nat Genet 2009;41(12):1350–3.

[45] Ji H, Ehrlich LI, Seita J, Murakami P, Doi A, Lindau P, et al. Comprehensive methylome map of lineage commitment from haematopoietic progenitors. Nature 2010;467 (7313):338–42.

[46] Shenker N, Flanagan JM. Intragenic DNA methylation: implications of this epigenetic mechanism for cancer research. Br J Cancer 2012;106(2):248–53.

[47] Maunakea AK, Nagarajan RP, Bilenky M, Ballinger TJ, D'Souza C, Fouse SD, et al. Conserved role of intragenic DNA methylation in regulating alternative promoters. Nature 2010;466(7303):253–7.

[48] Robertson KD. DNA methylation and human disease. Nat Rev Genet 2005;6 (8):597−610.

[49] Jones PA, Baylin SB. The fundamental role of epigenetic events in cancer. Nat Rev Genet 2002;3(6):415−28.

[50] Yang IV, Schwartz DA. Epigenetic control of gene expression in the lung. Am J Respir Crit Care Med 2011;183(10):1295−301.

[51] Ladd-Acosta C, Aryee MJ, Ordway JM, Feinberg AP. Comprehensive high-throughput arrays for relative methylation (CHARM). Curr Protoc Hum Genet 2010; Chapter 20:Unit 20.1.1−19.

[52] Schones DE, Zhao K. Genome-wide approaches to studying chromatin modifications. Nat Rev Genet 2008;9(3):179−91.

[53] Liu CG, Spizzo R, Calin GA, Croce CM. Expression profiling of microRNA using oligo DNA arrays. Methods 2008;44(1):22−30.

[54] Rakyan VK, Down TA, Balding DJ, Beck S. Epigenome-wide association studies for common human diseases. Nat Rev Genet 2011;12(8):529−41.

[55] Gervin K, Vigeland MD, Mattingsdal M, Hammero M, Nygard H, Olsen AO, et al. DNA methylation and gene expression changes in monozygotic twins discordant for psoriasis: identification of epigenetically dysregulated genes. PLOS Genet 2012;8(1):e1002454.

[56] Han J, Park SG, Bae JB, Choi J, Lyu JM, Park SH, et al. The characteristics of genome-wide DNA methylation in naive CD4^{+} T cells of patients with psoriasis or atopic dermatitis. Biochem Biophys Res Commun 2012;422(1):157−63.

[57] Hou R, Yin G, An P, Wang C, Liu R, Yang Y, et al. DNA methylation of dermal MSCs in psoriasis: identification of epigenetically dysregulated genes. J Dermatol Sci 2013;72(2):103−9.

[58] Roberson ED, Liu Y, Ryan C, Joyce CE, Duan S, Cao L, et al. A subset of methylated CpG sites differentiate psoriatic from normal skin. J Invest Dermatol 2012;132(3 Pt 1): 583−92.

[59] Zhang P, Zhao M, Liang G, Yin G, Huang D, Su F, et al. Whole-genome DNA methylation in skin lesions from patients with psoriasis vulgaris. J Autoimmun 2013;41: 17−24.

[60] Zhang P, Su Y, Chen H, Zhao M, Lu Q. Abnormal DNA methylation in skin lesions and PBMCs of patients with psoriasis vulgaris. J Dermatol Sci 2010;60(1):40−2.

[61] Nickoloff BJ, Wrone-Smith T. Superantigens, autoantigens, and pathogenic T cells in psoriasis. J Invest Dermatol 1998;110(4):459−60.

[62] Hong K, Chu A, Ludviksson BR, Berg EL, Ehrhardt RO. IL-12, independently of IFN-gamma, plays a crucial role in the pathogenesis of a murine psoriasis-like skin disorder. J Immunol 1999;162(12):7480−91.

[63] Ng A, Griffiths A, Cole T, Davison V, Griffiths M, Larkin S, et al. Congenital abnormalities and clinical features associated with Wilms' tumour: a comprehensive study from a centre serving a large population. Eur J Cancer 2007;43(9):1422−9.

[64] Wrone-Smith T, Nickoloff BJ. Dermal injection of immunocytes induces psoriasis. J Clin Invest 1996;98(8):1878−87.

[65] Chen M, Chen ZQ, Cui PG, Yao X, Li YM, Li AS, et al. The methylation pattern of p16INK4a gene promoter in psoriatic epidermis and its clinical significance. Br J Dermatol 2008;158(5):987−93.

[66] Ruchusatsawat K, Wongpiyabovorn J, Shuangshoti S, Hirankarn N, Mutirangura A. SHP-1 promoter 2 methylation in normal epithelial tissues and demethylation in psoriasis. J Mol Med (Berl) 2006;84(2):175−82.

[67] Zhang Q, Wang HY, Marzec M, Raghunath PN, Nagasawa T, Wasik MA. STAT3- and DNA methyltransferase 1-mediated epigenetic silencing of SHP-1 tyrosine phosphatase tumor suppressor gene in malignant T lymphocytes. Proc Natl Acad Sci USA 2005;102(19):6948−53.

[68] Sano S, Chan KS, Carbajal S, Clifford J, Peavey M, Kiguchi K, et al. Stat3 links activated keratinocytes and immunocytes required for development of psoriasis in a novel transgenic mouse model. Nat Med 2005;11(1):43−9.

[69] Zhang K, Zhang R, Li X, Yin G, Niu X. Promoter methylation status of p15 and p21 genes in HPP-CFCs of bone marrow of patients with psoriasis. Eur J Dermatol 2009; 19(2):141−6.

[70] Zhang RL, Niu XP, Li XH, Zhang KM, Yin GH. CFU-HPP colony formation of bone marrow hematopoietic proginitor cells in psoriatic patients and methylation of p16 gene promotor in CFU-HPP colony cells. Zhongguo Shi Yan Xue Ye Xue Za Zhi 2007; 15(4):780−4.

[71] Li B, Carey M, Workman JL. The role of chromatin during transcription. Cell 2007;128(4):707−19.

[72] Henikoff S, Shilatifard A. Histone modification: cause or cog? Trends Genet 2011;27 (10):389−96.

[73] Greer EL, Shi Y. Histone methylation: a dynamic mark in health, disease and inheritance. Nat Rev Genet 2012;13(5):343−57.

[74] Rea S, Eisenhaber F, O'Carroll D, Strahl BD, Sun ZW, Schmid M, et al. Regulation of chromatin structure by site-specific histone H3 methyltransferases. Nature 2000;406 (6796):593−9.

[75] Lachner M, O'Carroll D, Rea S, Mechtler K, Jenuwein T. Methylation of histone H3 lysine 9 creates a binding site for HP1 proteins. Nature 2001;410(6824): 116−20.

[76] Cao R, Wang L, Wang H, Xia L, Erdjument-Bromage H, Tempst P, et al. Role of histone H3 lysine 27 methylation in polycomb-group silencing. Science 2002;298(5595): 1039−43.

[77] Czermin B, Melfi R, McCabe D, Seitz V, Imhof A, Pirrotta V. Drosophila enhancer of Zeste/ESC complexes have a histone H3 methyltransferase activity that marks chromosomal polycomb sites. Cell 2002;111(2):185−96.

[78] Kuzmichev A, Nishioka K, Erdjument-Bromage H, Tempst P, Reinberg D. Histone methyltransferase activity associated with a human multiprotein complex containing the enhancer of Zeste protein. Genes Dev 2002;16(22):2893−905.

[79] Plath K, Fang J, Mlynarczyk-Evans SK, Cao R, Worringer KA, Wang H, et al. Role of histone H3 lysine 27 methylation in X inactivation. Science 2003;300(5616):131−5.

[80] Hu JL, Zhou BO, Zhang RR, Zhang KL, Zhou JQ, Xu GL. The N-terminus of histone H3 is required for de novo DNA methylation in chromatin. Proc Natl Acad Sci USA 2009;106(52):22187−92.

[81] Cedar H, Bergman Y. Linking DNA methylation and histone modification: patterns and paradigms. Nat Rev Genet 2009;10(5):295−304.

[82] Tovar-Castillo LE, Cancino-Diaz JC, Garcia-Vazquez F, Cancino-Gomez FG, Leon-Dorantes G, Blancas-Gonzalez F, et al. Under-expression of VHL and over-expression of HDAC-1, HIF-1alpha, LL-37, and IAP-2 in affected skin biopsies of patients with psoriasis. Int J Dermatol 2007;46(3):239−46.

[83] McLaughlin F, La Thangue NB. Histone deacetylase inhibitors in psoriasis therapy. Curr Drug Targets Inflamm Allergy 2004;3(2):213−19.

[84] Reilly CM, Regna N, Mishra N. HDAC inhibition in lupus models. Mol Med 2011;17 (5−6):417−25.

[85] Haigis MC, Guarente LP. Mammalian sirtuins—emerging roles in physiology, aging, and calorie restriction. Genes Dev 2006;20(21):2913−21.

[86] Zhang P, Su Y, Zhao M, Huang W, Lu Q. Abnormal histone modifications in PBMCs from patients with psoriasis vulgaris. Eur J Dermatol 2011;21(4):552−7.

[87] Blander G, Bhimavarapu A, Mammone T, Maes D, Elliston K, Reich C, et al. SIRT1 promotes differentiation of normal human keratinocytes. J Invest Dermatol 2009;129 (1):41–9.

[88] DeGregori J, Leone G, Miron A, Jakoi L, Nevins JR. Distinct roles for E2F proteins in cell growth control and apoptosis. Proc Natl Acad Sci USA 1997;94(14): 7245–50.

[89] Bartel DP. MicroRNAs: genomics, biogenesis, mechanism, and function. Cell 2004;116(2):281–97.

[90] Flynt AS, Lai EC. Biological principles of microRNA-mediated regulation: shared themes amid diversity. Nat Rev Genet 2008;9(11):831–42.

[91] Croce CM. Causes and consequences of microRNA dysregulation in cancer. Nat Rev Genet 2009;10(10):704–14.

[92] Jopling C. Liver-specific microRNA-122: biogenesis and function. RNA Biol 2012;9(2): 137–42.

[93] Papageorgiou N, Tousoulis D, Androulakis E, Siasos G, Briasoulis A, Vogiatzi G, et al. The role of microRNAs in cardiovascular disease. Curr Med Chem 2012;19(16): 2605–10.

[94] Metzker ML. Sequencing technologies—the next generation. Nat Rev Genet 2010;11 (1):31–46.

[95] Alvarez-Garcia I, Miska EA. MicroRNA functions in animal development and human disease. Development 2005;132(21):4653–62.

[96] He L, Hannon GJ. MicroRNAs: small RNAs with a big role in gene regulation. Nat Rev Genet 2004;5(7):522–31.

[97] Kloosterman WP, Lagendijk AK, Ketting RF, Moulton JD, Plasterk RH. Targeted inhibition of miRNA maturation with morpholinos reveals a role for miR-375 in pancreatic islet development. PLOS Biol 2007;5(8):e203.

[98] Miska EA. How microRNAs control cell division, differentiation and death. Curr Opin Genet Dev 2005;15(5):563–8.

[99] Reinhart BJ, Slack FJ, Basson M, Pasquinelli AE, Bettinger JC, Rougvie AE, et al. The 21-nucleotide let-7 RNA regulates developmental timing in *Caenorhabditis elegans*. Nature 2000;403(6772):901–6.

[100] Chen CZ, Li L, Lodish HF, Bartel DP. MicroRNAs modulate hematopoietic lineage differentiation. Science 2004;303(5654):83–6.

[101] Sonkoly E, Stahle M, Pivarcsi A. MicroRNAs and immunity: novel players in the regulation of normal immune function and inflammation. Seminars Cancer Biol 2008;18(2):131–40.

[102] Sonkoly E, Wei T, Janson PC, Saaf A, Lundeberg L, Tengvall-Linder M, et al. MicroRNAs: novel regulators involved in the pathogenesis of psoriasis? PLOS ONE 2007;2(7):e610.

[103] Bonness S, Bieber T. Molecular basis of atopic dermatitis. Curr Opin Allergy Clin Immunol 2007;7(5):382–6.

[104] Yi R, Poy MN, Stoffel M, Fuchs E. A skin microRNA promotes differentiation by repressing "stemness". Nature 2008;452(7184):225–9.

[105] Kubo M, Hanada T, Yoshimura A. Suppressors of cytokine signaling and immunity. Nat Immunol 2003;4(12):1169–76.

[106] Taganov KD, Boldin MP, Chang KJ, Baltimore D. NF-kappaB-dependent induction of microRNA miR-146, an inhibitor targeted to signaling proteins of innate immune responses. Proc Natl Acad Sci USA 2006;103(33):12481–6.

[107] Meisgen F, Xu N, Wei T, Janson PC, Obad S, Broom O, et al. MiR-21 is up-regulated in psoriasis and suppresses T cell apoptosis. Exp Dermatol 2012;21(4):312–14.

[108] Xia P, Fang X, Zhang ZH, Huang Q, Yan KX, Kang KF, et al. Dysregulation of miRNA146a versus IRAK1 induces IL-17 persistence in the psoriatic skin lesions. Immunol Lett 2012;148(2):151−62.
[109] Tili E, Michaille JJ, Cimino A, Costinean S, Dumitru CD, Adair B, et al. Modulation of miR-155 and miR-125b levels following lipopolysaccharide/TNF-alpha stimulation and their possible roles in regulating the response to endotoxin shock. J Immunol 2007;179(8):5082−9.
[110] Xu N, Brodin P, Wei T, Meisgen F, Eidsmo L, Nagy N, et al. MiR-125b, a microRNA downregulated in psoriasis, modulates keratinocyte proliferation by targeting FGFR2. J Invest Dermatol 2011;131(7):1521−9.
[111] Lovendorf MB, Zibert JR, Hagedorn PH, Glue C, Odum N, Ropke MA, et al. Comparison of microRNA expression using different preservation methods of matched psoriatic skin samples. Exp Dermatol 2012;21(4):299−301.
[112] Cheng Y, Ji R, Yue J, Yang J, Liu X, Chen H, et al. MicroRNAs are aberrantly expressed in hypertrophic heart: do they play a role in cardiac hypertrophy? Am J Pathol 2007;170(6):1831−40.
[113] Sheedy FJ, Palsson-McDermott E, Hennessy EJ, Martin C, O'Leary JJ, Ruan Q, et al. Negative regulation of TLR4 via targeting of the proinflammatory tumor suppressor PDCD4 by the microRNA miR-21. Nat Immunol 2010;11(2):141−7.
[114] Chan JA, Krichevsky AM, Kosik KS. MicroRNA-21 is an antiapoptotic factor in human glioblastoma cells. Cancer Res 2005;65(14):6029−33.
[115] Si ML, Zhu S, Wu H, Lu Z, Wu F, Mo YY. miR-21-mediated tumor growth. Oncogene 2007;26(19):2799−803.
[116] Zibert JR, Lovendorf MB, Litman T, Olsen J, Kaczkowski B, Skov L. MicroRNAs and potential target interactions in psoriasis. J Dermatol Sci 2010;58(3):177−85.
[117] Pivarcsi A, Meisgen F, Xu N, Stahle M, Sonkoly E. Changes in the level of serum microRNAs in patients with psoriasis after antitumour necrosis factor-alpha therapy. Br J Dermatol 2013;169(3):563−70.
[118] Lerman G, Avivi C, Mardoukh C, Barzilai A, Tessone A, Gradus B, et al. MiRNA expression in psoriatic skin: reciprocal regulation of hsa-miR-99a and IGF-1R. PLOS ONE 2011;6(6):e20916.
[119] Ichihara A, Jinnin M, Yamane K, Fujisawa A, Sakai K, Masuguchi S, et al. microRNA-mediated keratinocyte hyperproliferation in psoriasis vulgaris. Br J Dermatol 2011;165 (5):1003−10.
[120] Xu N, Meisgen F, Butler LM, Han G, Wang XJ, Soderberg-Naucler C, et al. MicroRNA-31 is overexpressed in psoriasis and modulates inflammatory cytokine and chemokine production in keratinocytes via targeting serine/threonine kinase 40. J Immunol 2013;190(2):678−88.

12

Epigenetics and Systemic Sclerosis

Nezam Altorok[1] and Amr H. Sawalha[2,3]

[1]Division of Rheumatology, Department of Internal Medicine, University of
Toledo Medical Center, Toledo, OH [2]Division of Rheumatology,
Department of Internal Medicine, University of Michigan, Ann Arbor, MI
[3]Center for Computational Medicine and Bioinformatics, University of
Michigan, Ann Arbor, MI

12.1 INTRODUCTION

Scleroderma is a term that encompasses most forms of thickened and sclerotic skin, including both localized (morphea, linear scleroderma, etc.) and systemic sclerosis (SSc) (limited, diffuse) variants. SSc is a complex multisystem autoimmune disease that is characterized by three pathological hallmarks: activation of the immune system, vascular injury, and fibrosis of the skin and internal organs [1]. There are two major subsets of SSc: diffuse cutaneous (dcSSc) and limited cutaneous (lcSSc) that are distinguished by the extent of skin thickening; lcSSc is characterized by skin thickening that is confined to the extremities distal to the elbows and knees with or without facial involvement, whereas dcSSc is characterized by skin thickening that involves areas proximal to the elbows and knees, including the trunk [2]. Besides the extent of skin involvement, the two subsets of SSc have different patterns of organ involvement, autoantibody profiles, and survival rates. For instance, patients with lcSSc are at risk for developing subcutaneous calcinosis, telangiectasia, malabsorption, digital ulcers, and pulmonary hypertension, whereas patients with dcSSc are at high risk for interstitial lung disease, renal failure, diffuse gastrointestinal disease, and

© 2015 Elsevier Inc. All rights reserved.

TABLE 12.1 Examples of Environmental Agents Linked to SSc

Occupational exposures	Welding; silica dusts; toxic oil; xenobiotics; pesticides; ultraviolet light exposure; organic solvents; epoxy resins; benzene; trichloroethylene; xylene; urea formaldehyde; and vinyl chloride
Infectious agents	Human cytomegalovirus
Diet	L-Tryptophan
Drugs	Methysergide; pentazocine; cocaine; talc; heroin; bleomycin; ethosuximide; vitamin K; and amphetamines

myocardial involvement. Anti-topoisomerase I (Scl-70) and anti-RNA polymerase antibodies are common in dcSSc, and anti-centromere antibodies are more common in the lcSSc subset.

Despite significant efforts, the identity of the initial trigger(s) of SSc remains a major challenge. Current hypotheses suggest a possible infectious or perhaps chemical agent that activates the immune system that, in turn, causes vascular injury/dysfunction and persistent activation of fibroblasts [3]. The end product of this interaction is deposition of collagens and extracellular matrix (ECM) glycoproteins in organs, which cause organ damage and dysfunction.

Over the last few years it became evident that substantial epigenetic aberrancies are present in SSc. These findings stem from candidate gene and epigenome-wide studies and are supported by the striking geographic clustering of SSc [4]. These observations suggest a role for an epigenetic program in the pathogenesis of SSc, driven by epigenetic–environmental factors. The environmental factors that are involved in the pathogenesis of SSc are by large uncharacterized. However, epidemiological and experimental data have linked a number of occupational exposures to the development of SSc (Table 12.1).

In this chapter, we briefly discuss the pathogenesis of SSc; we then explore evidence for epigenetic aberrancies in DNA methylation, histone code, and altered expression of microRNAs (miRNAs) across different cell types that are involved in the pathogenesis of SSc.

12.2 PATHOGENESIS OF SSc

The current paradigm suggests that the pathogenesis of SSc is based upon a complex interaction between activation of the immune system and vascular damage, in association with fibroblast activation, which leads to progressive tissue fibrosis [5].

1. *Activation of the immune system*

SSc is a connective tissue disorder that is characterized by chronic deregulation of the immune system. It appears that the most prominent effect of immune system deregulation occurs in the early phases of SSc, based on the observation that there are significant inflammatory cell infiltrates in the skin in the early phases of SSc [6]. In addition, there is significant upregulation of growth factors and cytokines in skin and sera samples, respectively, from patients with SSc [7]. Moreover, SSc is characterized by the presence of disease-specific circulating autoantibodies. These observations suggest that activation of the immune system is a key feature in SSc.

a. T lymphocytes in SSc

T lymphocytes contribute to the pathogenesis of SSc. Although the total number of peripheral blood T lymphocytes in SSc is not increased and in fact may be lower than that in healthy people, there is evidence for activation of circulating T lymphocytes in SSc [8]. In addition, there is evidence for T-lymphocyte infiltration in lung and skin tissues in the early phases of SSc [9].

b. B lymphocytes

B cells, among other immune cells, are activated in SSc, as manifested by the presence of circulating antibodies, hypergammaglobulinemia, stimulation of polyclonal B cells, and overexpression of CD 19 molecules on naïve and memory B lymphocytes [10]. Although the number of naïve B cells is increased in SSc, the number of memory B cells is reduced, but they are activated [11]. Human B lymphocytes are a source of transforming growth factor-β (TGF-β) and express receptors for TGF-β [12], and B lymphocytes secrete IL-6. TGF-β and IL-6 may activate fibroblasts and induce upregulation of collagen production.

c. Other immune cells

In SSc, several cell types contribute to activation of the immune system; for example, dendritic cells, macrophages, and natural killer cells play an important role in the production of type I interferon, which is upregulated in SSc [13].

2. *Vascular injury/dysfunction*

Vascular damage occurs early in the course of SSc as suggested by the presence of Raynaud's phenomenon. There is evidence for abnormal microvascular endothelial cell (MVEC) function and structure in SSc [14]. MVEC dysfunction leads to a host of changes in the blood vessels, including obliterative vasculopathy, that eventually results in a state of chronic tissue ischemia [3].

3. Role of fibroblasts in SSc

Fibroblasts play an important role in the pathogenesis of SSc, especially considering that fibroblasts are the most proximate cell for collagen production. In comparison to normal fibroblasts, SSc fibroblasts produce more collagen [15] and are characterized by increased proliferation and decreased apoptosis *in vitro* [16]. Moreover, SSc fibroblasts exhibit increased responsiveness to TGF-β [17], and in response overexpress α-smooth muscle actin (α-SMA), which is a marker of myofibroblasts. Additionally, fibroblasts play a role in activation of the immune system via production of numerous cytokines and chemokines and upregulation of adhesion and costimulatory molecules. Fibroblasts are frequently detected near small blood vessels surrounded by inflammatory cellular infiltrate in the early stages of SSc [18]. These observations highlight an important role for fibroblasts in the pathogenesis of SSc that goes beyond collagen production and expansion of ECM, to involve activation of the immune system.

The TGF-β signaling pathway is the most potent stimulus for myofibroblast differentiation as demonstrated by a robust fibrotic response upon exposure of fibroblasts to TGF-β, along with upregulation of matrix gene expression, and myofibroblast transformation [19]. Other fibrotic pathways are also important in SSc, such as the Wnt/β-catenin, Hedgehog, and Jagged−Notch signaling pathways. Collagen gene transcription in fibroblasts is modulated by several profibrotic cytokines and transcription factors. Friend leukemia integration-1 (Fli-1) is one of the transcription factors that repress expression of collagen [20]. Fli-1 is among the transcription factors that are underexpressed in SSc fibroblasts. SMAD3 is a profibrotic factor in the TGF-β downstream signaling cascade [21], whereas SMAD7 is an inhibitory factor that modulates TGF-β signaling [22]. There is convincing evidence suggesting that deregulation of these factors and pathways in SSc fibroblasts results in an imbalance that favors increased collagen expression and tissue fibrosis.

12.3 GENETIC FACTORS IN SSc

Genome-wide and candidate-gene association studies have identified several genetic susceptibility loci in SSc (*PTPN22, STAT4, IRF5, TNFSF4, SOX5, CD247, TBX21, CTGF, BANK1, FAM167A, HGF, C8orf13-BLK, KCNA5, NLRP1, CD226, IL2RA, IL12RB2, TLR2,* and *HIF1A,* as well as several loci in the HLA region) [23,24]. However, it appears that genetic factors account for a small proportion of SSc heritability [25]. Concordance rate calculations between twin pairs help in identifying the respective contributions of genetics and the environment in disease pathogenesis. Studies have demonstrated low concordance rates in SSc monozygotic twins, which are not different from the rates seen in dizygotic twins ($\sim 5\%$) [25]. Indeed, these observations underscore a prominent role for epigenetic—environmental factors in the pathogenesis of SSc.

12.4 EPIGENETIC ABERRANCIES IN SSc

In general, epigenetic mechanisms regulate several aspects of chromatin structure and function, including regulation of the chromatin configuration and accessibility of the transcriptional machinery to gene regulatory regions. In this section, we will explore the evidence supporting the role of aberrancies in the three epigenetic programs (DNA methylation, histone code modification, and altered expression of miRNAs) involved in the pathogenesis of SSc.

1. **Fibroblasts**
   It is not surprising to see that most of the studies that have evaluated epigenetics in SSc used dermal fibroblasts, because the skin is the most common tissue involved in SSc and is easily accessible for biopsy.
   a. *DNA methylation aberrancies in fibroblasts*
      The evidence is expanding regarding the role of DNA methylation aberrancies in the pathogenesis of SSc. Genome-wide methylation studies and studies that evaluated candidate-gene DNA methylation have provided new insights into the role of DNA methylation in the pathogenesis of SSc.
      1. *Altered DNA methylation maintenance factors in SSc*
         Similarly to the situation with other autoimmune diseases, the molecular mechanism by which DNA methylation is regulated in patients with SSc is still elusive, but there is evidence of altered levels of epigenetic maintenance mediators—specifically, increased expression levels of DNMT1 in cultured SSc fibroblasts, increased expression of methyl-CpG DNA-binding protein 1 (MBD-1), MBD-2, and methyl-CpG-binding protein 2 (MeCP-2) in SSc fibroblasts [26]. Theoretically, these observations may explain the ability of cultured fibroblasts to maintain SSc phenotype over multiple generations by cellular epigenetic inheritance.
      2. *TGF-β signaling pathway*
         TGF-β is considered one of the master-regulators of fibrosis. It is generally accepted that activation of the TGF-β signaling pathway in SSc leads to a cascade of fibroblast activation [27]

and promotes the transition of fibroblasts and precursor cells toward persistently activated fibroblast phenotype, and upregulation of collagen and ECM [28]. Genome-wide DNA methylation studies have shed light on altered DNA methylation in genes that are important in activation of the TGF-β signaling pathway. For instance, *ITGA9*, which encodes for α integrin 9, is hypomethylated and overexpressed in SSc fibroblasts compared to controls [29]. Integrins are a family of transmembrane receptors that bind extracellularly to the ECM and intracellularly to the cytoskeleton, thereby "integrating" the extracellular environment with the cell interior to control cell behavior [30]. There is an interesting bidirectional interaction between integrins and TGF-β signaling in fibrosis, with TGF-β inducing integrin expression and several integrins directly controlling TGF-β activation including regulation of TGF-β downstream signaling pathway components [31]. Upregulation of integrins has been demonstrated in SSc fibroblasts [32−34] and lung fibroblasts from patients with lung fibrosis [35]. There is evidence that integrins contribute to fibroblast activation, persistent myofibroblast phenotype [36], and activation of TGF-β in fibrotic diseases [37]. Moreover, in the same study, *ADAM12* was hypomethylated and overexpressed in SSc fibroblasts. *ADAM12* contributes to the process of fibrosis through enhancing TGF-β signaling [38−41]. Thus, in light of these observations, there appears to be a role for DNA methylation in upregulation of *ITGA9* and *ADAM12* that in turn contributes to persistent activation of the TGF-β pathway, which leads to tissue fibrosis in SSc.

3. *Epigenetic aberrancies in transcription factors that are involved in collagen gene expression*

As set forth, there is an imbalance between profibrotic and antifibrotic factors in SSc. There is evidence that levels of Fli-1, which is a transcription factor encoded by the *FLI1* gene, are significantly reduced in SSc fibrotic skin and cultured SSc fibroblasts compared with healthy controls [20]. Fli-1 is a negative regulator of collagen production by fibroblasts. Therefore, it appears that reduced levels of Fli-1 may be responsible for increased collagen synthesis and accumulation in patients with SSc. Of interest, studies have demonstrated heavy methylation of the promoter region of *FLI1* in SSc fibroblasts [26]. Indeed, exposure of SSc fibroblasts to 5-azacytidine (5-AZA), a universal demethylating agent (DNMT1 inhibitor), resulted in reduced type I collagen production *in vitro*. These observations demonstrate that DNA methylation aberrancies contribute to excessive collagen production in SSc fibroblasts. It is difficult to draw conclusions regarding the potential for 5-AZA as a treatment modality in fibrosis based on this evidence, as other profibrotic factors could be overexpressed due to the global demethylation effect of 5-AZA on the genome, and, hence, there is a risk of paradoxical activation of the fibrotic process or autoimmunity.

Furthermore, there is evidence for DNA methylation aberrancies in genes encoding transcription factors that are indirectly involved in collagen production. RUNX1 and RUNX2 are transcription factors that induce expression of SOX5 and SOX6, which leads to the induction of type II collagen expression [42,43]. RUNX3, another member of the RUNX family, is also likely to contribute to collagen synthesis in association with RUNX2 [44]. Hypomethylation of *RUNX1*, *RUNX2*, and *RUNX3* associated with overexpression of at least RUNX3 in dcSSc and lcSSc has been established [29]. These data indicate that alteration of DNA methylation could affect expression of transcription factors that play a role in collagen production by SSc fibroblasts.

4. *DNA methylation aberrancies in collagen and ECM-protein encoding genes*

Tissue fibrosis is the most prominent clinical manifestation in SSc. Fibrosis is the result of excessive production of collagen and ECM components, or defective remodeling of the ECM. Studies have confirmed hypomethylation and overexpression of two collagen genes (*COL23A1, COL4A2*) in dcSSc and lcSSc fibroblasts compared to control fibroblasts [29], in addition to hypomethylation of several collagen genes in each subset separately [29]. Moreover, *TNXB* was hypomethylated in dcSSc and lcSSc fibroblasts [29]. *TNXB* encodes a member of the tenascin family of ECM glycoproteins, which are involved in matrix maturation [45].

5. *The Wnt/β-catenin signaling pathway*

There is an increasing interest in the role of the Wnt/β-catenin signaling pathway as one of the profibrotic pathways in SSc. Studies have demonstrated persistent activation of the Wnt/β-catenin pathway as demonstrated by localization of β-catenin in fibroblast-like cells present in affected tissues [46]. Moreover, stimulation of normal fibroblasts with Wnt ligands results in β-catenin-mediated expression of collagen and other matrix genes, and enhanced myofibroblast differentiation and increased cell migration as expected in SSc [47,48]. In SSc, canonical Wnt signaling is activated by overexpression of Wnt proteins and by downregulation of the endogenous Wnt antagonists. The intensity and duration of Wnt/β-catenin signaling is normally tightly regulated by endogenous inhibitors. There is evidence of reduced expression of the endogenous Wnt antagonists, DKK1 and SFRP1, due to hypermethylation of the promoter region of *DKK1* and *SFRP1* in SSc fibroblasts [49]. On the other hand, there is evidence of hypomethylation of genes representative of the Wnt/β-catenin pathway in SSc—specifically, we demonstrated recently hypomethylation of *CTNNA2* and *CTNNB1* in dcSSc fibroblasts, and *CTNNA3* and *CTNND2* in lcSSc fibroblasts compared to control fibroblasts [29]. These findings suggest that DNA methylation aberrancies contribute to decreased expression of Wnt antagonists and increased expression of Wnt ligands and probably contribute to chronic activation of Wnt/β-catenin pathway signaling in SSc.

6. *Cadherins*

Cadherins are a group of transmembrane glycoproteins that mediate calcium-dependent homophilic cell-to-cell adhesion at adherens junctions [50]. Microarray studies have demonstrated overexpression of *CDH11*, which encodes cadherin-11, in fibroblasts from patients with SSc [51,52]. Moreover, Cdh11-deficient mice developed less fibrosis in bleomycin-induced fibrosis [53]. There is evidence for hypomethylation of *CDH11* in dcSSc fibroblasts in comparison to fibroblasts from healthy controls [29]. It is possible that hypomethylation of *CDH11* contributes to its overexpression, which facilitates the differentiation of resident tissue fibroblasts into myofibroblasts in SSc.

7. *The methylome in dcSSc versus lcSSc fibroblasts*

Recently, a genome-wide DNA methylation study demonstrated an interesting difference in DNA methylation aberrancies between dcSSc and lcSSc subsets in reference to healthy fibroblasts. The study demonstrated 3528 differentially methylated CpG sites in SSc, of which there were only 203 (~6%) CpG sites differentially methylated in both dcSSc and lcSSc. This finding suggests an interesting divergence of the DNA methylome at the genome-wide level between dcSSc and lcSSc that reflects heterogeneity at the epigenome level in scleroderma subsets [29]. Therefore, it is prudent to evaluate DNA methylation aberrancies and probably other epigenetic mechanisms in SSc in subset-specific approaches.

b. *Histone modification aberrancies in SSc fibroblasts*

We have discussed DNA hypermethylation and repression of *FLI1* in SSc fibroblasts early in this chapter. It is interesting to note that there is also significant reduction of histone H3 and H4 acetylation in the promoter region of the *FLI1* gene in SSc fibroblasts compared to healthy fibroblasts [26]. Moreover, trimethyl histone H3 on lysine 27 (i.e. H3K27me3), which is a potent repressor mark for target gene transcription, is increased in SSc fibroblasts in comparison with controls [54]. Altogether, these observations indicate that there are defects in the histone code in SSc, and that cross-talk between DNA methylation and histone modification changes can be involved in the pathogenesis of SSc, as demonstrated by *FLI1* repression in SSc fibroblasts.

c. *miRNA expression aberrancies in fibroblasts*

Briefly, miRNAs are small noncoding RNAs (generally 19–25 nucleotides in length) that play important regulatory roles mainly by cleavage or translational repression of targeted mRNAs. Many miRNAs are reported to be differentially expressed in SSc, suggesting that miRNA dysregulation plays a role in the pathogenesis of SSc.

1. *miRNA regulation of the TGF-β signaling pathway*

TGF-β mediates fibrosis positively by activating its downstream mediators, SMAD2 and SMAD3, but negatively via its inhibitory factor SMAD7. miR-21, which is upregulated in SSc fibroblasts [55], targets SMAD7. Overexpression of miR-21 in SSc fibroblasts decreases levels of SMAD7, whereas knockdown of miR-21 increases SMAD7 expression [56,57]. Therefore, miR-21 probably exerts a profibrotic effect by negatively regulating SMAD7 in SSc fibroblasts.

Altered expression of several other miRNAs in SSc with putative targets in the TGF-β downstream pathway (such as miR-145, miR-146, and miR-503) has also been demonstrated (Table 12.2).

2. *miRNAs directly target collagen genes in SSc*

miR-29 underexpression was reported in skin fibroblasts from patients with SSc, as well as fibroblasts from the mouse model of bleomycin-induced skin fibrosis [65]. It was demonstrated that induced expression of miR-29 in SSc fibroblasts reduces the expression of its target genes, and collagen type I and type III. Other potential targets for miRNA-29 include profibrotic molecules such as platelet-derived growth factor B (PDGF-B) and thrombospondin. Of significant interest, the stimulatory effects of TGF-β and PDGF-B on collagen synthesis were reduced by inducing the expression of miR-29 [65]. On the other hand, downregulation of miR-29 leads to further upregulation of TGF-β and PDGF-B. Taken together, these data argue for an antifibrotic role of miR-29 in SSc.

TABLE 12.2　Summary of Key Epigenetic Aberrancies Reported in SSc

Gene/pathway	Epigenetic defect	Cell type	Putative target/ mechanism of action in SSc	References
DNA METHYLATION				
COL4A2[§¶], *COL23A1*[§¶], *COL8A1*[§], *COL16A1*[§], *COL29A1*[§], *COL1A1*[¶], *COL6A3*[¶], *COL12A1*[¶]	Hypomethylation	Fibroblasts	Likely contributes to overexpression of collagen genes	[29]
PAX9[§¶]	Hypomethylation	Fibroblasts	Hypomethylation of *PAX9* may contribute to the process of fibrosis by overexpression of pro-α 2 chain of type I collagen	[29]
TNXB[§¶]	Hypomethylation	Fibroblasts	Unclear; possible overexpression of ECM glycoproteins	[29]
ITGA9[§¶]	Hypomethylation	Fibroblasts	Hypomethylation of *ITGA9* contributes to *ITGA9* overexpression in SSc. ITGA9 plays an integral role in myofibroblast differentiation and activation of TGF-β signaling pathway	[29]
RUNX1[§¶], *RUNX2*[§¶], *RUNX3*[§¶]	Hypomethylation	Fibroblasts	Indirectly induce expression of type II collagen by increasing expression of SOX5 and SOX6	[29]
ADAM12[§¶]	Hypomethylation	Fibroblasts	*ADAM12* overexpression in SSc contributes to fibrosis by inducing TGF-β signaling pathway	[29]

(*Continued*)

TABLE 12.2 (Continued)

Gene/pathway	Epigenetic defect	Cell type	Putative target/ mechanism of action in SSc	References
CTNNA2[§], CTNNB1[§] CTNNA3[¶], CTNND2[¶]	Hypomethylation	Fibroblasts	Unclear; hypomethylation of these genes that are components of the embryonic Wnt/β-catenin pathway may be contributing to the observed recapitulation of Wnt/β-catenin pathway in SSc	[29]
DKK1, SFRP1	Hypermethylation	Fibroblasts, PBMCs	Reduced expression of Wnt/β-catenin antagonists	[49]
PDGFC[§]	Hypomethylation	Fibroblasts	A profibrotic factor that is overexpressed in SSc	[29]
CDH11[§]	Hypomethylation	Fibroblasts	Overexpression of CDH11 induces myofibroblast differentiation	[29]
FLI1[§]	Hypermethylation	Fibroblasts	Overexpression of collagen genes in SSc	[26]
BMPRII[§]	Hypermethylation	MVECs	Failure of the inhibitory mechanism for cell proliferation and induction of apoptosis	[59]
NOS3	Hypermethylation	MVECs	Reduced NOS activity in MVECs; increased expression of proinflammatory and vasospastic genes	[60]
CD40L	Hypomethylation	CD4[+] T cells	Costimulatory molecule	[61]
CD70 (TNFRSF7)	Hypomethylation	CD4[+] T cells	Costimulatory molecule	[62]

(Continued)

2. IMMUNOLOGIC SKIN DISEASES

TABLE 12.2   (Continued)

Gene/pathway	Epigenetic defect	Cell type	Putative target/ mechanism of action in SSc	References
CD11a (ITGAL)	Hypomethylation	CD4+ T cells	Involved in costimulatory signaling	[63]
HISTONE MODIFICATIONS				
H3K27me3	Increased	Fibroblasts, murine dermal fibroblasts	May contribute to inhibition of collagen suppressor genes and, therefore, collagen deposition	[54]
Global H4 acetylation, H3K methylation	Increased H4 acetylation, decreased H3K methylation	B cells	Favors target gene expression in B cells; could be contributing to activation of genes in the immune system and antibody production	[64]
FLI1 H3 and H4 acetylation[§]	Reduced	Fibroblasts	Repression of *FLI1*; therefore, overexpression of collagen genes	[26]
microRNA				
miR-29[§]	Downregulated	Fibroblasts, murine dermal fibroblasts	Antifibrotic factor; probable target is type I and type III collagen	[55,65,66]
miR-21[§]	Overexpressed	Skin tissue, fibroblasts, murine dermal fibroblasts	Profibrotic factor; targets SMAD7. Upregulates canonical and noncanonical TGF-β signaling pathways	[55,56]
miR-142[§¶]	Overexpressed	Serum	Seems to be involved in regulating the expression of integrin αV	[67]
miR-196a[§¶]	Downregulated	Fibroblasts, serum, and hair shafts	Predicted target is type I collagen	[68,69]

(Continued)

TABLE 12.2   (Continued)

Gene/pathway	Epigenetic defect	Cell type	Putative target/ mechanism of action in SSc	References
miR-145[§]	Downregulated	Skin tissue, fibroblasts	Predicted target is SMAD3	[55]
miR-146[§]	Overexpressed	Skin tissue, fibroblasts	Predicted target is SMAD4	[55]
miR-152	Downregulated	MVECs	Predicted target is *DNMT1*	[70]
miR-503[§]	Overexpressed	Skin tissue, fibroblasts	Predicted target is SMAD7	[55]
miR-7	Overexpressed	Fibroblasts, skin, serum	Predicted target is type I collagen	[71]
miR let-7a	Downregulated	Fibroblasts, serum	Predicted target is type I collagen	[72]
miR-92-a[§]	Overexpressed	Fibroblasts, serum	Predicted target MMP-1	[57]
miR-150[§]	Downregulated	Fibroblasts, serum	Predicted target is integrin β3	[73]
miR-129-5p	Downregulated	Fibroblasts	Predicted target is type I collagen	[74]

The study design and analysis allows for distinction between [§]dcSSc and [¶]lcSSc; [§¶]the finding was reported in both dcSSc and lcSSc.
BMPRII, bone morphogenetic protein type II receptor; DNMT1, DNA (cytosine-5-)-methyltransferase 1; ECM, extracellular matrix; H3K27me3, trimethylation of histone H3 on lysine 27; MBD1, methyl-CpG-binding domain protein1; MVEC, microvascular endothelial cells; MMP-1, matrix metalloproteinase 1; NOS, nitric oxide synthetase; PBMCs, peripheral blood mononuclear cells; SMAD, intracellular proteins that transduce extracellular signals from TGF-β ligands; TGF-β, transforming growth factor-β.
Reproduced and modified with permission [58].

Moreover, several studies have demonstrated downregulation of other antifibrotic miRNAs (such as miR-196a, miR let-7a, and miR-129-5p), or increased expression of profibrotic miRNAs (such as miR-7) in SSc fibroblasts. The putative target for the aforementioned miRNAs is probably type I collagen [68,71,72,74] (Table 12.2).

2. MVECs

a. *DNA methylation alterations in nitric oxide synthesis*
It has been demonstrated that there are intrinsic defects in the mechanism of nitric oxide (NO) production by MVECs isolated from SSc patients [75]. NO is a potent vasodilator and an inhibitor

of smooth muscle cell growth. Also, NO has antithrombotic, antiplatelet, and antioxidation properties [76]. NO is produced partly by MVEC by the action of nitric oxide synthase (NOS). There is evidence for underexpression of *NOS3*, the gene encoding endothelial NOS in SSc-MVECs, and that the promoter region of *NOS3* is hypermethylated in SSc-MVEC compared to controls [60]. This finding indicates that the epigenetic program contributes to MVEC dysfunction in SSc.

b. *DNA methylation in MVEC apoptosis*

Enhanced MVEC apoptosis is one of the pathogenic manifestations of SSc. It has become apparent that MVEC apoptosis could be an initial element in the pathogenesis of SSc, and that MVEC apoptosis may even precede the onset of the fibrotic stage [77]. Bone morphogenetic proteins (BMPs) are a group of proteins that constitute morphogenetic signals and orchestrate tissue architecture through coordinating cell survival and differentiation. BMP signaling through bone morphogenic protein receptor II (BMPRII) favors MVEC survival and apoptosis resistance. There is evidence for reduced expression of *BMPRII* in SSc-MVECs in comparison with healthy controls which could be related to heavy methylation of the promoter region of *BMPRII* in SSc-MVECs compared to healthy controls [59]. In the same study [59], treatment of SSc-MVECs with 5-AZA normalized *BMPRII* expression levels and restored SSc-MVEC response to apoptosis to normal levels. Therefore, it seems that DNA methylation may play a role in MVEC response to apoptosis in SSc.

c. *miRNA aberrant expression in MVEC*

Most of the studies that evaluated miRNA expression in SSc have focused on dermal fibroblasts; very few studies have evaluated the extent of aberrant miRNA expression in SSc-MVEC. It appears that *miR-152* is downregulated in SSc-MVEC and the target for *miR-152* is *DNMT1* [70]. Forced expression of *miR-152* in MVEC led to increased expression levels of *DNMT1*, whereas inhibition of *miR-152* expression led to enhanced *DNMT1* expression and lower expression levels of *NOS3*. These data indicate that *miR-152* plays a role in the SSc-MVEC phenotype probably through DNA methylation.

3. Lymphocytes

a. *DNA methylation aberrancies in T lymphocytes*

It has been established that DNA methylation is a natural physiological process to maintain inactivation of one X chromosome in order to keep a balance among genes encoded on the X chromosome in males and females [78]. CD40L is a costimulatory molecule that is expressed predominantly on the surface of activated T cells. The main function of CD40L is to regulate B-cell function by engaging CD40 on the B-cell surface. Studies have demonstrated increased expression of *CD40L*, which is encoded on the X chromosome, in female SSc patients, associated with demethylation of the promoter region of *CD40L* on the inactive X chromosome in CD4$^+$ T cells. Moreover, there was no difference in CD40L expression between male patients with SSc and male controls [61]. The same observation of hypomethylation and overexpression of *CD40L* was reported in SLE [79]. These data indicate that there are defects in the epigenetic program, which leads to reactivation of genes that are located on the naturally silenced X chromosome in female patients with autoimmune diseases like SLE and SSc, which may explain female predominance in autoimmune diseases.

The CD70/CD27 axis has gained interest in autoimmune diseases because of its capacity to regulate immunity versus tolerance. CD70 is another costimulatory molecule that is expressed on activated lymphocytes and plays an important role in regulating B- and T-cell activation. *CD70* is overexpressed in SSc CD4$^+$ T cells, and there is evidence that demethylation of the *CD70* promoter region contributes to the overexpression of *CD70* in CD4$^+$ T cells [62]. Overall, these data suggest that DNA methylation aberrancies contribute to overexpression of costimulatory molecules, but it remains to be seen whether CD40L and CD70 signaling are involved in the pathogenesis of SSc to the same extent that these molecules are involved in the pathogenesis of other autoimmune diseases.

b. *Histone code modification in B lymphocytes*

Activation of the immune system is one of the pathological features of SSc. B cells play a special role in the pathogenesis of SSc, as suggested by the presence of disease-specific circulating autoantibodies in SSc. Very little is known about epigenetic aberrancies in SSc B lymphocytes. However, there is an evidence that B cells from patients with SSc are characterized by global H4 hyperacetylation and global H3 lysine 9 (H3K9) hypomethylation, associated with downregulation of histone deacetylase 2 (HDAC2) and HDAC7 compared to B cells from healthy controls [64]. The aforementioned modifications of the histone code favor permissive chromatin architecture for gene expression. It is not clear at this stage what could be the effect of these changes on B-lymphocyte function, but it is suggested that this histone code in SSc B-lymphocytes might enhance overexpression of autoimmunity-related genes in SSc [64].

12.5 WHAT MIGHT TRIGGER EPIGENETIC DYSREGULATION IN SSc?

If the contribution of epigenetic aberrancies to pathogenesis of auto-immune diseases, like SSc, is becoming increasingly clear, the trigger(s) that induce the defects in the epigenetic program are less so. We will present in this section some theories about the triggers that are largely driven by observational studies in SSc as well as data from epigenetic deregulation in general.

Much of the current interest about epigenetics and human diseases stems from the idea that the modifications in the epigenetic program are sensitive to environmental factors. However, the environmental—epigenetic triggers remain largely uncharacterized with few exceptions. One of the central obstacles hampering progress in identifying the environmental—epigenetic triggers in general is complicated by the temporality and causality issue; where epigenetic changes take place prior to the onset of disease, even in some cases, it appears that disease may occur one or two generations after the exposure [80,81]. With regard to triggers of epigenetic deregulation in SSc, it seems that external factors (e.g., exposure to organic solvents, silica exposure, UV light, toxins, diet, drugs, and infective factors, particularly human cytomegalovirus) (Table 12.1), and internal factors (e.g., hypoxia, oxidative stress, aging, and sex hormones) could be possible candidates [58].

a. *Occupational exposures*

The observation of geographical clustering of SSc and the epidemiological studies that linked SSc to exposure to occupational agents suggest that the environment plays a role in predisposition to SSc in susceptible hosts. However, the "causality" inference of occupational exposure in the pathogenesis of SSc is challenging; in most cases, it is hard to identify a single occupational agent due to the complexity of our environment, which is characterized by exposure to numerous chemical and toxic agents every day. Also, the "pathogenic environment" in epigenetics has not yet been characterized. Moreover, the timing of environmental exposure is difficult to identify, which makes recall of exposure even more difficult. These factors, in addition to the retrospective case—control design of the studies that have reported a link between SSc and specific occupational agents, make the identification of the environmental trigger of SSc a challenge.

b. *Diet and nutrition*

While there is so far very little evidence to suggest that a particular diet is specifically linked to predisposition to SSc, there is a piece of evidence that the susceptibility to chronic disease is influenced by persistent adaptations to prenatal and early postnatal nutrition [82]. The nutritional element in predisposing to SSc is not clear, but abnormalities in the availability of methyl donors (methionine and choline) and cofactors (folic acid, vitamin B_{12}, and pyridoxal phosphate) may contribute to aberrancies in DNA and histone methylation. It is also possible that the observed geo-epidemiology of SSc may be linked to undefined dietary patterns.

c. *Hypoxia*

The observation that Raynaud's phenomenon is usually the earliest clinical manifestation in the vast majority of patients with SSc [83], and that Raynaud's phenomenon usually precedes onset of the fibrotic phase by several months or years in patients with SSc is intriguing. Raynaud's phenomenon is characterized by vasospasm and reduced tissue perfusion to the distal extremities that causes intermittent tissue hypoxia and possible endothelial injury which perpetuate vascular dysfunction. Interestingly, tissue fibrosis starts in the distal extremities in most patients with SSc, which is the same site of reduced tissue perfusion in Raynaud's phenomenon. Hypoxia itself is a potent stimulus for the synthesis of collagen and its cross-linking enzyme lysyl hydroxylase, fibronectin, and fibrogenic cytokines [84]. Therefore, these observations suggest the hypothesis that transient tissue hypoxia due to decreased blood flow related to Raynaud's phenomenon might be the trigger for fibrosis through an epigenetic mechanism. Indeed, it has been established that hypoxia decreases global transcriptional activity and has a major effect on cellular phenotype through different mechanisms that include the hypoxia-inducible factor (HIF) transcription paradigm in eukaryotic cells through HIF-1, which is a critical transcription regulator of a majority of genes in response to hypoxia [85]. There is evidence that epigenetic pathways are also relevant in the adaptation to hypoxia [86]. Hypothetically, hypoxia evokes the anaerobic metabolism pathways which lead to lower levels of acetyl-CoA; therefore, it is possible that hypoxia may lead to a global decrease in histone acetylation levels [87]. Also, it seems that hypoxia may also induce HDAC upregulation, which induces a global decrease in H3K9 acetylation in various cells [88].

d. *Oxidation*

SSc is an oxidative stress state based on the observations that there are abnormalities in the NO/NOS axis and the presence of increased levels of oxidative biomarkers in SSc [89,90]. Oxidative stress leads to excessive generation of oxygen free-radicals and reactive oxygen species (ROS) [91]. ROS have been implicated in causing vascular injury and predisposing to autoimmunity in SSc [92]. Interestingly, there is a cross-talk between oxidative stress and fibrosis, where oxidative stress stimulates the accumulation of ECM proteins, and fibrosis generates more oxidative stress [91]. Moreover, in fibroblasts, TGF-β induces the NADPH oxidase enzyme NOX4, which catalyzes the reduction of oxygen to ROS. In turn, ROS act as signals to induce fibroblast activation and myofibroblast differentiation [93].

Interest in the role of oxidative stress in epigenetic regulation is growing, especially the role of oxidative stress in controlling DNA methylation. It was demonstrated recently that oxidative stress could contribute to impaired T-cell extracellular signal-related kinase (ERK) pathway signaling in SLE, which is another autoimmune disease that shares several clinical features with SSc that include, but are not limited to, the presence of Raynaud's phenomenon and the presence of circulating autoantibodies. There is evidence that oxidative stress disrupts ERK signaling in $CD4^+$ T cells *in vitro*, therefore reducing *DNMT1* expression and consequently causing demethylation and overexpression of methylation-sensitive genes that have been previously shown to be upregulated in patients with SLE, like *CD70* [94]. It remains to be seen whether the oxidative stress effect on ERK pathway signaling also applies to SSc, and whether antioxidants like *N*-acetylcysteine could have therapeutic benefit in treatment of autoimmune diseases by reversing the oxidative stress state.

12.6 CLINICAL RELEVANCE OF EPIGENETIC ABERRANCIES IN SSc

We have presented the evidence for epigenetic alterations in different programs involving several cell types in SSc. In this section, we will look at these aberrancies from a clinical perspective and discuss the value of the epigenetic alterations as diagnostic markers, and perhaps the potential use of the knowledge that we gained from studying epigenetic alterations in SSc as therapeutic strategies.

A potential diagnostic marker, *SOX2OT*, encodes for one of the long nonprotein coding RNAs (lncRNAs) that may exert a regulatory role on stem cell pluripotency [95]. It has been demonstrated that *SOX2OT is* hypermethylated across multiple CpG sites in dcSSc fibroblasts, but not lcSSc fibroblasts, in comparison to control fibroblasts [29]. This observation suggests that the methylation status of *SOX2OT* might potentially be a useful marker in differentiating SSc subsets if reproduced and validated in other studies.

There is evidence that expression levels of some miRNAs might correlate with specific features of SSc or with disease activity. For instance, serum levels of miR-142 correlate with SSc disease severity [67], and expression of miR-21, miR-29, miR-145, and miR-146 correlates with disease activity in SSc. Overall, further studies in this field are needed before miRNAs can be considered useful clinical biomarkers.

There is remarkable interest in finding a disease modifying agent to treat SSc, based on the unsatisfactory results from most therapeutic agents that have been used in treatment of SSc up to this date, either because of nonefficacy (mostly) or because of unacceptable side effect profile. Therefore, it is not surprising that there is interest in using epigenetic modifying agents in experimental settings in SSc. Trichostatin (TSA) is one of the HDAC inhibitors that is available for treatment of myelodysplastic disease. *In vitro* studies have shown that TSA attenuates expression of collagen I in dermal SSc fibroblasts [96]. Also, TSA was associated with lower fibrotic end points in an animal model of skin fibrosis [97]. Despite the fact that these observations suggest a possible role for TSA in the treatment of SSc, the "off-site" effect of TSA, as an agent with an ability to induce widespread changes in the chromatin, will perhaps limit TSA usefulness in SSc. Future studies to explore the use of specific miRNAs as potential treatment strategies in SSc would be of interest.

12.7 CONCLUSION

In recent years, the field of epigenetics in rheumatic diseases has grown dramatically and has become one of the paradigms in explaining the link between environmental exposures and disease susceptibility in genetically predisposed individuals. The data we have provided in this chapter suggest new approaches to understand the pathogenesis of a complex disease like SSc. We have explored several lines of evidence that confirm substantial epigenetic modifications in SSc, particularly in fibroblasts, MVECs, and in B and T cells. The evidence extends to include a role for epigenetic modifications in fundamental pathways that are involved in the process of fibrosis, such as TGF-β and downstream pathways, and the Wnt/β-catenin signaling pathway. The triggers for the epigenetic alterations in SSc are not clear, but it is reasonable to suggest a role for occupational exposures, nutritional factors, hypoxia, and oxidative stress as possible triggering mechanisms. It remains to be determined if epigenetic alterations could be used as biomarkers for disease activity or severity in SSc, or even as therapeutic strategies. To move the field forward, studies focused on uncovering the potential pathogenic triggers in SSc, and the mechanisms by which these triggers induce epigenetic alterations, are warranted. Ultimately, characterization of the "pathogenic" environment could lead to better understanding of the disease risk, and probably prevention.

References

[1] Abraham DJ, Varga J. Scleroderma: from cell and molecular mechanisms to disease models. Trends Immunol 2005;26(11):587−95; PubMed PMID: 16168711.

[2] LeRoy EC, Black C, Fleischmajer R, Jablonska S, Krieg T, Medsger Jr. TA, et al. Scleroderma (systemic sclerosis): classification, subsets and pathogenesis. J Rheumatol 1988;15(2):202−5; PubMed PMID: 3361530.

[3] Kahaleh B. The microvascular endothelium in scleroderma. Rheumatology 2008;47 (Suppl. 5):v14−15; PubMed PMID: 18784128.

[4] Mayes MD. Scleroderma epidemiology. Rheum Dis Clin North Am 2003;29 (2):239−54; PubMed PMID: 12841293.

[5] Varga J, Abraham D. Systemic sclerosis: a prototypic multisystem fibrotic disorder. J Clin Invest 2007;117(3):557−67; PubMed PMID: 17332883. Pubmed Central PMCID: PMC1804347.

[6] Kraling BM, Maul GG, Jimenez SA. Mononuclear cellular infiltrates in clinically involved skin from patients with systemic sclerosis of recent onset predominantly consist of mono-cytes/macrophages. Pathobiology 1995;63(1):48−56; PubMed PMID: 7546275.

[7] Yamamoto T. Autoimmune mechanisms of scleroderma and a role of oxidative stress. Self Nonself 2011;2(1):4−10; PubMed PMID: 21776329. Pubmed Central PMCID: PMC3136898.

[8] Gustafsson R, Totterman TH, Klareskog L, Hallgren R. Increase in activated T cells and reduction in suppressor inducer T cells in systemic sclerosis. Ann Rheum Dis 1990;49(1):40−5; PubMed PMID: 2138008. Pubmed Central PMCID: PMC1003962.

[9] Mavalia C, Scaletti C, Romagnani P, Carossino AM, Pignone A, Emmi L, et al. Type 2 helper T-cell predominance and high CD30 expression in systemic sclerosis. Am J Pathol 1997;151(6):1751−8; PubMed PMID: 9403725. Pubmed Central PMCID: PMC1858349.

[10] Hasegawa M, Fujimoto M, Takehara K, Sato S. Pathogenesis of systemic sclerosis: altered B cell function is the key linking systemic autoimmunity and tissue fibrosis. J Dermatol Sci 2005;39(1):1−7; PubMed PMID: 15885984.

[11] Sato S, Fujimoto M, Hasegawa M, Takehara K. Altered blood B lymphocyte homeo-stasis in systemic sclerosis: expanded naive B cells and diminished but activated memory B cells. Arthritis Rheum 2004;50(6):1918−27; PubMed PMID: 15188368.

[12] Kehrl JH, Roberts AB, Wakefield LM, Jakowlew S, Sporn MB, Fauci AS. Transforming growth factor beta is an important immunomodulatory protein for human B lymphocytes. J Immunol 1986;137(12):3855−60; PubMed PMID: 2878044.

[13] Wu M, Assassi S. The role of type 1 interferon in systemic sclerosis. Front Immunol 2013;4:266; PubMed PMID: 24046769. Pubmed Central PMCID: PMC3764426.

[14] Campbell PM, LeRoy EC. Pathogenesis of systemic sclerosis: a vascular hypothesis. Semin Arthritis Rheum 1975;4(4):351−68; PubMed PMID: 1135634.

[15] Leroy EC. Connective tissue synthesis by scleroderma skin fibroblasts in cell culture. J Exp Med 1972;135(6):1351−62; PubMed PMID: 4260235. Pubmed Central PMCID: PMC2139167.

[16] Gu YS, Kong J, Cheema GS, Keen CL, Wick G, Gershwin ME. The immunobiology of systemic sclerosis. Semin Arthritis Rheum 2008;38(2):132−60; PubMed PMID: 18221988.

[17] Ihn H, Yamane K, Kubo M, Tamaki K. Blockade of endogenous transforming growth factor beta signaling prevents up-regulated collagen synthesis in scleroderma fibro-blasts: association with increased expression of transforming growth factor beta receptors. Arthritis Rheum 2001;44(2):474−80; PubMed PMID: 11229480.

[18] Scharffetter K, Lankat-Buttgereit B, Krieg T. Localization of collagen mRNA in nor-mal and scleroderma skin by in-situ hybridization. Eur J Clin Invest 1988;18(1):9−17; PubMed PMID: 3130266.

[19] Varga J, Pasche B. Transforming growth factor beta as a therapeutic target in systemic sclerosis. Nat Rev Rheumatol 2009;5(4):200–6; PubMed PMID: 19337284. Pubmed Central PMCID: PMC3959159.

[20] Kubo M, Czuwara-Ladykowska J, Moussa O, Markiewicz M, Smith E, Silver RM, et al. Persistent down-regulation of Fli1, a suppressor of collagen transcription, in fibrotic scleroderma skin. Am J Pathol 2003;163(2):571–81; PubMed PMID: 12875977. Pubmed Central PMCID: PMC1868228.

[21] Mori Y, Chen SJ, Varga J. Expression and regulation of intracellular SMAD signaling in scleroderma skin fibroblasts. Arthritis Rheum 2003;48(7):1964–78; PubMed PMID: 12847691.

[22] Asano Y, Ihn H, Yamane K, Kubo M, Tamaki K. Impaired Smad7–Smurf-mediated negative regulation of TGF-beta signaling in scleroderma fibroblasts. J Clin Invest 2004;113(2):253–64; PubMed PMID: 14722617. Pubmed Central PMCID: PMC310747.

[23] Luo Y, Wang Y, Wang Q, Xiao R, Lu Q. Systemic sclerosis: genetics and epigenetics. J Autoimmun 2013;41:161–7; PubMed PMID: 23415078.

[24] Agarwal SK, Reveille JD. The genetics of scleroderma (systemic sclerosis). Curr Opin Rheumatol 2010;22(2):133–8; PubMed PMID: 20090527.

[25] Feghali-Bostwick C, Medsger Jr. TA, Wright TM. Analysis of systemic sclerosis in twins reveals low concordance for disease and high concordance for the presence of antinuclear antibodies. Arthritis Rheum 2003;48(7):1956–63; PubMed PMID: 12847690.

[26] Wang Y, Fan PS, Kahaleh B. Association between enhanced type I collagen expression and epigenetic repression of the FLI1 gene in scleroderma fibroblasts. Arthritis Rheum 2006;54(7):2271–9; PubMed PMID: 16802366.

[27] Ihn H. Autocrine TGF-beta signaling in the pathogenesis of systemic sclerosis. J Dermatol Sci 2008;49(2):103–13; PubMed PMID: 17628443.

[28] Blobe GC, Schiemann WP, Lodish HF. Role of transforming growth factor beta in human disease. N Engl J Med 2000;342(18):1350–8; PubMed PMID: 10793168.

[29] Altorok N, Tsou PS, Coit P, Khanna D, Sawalha AH. Genome-wide DNA methylation analysis in dermal fibroblasts from patients with diffuse and limited systemic sclerosis reveals common and subset-specific DNA methylation aberrancies. Ann Rheum Dis 2014; pii annrheumdis-2014-205303. [Epub ahead of print.] PubMed PMID: 24812288.

[30] Hynes RO. Integrins: bidirectional, allosteric signaling machines. Cell 2002;110 (6):673–87; PubMed PMID: 12297042.

[31] Margadant C, Sonnenberg A. Integrin-TGF-beta crosstalk in fibrosis, cancer and wound healing. EMBO Rep 2010;11(2):97–105; PubMed PMID: 20075988. Pubmed Central PMCID: PMC2828749.

[32] Asano Y, Ihn H, Yamane K, Jinnin M, Mimura Y, Tamaki K. Increased expression of integrin alpha(v)beta3 contributes to the establishment of autocrine TGF-beta signaling in scleroderma fibroblasts. J Immunol 2005;175(11):7708–18; PubMed PMID: 16301681.

[33] Asano Y, Ihn H, Yamane K, Jinnin M, Tamaki K. Increased expression of integrin alphavbeta5 induces the myofibroblastic differentiation of dermal fibroblasts. Am J Pathol 2006;168(2):499–510; PubMed PMID: 16436664. Pubmed Central PMCID: PMC1606497.

[34] Asano Y, Ihn H, Yamane K, Kubo M, Tamaki K. Increased expression levels of integrin alphavbeta5 on scleroderma fibroblasts. Am J Pathol 2004;164(4):1275–92; PubMed PMID: 15039216. Pubmed Central PMCID: PMC1615355.

[35] Horan GS, Wood S, Ona V, Li DJ, Lukashev ME, Weinreb PH, et al. Partial inhibition of integrin alpha(v)beta6 prevents pulmonary fibrosis without exacerbating inflammation. Am J Respir Crit Care Med 2008;177(1):56–65; PubMed PMID: 17916809.

[36] Carracedo S, Lu N, Popova SN, Jonsson R, Eckes B, Gullberg D. The fibroblast integrin alpha11beta1 is induced in a mechanosensitive manner involving activin A and

regulates myofibroblast differentiation. J Biol Chem 2010;285(14):10434−45; PubMed PMID: 20129924. Pubmed Central PMCID: PMC2856250.

[37] Munger JS, Huang X, Kawakatsu H, Griffiths MJ, Dalton SL, Wu J, et al. The integrin alpha v beta 6 binds and activates latent TGF beta 1: a mechanism for regulating pulmonary inflammation and fibrosis. Cell 1999;96(3):319−28; PubMed PMID: 10025398.

[38] Shi-Wen X, Renzoni EA, Kennedy L, Howat S, Chen Y, Pearson JD, et al. Endogenous endothelin-1 signaling contributes to type I collagen and CCN2 overexpression in fibrotic fibroblasts. Matrix Biol 2007;26(8):625−32; PubMed PMID: 17681742.

[39] Atfi A, Dumont E, Colland F, Bonnier D, L'Helgoualc'h A, Prunier C, et al. The disintegrin and metalloproteinase ADAM12 contributes to TGF-beta signaling through interaction with the type II receptor. J Cell Biol 2007;178(2):201−8; PubMed PMID: 17620406. Pubmed Central PMCID: PMC2064440.

[40] Skubitz KM, Skubitz AP. Gene expression in aggressive fibromatosis. J Lab Clin Med 2004;143(2):89−98; PubMed PMID: 14966464.

[41] Taniguchi T, Asano Y, Akamata K, Aozasa N, Noda S, Takahashi T, et al. Serum levels of ADAM12-S: possible association with the initiation and progression of dermal fibrosis and interstitial lung disease in patients with systemic sclerosis. J Eur Acad Dermatol Venereol 2013;27(6):747−53; PubMed PMID: 22540429.

[42] Kimura A, Inose H, Yano F, Fujita K, Ikeda T, Sato S, et al. Runx1 and Runx2 cooperate during sternal morphogenesis. Development 2010;137(7):1159−67; PubMed PMID: 20181744. Pubmed Central PMCID: PMC2835330.

[43] Zhao Q, Eberspaecher H, Lefebvre V, De Crombrugghe B. Parallel expression of Sox9 and Col2a1 in cells undergoing chondrogenesis. Dev Dyn 1997;209(4):377−86; PubMed PMID: 9264261.

[44] Yoshida CA, Yamamoto H, Fujita T, Furuichi T, Ito K, Inoue K, et al. Runx2 and Runx3 are essential for chondrocyte maturation, and Runx2 regulates limb growth through induction of Indian hedgehog. Genes Dev 2004;18(8):952−63; PubMed PMID: 15107406. Pubmed Central PMCID: PMC395853.

[45] Egging D, van Vlijmen-Willems I, van Tongeren T, Schalkwijk J, Peeters A. Wound healing in tenascin-X deficient mice suggests that tenascin-X is involved in matrix maturation rather than matrix deposition. Connect Tissue Res 2007;48(2):93−8; PubMed PMID: 17453911.

[46] Chilosi M, Poletti V, Zamo A, Lestani M, Montagna L, Piccoli P, et al. Aberrant Wnt/beta-catenin pathway activation in idiopathic pulmonary fibrosis. Am J Pathol 2003;162(5):1495−502; PubMed PMID: 12707032. Pubmed Central PMCID: PMC1851206.

[47] Carthy JM, Garmaroudi FS, Luo Z, McManus BM. Wnt3a induces myofibroblast differentiation by upregulating TGF-beta signaling through SMAD2 in a beta-catenin-dependent manner. PLOS ONE 2011;6(5):e19809; PubMed PMID: 21611174. Pubmed Central PMCID: PMC3097192.

[48] Wei J, Melichian D, Komura K, Hinchcliff M, Lam AP, Lafyatis R, et al. Canonical Wnt signaling induces skin fibrosis and subcutaneous lipoatrophy: a novel mouse model for scleroderma? Arthritis Rheum 2011;63(6):1707−17; PubMed PMID: 21370225. Pubmed Central PMCID: PMC3124699.

[49] Dees C, Schlottmann I, Funke R, Distler A, Palumbo-Zerr K, Zerr P, et al. The Wnt antagonists DKK1 and SFRP1 are downregulated by promoter hypermethylation in systemic sclerosis. Ann Rheum Dis 2014;73(6):1232−9; PubMed PMID: 23698475.

[50] Wheelock MJ, Johnson KR. Cadherins as modulators of cellular phenotype. Annu Rev Cell Dev Biol 2003;19:207−35; PubMed PMID: 14570569.

[51] Gardner H, Shearstone JR, Bandaru R, Crowell T, Lynes M, Trojanowska M, et al. Gene profiling of scleroderma skin reveals robust signatures of disease that are

imperfectly reflected in the transcript profiles of explanted fibroblasts. Arthritis Rheum 2006;54(6):1961−73; PubMed PMID: 16736506.

[52] Whitfield ML, Finlay DR, Murray JI, Troyanskaya OG, Chi JT, Pergamenschikov A, et al. Systemic and cell type-specific gene expression patterns in scleroderma skin. Proc Natl Acad Sci USA 2003;100(21):12319−24; PubMed PMID: 14530402. Pubmed Central PMCID: PMC218756.

[53] Schneider DJ, Wu M, Le TT, Cho SH, Brenner MB, Blackburn MR, et al. Cadherin-11 contributes to pulmonary fibrosis: potential role in TGF-beta production and epithelial to mesenchymal transition. FASEB J 2012;26(2):503−12; PubMed PMID: 21990376. Pubmed Central PMCID: PMC3290437.

[54] Kramer M, Dees C, Huang J, Schlottmann I, Palumbo-Zerr K, Zerr P, et al. Inhibition of H3K27 histone trimethylation activates fibroblasts and induces fibrosis. Ann Rheum Dis 2013;72(4):614−20; PubMed PMID: 22915621.

[55] Zhu H, Li Y, Qu S, Luo H, Zhou Y, Wang Y, et al. MicroRNA expression abnormalities in limited cutaneous scleroderma and diffuse cutaneous scleroderma. J Clin Immunol 2012;32(3):514−22; PubMed PMID: 22307526. Epub 2012/02/07. eng.

[56] Zhu H, Luo H, Li Y, Zhou Y, Jiang Y, Chai J, et al. MicroRNA-21 in scleroderma fibrosis and its function in TGF-beta-regulated fibrosis-related genes expression. J Clin Immunol 2013;33(6):1100−9; PubMed PMID: 23657402. Epub 2013/05/10. eng.

[57] Sing T, Jinnin M, Yamane K, Honda N, Makino K, Kajihara I, et al. microRNA-92a expression in the sera and dermal fibroblasts increases in patients with scleroderma. Rheumatology 2012;51(9):1550−6; PubMed PMID: 22661558. Epub 2012/06/05. eng.

[58] Altorok N, Almeshal N, Wang Y, Kahaleh B. Epigenetics, the holy grail in the pathogenesis of systemic sclerosis. Rheumatology 2014 [Epub ahead of print.] PubMed PMID: 24740406.

[59] Wang Y, Kahaleh B. Epigenetic repression of bone morphogenetic protein receptor II expression in scleroderma. J Cell Mol Med 2013;17(10):1291−9; PubMed PMID: 23859708.

[60] Wang Y, Kahaleh B. Epigenetic regulation in scleroderma: high-throughput DNA methylation profiling of Ssc fibroblasts and microvascular endothelial cells and the central role for Nos3 and Fli1 epigenetic repression in the emergence of Ssc cellular phenotype [Abstract]. American College of Rheumatology; Annual Scientific Meeting, 2007.

[61] Lian X, Xiao R, Hu X, Kanekura T, Jiang H, Li Y, et al. DNA demethylation of CD40l in CD4+ T cells from women with systemic sclerosis: a possible explanation for female susceptibility. Arthritis Rheum 2012;64(7):2338−45; PubMed PMID: 22231486.

[62] Jiang H, Xiao R, Lian X, Kanekura T, Luo Y, Yin Y, et al. Demethylation of TNFSF7 contributes to CD70 overexpression in CD4+ T cells from patients with systemic sclerosis. Clin Immunol 2012;143(1):39−44; PubMed PMID: 22306512.

[63] Wang Y, Shu Y, Wang Q, Zhao M, Liang G, Lu Q, et al. Demethylation of ITGAL (CD11a) regulatory sequences in CD4+T lymphocytes of systemic sclerosis. [Abstract #2905], American College of Rheumatology, San Diego, CA, 2013.

[64] Wang Y, Yang Y, Luo Y, Yin Y, Wang Q, Li Y, et al. Aberrant histone modification in peripheral blood B cells from patients with systemic sclerosis. Clin Immunol 2013;149 (1):46−54; PubMed PMID: 23891737.

[65] Maurer B, Stanczyk J, Jungel A, Akhmetshina A, Trenkmann M, Brock M, et al. MicroRNA-29, a key regulator of collagen expression in systemic sclerosis. Arthritis Rheum 2010;62(6):1733−43; PubMed PMID: 20201077. Epub 2010/03/05. eng.

[66] Fabbri M, Garzon R, Cimmino A, Liu Z, Zanesi N, Callegari E, et al. MicroRNA-29 family reverts aberrant methylation in lung cancer by targeting DNA methyltransferases 3A and 3B. Proc Natl Acad Sci USA 2007;104(40):15805−10; PubMed PMID: 17890317. Pubmed Central PMCID: PMC2000384. Epub 2007/09/25. eng.

[67] Makino K, Jinnin M, Kajihara I, Honda N, Sakai K, Masuguchi S, et al. Circulating miR-142-3p levels in patients with systemic sclerosis. Clin Exp Dermatol 2012;37 (1):34−9; PubMed PMID: 21883400. Epub 2011/09/03. eng.

[68] Honda N, Jinnin M, Kajihara I, Makino T, Makino K, Masuguchi S, et al. TGF-beta-mediated downregulation of microRNA-196a contributes to the constitutive upregulated type I collagen expression in scleroderma dermal fibroblasts. J Immunol 2012;188(7):3323−31; PubMed PMID: 22379029.

[69] Wang Z, Jinnin M, Kudo H, Inoue K, Nakayama W, Honda N, et al. Detection of hair-microRNAs as the novel potent biomarker: evaluation of the usefulness for the diagnosis of scleroderma. J Dermatol Sci 2013;72(2):134−41; PubMed PMID: 23890704. Epub 2013/07/31. eng.

[70] Wang YK, Omar RK, Kahaleh B. Down-regulated microRNA-152 induces aberrant DNA methylation in scleroderma endothelial cells by targeting DNA methyltransferase 1 [Abstract]. Arthritis Rheum 2010;62(Suppl. 10):1352.

[71] Kajihara I, Jinnin M, Yamane K, Makino T, Honda N, Igata T, et al. Increased accumulation of extracellular thrombospondin-2 due to low degradation activity stimulates type I collagen expression in scleroderma fibroblasts. Am J Pathol 2012;180(2):703−14; PubMed PMID: 22142808.

[72] Makino K, Jinnin M, Hirano A, Yamane K, Eto M, Kusano T, et al. The downregulation of microRNA let-7a contributes to the excessive expression of type I collagen in systemic and localized scleroderma. J Immunol 2013;190(8):3905−15; PubMed PMID: 23509348.

[73] Honda N, Jinnin M, Kira-Etoh T, Makino K, Kajihara I, Makino T, et al. miR-150 down-regulation contributes to the constitutive type I collagen overexpression in scleroderma dermal fibroblasts via the induction of integrin beta3. Am J Pathol 2013;182(1):206−16; PubMed PMID: 23159943. Epub 2012/11/20. eng.

[74] Nakashima T, Jinnin M, Yamane K, Honda N, Kajihara I, Makino T, et al. Impaired IL-17 signaling pathway contributes to the increased collagen expression in scleroderma fibroblasts. J Immunol 2012;188(8):3573−83; PubMed PMID: 22403442.

[75] Romero LI, Zhang DN, Cooke JP, Ho HK, Avalos E, Herrera R, et al. Differential expression of nitric oxide by dermal microvascular endothelial cells from patients with scleroderma. Vasc Med 2000;5(3):147−58; PubMed PMID: 11104297.

[76] Fish JE, Marsden PA. Endothelial nitric oxide synthase: insight into cell-specific gene regulation in the vascular endothelium. Cell Mol Life Sci 2006;63(2):144−62; PubMed PMID: 16416260.

[77] Sgonc R, Gruschwitz MS, Dietrich H, Recheis H, Gershwin ME, Wick G. Endothelial cell apoptosis is a primary pathogenetic event underlying skin lesions in avian and human scleroderma. J Clin Invest 1996;98(3):785−92; PubMed PMID: 8698871. Pubmed Central PMCID: PMC507489.

[78] Lyon MF. Gene action in the X-chromosome of the mouse (Mus musculus L.). Nature 1961;190:372−3; PubMed PMID: 13764598.

[79] Lu Q, Wu A, Tesmer L, Ray D, Yousif N, Richardson B. Demethylation of CD40LG on the inactive X in T cells from women with lupus. J Immunol 2007;179(9):6352−8; PubMed PMID: 17947713.

[80] Klip H, Verloop J, van Gool JD, Koster ME, Burger CW, van Leeuwen FE, et al. Hypospadias in sons of women exposed to diethylstilbestrol in utero: a cohort study. Lancet 2002;359(9312):1102−7; PubMed PMID: 11943257.

[81] Greer JM, McCombe PA. The role of epigenetic mechanisms and processes in autoimmune disorders. Biologics 2012;6:307−27; PubMed PMID: 23055689. Pubmed Central PMCID: PMC3459549.

[82] Lucas A. Programming by early nutrition: an experimental approach. J Nutr 1998;128(2 Suppl.):401S−6S; PubMed PMID: 9478036. Epub 1998/03/21.

[83] Belch JJ. Raynaud's phenomenon: its relevance to scleroderma. Ann Rheum Dis 1991;50(Suppl. 4):839−45; PubMed PMID: 1750795. Pubmed Central PMCID: PMC1033318.

[84] Loizos N, Lariccia L, Weiner J, Griffith H, Boin F, Hummers L, et al. Lack of detection of agonist activity by antibodies to platelet-derived growth factor receptor alpha in a subset of normal and systemic sclerosis patient sera. Arthritis Rheum 2009;60 (4):1145−51; PubMed PMID: 19333919.

[85] Shen C, Nettleton D, Jiang M, Kim SK, Powell-Coffman JA. Roles of the HIF-1 hypoxia-inducible factor during hypoxia response in *Caenorhabditis elegans*. J Biol Chem 2005;280(21):20580−8; PubMed PMID: 15781453.

[86] Johnson AB, Denko N, Barton MC. Hypoxia induces a novel signature of chromatin modifications and global repression of transcription. Mutat Res 2008;640(1−2):174−9; PubMed PMID: 18294659. Pubmed Central PMCID: PMC2346607. Epub 2008/02/26.

[87] Costa M, Davidson TL, Chen H, Ke Q, Zhang P, Yan Y, et al. Nickel carcinogenesis: epigenetics and hypoxia signaling. Mutat Res 2005;592(1−2):79−88; PubMed PMID: 16009382.

[88] Chen H, Yan Y, Davidson TL, Shinkai Y, Costa M. Hypoxic stress induces dimethylated histone H3 lysine 9 through histone methyltransferase G9a in mammalian cells. Cancer Res 2006;66(18):9009−16; PubMed PMID: 16982742. Epub 2006/09/20.

[89] Ogawa F, Shimizu K, Muroi E, Hara T, Hasegawa M, Takehara K, et al. Serum levels of 8-isoprostane, a marker of oxidative stress, are elevated in patients with systemic sclerosis. Rheumatology 2006;45(7):815−18; PubMed PMID: 16449367. Epub 2006/02/02.

[90] Andersen GN, Caidahl K, Kazzam E, Petersson AS, Waldenstrom A, Mincheva-Nilsson L, et al. Correlation between increased nitric oxide production and markers of endothelial activation in systemic sclerosis: findings with the soluble adhesion molecules E-selectin, intercellular adhesion molecule 1, and vascular cell adhesion molecule 1. Arthritis Rheum 2000;43(5):1085−93; PubMed PMID: 10817563. Epub 2000/05/19.

[91] Gabrielli A, Svegliati S, Moroncini G, Amico D. New insights into the role of oxidative stress in scleroderma fibrosis. Open Rheumatol J 2012;6:87−95; PubMed PMID: 22802906. Pubmed Central PMCID: PMC3395898.

[92] Herrick AL, Rieley F, Schofield D, Hollis S, Braganza JM, Jayson MI. Micronutrient antioxidant status in patients with primary Raynaud's phenomenon and systemic sclerosis. J Rheumatol 1994;21(8):1477−83; PubMed PMID: PMC7983650. Epub 1994/08/01.

[93] Hecker L, Vittal R, Jones T, Jagirdar R, Luckhardt TR, Horowitz JC, et al. NADPH oxidase-4 mediates myofibroblast activation and fibrogenic responses to lung injury. Nat Med 2009;15(9):1077−81; PubMed PMID: 19701206. Pubmed Central PMCID: PMC2743335.

[94] Li Y, Gorelik G, Strickland FM, Richardson BC. Oxidative stress, T cell DNA methylation and lupus. Arthritis Rheumatol 2014;66(6):1574−82; PubMed PMID: 24577881.

[95] van II D, Gordebeke PM, Khoshab N, Tiesinga PH, Buitelaar JK, Kozicz T, et al. Long non-coding RNAs in neurodevelopmental disorders. Front Mol Neurosci 2013;6:53; PubMed PMID: 24415997. Pubmed Central PMCID: PMC3874560.

[96] Hemmatazad H, Rodrigues HM, Maurer B, Brentano F, Pileckyte M, Distler JH, et al. Histone deacetylase 7, a potential target for the antifibrotic treatment of systemic sclerosis. Arthritis Rheum 2009;60(5):1519−29; PubMed PMID: 19404935.

[97] Huber LC, Distler JH, Moritz F, Hemmatazad H, Hauser T, Michel BA, et al. Trichostatin A prevents the accumulation of extracellular matrix in a mouse model of bleomycin-induced skin fibrosis. Arthritis Rheumat 2007;56(8):2755−64; PubMed PMID: 17665426.

Epigenetics of Allergic and Inflammatory Skin Diseases

Nina Poliak[1] and Christopher Chang[2]

[1]Division of Allergy and Immunology, Nemours/AI duPont Hospital
for Children, Wilmington, DE [2]Division of Rheumatology,
Allergy and Clinical Immunology, University of California
at Davis, Davis, CA

13.1 INTRODUCTION

Epigenetics has been considered as a potential mechanism involved in the development of allergic and inflammatory skin diseases [1]. Common inflammatory skin diseases are atopic dermatitis, psoriasis, mastocytosis, and urticaria. The pathogenesis and development of these diseases are caused by a complex interplay of multiple genetic and environmental factors, and are frequently linked through epigenetic mechanisms [1]. Many recent scientific studies have been conducted to identify the role of epigenetics in inflammatory skin diseases.

13.2 ATOPIC DERMATITIS

Atopic dermatitis (AD), also known as atopic eczema and eczema, is a chronic, relapsing, pruritic, allergic, and inflammatory skin disease. It is characterized by a dysfunctional skin barrier and dysregulation of the immune system [2]. Clinical findings of AD include erythema, xerosis, edema, erosions, excoriations, crusting, oozing, and lichenification [3]. Pruritus is one of the main symptoms of this condition.

© 2015 Elsevier Inc. All rights reserved.

13.2.1 Epidemiology of Atopic Dermatitis

Atopic dermatitis is very common and affects all age groups. Its prevalence has increased globally in the last few decades. It is more common in industrialized countries; among other factors this is presumed to be due to the modern lifestyle [4]. Its prevalence is up to 25% in children and 2−3% in adults [2]. AD usually starts in infancy, frequently between 3 and 6 months of age. Most patients, approximately 60%, develop the first presentation of AD in the first year of life. 90% of cases develop by 5 years of age [2]. A majority of these patients experience resolution of atopic dermatitis by adulthood, while 10−30% continue to have AD as adults [2]. A small number of patients may have their first presentation of atopic dermatitis as adults. Children with AD are prone to develop other allergic diseases such as asthma, food allergies, and allergic rhinitis/rhinoconjunctivitis. The combination of AD, asthma, and allergic rhinitis is termed the "atopic march," with atopic dermatitis usually signaling the start of this triad.

Atopic dermatitis is the commonest inflammatory skin disease in children [5] and its prevalence continue to rise worldwide. Its prevalence varies between and within countries [5]. Like most allergic diseases, the prevalence of AD has been increasing in the developed, industrialized world. It has been postulated that the prevalence in the developing world may catch up with the prevalence in the developed world in the future [5]. The fact that AD is so common provides an opportunity for investigation of epigenetic mechanisms for this disease. Given its prevalence, large population-based studies can be conducted to investigate gene−environment interactions and the identification of environmental factors that cause the differences in prevalence, severity, and management response of AD even within the same community/country.

The International Study of Asthma and Allergies in Childhood (ISAAC) is the biggest and only allergy study that has a global approach [6] and provides information about trends in prevalence. ISAAC includes almost 2 million children from 106 countries. ISAAC Phase One demonstrated the highest prevalence of AD in the United Kingdom, Finland, Sweden, Ireland, Nigeria, and New Zealand [7]. The lowest prevalence was reported in Iran, Georgia, Indonesia, China, Taiwan, and Albania [7]. The fact that the prevalence of AD varies not only between but also within countries suggests that environmental factors rather than genetic factors are the main drivers of change in AD burden [8,9]. ISAAC Phase Three demonstrated that the prevalence of AD continues to rise in most developing countries [10]. Table 13.1 demonstrates worldwide prevalence based on ISAAC Phase One data.

An excellent example of environmental influences that changed AD prevalence in a genetically similar population over a short period of

TABLE 13.1 Worldwide Prevalence of AD per ISAAC Phase One [7]

Countries with highest prevalence of atopic dermatitis	Countries with lowest prevalence of atopic dermatitis
United Kingdom	Iran
Nigeria	Georgia
Finland	Taiwan
Sweden	Indonesia
Ireland	China
New Zealand	Albania

time (approximately 30 years) is the prevalence of AD in the German population before and after reunification. East Germany's preschoolers usually had a low prevalence of AD before reunification. After reunification, however, East Germany saw an increase in newly diagnosed AD cases in children up to 6 years of age. The prevalence of AD in East Germany in 1991 was 16%, and in 1997 the prevalence in the same preschooler population was 23.4% [11]. In contrast, the prevalence in West Germany's preschoolers stayed the same. Similar data can be found in migrant populations who move from areas of low disease prevalence to areas with high disease prevalence, and they typically adopt the prevalence of their new environment [12].

Other environmental risk factors for AD have been studied. Climate, urban versus rural living, diet, breastfeeding and delayed weaning, obesity and physical exercise, pollution, tobacco smoke, microbial exposure, day care influences, farm environment and animal exposure, and infections and antibiotics exposure prenatally and postnatal are all risk factors that have been investigated.

13.2.2 Genetics of Atopic Dermatitis

The complex pathogenesis of AD involving genetic, environmental, and immunologic factors has long been recognized. Numerous studies have underlined the genetic factors of this disease. A family history of AD is one of the major risk factors. Seventy percent of AD patients have a positive family history [13]. The risk of developing AD is two- to threefold higher in children with one atopic parent and three- to fivefold higher if both parents are atopic [2]. The risk is more predictive if there is a maternal history [14]. Recent twin studies have provided additional evidence for the heritability of AD. These studies showed that there is a sevenfold increased risk of atopic dermatitis in the co-twin of an affected monozygotic twin; Thomsen et al. showed that there is a threefold increased risk in the co-twin of an affected dizygotic twin in relation to the general population [15]. This study estimated that genetic factors account for 82—84% of the risk of developing AD while environmental factors account for 16—18%.

A dysfunctional skin barrier is a hallmark of eczema. Filaggrin (filament-aggregating protein) is involved in the development and differentiation of keratinocytes. Thus, filaggrin is a protein that plays a key role in the terminal differentiation of the epidermis and the stratum corneum. Filaggrin contributes to epidermal hydration, as filaggrin breakdown products are a component of natural moisturizing factors. Filaggrin deficiency also increases transepidermal water loss [16]. Filaggrin mutations are common in AD. Loss-of-function mutations of filaggrin have been considered to be an important risk factor for the development of AD in patients with allergic sensitization, and are also associated with early onset and a more persistent course of atopic dermatitis. The null mutation of filaggrin is found in 50% of moderate-to-severe cases of AD but in only approximately 15% of mild-to-moderate atopic dermatitis [17]. Whether the filaggrin mutation alone can induce epidermal barrier dysfunction, and subsequently AD, is controversial. This controversy has arisen from the fact that 9% of Europeans have two variants of filaggrin null mutation [18]. Additionally, a significant number of patients with AD do not have this mutation, and 40% of people with filaggrin loss-of-function mutations do not develop atopic dermatitis. On the other hand, filaggrin is a candidate gene involved in other types of allergy including asthma and food allergy. Filaggrin null mutations increase the risk of food allergies because skin and mucosal barriers become more permeable to food allergens. Filaggrin null mutations are also associated with the risk of peanut allergy, a condition that is also associated with some cases of AD.

Another gene that is associated with increased incidence and increased severity of AD is the serine protease inhibitor Kazal-type 5 (SPINK5) gene. SPINK5 is a protease inhibitor protein. Its gene is located on chromosome 5q31. SPINK5 is expressed in the thymus. Its defects have been suggested to cause abnormal maturation of T lymphocytes and accentuated Th2 responses, including eosinophilia and increased IgE level. A recent report suggested that SPINK5 polymorphisms in Japanese children are associated with increased severity of AD and food allergy [19]. SPINK5 encodes the protease inhibitor lymphoepithelial Kazal-type-related inhibitor (LEKTI) that is expressed in epithelium and mucous membranes. While SPINK5 regulates proteolysis in keratinocyte differentiation and generates normal epithelium, LEKTI is involved in maintaining normal skin permeability. SPINK5 polymorphisms are associated with the incidence and severity of AD in some populations [20].

Genome-wide linkage analysis or genome-wide association studies (GWAS) are the methods of choice for gene identification in complex diseases. Genes that are involved in epidermal structural development and the immune system are important for understanding the etiology of AD. GWAS have identified multiple candidate regions on multiple chromosomes that have been associated with AD. For instance, GWAS have identified two genetic loci for AD on chromosomes 1p21 and 11q13. The chromosome 1 locus encompasses filaggrin and supports the important role of filaggrin as a major susceptibility gene [16]. Chromosome 11 encompasses genes for allergic rhinitis and asthma, diseases associated with AD. Another gene described in GWAS is *HRH4*, encoding the histamine H4 receptor, which is important for pruritus.

13.2.3 Epigenetics of Atopic Dermatitis

Although several susceptibility genes for AD have been described in the past, only 14.4% of the estimated total heritability can be explained to date [21]. Epigenetic mechanisms represent another source of hidden heritability and may play a role in the differential expression of phenotypes of AD. Epigenetics contributes to phenotype plasticity and is linked to the diversity, different severity, and distinct pathophysiology of AD. The most commonly known epigenetic mechanisms are DNA methylation, chromatin modification, such as histone acetylation and deacetylation, and microRNA regulation. In 2006, Nakamura et al. investigated the DNA methylation profiles in patients with AD. They described a significantly reduced expression of DNA methyltransferase 1 (DNMT1) in peripheral blood mononuclear cells of AD affected individuals with high serum IgE levels [22]. Their report was the first report of DNMT1 expression in AD patients. DNMT1 is the key enzyme for maintenance of DNA methylation patterns during cell division [22]. DNMT1 also has a *de novo* DNA methylation capability, and dysfunction of this enzyme may lead to changed DNA methylation and changed expression of AD-related genes. Nakamura et al. measured the levels of DNMT1 by measuring messenger RNA (mRNA) expression in patients with AD. DNMT1 mRNA levels were significantly lower in AD patients with high IgE levels in comparison to healthy controls.

In 2012, Liang et al. investigated 10 patients with AD and 10 healthy controls while looking at the high-affinity IgE receptor gamma subunit (FceR1G) promoter in monocytes and dentritic cells (DCs). It is well known that overexpression of the high-affinity IgE receptor on monocytes and dendritic cells contributes to the pathogenesis of AD [23]. Liang et al. showed that monocytes from individuals suffering from AD demonstrated a global hypomethylation as well as locus-specific hypo-methylation of the *FceR1G* promoter in comparison to healthy controls [23]. Additionally, hypomethylation of *FceR1G* is correlated with its overexpression [23]. Liang et al. confirmed the relationship between methylation and expression of *FceR1G*. Furthermore, they showed that treating healthy monocytes with 5-azacytidine caused a reduction in methylation levels and an induction in *FceR1G* transcription and surface expression [23].

In 2012, Ziyab et al. analyzed the methylation levels of CpG sites across the filaggrin (*FLG*) gene in DNA taken from the peripheral blood of 245 individuals. Thirty-seven of those 245 individuals suffered from AD. Although there were no significant differences between carriers and noncarriers of *FLG* mutations, it was reported that there was a potential interaction between filaggrin variants and methylation at the single CpG site in the *FLG* gene body [24]. Ziyab et al. suggested that their study shows that *FLG* loss-of-function variants and AD is modu-lated by DNA methylation [24]. This provided evidence of the impor-tance of the filaggrin genomic region in the manifestation of AD.

Wang et al. further investigated the effects of cigarette smoke and DNA methylation on AD in children [25]. Their goal was to evaluate whether smoke exposure can lead to differential DNA methylation pat-terns in genes that play a role in the development of atopy. A total of 261 mother and child pairs were recruited, who completed study ques-tionnaires regarding smoking history. In addition, prenatal smoke inha-lation was monitored through measurement of cord blood cotinine levels. The investigators subsequently evaluated the blood of seven sub-jects at the age of 2 years for DNA methylation patterns using an Illumina Infinium Assay Methylation Protocol to identify candidate genes and found that thymic stromal lymphopoietin (*TSLP*) showed a differential pattern. The study was then expanded to 150 subjects who provided blood at 2 years of age. The study population was stratified into high and low exposure groups and the results confirmed that DNA methylation of the *TLSP* promoter was higher in the low exposure group compared with the high exposure group (44.2% vs. 29.6%). Wang et al. also found that the degree of *TSLP* promoter methylation was lower in the atopic dermatitis group compared with nonatopic dermati-tis patients ($25.10 \pm 6.53\%$ vs. $30.41 \pm 10.65\%$, $P = 0.006$). The TSLP pro-tein levels also correlated inversely with the methylation status of the *TSLP* promoter.

A more recent study conducted by Rodriguez et al. in 2014 investi-gated 28 atopic dermatitis-affected and 29 healthy individuals for meth-ylation changes of DNA derived from whole blood, T cells, B cells, and lesional and nonlesional epidermis. This study showed that there were significant methylation differences between lesional epidermis and epidermis of healthy controls for various CpG sites, which correlated with altered transcript levels of genes relevant for epidermal differentiation and the innate immune response [26]. The DNA methylation was significantly discordant in skin and blood samples suggesting that blood is not an ideal surrogate for skin tissue [26]. Further research on epigenetics and atopic dermatitis will clarify and improve our understanding of the mechanisms of AD.

Sonkoly et al. studied the microRNA profiles of patients with atopic dermatitis. They generated microRNA heat maps which suggested the particular importance of miR-155 in the pathogenesis of atopic dermatitis. miR-155 was shown to modulate T-cell specific responses through downregulation of cytotoxic T-cell lymphocyte antigen 4 (CTLA-4) [27]. This was accompanied by an increased proliferation rate. miR-155 was preferentially expressed in infiltrating immune cells and could be upregulated *in vivo* in PMBCs by T-cell activators and by external stimulants including superantigens and allergens *in vivo*.

Other miRNAs that are upregulated in AD include miR-21, miR-142-3p, miR-142-5p, miR-146a, and miR-223 [28]. The targets of these miRNAs include interleukin-12p35 (miR-21), STAT 1 (miR-146a), and insulin growth factor 1 receptor (IGF1R) for miR-223. The targets of miR-21 and miR-223 include regulation of eosinophil development. miRNAs that are downregulated in AD included miR-365 and miR-375. miR-375 as well as the let family of miRNAs including let 7a, b, c, and d are all downregulated in AD and play a role in regulation of IL-13 expression. These findings again illustrate how microRNA analysis can contribute to an understanding of the pathogenesis of diseases such as atopic dermatitis.

13.2.4 Pathogenesis of Atopic Dermatitis

The pathogenesis of AD is a complex interplay of abnormal skin barriers, innate and adaptive immune systems, different T-cell subpopulations and their cytokines, mast cells, eosinophils, different pathways, and many more unexplored factors. Figure 13.1 illustrates the possible pathogenesis of atopic dermatitis.

Th2 cells are critical cells in the pathogenesis of AD [29]. Atopic dermatitis is a Th2/Th22-dominant disease in the acute form and a Th1/Th17-dominant condition in the chronic state. When allergens and pathogens penetrate into impaired and defective skin of AD, they stimulate B cells to produce IgE antibodies that bind to allergens. Cross-linking of antigen-bound IgE on the surface of mast cells and basophils leads to release of mediators that stimulate an immunologic response involving T cells and DCs. Keratinocytes secrete TSLP, an important

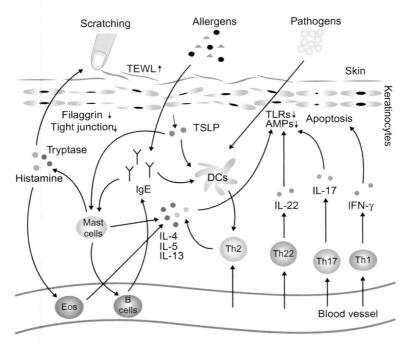

FIGURE 13.1   The pathogenesis of atopic dermatitis. *Reproduced with permission* [4].

cytokine in allergic diseases, which also activates DCs. The activated DCs induce the Th2 polarization seen in acute AD. Th2 cells are critical for induction of isotype class switching to IgE synthesis. IL-4, IL-5, and IL-13 are the main Th2 cytokines, and these cytokines are increased in lesional and nonlesional skin in the acute phase of AD [30]. The cytokines IL-4, IL-5, and IL-13 suppress the expression of Toll-like receptors (TLRs) and antimicrobial peptides (AMPs). AMPs expressed on keratinocytes play an important role in the clearance of pathogens. Mutations and malfunction of pattern recognition receptors (PRRs), such as TLRs and nucleotide-binding oligomerization domain-like receptors (NLRs), and AMPs, are associated with increased susceptibility to skin infections caused by *Staphylococcus aureus* and *Malassezia furfur*. Downregulation of TLRs, NLRs, and AMPs plays a significant role in the initiation and worsening of atopic dermatitis.

Th22 lymphocytes are also involved in pathogenesis of AD. Th22 cells secrete IL-22, which downregulates AMPs and TLRs. IL-22 also downregulates filaggrin and profilaggrin processing enzyme expression and exaggerates the epidermal barrier dysfunction in AD [31]. In the chronic stage of AD, Th1 cells become important elements through secretion of IFN-gamma. Th1 cells are characterized by production of IFN-gamma, IL-12, IL-2, TGF-Beta 1, and chemokine CCL20. Cytokines IFN-gamma, IL-12, and CCL20 are increased in chronic AD [32]. These cytokines are responsible for tissue remodeling and fibrosis in chronic AD. INF-gamma also exacerbates apoptosis of keratinocytes, and promotes inflammation and tissue remodeling [33].

Th17 cells also play a role in AD. These cells produce high levels of IL-17A and IL-17F [34]. AD is considered to be an IL-17-mediated disease. The levels of Th17 and IL-17 vary depending on the phase of AD. IL-17 may also stimulate B cells to secrete antigen-specific IgE [35] and subsequently participates in the pathogenesis of acute AD. IL-17 also contributes to the downregulation of AMPs and TLRs. Mast cells and eosinophils contribute to inflammation of AD by secreting Th2 cytokines, IL-4, IL-5, and IL-13. Additionally, mast cells secrete histamine and tryptase, which exacerbate pruritus. Eosinophils contribute to AD by secreting many cytokines and chemokines, especially IL-16. IL-16 levels are much higher in patients with atopic dermatitis and it also promotes exacerbation of AD [36].

13.2.5 Phenotypes and Diagnosis of Atopic Dermatitis

Atopic dermatitis is a heterogeneous disease. It represents a wide spectrum of phenotypes. AD can be divided into acute and chronic. It can be separated by the age of onset: early onset, late onset, and very late onset. It can be separated further into AD occurring in infancy, childhood, and adolescence. Atopic dermatitis can be categorized by the presence of filaggrin mutations because the AD that is associated with filaggrin null mutations represents an early onset and more severe form of the disease. Eczema can be also divided into mild, moderate, and severe. Another two distinct phenotypes are intrinsic and extrinsic AD. Extrinsic AD, also known as allergic or classical AD, is usually present early in life, and has typical clinical features of atopy, as well as high levels of total and specific IgEs. The extrinsic phenotype is associated with food allergies, environmental allergies, and respiratory diseases. The extrinsic phenotype is associated with Th2 dominance and presents with high levels of Th2 cells secreting IL-4, IL-5, and IL-13. Intrinsic AD accounts for 10−45% of patients. This phenotype represents a milder disease, is not associated with atopy, affects more female patients, and usually has normal IgE levels. Clinically, extrinsic and intrinsic AD are very similar [37]. In Table 13.2 are detailed the various phenotypes of atopic dermatitis.

The diagnosis of atopic dermatitis is made clinically. It is based on the patient's history, morphology of the skin lesions, distribution of

TABLE 13.2   Phenotypes of Atopic Dermatitis

Acute atopic dermatitis	Chronic atopic dermatitis
Pruritic erythematous papules with excoriation, serious exudates, microvesiculation	Dry, erythematous skin, lichenification, excoriations
Th2 dominant response	Th1 dominant response
Extrinsic Atopic Dermatitis (55–90% of AD)	**Intrinsic Atopic Dermatitis** (10–45% of AD)
High serum IgE and allergen-specific IgE associated with allergic triggers	Normal IgE level
Susceptibility to asthma and rhinitis	No association with allergic triggers
High levels of Th2, IL-5, IL-13	Little susceptibility to asthma and rhinitis
Eosinophilia	Normal levels of Th2 and cytokines
More severe clinical manifestations	Mild eosinophilia
Filaggrin mutation presence	Milder clinical manifestations
	No filaggrin mutation
Filaggrin mutation	**Absence of filaggrin mutation**
Early onset	Later onset
More severe and persistent	Milder clinical presentation
Susceptibility to asthma	No increased risk for asthma
Associated with ichthyosis, keratosis pilaris, palmar hyperlinearity	
Early onset atopic dermatitis	**Adult onset atopic dermatitis**

Atopic dermatitis presents with a wide spectrum of different phenotypes.

lesions, and associated clinical features. There are major and minor features of AD. Essential features of AD are pruritus, typical rash, age-specific distribution of the rash, and a chronic and relapsing history. Important features that support the diagnosis of AD are the early age of its onset, personal and/or family history of atopy, immunoglobulin E (IgE) reactivity, and xerosis. There are also associated features of AD that help to suggest the diagnosis, but they are nonspecific. They include dermographism, delayed blanch response, keratosis pilaris, hyperlinear palms, ocular and periorbital changes, and inflammation around the lips.

The differential diagnosis of AD is broad, because many other skin diseases can present with a similar erythematous, scaly rash. Other conditions that should be considered while diagnosing atopic dermatitis are seborrheic dermatitis, contact dermatitis, ichthyosis, cutaneous T-cell lymphoma, psoriasis, immunodeficiencies associated with dermatitis, and photosensitivity dermatitis.

13.2.6 Biomarkers of Atopic Dermatitis

Currently there are no reliable biomarkers or tests that can differentiate AD from other conditions [2]. The most commonly utilized laboratory test is an elevated total and/or allergen-specific serum IgE level. Patients with AD tend to have higher levels of IgE. However, this is not universally seen [2]. Some patients develop an elevated IgE level later in the course of disease [2], while some studies suggest that elevated IgE level is a secondary phenomenon. Sometimes, unexplained elevated IgE levels may be found in the general population as well. Additionally, IgE may also be elevated in several nonatopic conditions.

Increases in peripheral eosinophils and tissue mast cells have been evaluated, but these associations are nonspecific. There are several novel cytokines, chemokines, and T-lymphocyte subsets (e.g., macrophage-derived chemoattractant (MDC), IL-12, IL-16, IL-18, IL-31, thymus and activation-regulated chemokine (TARC), and CD30) that have been evaluated as potential biomarkers, but they have not shown an appropriate sensitivity or specificity in order to be used for diagnosis or monitoring of AD [38,39]. There are also no reliable markers for the prognosis of AD, but high total serum IgE levels and filaggrin gene null mutations seem to predict a more severe course of disease.

13.2.7 Associated Comorbidities of Atopic Dermatitis

Atopic dermatitis has been associated with several other conditions, such as asthma, allergic rhinitis/rhinoconjunctivitis, and food allergies. AD is usually the start of the "atopic march." However, the progression from AD to other atopic conditions does not occur in all patients. Atopic dermatitis and food allergies are highly correlated. The estimated prevalence of food allergies in patients with AD has ranged from 20% to 80% [40]. Generally, food allergies are more likely present in patients with early onset and increased severity of AD. Food allergens may lead to rapid IgE-mediated reactions or to late eczematous reactions. The most common products causing food allergies in the United States are peanuts, tree nuts, cows' milk, soy, eggs, wheat, seafood, and shellfish [24]. Some studies suggest that food allergies might be an exacerbating factor for AD in the pediatric population. Food allergies as an exacerbating factor for atopic dermatitis are more likely to exist in infants and children with moderate-to-severe AD [41]. Some studies suggest that elimination of foods triggering immediate reactions improves AD in infants and children.

It was demonstrated in a trial of 55 children with AD and egg sensitivity that children in the egg exclusion group demonstrated improvement in the disease after 4 weeks of egg exclusion. Food allergen-specific serum IgE tests or skin prick tests can help identify sensitization to specific foods, while the double-blind, placebo-controlled food challenge is the gold standard for the diagnosis of food allergies.

There are two main risk factors for the development of AD: family history and filaggrin null mutations [2]. It would be interesting to evaluate whether epigenetic changes that increase the chance of developing AD also increase the risk of developing other atopic diseases. These kinds of studies have not yet been done but they do provide us with a potentially productive area for research. Epigenetic studies can be conducted with either large populations of patients or by comparing discordant monozygotic twins. Each of these approaches present very different but substantial challenges.

13.2.8 Treatment of Atopic Dermatitis

Atopic dermatitis is a heterogeneous disease with different phenotypes and severity. Subsequently, the treatment of AD is complicated and involves different agents in order to address different aspects and pathways of pathogenesis. Treatment response is also very variable. Variability in response to treatment might be also related to different aspects of epigenetics and to the personal epigenetics of each patient. Some patients respond well and some patients are resistant to the same treatment plan.

Topical agents are the mainstay of AD therapy, while more severe cases require systemic therapy. The use of topical moisturizers is an important therapeutic concept. Topical moisturizers address xerosis and transepidermal water loss. For moderate and severe AD, wet-wrap therapy is used to quickly reduce the severity of the disease and flare-ups. Response to wet-wrap therapy is also variable, suggesting a role of epigenetic mechanisms in the treatment response.

Topical corticosteroids (TCS) belong to the anti-inflammatory arm of the treatment protocol. These agents affect T lymphocytes, monocytes, dendritic cells, and macrophages through interference with antigen processing and suppression of inflammatory cytokine release. Application of TCS also reduces S. aureus bacterial load on atopic dermatitis lesions, most likely because TCS decrease the release of inflammatory cytokines and inhibit AMP production.

Topical calcineurin inhibitors (TCIs) belong to a secondary class of anti-inflammatory therapy. They inhibit calcineurin-dependent T-cell activation, blocking the production of anti-inflammatory cytokines and mediators. They also affect mast-cell activation and epidermal dendritic cells. The most commonly used TCIs are used in the form of topical tacrolimus ointment and pimecrolimus cream.

If AD is severe and cannot be controlled by topical therapy, systemic therapy may be considered. Such therapy is directed to decrease inflammation by suppressing or modulating immune responses. Systemic therapy includes use of systemic corticosteroids, cyclosporine, azathioprine, mycophenolate mofetil, methotrexate, alitretinoin, interferon, intravenous immunoglobulin, and biologics (anti-CD20, anti-IL-5, anti-IgE). Systemic corticosteroids are the most common systemic therapy and are often used to treat severe AD exacerbations and pruritus. Epigenetic studies may help to define the characteristics of patient-specific responses to the different treatment modalities available to us as well as to define future pharmacological targets involved in the pathogenesis of atopic dermatitis.

13.3 PSORIASIS

Psoriasis is a common chronic immune-mediated inflammatory skin disease. It affects mainly the skin but joints can be affected as well. It presents with thickened, inflamed, and scaly skin patches. Psoriasis is characterized by abnormal keratinocyte proliferation, vascular hyperplasia, and infiltration of inflammatory cells into the dermis and epidermis. Psoriasis is a T-cell-mediated autoimmune disease. Psoriasis and atopic dermatitis have many similar features. Both of these conditions are inflammatory skin disorders in which genetic and environmental factors play important roles.

13.3.1 Epidemiology of Psoriasis

Depending on ethnicity and geographic area, the prevalence of psoriasis ranges between 1% and 11.8%. Prevalence rates vary between people of different ethnic backgrounds; psoriasis is most common in whites [42]. The incidence in white individuals is estimated to be 60 cases per 100,000 individuals per year [42]. In the United States, psoriasis affects approximately 3% of the population [43]. The disease affects both children and adults. In children, the prevalence is estimated to be around 0.71% [44]. Prevalence increases with age to 1.2% by age 18 [44]. One-third of patients develop psoriasis in childhood and there is no gender bias in children [44].

13.3.2 Genetics of Psoriasis

Psoriasis has a strong genetic component with an estimated heritability of 66%. Population studies demonstrate that the incidence of psoriasis is greater in first- and second-degree relatives of patients than in the general population [42]. The concordance rate among monozygotic twins is 35−72%. The fact that the concordance rate in monozygotic twins is not 100% suggests that factors other than genetics play a role in the development of psoriasis. Moreover, several twin studies have shown that concordance rates in monozygotic twins vary greatly between different geographic areas. This suggests the role of environmental and lifestyle factors such as ultraviolet radiation/solar exposure and diet in the manifestation of psoriasis. The interaction between environment and genetics most likely occurs through epigenetic mechanisms.

Several genome-wide linkage analyses have been performed. At least nine chromosomal loci have been identified in association with psoriasis [45]; these loci are termed psoriasis susceptibility 1 through 9 (*PSORS1−9*). *PSORS1* is a major gene that accounts for 35−50% of the heritability of psoriasis. *PSORS1* is located within the major histocompatibility complex (MHC) on chromosome 6p [45]. Psoriasis vulgaris and guttate psoriasis are associated with *PSORS1* while other psoriasis variants such as late-onset psoriasis vulgaris and palmoplantar pustulosis are not [46].

PSORS2 is located on chromosome 17q and its polymorphism causes loss of binding to the RUNX1 transcription factor [47]. *PSORS4* is important for epidermal differentiation and *PSORS8* is located on chromosome 16q and overlaps with a Crohn's disease locus. Genome-wide association analyses have found variants in the gene encoding the interleukin-23 receptor (IL-23R) as well as variants in the untranslated region of the interleukin-12B (*IL12B*) gene to be indicators of psoriasis risk [48].

CDKAL1 is another gene associated with psoriasis as well as Crohn's disease and diabetes mellitus type 2. This is an interesting finding given the fact that Crohn's disease and diabetes mellitus type 2 are associated with the moderate-to-severe form of psoriasis. Psoriasis and AD share certain susceptibility loci such as 1q21, 17q25, and 20p [49]. *PSORS4* and *ATOD2* are both located on 1q21 and both are susceptibility loci for psoriasis and AD [49]. Both loci are involved in epidermal differentiation. In a very recent study, Sheng et al. identified three new susceptibility loci for psoriasis: *NFKB1* on chromosome 4q, *CD27-LAG3* on chromosome 12p, and *IKZF3* on chromosome 17q [50].

13.3.3 Epigenetics of Psoriasis

There have been several studies conducted on the epigenetic mechanisms involved in psoriasis. In 2010, Zhang et al. studied 30 patients with psoriasis vulgaris. They showed that genomic DNA in peripheral blood mononuclear cells of psoriatic patients was aberrantly hypermethylated in comparison with healthy controls. They also found that methyltransferase enzyme DNMT1 mRNA expression was increased [51]. In 2011, the same group investigated global histone H4 in another 30 patients with psoriasis vulgaris and 20 healthy controls. They showed that, compared with normal controls, global histone H4 hypoacetylation was observed in PBMCs from psoriasis vulgaris patients [52]. They also showed a negative correlation between the degree of histone H4 acetylation and disease activity in patients [52].

The role of CD4$^+$ T cells in diseases with immune dysregulation such as psoriasis has driven research into exploring aberrant expression of CD4$^+$ T-cell-related cytokines and mediators. Similarly, CD4$^+$ T cells are a popular target for investigating epigenetic regulation mechanisms such as DNA methylation. Park et al. studied genome-wide DNA methylation of CD4$^+$ cells in patients with psoriasis compared to healthy controls using MeDIP-seq methodology and found that DNA methylation is globally enhanced in the psoriasis patients [53]. The authors targeted genes that have reduced expression in CD4$^+$ cells, with one of the candidate genes being phosphatidic acid phosphatase type 2 domain containing 3 (*PPAPDC3*). Bisulfate sequencing of the transcription start region demonstrated hypermethylation in CD4$^+$ cells which was associated with reduced expression of the gene. The authors concluded that DNA methylation of specific relevant genes is an epigenetic mechanism that may play a role in the pathogenesis of psoriasis. Figure 13.2 illustrates a protocol for the use of MeDip-seq to study DNA methylation.

Another study on whole genome DNA methylation also using MeDIP-seq technology was conducted by Zhang et al. [54]. The authors identified differentially methylated regions (DMRs) in the flanking regions of two genes, namely programmed cell death 5 (*PDCD5*) and tissue inhibitor of metalloproteinase 2 (*TIMP2*), which play roles in apoptosis and inhibition of matrix metalloproteinases. Both of these genes are known to play a significant role in various cellular and physiologic functions such as wound healing, tumor cell invasion, and angiogenesis. Figure 13.3 illustrates an analysis of DMRs in nonaffected skin and affected skin in psoriasis patients compared to healthy patients (*courtesy*: Qianjin Lu, Second Xiangya Hospital, Changsha, PR China).

DNA methylation pattern in genome-wide detected by MeDIP-seq

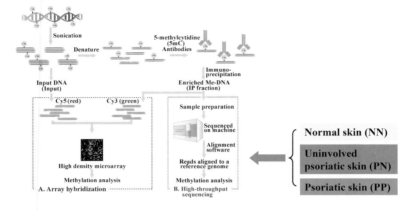

FIGURE 13.2   An example of the use of MeDIP-seq to study DNA methylation in psoriasis.

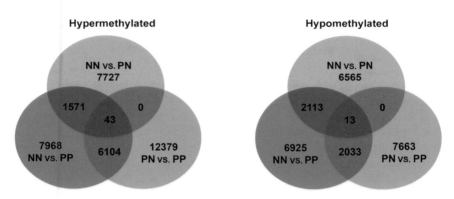

FIGURE 13.3   **DMRs in psoriasis.** NN, nonaffected skin in normal subjects; PN, nonaffected skin in psoriasis patients; PP, affected skin in psoriasis patients. *Courtesy of Professor Qianjin Lu, Second Xiangya Hospital, Changsha, PR China.*

Several microRNAs (miRs) have been implicated in the pathogenesis of psoriasis. miR-203 is the first skin specific miR described. miR-203 suppresses the gene expression of suppressor of cytokine signaling-3 (*SOCS-3*) which is a negative regulator of the STAT3 pathway. miR-203 is overexpressed in psoriatic skin lesions [44]. Xu et al. showed that miR-31 also contributes to inflammation in psoriasis and is overexpressed the disease [55]. Other miRs investigated in psoriasis include miR-146a, investigated by Yang et al. [56]; miR-125b, described by Tili et al. [57]; and miR-221 and 222, evaluated by Zibert et al. [58].

Because psoriasis has some similarities with AD, contact dermatitis, drug eruption, and lymphoma, there is a need for a serum biomarker that can distinguish between psoriasis and other inflammatory skin diseases. Thus, Koga et al. measured serum levels of 6 different miRs in 15 healthy participants, 15 AD patients, and 15 psoriasis vulgaris patients. They found that serum levels of miR-125b, miR-203, miR-146a, and miR-205 were significantly decreased in psoriasis vulgaris patients in comparison to healthy participants [59]. They also found that levels of miR-203 demonstrated the strongest downregulation in psoriasis patients and thus might serve as a biomarker for psoriasis [59].

In 2007, Sonkoly et al. investigated the microRNA profiles of patients with psoriasis. They found that multiple miRNAs were upregulated and downregulated in psoriasis affected skin compared with healthy skin. In particular, they also confirmed the role of miR-203 in the pathogenesis of psoriasis and their analysis of 21 different organs and tissues revealed that miR-203 was preferentially expressed in keratinocytes [44]. They also observed that the upregulation of miR-203 was concurrent with a downregulation of the *SOCS-3* gene. As alluded to above, the *SOCS-3* gene is involved in the inflammatory response as well as in keratinocyte function. Additionally, Koga et al. suggested that the combination of miR-146a and miR-203 levels can be a diagnostic marker for psoriasis [59]. They investigated this hypothesis in a study that included 10 psoriasis patients, 9 normal participants, and 7 patients with cutaneous lymphoma. In this study, they showed that the combination of miRNA levels may be more reliable to distinguish psoriasis patients from healthy participants than levels of each miRNA alone [59].

Recently, Zhao et al. looked at miR-210 expression in the T cells of 18 psoriasis vulgaris patients and 18 healthy participants [60]. They found that miR-210 was significantly increased in psoriasis vulgaris CD4$^+$ T cells compared with healthy participants. However, they found no significant correlation between miR-210 expression levels and severity score. They also evaluated miR-210 expression and *FOXP3*. They confirmed that miR-210 suppresses *FOXP3* expression and subsequently leads to immune dysfunction. This study suggests that the gene *FOXP3* is one of the targets of miR-210 in CD4$^+$ T cells. Several prior studies showed that FOXP3 is required for development and function of T regulatory (Treg) cells. Treg cells are important for maintaining peripheral tolerance, preventing autoimmune diseases and limiting chronic inflammatory diseases [60]. This study demonstrates that miR-210 is involved in regulation of Treg cells through FOXP3 and consequently is involved in the immune response. Additionally, this study showed that overexpression of miR-210 leads to increased inflammatory cytokine (IFN-gamma and IL-17) expression and decreased regulatory cytokine (IL-10

(A)

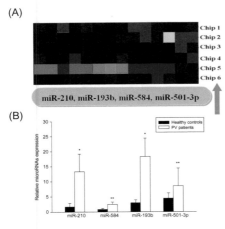

(A) Chips 1–3 for PBMCs of PV patients

Chips 4–6 for PBMCs of Healthy controls

Expression of miR-210, miR-193b, miR-584, and miR-501-3p was upregulated in PV PBMCs compared with healthy controls.

(B)

(B) Validation of miR-210, miR-193b, miR-584, and miR-501-3p by RT-qPCR

FIGURE 13.4   **miRNA in psoriasis.** PV, psoriasis vulgaris. Upregulated miRNAs in psoriasis include miR-210, miR-193b, miR-584, and miR-501-3p. *Courtesy of Professor Qianjin Lu, Second Xiangya Hospital, Changsha, PR China.*

and TGF-beta) expression in CD4$^+$ T cells of patients with psoriasis vulgaris [60]. Finally, this study suggested that miR-210 can be a potential therapeutic target for psoriasis vulgaris (Figure 13.4).

Another very recent study by Lovendorf et al., which was published in June 2014, suggests that miR-223 and miR-143 can serve as biomarkers for disease activity in psoriasis [61]. This group measured miR-223 and miR-143 in PBMCs from patients with psoriasis and healthy controls. They found that miR-223 and miR-143 were significantly upregulated in the PBMCs from patients with psoriasis compared to healthy individuals [61]. Additionally, after 3–5 weeks of treatment with methotrexate accompanied by a decrease in psoriasis severity, miR-223 and miR-143 were found to be significantly downregulated in the PBMCs from patients with psoriasis [61]. A more comprehensive discussion of epigenetics in psoriasis can be found in Chapter 11.

13.3.4 Pathogenesis of Psoriasis

The pathogenesis of psoriasis is still not completely clarified. In the past, psoriasis was considered to be a disease of keratinocytes. The current opinion is that psoriasis is an inflammatory disease driven by cross-talk between T cells, dendritic cells, connective tissue fibroblasts, and keratinocytes. It is a T-cell-mediated disease and different pathways/mechanisms are involved in the pathogenesis. Innate immune cells, myeloid dendritic cells, and cell-mediated adaptive immune system pathways play an important role in psoriasis. Activation of Th1,

Th17, and Th22 results in the production of specific cytokines such as INF-gamma, IL-2, IL-12, IL-22, IL-17, and IL-23 [62]. These cytokines activate keratinocytes [62] and induce the production of AMPs (cathelicidin, beta-defensins, and LL-37), proinflammatory cytokines, chemokines, and S100 proteins. These mediators continue to activate the proinflammatory disease cycle and lead to development of the inflammatory infiltrates. While AD is a primarily Th2-mediated disease, psoriasis is primarily a Th1-mediated inflammatory skin disease.

13.3.5 Phenotypes and Diagnosis of Psoriasis

Psoriasis appears as several different clinical variants. Psoriasis phenotypes include psoriasis vulgaris, inverse psoriasis, generalized pustular psoriasis, palmoplantar psoriasis, and guttate psoriasis. Psoriasis vulgaris accounts for 90% of all cases. Guttate psoriasis usually affects children and adolescents after a beta-hemolytic streptococcal infection. Psoriasis is diagnosed clinically. It manifests as raised, erythematous, well-demarcated plaques with adherent silvery scales [48].

13.3.6 Treatment of Psoriasis

The treatment of psoriasis is a stepwise approach. However, the classic systemic treatments often are not completely effective. Thus, we will review newer strategies such as biological agents. Such agents show high effectiveness in psoriasis. They can be placed, based on the mechanism of action, into two main classes: T-cell targeted therapies (e.g., alefacept) and anticytokine therapies (e.g., antitumor necrosis factor therapies, monoclonal antibodies against IL-12 and IL-23) [48].

Alefacept is the first biological agent developed for psoriasis treatment. It is a medication targeting Th1 cells. Medications that target Th1 cell pathways, the key players in the pathogenesis of psoriasis, tend to work well in psoriasis. Alefacept blocks the interaction between LFA-3 on antigen-presenting cells (dendritic cells) and CD2 on T cells.

Anticytokine therapies include infliximab, adalimumab, etanercept, and a monoclonal antibody against IL-12 and IL-23 (ustekinumab). Blockade of IL-12 and IL-23 is achieved when antibodies block the p40 chain, the common chain of these cytokines [48]. Anti-TNF strategies include three different variants: a complete monoclonal anti-TNF-alpha antibody, a human p75 TNF-receptor Fc fusion protein, and a humanized chimeric anti-TNF-alpha monoclonal antibody [48]. These biologic regulatory modulators are potential targets for epigenetic studies. Identification of specific DNA methylation and histone modification patterns as well as specific relevant miRNA molecules involved in psoriasis will help guide these types of epigenetic studies.

13.4 MASTOCYTOSIS

Mastocytosis is a disorder characterized by accumulation of pathologic mast cells in tissues [63]. It is associated with symptoms of mast-cell activation. Skin, bone marrow, and gastrointestinal tract are the most commonly affected tissues. Liver, spleen, and lymph nodes are additional tissues that can be affected by mastocytosis.

13.4.1 Epidemiology of Mastocytosis

Mastocytosis is a rare disease and affects both children and adults. There are no epidemiologic studies to precisely define the incidence, point prevalence, or cumulative prevalence of mastocytosis because it is rare and often not recognized [64]. The situation of poor epidemiological data will change in the near future because several centers of excellence under the umbrella of the European Competence Network on Mastocytosis have been established in Europe, and more firm criteria for the diagnosis of mastocytosis were defined in 2001 [64]. In the United States, there are estimated to be up to 200,000 patients affected with mastocytosis, leading to an estimate in the prevalence of mastocytosis of approximately 1 in 10,000 persons [64]. Mastocytosis has bimodal distribution affecting both children and adults [65]. Approximately 55% of cases present before 2 years of age, 10% of patients are between 2 and 15 years of age, and 35% of cases are seen in adult patients [65]. There is no gender bias noted [66].

13.4.2 Genetics of Mastocytosis

KIT, the tyrosine kinase receptor for stem cell factor, is critical for the proliferation, survival, differentiation, and homing of mast cells [67]. KIT is found on the surface of mast cells. Gain of function mutation in c-KIT results in ligand-independent receptor activity. Patients with mastocytosis frequently have the c-KIT mutation at codon 816 [67].

D816V c-KIT mutation is detectable in more than 80% of patients with systemic mastocytosis in bone marrow mast cells and in approximately 40% of children with mastocytosis in lesional skin. The mutation leads to activation of the tyrosine kinase domain resulting in persistent cell proliferation. In patients with indolent disease, C-KIT mutation is usually confined to mast cells, while in patients with an advanced disease this mutation involves multiple other myeloid cells as well as lymphoid lineages. Patients with multilineage involvement usually have a more aggressive disease.

13.4.3 Epigenetics of Mastocytosis

To date, there have been only two studies on the epigenetics of mastocytosis. Both studies have explored the role of miRNAs in mast cells. In 2011, Mayoral et al. examined the role of miR-221 in mast cell biology [68]. They used a mouse model and constructed a robust lentiviral system for manipulating miR-221 expression (overexpression vs. depletion) in mouse mast cells. They found that miR-221 does not affect mast cell differentiation, but does affect differentiated mast cells. They suggested that miR-221 has dual role in mast cells based on the state of the mast cells. In resting cells, miR-221 partially regulates normal cell homeostasis through regulation of the cell cycle and the actin cytoskeleton. In the activated mast cell state, miR-221 is involved in cytokine production and degranulation of preformed and *de novo* produced mediators in response to antigen stimulation.

In 2014, Yamada et al. similarly investigated the relationship between miRNAs and mast cell degranulation. They demonstrated that miRNA, especially miR-142-3p, are involved in mast cell degranulation by showing that silencing Dicer, a major enzyme in miRNA biogenesis, attenuates degranulation of human mast cells, specifically the LAD2 mast cell line [69]. They also discovered that overexpression of miR-142-3p enhances FceRI-mediated degranulation and that miR-142-3p rescues the decrease of degranulation by silencing Dicer [69]. Similar effects were shown by this group in bone marrow-derived mast cells. The Yamada et al. study suggests that miR-142-3p is a potential therapeutic target in mastocytosis or allergies. There is extensive potential for further investigations on the epigenetics of mastocytosis. The reason for the difference in presentation of mastocytosis between the pediatric and adult populations is not clear, and an epigenetic mechanism might be one of the reasons. Additionally, the heterogeneity in the phenotypes of mastocytosis might also be attributed to epigenetic mechanisms. Epigenetic investigations can help us better understand the cause of this significantly different disease presentation and course in different age groups, as well as to better understand the pathogenesis of the disease.

13.4.4 Phenotypes of Mastocytosis

The WHO classification divides mastocytosis into seven subtypes. Mastocytosis is initially separated into two main subtypes: cutaneous and systemic phenotypes. The systemic subtype is further divided into six subtypes [63]. Table 13.3 demonstrates the various phenotypes of mastocytosis.

The mastocytosis that presents in any extracutaneous tissue and exhibits multifocal or diffuse organ involvement is termed systemic

TABLE 13.3   Classification of Mastocytosis

Cutaneous mastocytosis	Systemic mastocytosis
	1a. ISM (indolent systemic mastocytosis) 1b. Smoldering SM 2a. SM-AHNMD 2b. ASM (aggressive SM) 3. MCL (mast cell leukemia) 4. MCS (mast cell sarcoma) 5. Extracutaneous mastocytosis

mastocytosis (SM). SM can occur in the form of indolent systemic mastocytosis (ISM), meaning that there is no other hematologic disease and there is no end-organ damage consistent with mast cell infiltration. The more severe form of ISM is termed smoldering systemic mastocytosis [63]. Smoldering SM is characterized by increased mast cell burden represented by more than 30% infiltration in bone marrow biopsy and a tryptase level of more than 200 ng/mL [63]. Approximately 20% of patients with SM have a second myeloproliferative or myelodysplastic disease of bone marrow. These patients have so-called SM associated with a hematologic nonmast cell clonal disease (SM-AHNMD) [70]. Approximately 5% of the patients show evidence of end-organ damage by mast cell infiltration and this form of SM is termed aggressive SM (ASM). Mast cell leukemia is a subset of SM that is diagnosed when bone marrow aspirate has more than 20% of mast cells. MCS is a malignant form of solid mast cell tumor associated with malignant and destructive infiltrates and metastatic potential [71]. Extracutaneous mastocytosis is a rare benign mast cell tumor with only a few cases described in the literature.

13.4.5 Diagnosis of Mastocytosis

Thorough skin examination and detection of a typical lesion are the first steps in diagnosing mastocytosis. Typical lesions are a few millimeters to a few centimeters in diameter and have a brownish hyperpigmentation with hyperemic component. The lesions are fixed and usually are nonpruritic. Stroking of the lesion can cause a wheal localized to the lesion, and this phenomena is termed Darier sign. Although skin lesions have a typical appearance, skin biopsy confirms the diagnosis. Typical biopsy shows perivascular mast cell infiltrates in the upper dermis. The presence of multifocal dense aggregates of mast cells with more than 15 cells per aggregate in an extracutaneous tissue, especially bone marrow, is a major diagnostic criterion for diagnosis of mastocytosis [72].

In children symptoms include skin lesions of urticaria pigmentosa (UP) or mastocytoma. Mastocytosis diagnosed in infancy has a good prognosis because it is limited to skin and does not show any evidence of pathological mast cell accumulation in other tissues. Most pediatric patients experience resolution of symptoms by adolescence. In adults, however, mastocytosis is usually associated with mast cell accumulation in bone marrow and has a persistent clinical course [70].

13.4.6 Biomarkers of Mastocytosis

Histamine is the major chemical released during mast cell degranulation. Normal plasma levels of histamine are <1.0 ng/mL. Histamine is increased in patients with mastocytosis. The increased levels are usually not correlated with mast cell load, but a correlation is found between increased histamine levels in blood and urine and the extent of bone involvement in patients with mastocytosis.

Tryptase is an enzyme that is increased in mastocytosis. Mast cells are the major source of the neutral protease tryptase but basophils and myeloid progenitors also release small amounts of tryptase. A normal tryptase level in healthy individuals is 5 ng/mL. A serum tryptase level that is >20 ng/mL is suggestive of mastocytosis. Because elevated tryptase levels are seen in other diseases as well, tryptase levels cannot be used as the sole criterion for diagnosis of mastocytosis. CD25 and CD2 are other markers of mastocytosis. Normal or reactive mast cells do not express these markers. CD25 is a more specific and sensitive marker than CD2 [73].

13.5 URTICARIA

Urticaria is an inflammatory disease, characterized by the development of wheals (hives), angioedema, or both [74]. It is a common condition that can be classified, based on the duration of the urticarial episodes, into acute and chronic. Urticaria with a duration of urticarial episodes of <6 weeks is termed acute urticaria. Chronic urticaria (CU) is defined when the symptoms last more than 6 weeks. Some patients with recurrent urticaria do not fit completely into the acute or chronic classification because they may have an episode that lasts <6 weeks but then recurs frequently. These patients should be occasionally re-evaluated until a diagnosis is established. Acute urticaria should be differentiated from anaphylaxis [75]. If, during the urticarial episodes, symptoms occur in organs other than the skin, for instance in those of the respiratory, gastrointestinal, cardiac, or nervous systems, anaphylaxis should be considered [75].

13.5.1 Epidemiology of Urticaria

Urticaria is a very common condition. It has been estimated that 10–20% of the population may develop an episode of urticaria at some point in their lives [76]. The incidence of CU has been reported at 1.4% per year. The prevalence of CU in the general population has ranged from 0.5% to 5% [75].

13.5.2 Epigenetics of Urticaria

A thorough search of the literature has not revealed any significant studies involving epigenetics and urticaria. This provides us with the evidence that epigenetics is still a very young science and more research is needed to investigate the role of epigenetics in urticaria.

13.5.3 Pathogenesis of Urticaria

The pathogenesis of urticaria is not completely clear and several theories exist. Mast cells and basophils are the primary effector cells in the pathogenesis of urticaria [77]. Mast cells produce a wide variety of proinflammatory and vasodilatory substances. They release histamine and produce leukotriene C4 and prostaglandin D2 during the immediate phase. These immediately released products are responsible for erythema, swelling, and pruritus. During the delayed secretion phase, TNF-alpha, IL-4, and IL-5 are produced and secreted. These cytokines lead to an influx of inflammatory cells. Lesions of acute urticaria consist of subcutaneous edema but only rare inflammatory cells. Lesions of CU demonstrate, in addition to subcutaneous edema, perivascular inflammatory infiltrates consisting of $CD4^+$ and $CD8^+$ T lymphocytes, basophils, eosinophils, and neutrophils.

Two main mechanisms have been investigated as potential pathogenetic explanations for CU. One theory suggests that CU is caused by IgE/antigen interactions that stimulate the high-affinity receptor for IgE, FcεRI [78]. Development of autoantibodies to FcεRI or to IgE located on mast cells and basophils and binding of these antibodies to FcεRI on basophils and mast cells leads to release of numerous mediators, such as histamine, LTC4, platelets-activating factor, IL-4, IL-5, and IL-13, that cause symptoms of urticaria and angioedema [78,79]. Another hypothesis suggests that dysregulation of intracellular signaling pathways involving Syk, SHIP-1, or SHIP-2 in basophils and mast cells [80] may play a significant role in the pathogenesis of chronic urticaria.

13.5.4 Phenotypes of Urticaria

Based on possible etiology and possible triggers, CU has been classified into different subtypes. Autoimmune urticaria is a subset of urticaria where specific IgG antibodies against the FcεRI alpha subunit component of the high-affinity IgE receptor and IgG antibodies against IgE are present. About 30—50% of patients with CU have antibodies against the FcεRI alpha subunit of the IgE receptor and 5—10% produce IgG antibodies against IgE [75]. In the past, the autologous serum skin test (ASST) and the autologous plasma skin test were utilized but currently the usefulness of these tests is unclear and they are not recommended routinely.

Physical urticaria is another category of CU that can be distinguished from other subtypes not only by triggers but also duration of the urticarial episodes itself. Physical urticaria lesions usually last from 30 min to 2 h while lesions of other types of CU generally last the entire day. Physical urticaria can further be classified based on physical triggers.

Cold urticaria is more common in colder climates and represents 5—30% of physical urticaria cases [76]. Patients with this type of urticaria typically develop wheals on cold-exposed areas. Oropharyngeal edema on ingestion of cold beverages has been reported [81]. Systemic reactions in patients with cold urticaria have been reported and were associated with cold water swimming fatalities [82]. Cold urticaria is usually diagnosed with an ice cube test. During this test, an ice cube is usually placed on the skin for several minutes and a wheal develops on rewarming, determining the test to be positive.

Cholinergic urticaria accounts for 30% of physical urticarias and 3—5% of CU [76]. This type of CU is more common in teenagers and young adults and is triggered by exercise, warm water, and stress. Other cholinergic-mediated symptoms such as wheezing may occur along with urticaria. The lesions of the cholinergic urticaria are usually smaller and macular initially. Cholinergic urticaria can be diagnosed by exercise challenge or partial body immersion in hot water [76]. Delayed pressure urticarial accounts for 2% of CU [76]. This type of urticaria occurs usually approximately 4—6 h after a pressure stimulus.

13.5.5 Diagnosis of Urticaria

The most important diagnostic tool in evaluating patients with urticaria is a detailed history [76]. After a thorough history and physical examination, no diagnostic testing might be necessary for some patients [75]. A thorough history is particularly valuable in diagnosing physical urticarias. Autoantibodies to FcεRI and activation markers on basophils can be used to diagnose autoimmune urticaria [76]. If a trigger for urticaria is known, targeted laboratory testing can be done but in most cases triggers are not identified. In addition, thyroid stimulating hormone and thyroid antibodies can be tested in refractory cases. Increased levels of anti-thyroglobulin or anti-thyroid antibodies in euthyroid subjects are commonly detected, although the clinical significance of this is unclear [75].

13.6 CONCLUSION

Atopic dermatitis, psoriasis, mastocytosis, and urticaria are multifactorial diseases that involve immunological, genetic, and environmental factors. The prevalence of the allergic skin diseases, especially the prevalence of atopic dermatitis, has increased in recent decades much more rapidly than can be explained by genetics alone. Any change in the genetics of a population would take several generations to occur. Epigenetic changes can cause a much faster rise in disease prevalence. These changes, similarly to genetic changes, can be passed from one generation to the next [83]. Additionally, studies conducted with monozygotic twins who presented with discordant development of the diseases have shown that genetic factors are only partially responsible. Epigenetic mechanisms represent another source of hidden heritability. Epigenetics can be linked to the diversity and different severity of these diseases, and contributes to phenotype plasticity.

The most well known epigenetic mechanisms are DNA methylation, chromatin modifications (such as histone acetylation and deacetylation), and microRNA regulation. DNA methylation was the first to be described and extensively studied [83]. Environmental factors such as diet and nutrition, including dietary methyl donors and cofactors, such as folic acid, vitamins B12, B6, B2, and zinc, are necessary for DNA methylation and may modify the genetic risk of skin diseases.

The pathogenesis of the described inflammatory and allergic skin diseases is not completely understood. The fact that the treatment success rate is not 100% for any of the aforementioned diseases suggests that some pathways of pathogenesis have not been explored and thus are not affected by the treatment. Epigenetics provides us with additional areas for research of unexplored pathophysiologic pathways.

Most of the allergic and inflammatory skin diseases do not have good biomarkers that can be used to diagnose, monitor disease activity, and measure response to treatment. Several studies propose that miRNA will become the potential biomarker of inflammatory skin diseases. miRNAs are small noncoding RNA molecules that negatively regulate gene expression at the posttranscriptional level through degradation [80]. The regulation and expression of miRNAs has been extensively studied in psoriasis, but not in mastocytosis or urticaria.

Epigenetics involves changes in gene expression that are heritable and can be passed on to several future generations, but they are also reversible. Additionally, epigenetic changes can be limited to certain cell types [84]. Therefore drugs that are focused on epigenetic mechanisms can lead to more targeted and cell- or organ-specific effects. Additionally, the patient's epigenome will become a biomarker in determining the late risk of developing allergic or inflammatory skin diseases. This will be potentially valuable in the prevention of diseases and early intervention. In summary, epigenetics is a growing and promising field that will open many avenues for understanding the inheritance, pathogenesis, heterogeneity, and treatment of allergic and inflammatory skin diseases.

List of Abbreviations

AMP   antimicrobial peptides
ASM   aggressive systemic mastocytosis
CU   chronic urticaria
DNMT1   DNA methyltransferase 1
FcɛR1   high-affinity IgE receptor
FLG   filaggrin
GWAS   genome-wide linkage analysis
ISAAC   International Study of Asthma and Allergies in Childhood
ISM   indolent systemic mastocytosis
LEKTI   lymphoepithelial Kazal-type-related inhibitor
MCL   mast cell leukemia
MCS   mast cell sarcoma
miRNA   microRNA
NLR   nucleotide-binding oligomerization domain-like receptors
PSORS   psoriasis susceptibility
SM   systemic mastocytosis
SM-AHNMD   systemic mastocytosis associated with a hematologic nonmast cell clonal disease
SPINK5   serine protease inhibitor Kazal-type 5
TARC   thymus and activation-regulated chemokine
TCI   topical calcineurin inhibitor
TCS   topical corticosteroids
TLR   Toll-like receptor

References

[1] Tezza G, Mazzei F, Boner A. Epigenetics of allergy. Early Hum Dev 2013;89(Suppl. 1): S20—1. http://dx.doi.org/10.1016/S0378-3782(13)70007-0.
[2] Eichenfield LF, et al. Guidelines of care for the management of atopic dermatitis: Section 2. Management and treatment of atopic dermatitis with topical therapies. J Am Acad Dermatol 2014;71:116—32. http://dx.doi.org/10.1016/j.jaad.2014.03.023.
[3] Totri CR, Diaz L, Eichenfield LF. 2014 update on atopic dermatitis in children. Curr Opin Pediatr 2014;26(4):466—71. http://dx.doi.org/10.1097/MOP.0000000000000109.

[4] Mu Z, Zhao Y, Liu X, Chang C, Zhang J. Molecular biology of atopic dermatitis. Clin Rev Allergy Immunol 2014;47:193−218. http://dx.doi.org/10.1007/s12016-014-8415-1.

[5] Flohr C, Mann J. New insights into the epidemiology of childhood atopic dermatitis. Allergy 2014;69:3−16. http://dx.doi.org/10.1111/all.12270.

[6] Flohr C. Recent perspectives on the global epidemiology of childhood eczema. Allergol Immunopathol 2011;39:174−82. http://dx.doi.org/10.1016/j.aller.2011.02.004.

[7] Worldwide variation in prevalence of symptoms of asthma, allergic rhinoconjunctivitis, and atopic eczema: ISAAC. The International Study of Asthma and Allergies in Childhood (ISAAC) Steering Committee. Lancet 1998;351:1225−32.

[8] Flohr C, et al. The role of atopic sensitization in flexural eczema: findings from the International Study of Asthma and Allergies in Childhood Phase Two. J Allergy Clin Immunol 2008;121:141−7, e144. http://dx.doi.org/10.1016/j.jaci.2007.08.066.

[9] Shaw TE, Currie GP, Koudelka CW, Simpson EL. Eczema prevalence in the United States: data from the 2003 National Survey of Children's Health. J Invest Dermatol 2011;131:67−73. http://dx.doi.org/10.1038/jid.2010.251.

[10] Williams H, et al. Is eczema really on the increase worldwide? J Allergy Clin Immunol 2008;121:947−54, e915. http://dx.doi.org/10.1016/j.jaci.2007.11.004.

[11] Schafer T, et al. The excess of atopic eczema in East Germany is related to the intrinsic type. Br J Dermatol 2000;143:992−8.

[12] Flohr C. Is there a rural/urban gradient in the prevalence of eczema? Br J Dermatol 2010;162:951. http://dx.doi.org/10.1111/j.1365-2133.2010.09786.x.

[13] Wen HJ, et al. Predicting risk for early infantile atopic dermatitis by hereditary and environmental factors. Br J Dermatol 2009;161:1166−72. http://dx.doi.org/10.1111/j.1365-2133.2009.09412.x.

[14] Ruiz RG, Kemeny DM, Price JF. Higher risk of infantile atopic dermatitis from maternal atopy than from paternal atopy. Clin Exp Allergy 1992;22:762−6.

[15] Thomsen SF, et al. Importance of genetic factors in the etiology of atopic dermatitis: a twin study. Allergy Asthma Proc 2007;28:535−9. http://dx.doi.org/10.2500/aap2007.28.3041.

[16] Irvine AD, McLean WH, Leung DY. Filaggrin mutations associated with skin and allergic diseases. N Engl J Med 2011;365:1315−27. http://dx.doi.org/10.1056/NEJMra1011040.

[17] Brown SJ, McLean WH. Eczema genetics: current state of knowledge and future goals. J Invest Dermatol 2009;129:543−52. http://dx.doi.org/10.1038/jid.2008.413.

[18] Palmer CN, et al. Common loss-of-function variants of the epidermal barrier protein filaggrin are a major predisposing factor for atopic dermatitis. Nat Genet 2006;38:441−6. http://dx.doi.org/10.1038/ng1767.

[19] Kusunoki T, et al. Effect of eczema on the association between season of birth and food allergy in Japanese children. Pediatr Int 2013;55:7−10. http://dx.doi.org/10.1111/j.1442-200X.2012.03725.x.

[20] Nishio Y, et al. Association between polymorphisms in the SPINK5 gene and atopic dermatitis in the Japanese. Genes Immun 2003;4:515−17. http://dx.doi.org/10.1038/sj.gene.6363889.

[21] Ellinghaus D, et al. High-density genotyping study identifies four new susceptibility loci for atopic dermatitis. Nat Genet 2013;45:808−12. http://dx.doi.org/10.1038/ng.2642.

[22] Nakamura T, et al. Expression of DNMT-1 in patients with atopic dermatitis. Arch Dermatol Res 2006;298:253−6. http://dx.doi.org/10.1007/s00403-006-0682-0.

[23] Liang Y, et al. Demethylation of the FCER1G promoter leads to FcepsilonRI overexpression on monocytes of patients with atopic dermatitis. Allergy 2012;67:424−30. http://dx.doi.org/10.1111/j.1398-9995.2011.02760.x.

[24] Ziyab AH, et al. Interplay of filaggrin loss-of-function variants, allergic sensitization, and eczema in a longitudinal study covering infancy to 18 years of age. PLOS ONE 2012;7:e32721. http://dx.doi.org/10.1371/journal.pone.0032721.

[25] Wang IJ, Chen SL, Lu TP, Chuang EY, Chen PC. Prenatal smoke exposure, DNA methylation, and childhood atopic dermatitis. Clin Exp Allergy 2013;43:535–43. http://dx.doi.org/10.1111/cea.12108.

[26] Rodriguez E, et al. An integrated epigenetic and transcriptomic analysis reveals distinct tissue-specific patterns of DNA methylation associated with atopic dermatitis. J Invest Dermatol 2014;134:1873–83. http://dx.doi.org/10.1038/jid.2014.87.

[27] Sonkoly E, et al. MiR-155 is overexpressed in patients with atopic dermatitis and modulates T-cell proliferative responses by targeting cytotoxic T lymphocyte-associated antigen 4. J Allergy Clin Immunol 2010;126:581–9, e581–620. http://dx.doi.org/10.1016/j.jaci.2010.05.045 S0091-6749(10)00966-8 [pii].

[28] Lu TX, Rothenberg ME. Diagnostic, functional, and therapeutic roles of microRNA in allergic diseases. J Allergy Clin Immunol 2013;132:3–13 [quiz 14]. http://dx.doi.org/10.1016/j.jaci.2013.04.039 S0091-6749(13)00686-6 [pii].

[29] Ong PY, Leung DY. Immune dysregulation in atopic dermatitis. Curr Allergy Asthma Rep 2006;6:384–9.

[30] Neis MM, et al. Enhanced expression levels of IL-31 correlate with IL-4 and IL-13 in atopic and allergic contact dermatitis. J Allergy Clin Immunol 2006;118:930–7. http://dx.doi.org/10.1016/j.jaci.2006.07.015.

[31] Gutowska-Owsiak D, Schaupp AL, Salimi M, Taylor S, Ogg GS. Interleukin-22 downregulates filaggrin expression and affects expression of profilaggrin processing enzymes. Br J Dermatol 2011;165:492–8. http://dx.doi.org/10.1111/j.1365-2133.2011.10400.x.

[32] Yamanaka K, Mizutani H. The role of cytokines/chemokines in the pathogenesis of atopic dermatitis. Curr Probl Dermatol 2011;41:80–92. http://dx.doi.org/10.1159/000323299.

[33] Gros E, Petzold S, Maintz L, Bieber T, Novak N. Reduced IFN-gamma receptor expression and attenuated IFN-gamma response by dendritic cells in patients with atopic dermatitis. J Allergy Clin Immunol 2011;128:1015–21. http://dx.doi.org/10.1016/j.jaci.2011.05.043.

[34] Park H, et al. A distinct lineage of CD4 T cells regulates tissue inflammation by producing interleukin 17. Nat Immunol 2005;6:1133–41. http://dx.doi.org/10.1038/ni1261.

[35] Milovanovic M, Drozdenko G, Weise C, Babina M, Worm M. Interleukin-17A promotes IgE production in human B cells. J Invest Dermatol 2010;130:2621–8. http://dx.doi.org/10.1038/jid.2010.175.

[36] Masuda K, Katoh N, Okuda F, Kishimoto S. Increased levels of serum interleukin-16 in adult type atopic dermatitis. Acta Derm Venereol 2003;83:249–53.

[37] Pugliarello S, Cozzi A, Gisondi P, Girolomoni G. Phenotypes of atopic dermatitis. J Dtsch Dermatol Ges 2011;9:12–20. http://dx.doi.org/10.1111/j.1610-0387.2010.07508.x.

[38] Aral M, et al. The relationship between serum levels of total IgE, IL-18, IL-12, IFN-gamma and disease severity in children with atopic dermatitis. Mediators Inflamm 2006;2006:73098. http://dx.doi.org/10.1155/MI/2006/73098.

[39] Di Lorenzo G, et al. Serum levels of soluble CD30 in adult patients affected by atopic dermatitis and its relation to age, duration of disease and scoring atopic dermatitis index. Mediators Inflamm 2003;12:123–5. http://dx.doi.org/10.1080/0962935031000097736.

[40] Werfel T, Breuer K. Role of food allergy in atopic dermatitis. Curr Opin Allergy Clin Immunol 2004;4:379–85.

[41] Katta R, Schlichte M. Diet and dermatitis: food triggers. J Clin Aesthet Dermatol 2014;7:30−6.

[42] Griffiths CE, Barker JN. Pathogenesis and clinical features of psoriasis. Lancet 2007;370:263−71. http://dx.doi.org/10.1016/S0140-6736(07)61128-3.

[43] Rachakonda TD, Schupp CW, Armstrong AW. Psoriasis prevalence among adults in the United States. J Am Acad Dermatol 2014;70:512−16. http://dx.doi.org/10.1016/j.jaad.2013.11.013.

[44] Sonkoly E, et al. MicroRNAs: novel regulators involved in the pathogenesis of psoriasis? PLOS ONE 2007;2:e610. http://dx.doi.org/10.1371/journal.pone.0000610.

[45] Capon F, Trembath RC, Barker JN. An update on the genetics of psoriasis. Dermatol Clin 2004;22:339−47, vii. http://dx.doi.org/10.1016/S0733-8635(03)00125-6.

[46] Allen MH, et al. The major psoriasis susceptibility locus PSORS1 is not a risk factor for late-onset psoriasis. J Invest Dermatol 2005;124:103−6. http://dx.doi.org/10.1111/j.0022-202X.2004.23511.x.

[47] Helms C, et al. A putative RUNX1 binding site variant between SLC9A3R1 and NAT9 is associated with susceptibility to psoriasis. Nat Genet 2003;35:349−56. http://dx.doi.org/10.1038/ng1268.

[48] Nestle FO, Kaplan DH, Barker J. Psoriasis. N Engl J Med 2009;361:496−509. http://dx.doi.org/10.1056/NEJMra0804595.

[49] Christensen U, et al. Linkage of atopic dermatitis to chromosomes 4q22, 3p24 and 3q21. Hum Genet 2009;126:549−57. http://dx.doi.org/10.1007/s00439-009-0692-z.

[50] Sheng Y, et al. Sequencing-based approach identified three new susceptibility loci for psoriasis. Nat Commun 2014;5:4331. http://dx.doi.org/10.1038/ncomms5331.

[51] Zhang P, Su Y, Chen H, Zhao M, Lu Q. Abnormal DNA methylation in skin lesions and PBMCs of patients with psoriasis vulgaris. J Dermatol Sci 2010;60:40−2. http://dx.doi.org/10.1016/j.jdermsci.2010.07.011.

[52] Zhang P, Su Y, Zhao M, Huang W, Lu Q. Abnormal histone modifications in PBMCs from patients with psoriasis vulgaris. Eur J Dermatol 2011;21:552−7. http://dx.doi.org/10.1684/ejd.2011.1383.

[53] Park GT, Han J, Park SG, Kim S, Kim TY. DNA methylation analysis of CD4$^+$ T cells in patients with psoriasis. Arch Dermatol Res 2014;306:259−68. http://dx.doi.org/10.1007/s00403-013-1432-8.

[54] Zhang P, et al. Whole-genome DNA methylation in skin lesions from patients with psoriasis vulgaris. J Autoimmun 2013;41:17−24. http://dx.doi.org/10.1016/j.jaut.2013.01.001S0896-8411(13)00002-4 [pii].

[55] Xu N, et al. MicroRNA-31 is overexpressed in psoriasis and modulates inflammatory cytokine and chemokine production in keratinocytes via targeting serine/threonine kinase 40. J Immunol 2013;190:678−88. http://dx.doi.org/10.4049/jimmunol.1202695.

[56] Yang L, et al. miR-146a controls the resolution of T cell responses in mice. J Exp Med 2012;209:1655−70. http://dx.doi.org/10.1084/jem.20112218.

[57] Tili E, et al. Modulation of miR-155 and miR-125b levels following lipopolysaccharide/TNF-alpha stimulation and their possible roles in regulating the response to endotoxin shock. J Immunol 2007;179:5082−9.

[58] Zibert JR, et al. MicroRNAs and potential target interactions in psoriasis. J Dermatol Sci 2010;58:177−85. http://dx.doi.org/10.1016/j.jdermsci.2010.03.004.

[59] Koga Y, et al. Analysis of expression pattern of serum microRNA levels in patients with psoriasis. J Dermatol Sci 2014;74:170−1. http://dx.doi.org/10.1016/j.jdermsci.2014.01.005.

[60] Zhao M, et al. Up-regulation of microRNA-210 induces immune dysfunction via targeting FOXP3 in CD4(+) T cells of psoriasis vulgaris. Clin Immunol 2014;150:22−30. http://dx.doi.org/10.1016/j.clim.2013.10.009.

[61] Lovendorf MB, Zibert JR, Gyldenlove M, Ropke MA, Skov L. MicroRNA-223 and miR-143 are important systemic biomarkers for disease activity in psoriasis. J Dermatol Sci 2014;75:133−9. http://dx.doi.org/10.1016/j.jdermsci.2014.05.005.

[62] Tollefson MM. Diagnosis and management of psoriasis in children. Pediatr Clin North Am 2014;61:261−77. http://dx.doi.org/10.1016/j.pcl.2013.11.003.

[63] Akin C. Mastocytosis. Immunol Allergy Clin North Am 2014;34:xvii−xviii. http://dx.doi.org/10.1016/j.iac.2014.02.005.

[64] Brockow K. Epidemiology, prognosis, and risk factors in mastocytosis. Immunol Allergy Clin North Am 2014;34:283−95. http://dx.doi.org/10.1016/j.iac.2014.01.003.

[65] Frieri M. Mechanisms of disease for the clinician: systemic lupus erythematosus. Ann Allergy Asthma Immunol 2013;110:228−32. http://dx.doi.org/10.1016/j.anai.2012.12.010.

[66] Kettelhut BV, Metcalfe DD. Pediatric mastocytosis. J Invest Dermatol 1991;96:15S−18S discussion 18S, 60S−65S. http://dx.doi.org/10.1111/1523-1747.ep12468942.

[67] Cruse G, Metcalfe DD, Olivera A. Functional deregulation of KIT: link to mast cell proliferative diseases and other neoplasms. Immunol Allergy Clin North Am 2014;34:219−37. http://dx.doi.org/10.1016/j.iac.2014.01.002.

[68] Mayoral RJ, et al. MiR-221 influences effector functions and actin cytoskeleton in mast cells. PLOS ONE 2011;6:e26133. http://dx.doi.org/10.1371/journal.pone.0026133.

[69] Yamada Y, Kosaka K, Miyazawa T, Kurata-Miura K, Yoshida T. miR-142-3p enhances FcepsilonRI-mediated degranulation in mast cells. Biochem Biophys Res Commun 2014;443:980−6. http://dx.doi.org/10.1016/j.bbrc.2013.12.078.

[70] Soucie E, Brenet F, Dubreuil P. Molecular basis of mast cell disease. Mol Immunol 2015;63:55−60. http://dx.doi.org/10.1016/j.molimm.2014.03.013.

[71] Ryan RJ, et al. Mast cell sarcoma: a rare and potentially under-recognized diagnostic entity with specific therapeutic implications. Mod Pathol 2013;26:533−43. http://dx.doi.org/10.1038/modpathol.2012.199.

[72] Akin C. Mast cell activation disorders. J Allergy Clin Immunol Practice 2014;2:252−257, e251 [quiz 258]. http://dx.doi.org/10.1016/j.jaip.2014.03.007.

[73] Brockow K, Metcalfe DD. Mastocytosis. Chem Immunol Allergy 2010;95:110−24. http://dx.doi.org/10.1159/000315946.

[74] Darlenski R, Kazandjieva J, Zuberbier T, Tsankov N. Chronic urticaria as a systemic disease. Clin Dermatol 2014;32:420−3. http://dx.doi.org/10.1016/j.clindermatol.2013.11.009.

[75] Bernstein JA, et al. The diagnosis and management of acute and chronic urticaria: 2014 update. J Allergy Clin Immunol 2014;133:1270−7. http://dx.doi.org/10.1016/j.jaci.2014.02.036.

[76] Khan DA. Chronic urticaria: diagnosis and management. Allergy Asthma Proc 2008;29:439−46. http://dx.doi.org/10.2500/aap.2008.29.3151.

[77] Saini SS. Basophil responsiveness in chronic urticaria. Curr Allergy Asthma Rep 2009;9:286−90.

[78] Zhang M, Murphy RF, Agrawal DK. Decoding IgE Fc receptors. Immunol Res 2007;37:1−16.

[79] Altman K, Chang C. Pathogenic intracellular and autoimmune mechanisms in urticaria and angioedema. Clin Rev Allergy Immunol 2013;45:47−62. http://dx.doi.org/10.1007/s12016-012-8326-y.

[80] Ambros V. The functions of animal microRNAs. Nature 2004;431:350−5. http://dx.doi.org/10.1038/nature02871.

[81] Mathelier-Fusade P, Aissaoui M, Bakhos D, Chabane MH, Leynadier F. Clinical predictive factors of severity in cold urticaria. Arch Dermatol 1998;134:106−7.

[82] Delore P, Gerin P, Chapuy A. [Drowning and urticaria caused by cold]. J Med Lyon 1956;37:497–503.
[83] Begin P, Nadeau KC. Epigenetic regulation of asthma and allergic disease. Allergy Asthma Clin Immunol 2014;10:27. http://dx.doi.org/10.1186/1710-1492-10-27.
[84] Bell CG, Beck S. The epigenomic interface between genome and environment in common complex diseases. Brief Funct Genomics 2010;9:477–85. http://dx.doi.org/10.1093/bfgp/elq026.

Epigenetics and Other Autoimmune Skin Diseases

Ming Zhao, Ruifang Wu, and Qianjin Lu

Department of Dermatology, The Second Xiangya Hospital of Central South University, Hunan Key Laboratory of Medical Epigenomics, Changsha, Hunan, PR China

Heritable changes in gene expression may occur without alteration of DNA sequence, in the forms of DNA methylation, histone modifications, and microRNA (miRNA), collectively known as epigenetics [1,2]. DNA methylation, which is an important epigenetic mechanism involved in the regulation of gene transcription, refers to the addition of a methyl group to the 5-C position of cytosine bases, typically at cytosine phosphate guanine (CpG) dinucleotides, and is catalyzed by DNA methyltransferases (DNMTs) [1,2]. CpG islands are CG-rich sequences found in or near approximately 40% of promoters of mammalian genes and serve as promoters for their associated genes [3]. In promoter regions, DNA methylation silences genes expression by interfering with transcription activators binding and/or recruitment of chromatin repressor complexes through the binding of methyl-CpG-binding proteins (MBDs), which leads to tightly condensed chromatin around the gene promoter [1,4,5].

Histone modifications regulate gene expression by altering the chromatin conformation [6,7]. Posttranslational modifications, including acetylation, methylation, phosphorylation, sumoylation, and ubiquitination to the N-terminal tail regions of histones, which protrude from nucleosomal core particles, can change local chromatin conformation and affect the accessibility of genes [6,7]. Modifications associated with loosening of local chromatin such as histone acetylation can activate gene expression, while histone deacetylation has been proved to play an important role in gene inactivation by modifying the chromatin into repressive configurations [8].

© 2015 Elsevier Inc. All rights reserved.

miRNAs are a group of small noncoding RNA molecules of 21–23 nucleotides that negatively regulate gene expression at the posttranscriptional level through degradation and translational inhibition of target messenger RNAs (mRNAs) [9]. They are first transcribed from a huge double-stranded primary transcript, known as pri-miRNA, by RNA polymerase II [10]. Then the RNase III enzyme Drosha converts the pri-miRNA into an approximately 70 nt stem-loop structure called pre-miRNA, which is imported to the cytoplasm by Exportin 5 and is further processed by Dicer enzyme into an unstable miRNA duplex. After strand separation, the unstable miRNA duplex becomes mature miRNA, which is incorporated into the RNA-induced silencing complex (RISC) to regulate protein translation [2,10].

Due to the important role of epigenetic mechanisms in regulating gene expression, a failure to maintain epigenetic homeostasis may lead to aberrant gene expression, contributing to a wide range of diseases [11]. Epigenetic mechanisms are essential for normal development and function of the immune system. Similarly, aberrant epigenetic mechanisms in the immune response process, due to some factors such as environmental influences, lead to dysregulated gene expression, contributing to immune dysfunction and the development of autoimmune disorders. The interactions between genetically determined susceptibility and environmental factors are implicated in systemic autoimmune diseases such as rheumatoid arthritis and scleroderma, as well as in organ-specific autoimmunity. The skin is exposed to a wide variety of environmental agents, including UV radiation, mineral dust, and pathogens, and is prone to the development of autoimmune conditions such as psoriasis, pemphigus, and some forms of vitiligo, depending on environmental and genetic influences [11]. In fact, continued efforts in the field of epigenetics have demonstrated that alterations to epigenetic modification are indeed involved in many common autoimmune skin disease [11,12]. In this chapter, we review the advanced discovery of the involvement of epigenetic mechanisms in four kinds of autoimmune skin diseases: vitiligo, alopecia areata (AA), dermatomyositis (DM), and pemphigus vulgaris (PV).

14.1 EPIGENETICS IN VITILIGO

Vitiligo is an idiopathic, acquired, progressive, multifactorial, depigmenting disorder with devastating psychological and social consequences [13]. It is clinically characterized by white macules or patches in the skin, often overlying hair and mucous membranes, due to chronic and progressive loss of functional melanocytes from the involved areas

[13,14]. As the most common depigmenting disorder, vitiligo affects an estimated 0.5−1% of the population worldwide, with no predilection for gender, race, or geography [15]. Manifestations can appear at almost any age; however, 50% of all vitiligo patients present before 20 years of age and 25% of cases present before 14 years of age [15,16]. Although the classification of vitiligo is complex and much debated, there are two main types recognized: segmental vitiligo (SV) and nonsegmental vitiligo (NSV, also referred to as generalized vitiligo), which accounts for roughly 90% of total vitiligo cases [17].

The etiology of vitiligo is still obscure and many different hypotheses have been proposed, the most compelling of which involves a combination of environmental and genetic factors, contributing to autoimmune melanocyte destruction [18,19]. Large-scale epidemiological surveys have shown that about 15−20% of vitiligo patients have one or more affected first-degree relatives and the familial aggregation takes a non-Mendelian pattern that suggests a polygenic, multifactorial inheritance of vitiligo [19−21]. Many studies performed in vitiligo patients' close relatives suggest a significant increased disease risk with decreasing genetic distance from an affected proband [20,22]. For example, Alkhateeb et al. [20] have found that the risk to a patient's first-degree relatives is about 7.1% in Caucasians, 6.1% in Indo-Pakistanis, and 4.8% in Hispanics, with lower risks to more distant relatives. In addition, their studies also show that the concordance for generalized vitiligo in monozygotic twins is 23%, which is more than 60 times the general population risk of 0.38% [20]. The application of genome-wide association studies (GWAS) to investigate the genetics of vitiligo in recent years has resulted in the identification of multiple vitiligo susceptibility genes [23,24]. All the evidence strongly supports genetic factors contributing to one's risk for vitiligo.

Nevertheless, as mentioned above, monozygotic twins share all of their genes identically, and the limited concordance thus suggests that environmentally induced epigenetic factors must also play an important role in the pathogenesis of vitiligo, perhaps being even more important than genetics. Numerous factors including stress, pregnancy, severe sunburn, physical trauma, and infections have been described in association with the onset of vitiligo, but the mechanism by which these factors exert their effect is unclear [21].

In 1996, Sreekumar et al. [25] discovered that the DNA methylation inhibitor 5-azacytidine (5azaC) could induce autoimmune vitiligo in vitiligo-susceptible but normally pigmented chicken strains (BL), which are parental control strains of the Smyth Line chicken model, due to loss of skin melanocytes. In this model, both T cells and B cells were involved in the disease, and the chickens developed antibodies to the melanocyte-specific protein TRP-1 that cross-reacts with mouse and human melanocytes. It is possible that hypomethylation of DNA caused by 5azaC turned on a certain gene that induces autoreactivity in a certain T-cell or (and) B-cell subpopulation, suggesting a role for demethylation of immune genes in the pathogenesis of vitiligo in genetically susceptible individuals.

To investigate whether epigenetic changes are also involved in human vitiligo, our group [26] examined genomic and gene-specific DNA methylation levels as well as mRNA levels of DNMTs, methyl-DNA-binding domain proteins (MBDs), and interleukin-10 (IL-10) in peripheral blood mononuclear cells (PBMCs) from vitiligo patients and controls (Table 14.1). We chose PBMCs as objects because evidence has shown that PBMCs of vitiligo patients overproduce proinflammatory cytokines such as IL-1b, IL-6, IL-8, and tumor necrosis factor (TNF)-α [30] and infiltrate the periphery of vitiligo lesions [31], suggesting that PBMCs may contribute to the development of vitiligo. According to our study, the mean genomic DNA methylation levels were significantly higher in vitiligo patients than those in healthy controls, which is different from the global hypomethylation observed in $CD4^+$ T cells from patients with systemic lupus erythematosus (SLE) and systemic sclerosis (SSc) [28]. Why is the result not the same? In our opinion, vitiligo is an organ-specific autoimmune disease, not a systemic autoimmune condition like SLE. Therefore, it could be that disease-specific methylation defects in PBCMs from patients with vitiligo, which have unique effects on immune activation, determine the tissue specificity of the disorder. Similarly to vitiligo, other organ-specific antoimmune diseases such as psoriasis and type 1 diabetes mellitus (T1DM) also show global genome hypermethylation, indicating that hypermethylation may be important in organ-specific immune activation. In addition, DNMT1, an enzyme maintaining DNA methylation, was significantly increased in PBMCs of vitiligo patients. Some MBD proteins including MBD1, MBD3, MBD4, and MeCP2 were also significantly upregulated in PBMCs from vitiligo patients. What is more, MBD1 and MBD3 expression was positively correlated with overall methylation levels in PBMCs of vitiligo patients, which indicates the increased expression of MBD1 and MBD3 in PBMCs, which may contribute to DNA hypermethylation and the development of vitiligo [26].

It has been widely believed that vitiligo, which is commonly associated with other autoimmune diseases such as autoimmune thyroid disease [32] and Addison's disease [33], is preferentially an autoimmune disease. The vitiligo susceptibility genes that were identified by GWAS almost universally are involved in immune regulation and immune targeting of melanocytes [23,24]. Recent studies have found that different circulating autoantibodies to melanocytes and melanocytic protein components, which are uncommon in healthy individuals, are detectable in the serum

TABLE 14.1 Epigenetic Changes in the Four Kinds of Autoimmune Skin Diseases

Disease	Genomic DNA methylation level	DNMTs	MBDs	Histone modification	Reference
Vitiligo	Global hypermethylation	DNMT1 was significantly increased in PBMCs of vitiligo patients	MBD1, MBD3, MBD4, and MeCP2 were upregulated in PBMCs of vitiligo patients	—	[26]
Alopecia areata	Global hypermethylation	DNMT1 was upregulated in PBMCs of AA patients	MBD1 and MBD4 were increased in AA PBMCs	H3 acetylation was increased, H3K4 methylation was decreased in AA PBMCs	[27]
Dermatomyositis	No significant change	—	MBD2 and MeCP2 were significantly increased in DM patients	—	[28]
Pemphigus	Global hypermethylation	DNMT1 was upregulated in PBMCs of PV patients	—	Histone H3/H4 acetylation and H3K4/H3K27 methylation levels were decreased in PV PBMCs	[29]

DNMT1, DNA methyltransferase1; MBDs, methyl-DNA-binding domain proteins; AA, alopecia areata; DM, dermatomyositis; PV, pemphigus vulgaris.

of many but not all vitiligo patients [34,35]. What is more important, the discoveries of T-cell infiltration in the margins of active vitiligo skin lesions [36] and circulating skin-homing melanocyte-specific cytotoxic T lymphocytes [37] suggest the involvement of T cells in the pathogenesis of vitiligo. Recently, Zhou et al. [38] reported that patients with NSV have number and functional defects in invariant natural killer T cells and $CD4^+$ $CD25^+$ $Foxp3^+$ regulatory T cells (Tregs). The decrease in the number of properly functioning Tregs can affect immune balance and is accompanied by reduced transforming growth factor (TGF-β) in the serum of vitiligo patients [39]. Tregs synthesize IL-10 and TGF-β, which in turn contributes to the differentiation and function of Tregs. In our study, we demonstrated that mRNA transcription of IL-10 is lower in vitiligo PBMCs than in healthy controls [26]. Our previous work showed that $CD4^+$ T cells of SLE patients express high levels of IL-10 mRNA and protein, and that IL-10 expression levels are inversely correlated with DNA hypomethylation of IL-10 intron 4 in SLE $CD4^+$ T cells, which suggest that IL-10 is a methylation-sensitive gene. In vitiligo, we found that the enhancer region within intron 4 of IL-10 containing 8 CG pairs is hypermethylated in vitiligo PBMCs and the average methylation status of this region is negatively correlated with IL-10 mRNA expression, suggesting that DNA hypermethylation may contribute to decreased IL-10 expression in vitiligo [26]. All these results show that altered DNA methylation in PBMCs of vitiligo patients may contribute to the pathogenesis of disease by affecting the expression of autoimmunity-related genes.

Besides DNA methylation, a study of miRNAs from vitiligo patients' serum has been reported recently (Table 14.2). In this study, serum miRNA expression profiles were performed in patients with NSV and healthy controls by miRNA arrays [40]. Shi et al. identified 31 differently expressed miRNAs, among which 12 miRNAs had more significant changes (>threefold) between the two groups. Specifically, serum miR-16 [42,43], miR-19b [41], and miR-720 appeared to be the best serum biomarkers distinguishing NSV from healthy controls, with a 0.975 estimated area under the curve (AUC). In addition, Shi et al. explored whether the changed serum miRNAs could be associated with disease severity, and found miR-574-3p expression significantly downregulated in vitiligo patients with body surface area (BSA) greater than 50% compared with vitiligo patients with 30% < BSA < 50% and healthy controls, suggesting the level of miR-574-3p may be related to disease severity. A similar study was performed on a mouse autoimmune vitiligo model, in which melanocyte autoreactive $CD4^+$ T cells from $Tyrp1^{B-W}Rag1^{-/-}$TRP-1-specific CD4 transgenic mice were adoptively transferred into $Rag1^{-/-}$ host mice [44]. After the adoptive transferring of TRP-1-specific $CD4^+$ T cells, the Rag1 knockout mice developed severe vitiligo. The serum was collected at 8 weeks after the transfer of

TABLE 14.2 Potential miRNAs Involved in Vitiligo and Dermatomyositis

miRNAs	Changed directions	Subjects	Biological function	References
VITILIGO				
miR-574-3p	Downregulated	NSV patients and healthy controls	May be associated with NSV lesion severity	[40]
miR-19b	Upregulated	NSV patients and healthy controls	Potential serum biomarkers; regulate Th1 responses by supporting IFN-γ production and suppressing inducible Tregs	[40,41]
miR-16	Upregulated	NSV patients and healthy controls	Potential serum biomarkers; might be implicated in the common inflammation process in autoimmune disorders	[40,42,43]
miR-146	Upregulated	Rag1$^{-/-}$ host mice transferred with melanocyte autoreactive CD4$^+$ T cells	May contribute to the prolonged production of TNF-α and significantly upregulate phagocytic activity	[43,44]
miR-191	Upregulated	Rag1$^{-/-}$ host mice transferred with melanocyte autoreactive CD4$^+$ T cells	Involved in the proliferation or survival of melanocytes and melanoma cells	[44,45]
DERMATOMYOSITIS				
miR-7	Downregulated	DM patients	A potential marker for the diagnosis or evaluation of disease activity in DM patients and may target inflammatory molecules including FGF11 and CCL16	[46]
miR-206	Downregulated	DM patients	Regulates the proportion of Th17 cells mediated by KLF4	[47]
miR-223	Downregulated	Gottron's papules of DM patients	Involved in keratinocyte proliferation via targeting PKCε	[48]

T cells, when depigmentation lesions were larger than 50% of the BSA. Similarly, 20 miRNAs showed significantly different expression levels (more than twofold changes) between vitiligo and control mice. Among these miRNAs, miR-146a [43] and miR-191 [45] were differentially expressed in the serum of both vitiligo mouse and NSV patient, with an upregulated expression pattern. These studies suggested the possibility of the involvement of miRNAs in NSV development and may provide a new approach based on miRNAs for differential diagnosis and determining therapy response of NSV.

14.2 EPIGENETICS IN ALOPECIA AREATA

Alopecia areata (AA), a common hair loss disorder which affects approximately 1−2% of the general population in all age groups [49], is characterized by well-circumscribed patches of hair loss, which can progress to AA totalis, affecting the whole scalp, and AA universalis, affecting the whole body [50]. Typically, one or more discrete round or oval, nonscarring, circumscribed hair loss areas of varying sizes are found. Episodes of hair loss generally have a sudden onset and a recurrent course, and the isolated hairless patches can subsequently extend centrifugally and may coalesce [50].

Despite extensive research in AA, the exact etiology is still wrapped in mist. Numerous evidence indicates that AA is a chronic inflammatory organ-specific autoimmune disease due to a T-cell-mediated autoimmune response against anagen-stage hair follicles [51,52]. Previous studies have shown that there is a strong association between AA and some autoimmune diseases, especially Hashimoto's thyroiditis, vitiligo, and pernicious anemia [52,53], providing a basis on which this autoimmune etiology is proposed. It has been demonstrated that the level of hair follicle-specific IgG is increased in the peripheral blood of AA patients and it can be found deposited around the periphery of hair follicles adjacent to the border of the active lesions [54]. Recent evidence indicates that the hair follicle may be an immune-privileged site with low levels of major histocompatibility complex (MHC) expression, while the immune privilege collapse caused by both genetic predisposition and environmental factors induces the infiltration of the hair follicle by T lymphocytes, during which the CD8$^+$ T cells act as the main effectors with the help of CD4$^+$ T cells [55,56].

In fact, genetic factors leading to AA have been studied. A number of candidate-gene association studies have been performed over the past two decades, and several of these have indicated a significant association between AA and some human leukocyte antigen (HLA) genes such as HLA-DQB1*03 [57] and HLA-DQB1*0604 [58]. Furthermore, the MHC class I chain-related gene A (MICA(*)6) is also associated with AA [59]. Recently, a GWAS in a sample of 1054 cases and 3278 controls identified 139 single nucleotide polymorphisms (SNPs) that are significantly associated with AA [60]. The study showed an association with genomic regions containing several genes controlling the activation and proliferation of Tregs, such as cytotoxic T lymphocyte-associated antigen 4 (CTLA4), IL-2/IL-21, IL-2 receptor A (IL-2RA; CD25), and Eos (also known as Ikaros family zinc finger 4; IKZF4), as well as the HLA region.

However, the result from a limited twin study showed a concordance rate for AA in monozygotic twins of 55% [61], suggesting the involvement of environmental factors in the pathology of AA. Many environmental triggers, such as viral infection, psychological stress, and oxidative stress, have been proposed to activate pathological immune responses in AA, although most of them need to be further confirmed [61,62]. Given that environmental factors often lead to autoimmune disease-relevant changes in cell function by triggering alterations in the epigenetic regulation of gene expression, investigators hypothesize that epigenetic mechanisms may contribute to the pathogenesis of AA. To explore whether patterns of DNA methylation and histone modifications are altered in patients with AA, our group [27] compared the global DNA methylation levels, and histone H3 and H4 acetylation levels, as well as the methylation levels of histone H3 lysine 4 and lysine 9 (H3K4 and H3K9) in PBMCs from patients with AA and healthy controls (Table 14.1). We identified a significant increase in DNA methylation in PBMCs of patients with AA, which is similar for patients with psoriasis and vitiligo exhibiting T-cell hypermethylation. We also found that the expression of some DNMT and MBD genes was significantly changed. DNMT1 transcript levels were upregulated in PBMCs of AA patients compared with healthy controls and the increase in expression correlated with the level of hypermethylation in AA patients. MBD1 and MBD4 expression was increased in AA PBMCs. MBDs, like DNMTs, are involved in regulating DNA methylation levels and patterns, and suppress gene transcription in a methylation-dependent manner by binding methylated DNA and preventing transcription factors from binding to gene regulatory elements [63].

In addition, we found that H3 acetylation was increased, whereas H3K4 methylation was decreased in PBMCs of patients with AA. Correlation analysis showed that H3 acetylation levels were positively correlated with AA disease severity and with the expression level of RANTES mRNA. RANTES protein is secreted by PBMCs and has been previously demonstrated to be overexpressed in AA patients [64].

The histone modification regulators have also been found changed in AA PBMCs. For example, P300 was upregulated while HDAC2 and HDAC7 were downregulated in AA PBMCs, consistent with the observed hyperacetylation of histone H3 [27]. These findings provide novel insights into the pathogenesis of AA and help elucidate the role of epigenetic changes in the development of AA.

14.3 EPIGENETICS IN DERMATOMYOSITIS

Dermatomyositis (DM), a subtype of rare idiopathic inflammatory myopathies (IIMs), is typically characterized by skin manifestations accompanying or preceding muscle weakness [65], although amyopathic DM (DM with subclinical or absent myopathy) also occurs. Progressive, symmetric, proximal muscle weakness without skin changes is called polymyositis (PM) [65]. The typical features of DM are subacute onset, symmetric proximal muscle weakness, electromyographic and muscle alteration, characteristic skin lesions (Gottron's papules, heliotrope rash, V-neck sign, or shawl sign), and elevated serum muscle enzymes such as creatine kinase (CK) or lactate dehydrogenase (LDH) [66]. According to the literature, there is a strong female predilection and a bimodal peak; the smaller peak is seen in children and the larger one in adults between the ages of 40 and 60 [67].

Patients with DM have an increased risk of cancer and this disease sometimes overlaps with other connective tissue disease (CTD) such as SLE, rheumatoid arthritis, and SSc [68]. Although the pathogenic mechanism of DM still remains unclear, mounting evidence is building for a model of autoimmune-mediated inflammation of muscle and skin [65,69]. Immunofluorescence studies revealed the involvement of complement components in the development of vascular pathology associated with DM [70]. Other studies have found that inflammatory infiltration is composed mainly of macrophages, B cells, and CD4[+] T cells in the perivascular and perimysial regions [71,72]. In addition, many antibodies including both myositis-associated autoantibodies (MAAs), such as antinuclear antibodies (ANA), and myositis-specific autoantibodies (MSAs), such as anti-Jo-1 antibodies and anti-Mi-2 antibodies, have been detected in the serum of DM patients [73,74]. Kubo et al. [75] have identified the antihistone antibodies (AHAs) as one of the serologic abnormalities observed in DM/PM.

An inherited predisposition has been demonstrated to be involved in the pathogenesis of DM, and the HLA system is reported to confer susceptibility to DM in both adults and children [76]. Additionally, environmental factors such as various infections and ultraviolet light seem to play a role in the etiology of DM [76,77]. Accordingly, the question of whether epigenetic changes induced by environmental agents contribute to the pathogenesis of DM has intrigued researchers.

DNA hypomethylation is a common epigenetic change, which has been proved to contribute to several systemic autoimmune diseases such as SLE and SSc. To explore whether impaired DNA methylation occurs in DM, Lei et al. [28] measured global DNA methylation levels as well as mRNA levels of DNMTs and MBDs in CD4$^+$ T cells from DM patients and healthy controls. Although no significant change in global DNA methylation levels was observed in DM patients, they found significant increases in the mRNA levels of two MBD genes, *MBD2* and *MeCP2* (Table 14.1).

In a recent study on the relationship between miRNAs and DM, several miRNAs were found to be up- or downregulated in DM skin, using miRNA array analysis [46]. miRNA-7 (miR-7), which has been confirmed to be upregulated in SSc skin, has been found to be the most downregulated miRNA in DM skin (Table 14.2). Oshikawa et al. further measured the serum level of miR-7 and found a specifically decreased expression in patients with DM compared with normal subjects or patients with other autoimmune diseases such as SLE and SSc, suggesting the possibility of serum miR-7 level being used as a diagnostic marker for DM. However, they did not find a significant association of serum miR-7 levels with clinical and laboratory findings of DM patients. Moreover, the mechanism inducing the downregulation of miR-7 expression in the skin and sera of DM patients and the role of decreased miR-7 in the pathogenesis of DM are still unknown. The researchers speculated that miR-7 expression is decreased in the infiltrated lymphocytes or fibroblasts of DM skin, which may result in the increased production of inflammatory molecules, leading to skin inflammation. Further studies are needed to clarify these points.

IL-17, a key cytokine of Th17 cells, has been detected in the inflammatory infiltrates of patients with DM [78]. In a new study, Tang et al. [47] investigated the association between frequency of Th17 cells and the expression of miR-206 in the peripheral blood of DM patients. They identified CD3$^+$CD8$^-$IL-17$^+$ cells to distinguish the Th17 cells from PBMCs and found increased Th17 cells and enhanced expression of IL-17 mRNA in PBMCs from patients with DM. At the same time, the mRNA levels of RORC and KLF4, two positive regulators of Th17 differentiation, were also increased significantly in the PBMCs of DM patients. As expected, the augmented expression of KLF4 was accompanied by the attenuated expression of miR-206, a miRNA picking KLF4 as one of its multiple targets. Furthermore, this study showed a negative correlation between the percentages of Th17 cells and the expression of miR-206 in DM patients. These findings collectively suggest that the augmented expression of KLF4 mRNA may be caused by the attenuated expression of miR-206, and the high level of KLF4 mRNA evokes the proportion of Th17 cells found in DM patients (Table 14.2).

Gottron's papules, one of the skin manifestations found, are of great diagnostic value because they are specific to DM. In another study, Inoue et al. [48] performed miRNA polymerase chain reaction (PCR) array analysis to explore the expression pattern of miRNAs in Gottron's papules of DM patients and then evaluated the role of miRNAs in the pathogenesis of Gottron's papules. Among the several miRNAs with changed expression, they identified miR-223, which was detected in normal skin but not in DM skin. The results of quantitative real-time PCR analysis also showed that the expression of miR-223 was significantly decreased in DM compared with normal skin. Further research found increased protein expression of PKCε, one of the putative target genes of miR-223, in the hyperproliferated epidermis of Gottron's papules in DM patients. A specific inhibitor of miR-223 can induce cell proliferation while knockdown of PKCε by a specific small interfering RNA (siRNA) can decrease cell numbers, demonstrating that miR-223 is involved in keratinocyte proliferation via targeting PKCε. Taken together, these results suggest that the hyperproliferation of keratinocytes in the epidermis of Gottron's papules may be induced by decreased miR-223 expression as well as by subsequently increased protein levels of PKCε [48] (Table 14.2). In this study, the researchers also determined the serum level of miR-223 in DM patients and found no significant differences in the serum miR-223 level between DM patients and controls, although it was decreased in DM. However, a significantly decreased level of serum miR-223 was observed in patients with clinically amyopathic DM (CADM), which was characterized by skin lesions without muscle weakness, indicating that the decreased serum miR-223 level may be associated with cutaneous involvement of the disease. In addition, the results also showed that patients with decreased serum miR-223 levels tend to have more severe symptoms. Thus, the serum miR-223 level might serve as new biomarker for CADM, whose diagnosis is sometimes difficult, especially in the absence of myositis or lung involvement.

14.4 EPIGENETICS IN PEMPHIGUS

Pemphigus encompasses a group of potentially life-threatening, antibody-mediated autoimmune mucocutaneous diseases characterized by the presence of IgG autoantibodies directed against the extracellular domains of cadherin-type epithelial cell adhesion molecules—the desmogleins (Dsg)—leading to loss of cell−cell contact (acantholysis) with consequent intraepithelial vesiculation formation [79,80]. There are different clinical forms of pemphigus, the two main subtypes being pemphigus vulgaris (PV) with its variant pemphigus vegetans (PVeg), and pemphigus foliaceus (PF) with its variant pemphigus erythematosus (PE). Less common forms include pemphigus herpetiformis, IgA pemphigus, and paraneoplastic pemphigus [81]. Since PV is the most common form of pemphigus worldwide, it will be the focus of this chapter.

Epidemiological data of pemphigus have shown different distributions depending on different ethnic groups and regions. PV occurs with equal frequency in both genders and has a mean age of onset of 50−60 years, although it can also be seen in children and the elderly. A higher prevalence has been noted in Jewish people and those of Mediterranean descent [82].

It has been demonstrated that both humoral and cellular autoimmunities are important in the pathogenesis of this disease. Antibodies presented in PV are most commonly directed against Dsg3, whereas that in PF is Dsg1, although some PV patients also have autoantibodies to Dsg1 and the proportion of Dsg1 and Dsg3 antibodies appears to be related to clinical severity [80,83,84]. It has been proved that both IgG1 and IgG4 autoantibodies to Dsg3 are present in patients with pemphigus, but some evidence suggests that the IgG4 antibodies are pathogenic [85]. Apoptosis has been proposed to contribute to IgG-induced acantholysis [81]. Moreover, peripheral $CD4^+$ T-cell responses, and occasionally $CD8^+$ T-cell responses, to the ectodomain of Dsg3 were also identified in PV patients by several independent investigators [86−88]. Dsg3-reactive Th1 [86] and Th2 [88,89] cells recognized portions of the extracellular domain of Dsg3 in the context of PV-associated HLA class II alleles. Eming et al. [90] showed that Dsg3-specific autoreactive Th1 and Th2 cells were detected at similar frequencies in acute onset PV. Veldman et al. [91] showed that Dsg3-reactive Th2 cells were detected at similar frequencies in acute onset, chronic active, and remittent PV, while the number of autoreactive Th1 cells exceeded that of Th2 cells in chronic active PV. Autoreactive Th1 and Th2 cells may be involved in the regulation of the production of pathogenic autoantibodies by B cells in PV since sera of patients with PV contain Th1-regulated IgG1 and Th2-regulated IgG4 autoantibodies directed against Dsg3 [85,92].

Although the autoimmune mechanisms have been established, the precise initiating factors are still indistinct. There is considerable evidence to suggest that both genetic susceptibility and environmental elements contribute to trigger the immune response against self antigens. According to population studies, PV is strongly associated with the serotypes HLA-DR4 and HLA-DRw6 [93,94]. In addition, Miyagawa et al. [95] have proved that Asian alleles of the HLA-B15 family, including the allele B*1507, were significantly increased in comparison with normal controls, especially in Japanese patients with PV.

Besides genetic factors, several studies have implicated environmental factors as contributing to the pathogenesis of pemphigus, such as virus infection [96] and exposure to pesticides and sunlight [97], which are associated with epigenetic modifications. For example, Epstein—Barr virus (EBV), which has been implicated as being a cause of pemphigus, increases DNA methylation of the E-cadherin gene and p16^{INK4A} and represses *BIM* gene transcription, initially involving epigenetic modification via H3K27 trimethylation during lymphomagenesis [98]. Sun exposure can induce a distinct DNA hypermethylation pattern in UVB-exposed epidermal skin [99], which has been reported to exacerbate PV [97]. Moreover, pesticides, air pollutants, industrial chemicals, and heavy metals have also been reported to change gene expression through histone modifications and DNA methylation [100]. These studies suggested that alterations of epigenetic modifications may be involved in the development of PV. But unfortunately, the role and mechanism of epigenetic modifications in the development of PV has been poorly reported. Recently, our group measured global DNA methylation levels, global histone H3 and H4 acetylation levels, and global methylation levels of histone H3 lysine 4 and lysine 27 (H3K4 and H3K27) in PBMCs from PV patients and healthy controls [29] (Table 14.1). At the same time, the expression of several epigenetic modifiers was detected too. We found that global DNA methylation levels were significantly increased accompanied by upregulated DNMT1 mRNA levels in PBMCs of PV patients compared with controls, similarly to previous findings in vitiligo [26] and AA [27], both of which are organ-specific autoimmune disorders. In contrast, genomic DNA hypomethylation and decreased DNMT1 mRNA levels were identified in systemic autoimmune diseases such as SLE and SSc [28]. Our study also showed increased 5-methylcytosine content in keratinocytes of PV lesions compared with normal skin, indicating global DNA hypermethylation in PV lesional skin. In addition, global histone H3/H4 acetylation and H3K4/H3K27 methylation levels were decreased significantly in PBMCs of patients relative to those of healthy controls, suggesting the involvement of aberrant histone modification in the pathogenesis of PV. Interestingly, the changes in global H4 acetylation and global H3K27 methylation were negatively correlated with the levels of autoantibody against Dsg3 in serum of patients with PV. Moreover, our study also showed an increased expression of HDAC1 and HDAC2, which is inversely correlated with H3 and H4 acetylation levels in PV PBMCs. These findings collectively suggest a potentially important role of epigenetic changes in the pathogenesis of PV.

14.5 CONCLUSION

In addition to SLE, systemic scleroderma, and psoriasis, there is evidence that strongly supports the notion that epigenetic changes are involved in some autoimmune skin diseases including vitiligo, AA, DM, and PV, although the detailed epigenetic mechanisms remain to be further elucidated. In the future, clarifying the up- or downstream events of the implicated epigenetic factors may lead to further understanding of the complexity of these diseases. More critically, as epigenetic changes are early events and can be interfered with and reversed, the diagnostic markers and therapies based on epigenetic modifications will have broader prospects of application in the future.

References

[1] Cheng JB, Cho RJ. Genetics and epigenetics of the skin meet deep sequence. J Invest Dermatol 2012;132(3 Pt 2):923–32.

[2] Lu Q, Renaudineau Y, Cha S, Ilei G, Brooks WH, Selmi C, et al. Epigenetics in autoimmune disorders: highlights of the 10th Sjogren's Syndrome Symposium. Autoimmun Rev 2010;9(9):627–30.

[3] Antequera F, Bird A. Number of CpG islands and genes in human and mouse. Proc Natl Acad Sci USA 1993;90(24):11995–9.

[4] Attwood JT, Yung RL, Richardson BC. DNA methylation and the regulation of gene transcription. Cell Mol Life Sci 2002;59(2):241–57.

[5] Fuks F, Hurd PJ, Wolf D, Nan X, Bird AP, Kouzarides T. The methyl-CpG-binding protein MeCP2 links DNA methylation to histone methylation. J Biol Chem 2003;278 (6):4035–40.

[6] Jenuwein T, Allis CD. Translating the histone code. Science 2001;293(5532):1074–80.

[7] Ellis L, Atadja PW, Johnstone RW. Epigenetics in cancer: targeting chromatin modifications. Mol Cancer Ther 2009;8(6):1409–20.

[8] Berger SL. The complex language of chromatin regulation during transcription. Nature 2007;447(7143):407–12.

[9] Ambros V. The functions of animal microRNAs. Nature 2004;431(7006):350–5.

[10] Lee Y, Kim M, Han J, Yeom KH, Lee S, Baek SH, et al. MicroRNA genes are transcribed by RNA polymerase II. EMBO J 2004;23(20):4051–60.

[11] Strickland FM, Richardson BC. Epigenetics in human autoimmunity. Epigenetics in autoimmunity—DNA methylation in systemic lupus erythematosus and beyond. Autoimmunity 2008;41(4):278–86.

[12] Hewagama A, Richardson B. The genetics and epigenetics of autoimmune diseases. J Autoimmun 2009;33(1):3–11.

[13] Guerra L, Dellambra E, Brescia S, Raskovic D. Vitiligo: pathogenetic hypotheses and targets for current therapies. Curr Drug Metab 2010;11(5):451–67.

[14] Dell'Anna ML, Cario-Andre M, Bellei B, Taieb A, Picardo M. In vitro research on vitiligo: strategies, principles, methodological options and common pitfalls. Exp Dermatol 2012;21(7):490–6.

[15] Silverberg NB, Travis L. Childhood vitiligo. Cutis 2006;77(6):370–5.

[16] Kakourou T. Vitiligo in children. World J Pediatr 2009;5(4):265–8.

[17] Spritz RA. The genetics of generalized vitiligo: autoimmune pathways and an inverse relationship with malignant melanoma. Genome Med 2010;2(10):78.

[18] Halder RM, Chappell JL. Vitiligo update. Semin Cutan Med Surg 2009;28(2):86—92.
[19] Spritz RA. The genetics of generalized vitiligo and associated autoimmune diseases. Pigment Cell Res 2007;20(4):271—8.
[20] Alkhateeb A, Fain PR, Thody A, Bennett DC, Spritz RA. Epidemiology of vitiligo and associated autoimmune diseases in Caucasian probands and their families. Pigment Cell Res 2003;16(3):208—14.
[21] Korsunskaya IM, Suvorova KN, Dvoryankova EV. Modern aspects of vitiligo pathogenesis. Dokl Biol Sci 2003;388:38—40.
[22] Sun X, Xu A, Wei X, Ouyang J, Lu L, Chen M, et al. Genetic epidemiology of vitiligo: a study of 815 probands and their families from south China. Int J Dermatol 2006;45 (10):1176—81.
[23] Birlea SA, Gowan K, Fain PR, Spritz RA. Genome-wide association study of generalized vitiligo in an isolated European founder population identifies SMOC2, in close proximity to IDDM8. J Invest Dermatol 2010;130(3):798—803.
[24] Quan C, Ren YQ, Xiang LH, Sun LD, Xu AE, Gao XH, et al. Genome-wide association study for vitiligo identifies susceptibility loci at 6q27 and the MHC. Nat Genet 2010;42(7):614—18.
[25] Sreekumar GP, Erf GF, Smyth JJ. 5-Azacytidine treatment induces autoimmune vitiligo in parental control strains of the Smyth Line chicken model for autoimmune vitiligo. Clin Immunol Immunopathol 1996;81(2):136—44.
[26] Zhao M, Gao F, Wu X, Tang J, Lu Q. Abnormal DNA methylation in peripheral blood mononuclear cells from patients with vitiligo. Br J Dermatol 2010;163(4):736—42.
[27] Zhao M, Liang G, Wu X, Wang S, Zhang P, Su Y, et al. Abnormal epigenetic modifications in peripheral blood mononuclear cells from patients with alopecia areata. Br J Dermatol 2012;166(2):226—73.
[28] Lei W, Luo Y, Lei W, Luo Y, Yan K, Zhao S, et al. Abnormal DNA methylation in CD4+ T cells from patients with systemic lupus erythematosus, systemic sclerosis, and dermatomyositis. Scand J Rheumatol 2009;38(5):369—74.
[29] Zhao M, Huang W, Zhang Q, Gao F, Wang L, Zhang G, et al. Aberrant epigenetic modifications in peripheral blood mononuclear cells from patients with pemphigus vulgaris. Br J Dermatol 2012;167(3):523—31.
[30] Zailaie MZ. Decreased proinflammatory cytokine production by peripheral blood mononuclear cells from vitiligo patients following aspirin treatment. Saudi Med J 2005;26(5):799—805.
[31] Wankowicz-Kalinska A, van den Wijngaard RM, Tigges BJ, Westerhof W, Ogg GS, Cerundolo V, et al. Immunopolarization of CD4+ and CD8+ T cells to Type-1-like is associated with melanocyte loss in human vitiligo. Lab Invest 2003;83(5):683—95.
[32] Schallreuter KU, Lemke R, Brandt O, Schwartz R, Westhofen M, Montz R, et al. Vitiligo and other diseases: coexistence or true association? Hamburg study on 321 patients. Dermatology 1994;188(4):269—75.
[33] Zelissen PM, Bast EJ, Croughs RJ. Associated autoimmunity in Addison's disease. J Autoimmun 1995;8(1):121—30.
[34] Kemp EH, Gavalas NG, Gawkrodger DJ, Weetman AP. Autoantibody responses to melanocytes in the depigmenting skin disease vitiligo. Autoimmun Rev 2007;6(3):138—42.
[35] Abu TM, Pramod K, Ansari SH, Ali J. Current remedies for vitiligo. Autoimmun Rev 2010;9(7):516—20.
[36] Le Poole IC, Wankowicz-Kalinska A, van den Wijngaard RM, Nickoloff BJ, Das PK. Autoimmune aspects of depigmentation in vitiligo. J Investig Dermatol Symp Proc 2004;9(1):68—72.
[37] Ogg GS, Rod DP, Romero P, Chen JL, Cerundolo V. High frequency of skin-homing melanocyte-specific cytotoxic T lymphocytes in autoimmune vitiligo. J Exp Med 1998;188(6):1203—8.

[38] Zhou L, Li K, Shi YL, Hamzavi I, Gao TW, Henderson M, et al. Systemic analyses of immunophenotypes of peripheral T cells in non-segmental vitiligo: implication of defective natural killer T cells. Pigment Cell Melanoma Res 2012;25(5):602−11.

[39] Basak PY, Adiloglu AK, Ceyhan AM, Tas T, Akkaya VB. The role of helper and regulatory T cells in the pathogenesis of vitiligo. J Am Acad Dermatol 2009;60(2):256−60.

[40] Shi YL, Weiland M, Li J, Hamzavi I, Henderson M, Huggins RH, et al. MicroRNA expression profiling identifies potential serum biomarkers for non-segmental vitiligo. Pigment Cell Melanoma Res 2013;26(3):418−21.

[41] Jiang S, Li C, Olive V, Lykken E, Feng F, Sevilla J, et al. Molecular dissection of the miR-17-92 cluster's critical dual roles in promoting Th1 responses and preventing inducible Treg differentiation. Blood 2011;118(20):5487−97.

[42] Paraskevi A, Theodoropoulos G, Papaconstantinou I, Mantzaris G, Nikiteas N, Gazouli M. Circulating microRNA in inflammatory bowel disease. J Crohns Colitis 2012;6(9):900−4.

[43] Pauley KM, Satoh M, Chan AL, Bubb MR, Reeves WH, Chan EK. Upregulated miR-146a expression in peripheral blood mononuclear cells from rheumatoid arthritis patients. Arthritis Res Ther 2008;10(4):R101.

[44] Shi YL, Weiland M, Lim HW, Mi QS, Zhou L. Serum miRNA expression profiles change in autoimmune vitiligo in mice. Exp Dermatol 2014;23(2):140−2.

[45] Mueller DW, Rehli M, Bosserhoff AK. miRNA expression profiling in melanocytes and melanoma cell lines reveals miRNAs associated with formation and progression of malignant melanoma. J Invest Dermatol 2009;129(7):1740−51.

[46] Oshikawa Y, Jinnin M, Makino T, Kajihara I, Makino K, Honda N, et al. Decreased miR-7 expression in the skin and sera of patients with dermatomyositis. Acta Derm Venereol 2013;93(3):273−6.

[47] Tang X, Tian X, Zhang Y, Wu W, Tian J, Rui K, et al. Correlation between the frequency of Th17 cells and the expression of microRNA-206 in patients with dermatomyositis. Clin Dev Immunol 2013;2013:345347.

[48] Inoue K, Jinnin M, Yamane K, Makino T, Kajihara I, Makino K, et al. Down-regulation of miR-223 contributes to the formation of Gottron's papules in dermatomyositis via the induction of PKCε. Eur J Dermatol 2013;23(2):160−7.

[49] Wasserman D, Guzman-Sanchez DA, Scott K, McMichael A. Alopecia areata. Int J Dermatol 2007;46(2):121−31.

[50] Forstbauer LM, Brockschmidt FF, Moskvina V, Herold C, Redler S, Herzog A, et al. Genome-wide pooling approach identifies SPATA5 as a new susceptibility locus for alopecia areata. Eur J Hum Genet 2012;20(3):326−32.

[51] Todes-Taylor N, Turner R, Wood GS, Stratte PT, Morhenn VB. T cell subpopulations in alopecia areata. J Am Acad Dermatol 1984;11(2 Pt 1):216−23.

[52] Gilhar A, Kalish RS. Alopecia areata: a tissue specific autoimmune disease of the hair follicle. Autoimmun Rev 2006;5(1):64−9.

[53] Seetharam KA. Alopecia areata: an update. Indian J Dermatol Venereol Leprol 2013;79(5):563−75.

[54] Wang E, McElwee KJ. Etiopathogenesis of alopecia areata: why do our patients get it? Dermatol Ther 2011;24(3):337−47.

[55] McElwee KJ, Gilhar A, Tobin DJ, Ramot Y, Sundberg JP, Nakamura M, et al. What causes alopecia areata? Exp Dermatol 2013;22(9):609−26.

[56] Gilhar A, Landau M, Assy B, Shalaginov R, Serafimovich S, Kalish RS. Mediation of alopecia areata by cooperation between CD4+ and CD8+ T lymphocytes: transfer to human scalp explants on Prkdc(scid) mice. Arch Dermatol 2002;138(7):916−22.

[57] Colombe BW, Lou CD, Price VH. The genetic basis of alopecia areata: HLA associations with patchy alopecia areata versus alopecia totalis and alopecia universalis. J Investig Dermatol Symp Proc 1999;4(3):216−19.

[58] Xiao FL, Zhou FS, Liu JB, Yan KL, Cui Y, Gao M, et al. Association of HLA-DQA1 and DQB1 alleles with alopecia areata in Chinese Hans. Arch Dermatol Res 2005;297 (5):201–9.

[59] Barahmani N, de Andrade M, Slusser JP, Zhang Q, Duvic M. Major histocompatibility complex class I chain-related gene A polymorphisms and extended haplotypes are associated with familial alopecia areata. J Invest Dermatol 2006;126 (1):74–8.

[60] Petukhova L, Duvic M, Hordinsky M, Norris D, Price V, Shimomura Y, et al. Genome-wide association study in alopecia areata implicates both innate and adaptive immunity. Nature 2010;466(7302):113–17.

[61] Jackow C, Puffer N, Hordinsky M, Nelson J, Tarrand J, Duvic M. Alopecia areata and cytomegalovirus infection in twins: genes versus environment? J Am Acad Dermatol 1998;38(3):418–25.

[62] Panconesi E, Hautmann G. Psychophysiology of stress in dermatology. The psychobiologic pattern of psychosomatics. Dermatol Clin 1996;14(3):399–421.

[63] Hendrich B, Bird A. Identification and characterization of a family of mammalian methyl-CpG binding proteins. Mol Cell Biol 1998;18(11):6538–47.

[64] Kuwano Y, Fujimoto M, Watanabe R, Ishiura N, Nakashima H, Ohno Y, et al. Serum chemokine profiles in patients with alopecia areata. Br J Dermatol 2007;157 (3):466–73.

[65] Dalakas MC, Hohlfeld R. Polymyositis and dermatomyositis. Lancet 2003;362 (9388):971–82.

[66] Callen JP. Dermatomyositis. Lancet 2000;355(9197):53–7.

[67] Oddis CV, Conte CG, Steen VD, Medsger TJ. Incidence of polymyositis–dermatomyositis: a 20-year study of hospital diagnosed cases in Allegheny County, PA 1963–1982. J Rheumatol 1990;17(10):1329–34.

[68] Bronner IM, van der Meulen MF, de Visser M, Kalmijn S, van Venrooij WJ, Voskuyl AE, et al. Long-term outcome in polymyositis and dermatomyositis. Ann Rheum Dis 2006;65(11):1456–61.

[69] Reddy BY, Hantash BM. Cutaneous connective tissue diseases: epidemiology, diagnosis, and treatment. Open Dermatol J 2009;3(1):22–31.

[70] Mendell JR, Garcha TS, Kissel JT. The immunopathogenic role of complement in human muscle disease. Curr Opin Neurol 1996;9(3):226–34.

[71] Arahata K, Engel AG. Monoclonal antibody analysis of mononuclear cells in myopathies. I: Quantitation of subsets according to diagnosis and sites of accumulation and demonstration and counts of muscle fibers invaded by T cells. Ann Neurol 1984;16 (2):193–208.

[72] Engel AG, Arahata K. Monoclonal antibody analysis of mononuclear cells in myopathies. II: Phenotypes of autoinvasive cells in polymyositis and inclusion body myositis. Ann Neurol 1984;16(2):209–15.

[73] Mammen AL. Dermatomyositis and polymyositis: clinical presentation, autoantibodies, and pathogenesis. Ann NY Acad Sci 2010;1184:134–53.

[74] Hengstman GJ, van Brenk L, Vree EW, van der Kooi EL, Borm GF, Padberg GW, et al. High specificity of myositis specific autoantibodies for myositis compared with other neuromuscular disorders. J Neurol 2005;252(5):534–7.

[75] Kubo M, Ihn H, Yazawa N, Sato S, Kikuchi K, Tamaki K. Prevalence and antigen specificity of anti-histone antibodies in patients with polymyositis/dermatomyositis. J Invest Dermatol 1999;112(5):711–15.

[76] Tansley SL, McHugh NJ, Wedderburn LR. Adult and juvenile dermatomyositis: are the distinct clinical features explained by our current understanding of serological subgroups and pathogenic mechanisms? Arthritis Res Ther 2013;15(2):211.

[77] Okada S, Weatherhead E, Targoff IN, Wesley R, Miller FW. Global surface ultraviolet radiation intensity may modulate the clinical and immunologic expression of autoimmune muscle disease. Arthritis Rheum 2003;48(8):2285–93.

[78] Chevrel G, Page G, Granet C, Streichenberger N, Varennes A, Miossec P. Interleukin-17 increases the effects of IL-1 beta on muscle cells: arguments for the role of T cells in the pathogenesis of myositis. J Neuroimmunol 2003;137(1–2):125–33.

[79] Nishikawa T, Hashimoto T, Shimizu H, Ebihara T, Amagai M. Pemphigus: from immunofluorescence to molecular biology. J Dermatol Sci 1996;12(1):1–9.

[80] Amagai M, Klaus-Kovtun V, Stanley JR. Autoantibodies against a novel epithelial cadherin in pemphigus vulgaris, a disease of cell adhesion. Cell 1991;67(5):869–77.

[81] Kanwar AJ, De D. Pemphigus in India. Indian J Dermatol Venereol Leprol 2011;77 (4):439–49.

[82] Yeh SW, Ahmed B, Sami N, Razzaque AA. Blistering disorders: diagnosis and treatment. Dermatol Ther 2003;16(3):214–23.

[83] Harman KE, Gratian MJ, Seed PT, Bhogal BS, Challacombe SJ, Black MM. Diagnosis of pemphigus by ELISA: a critical evaluation of two ELISAs for the detection of antibodies to the major pemphigus antigens, desmoglein 1 and 3. Clin Exp Dermatol 2000;25(3):236–40.

[84] Harman KE, Gratian MJ, Bhogal BS, Challacombe SJ, Black MM. A study of desmoglein 1 autoantibodies in pemphigus vulgaris: racial differences in frequency and the association with a more severe phenotype. Br J Dermatol 2000;143(2):343–8.

[85] Bhol K, Natarajan K, Nagarwalla N, Mohimen A, Aoki V, Ahmed AR. Correlation of peptide specificity and IgG subclass with pathogenic and nonpathogenic autoantibodies in pemphigus vulgaris: a model for autoimmunity. Proc Natl Acad Sci USA 1995;92(11):5239–43.

[86] Hertl M, Amagai M, Sundaram H, Stanley J, Ishii K, Katz SI. Recognition of desmoglein 3 by autoreactive T cells in pemphigus vulgaris patients and normals. J Invest Dermatol 1998;110(1):62–6.

[87] Wucherpfennig KW, Yu B, Bhol K, Monos DS, Argyris E, Karr RW, et al. Structural basis for major histocompatibility complex (MHC)-linked susceptibility to autoimmunity: charged residues of a single MHC binding pocket confer selective presentation of self-peptides in pemphigus vulgaris. Proc Natl Acad Sci USA 1995;92(25):11935–9.

[88] Lin MS, Swartz SJ, Lopez A, Ding X, Fernandez-Vina MA, Stastny P, et al. Development and characterization of desmoglein-3 specific T cells from patients with pemphigus vulgaris. J Clin Invest 1997;99(1):31–40.

[89] Rizzo C, Fotino M, Zhang Y, Chow S, Spizuoco A, Sinha AA. Direct characterization of human T cells in pemphigus vulgaris reveals elevated autoantigen-specific Th2 activity in association with active disease. Clin Exp Dermatol 2005;30(5):535–40.

[90] Eming R, Budinger L, Riechers R, Christensen O, Bohlen H, Kalish R, et al. Frequency analysis of autoreactive T-helper 1 and 2 cells in bullous pemphigoid and pemphigus vulgaris by enzyme-linked immunospot assay. Br J Dermatol 2000;143 (6):1279–82.

[91] Veldman C, Stauber A, Wassmuth R, Uter W, Schuler G, Hertl M. Dichotomy of autoreactive Th1 and Th2 cell responses to desmoglein 3 in patients with pemphigus vulgaris (PV) and healthy carriers of PV-associated HLA class II alleles. J Immunol 2003;170(1):635–42.

[92] Spaeth S, Riechers R, Borradori L, Zillikens D, Budinger L, Hertl M. IgG, IgA and IgE autoantibodies against the ectodomain of desmoglein 3 in active pemphigus vulgaris. Br J Dermatol 2001;144(6):1183–8.

[93] Brautbar C, Moscovitz M, Livshits T, Haim S, Hacham-Zadeh S, Cohen HA, et al. HLA-DRw4 in pemphigus vulgaris patients in Israel. Tissue Antigens 1980;16(3):238–43.

[94] Szafer F, Brautbar C, Tzfoni E, Frankel G, Sherman L, Cohen I, et al. Detection of disease-specific restriction fragment length polymorphisms in pemphigus vulgaris linked to the DQw1 and DQw3 alleles of the HLA-D region. Proc Natl Acad Sci USA 1987;84(18):6542—5.

[95] Miyagawa S, Niizeki H, Yamashina Y, Kaneshige T. Genotyping for HLA-A, B and C alleles in Japanese patients with pemphigus: prevalence of Asian alleles of the HLA-B15 family. Br J Dermatol 2002;146(1):52—8.

[96] Sagi L, Sherer Y, Trau H, Shoenfeld Y. Pemphigus and infectious agents. Autoimmun Rev 2008;8(1):33—5.

[97] Orion E, Barzilay D, Brenner S. Pemphigus vulgaris induced by diazinon and sun exposure. Dermatology 2000;201(4):378—9.

[98] Paschos K, Allday MJ. Epigenetic reprogramming of host genes in viral and microbial pathogenesis. Trends Microbiol 2010;18(10):439—47.

[99] Nandakumar V, Vaid M, Tollefsbol TO, Katiyar SK. Aberrant DNA hypermethylation patterns lead to transcriptional silencing of tumor suppressor genes in UVB-exposed skin and UVB-induced skin tumors of mice. Carcinogenesis 2011;32 (4):597—604.

[100] Edwards TM, Myers JP. Environmental exposures and gene regulation in disease etiology. Environ Health Perspect 2007;115(9):1264—70.

NONIMMUNOLOGIC SKIN DISEASES

15

Epigenetics and Infectious Skin Disease

Jack L. Arbiser[1,2] and Michael Y. Bonner[1]

[1]Department of Dermatology, Emory School of Medicine, Winship Cancer Institute, Atlanta, GA [2]Department of Dermatology, Atlanta Veterans Affairs Medical Center, Decatur, GA

15.1 HERPES SIMPLEX VIRUSES

The herpesvirus family includes at least eight members that infect humans [7]. In both humans and lower animals, herpesvirus infection is associated with malignancy [8−10]. In humans, HSV1 and -2 and VZV (varicella zoster) are associated with acute infections and lifelong latency [11,12]. Epstein−Barr virus (EBV) and HHV8 (KSHV) are associated with human malignancy, with EBV accounting for the vast majority of human malignancies. While EBV infections rarely cause skin disorders (oral hairly leukoplakia, cutaneous leiomyosarcoma, CD56 lymphoid malignancies), the mechanism of how the EBV signals is likely conserved among other herpesviruses that cause malignancy [13−16].

EBV was originally isolated from African Burkitt's lymphoma, a malignancy that is common in children in East Africa [17]. Burkitt's lymphoma is also of historic interest because it was one of the first malignancies associated with a specific chromosomal translocation, that of the immunoglobulin promoter onto the *c-myc* oncogene [17,18]. The requirement for both viral infection and chromosomal translocation indicates that the genesis of Burkitt's lymphoma is a multistep process that is not dependent on a single event. In the United States, Burkitt's lymphoma is usually not associated with EBV infection except in the presence of HIV infection or other immunosuppression [19]. In East Africa, EBV-associated Burkitt's lymphoma is linked to high levels of untreated malaria infection, and treatment of malaria may reduce the rate of development of Burkitt's lymphoma [20].

© 2015 Elsevier Inc. All rights reserved.

Since the discovery of Burkitt's lymphoma, additional malignancies have been associated with EBV infection. These include nasopharyngeal carcinoma, subsets of gastric carcinoma, cutaneous leiomyosarcoma, Hodgkin's disease, and hematodermic CD56 positive lymphoma [20–23]. The development of tumors results from the development of viral latency, and in the case of Burkitt's lymphoma, the most well-studied, three types of latency have been described [24]. Type 1 latency is characterized by EBNA-1 protein and EBER noncoding RNAs. Type 2 latency is characterized by the expression of additional viral proteins, including EBNA, LMP-1, LMP-2A, LMP-2B, and EBER. Type 3 latency is characterized by a full panel of viral proteins, including EBNA-2. EBNA 3A, 3B, and 3C in addition to the proteins are expressed in latency type 2. Notably, viral proteins associated with viral coats are suppressed in latent states, making immune evasion possible.

Analysis of signaling in EBV-expressing Burkitt's cell lines allowed the elucidation of signaling events common to the majority of EBV-induced tumors [25]. Several years ago, we hypothesized the existence of a common signaling phenotype, termed the reactive oxygen-driven tumor [17,26]. The characteristics of this tumor are that it is caused by a reactive oxygen inducing carcinogen, and chronic reactive oxygen results in upregulation of DNA methyltransferase 1 (DNMT1) in EBV-infected lymphocytes, which hypermethylates (silences) the tumor suppressor p16ink4a [27]. In addition, these tumors have wild-type p53, because p53 is oxidatively inactivated by superoxide, thus removing the selective pressure for p53 mutation [28,29]. Reactive oxygen also inactivates other tumor suppressors such as PTEN and I-κB , leading to the constitutive activation of Akt and NF-κB, respectively [30,31]. Early studies of EBV-induced Burkitt's lymphoma revealed that these lymphomas tended to have hypermethylation of p16ink4a and wild-type p53, indicating that these tumors might be using the reactive oxygen-driven tumor signaling phenotype. In contrast, most sporadic Burkitt's lymphoma has mutant p53 and is EBV negative. In order to determine this, we examined a panel of Burkitt's lymphoma cell lines and discovered a tight correlation between EBV status and presence of high levels of reactive oxygen [17]. We also demonstrated high levels of mitogen-activated protein (MAP) kinase activation in the same cell lines that have elevated reactive oxygen. This was true in both type 1 and 3 latencies. We found that by blocking IL-10 in type 1 latency, we could reduce levels of reactive oxygen, indicating that the immunosuppressive molecule IL-10 can activate reactive oxygen through an autocrine loop. In type 3 latency, we demonstrated that the viral oncoproteins LMP1 and EBNA2 can induce reactive oxygen. Treatment with an antioxidant, ebselen, led to the decrease of NF-κB signaling, and a compensatory increase in MAP kinase signaling. The phenotype of this would be a fast growing tumor cell that is highly sensitive to apoptosis inducing agents [17]. Blockade of reactive oxygen could also decrease the production of immunosuppressive IL-10 and thus augment immune responses to this virally induced tumor. Of interest, virtually every EBV-induced tumor, regardless of anatomic site, demonstrates hypermethylation of p16ink4a, indicating that this is a common phenomenon. We demonstrated p16ink4a hypermethylation in an EBV-induced cutaneous lymphoma.

Finally, we have demonstrated that elevated reactive oxygen signaling can be targeted in humans. Oral hairy leukoplakia is an example of type 2 latency. Given that all forms of latency are associated with elevated reactive oxygen, we treated an HIV positive patient with oral hairy leukoplakia with gentian violet, a topically available NADPH oxidase inhibitor [13]. After three applications of gentian violet, the oral hairy leukoplakia completely resolved. A final way of eliminating latent cells is through induction of lytic infection, which can be done with histone deacetylase (HDAC) inhibitors, including the FDA approved drug valproic acid (VA) [13,32−34].

15.1.1 KSHV (HHV8)

Epidemiological studies of Kaposi's sarcoma (KS) suggested that KS was caused by an infectious agent [35]. Intensive studies of KS lesions led to the discovery of KSHV in 1994 [36], and subsequently this virus was discovered in hematologic disorders, for example, Castleman's disease and primary effusion lymphoma [37,38]. KSHV differs substantially from EBV in that it does not have an immortalizing function and has a tropism for lymphatic endothelial cells [39]. While KS expresses lymphatic markers, it is possible that KSHV can convert hemogenic endothelial cells into lymphatic endothelial cells through induction of the transcription factor Prox1.

The role of mammalian target of rapamycin (MTORC1) in KS was discovered when patients with posttransplantion KS responded to oral rapamycin. Subsequently, KS was found to have activation of this pathway, and infection of lymphatic but not hemogenic endothelial cells led to activation of mTORC1. The sensitivity of KS-infected endothelial cells to rapamycin was found to be conferred by a latency associated protein, ORF45, which induced mTORC1 activation through a MAP kinase/RSK pathway [40]. Of interest, notch pathways are upregulated in KS lesions, implicating a reactive oxygen pathway in these cells as well [41]. We have treated localized KS lesions with destructive modalities including cryotherapy and electrodesiccation followed by the NADPH oxidase inhibitor gentian violet, with good results (unpublished data).

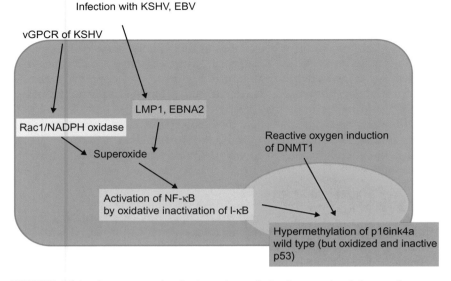

FIGURE 15.1   Common mode of epigenetic regulation in tumorigenic herpesviruses.

HDACs modify chromatin structure, and several HDAC inhibitors have been used in humans [16]. These include VA, which was initially developed as an antiseizure medicine and is widely used. Treatment of AIDS KS patients with clinically relevant doses of VA resulted in the increased expression of lytic proteins, but no major regression of KS lesions was noted. No major toxicities were observed (Figure 15.1) [32].

15.2 HSV1 AND -2 AND VZV

HSV1 and -2 are ubiquitous pathogens of humans, with approximately 90% of humans showing evidence of past infection with HSV1 and 10–20% that with HSV2. The prevalence of VZV infection was nearly universal but has dropped recently because of emphasis on childhood vaccination. All three of these viruses begin with epithelial infections and then exist in a lifelong latent state in neurons. Both clinically evident reactivation and asymptomatic viral shedding exist, in particular in HSV1 and -2, and it is likely that the differences between asymptomatic viral shedding and clinically evident infection reflect an interplay between the magnitude of neuronal activation versus immune control of infection.

HSV can bind to several cellular receptors through the promiscuous ligand glycoprotein D [34], these include nectin-1 and heparan sulfate proteoglycans. Viral protein VP16 interacts with host cell proteins Oct1 and

HCF1 to begin transcription of immediate early viral genes. In epithelial cells, this leads to productive viral infection that gives rise to mature virions which can then infect neighboring cells. Virions also infect neurons and spread retrograde to the nucleus, leading to the onset of the latency program. In the neuron, the major transcript is the latency associated transcript (LAT), and few, if any, other viral products are transcribed. Productive infection is halted primarily by the cytotoxic activity of CD8 cytotoxic lymphocytes, and patients with immune deficits of CD8 lymphocyte-based immunity have a more severe course. These include HIV infection, patients on chemotherapy, early infancy, and genetic immunodeficiencies.

Viral nucleic acids and double stranded RNA are sensed by TLR3 and TLR9, and patients who are deficient in TLR3 are highly susceptible to severe HSV infection. The activation of TLR signaling leads to a concomitant increase in interferon alpha and gamma. CD8 cells also limit infection through induction of granzymes A and B and induction of CD95.

The main difference between HSV and VZV is the frequency of reactivation from latency [34]. VZV usually reactivates once in a lifetime as an episode of shingles, while HSV often reactivates and/or sheds asymptomatically. The clinical settings associated with HSV reactivation include ultraviolet exposure, menstruation, and surgical procedures.

Of therapeutic relevance in HSV and VZV infection are epigenetic methods that either (i) induce viral replication that could be combined with antiviral therapy with the hope of eliminating latency or (ii) prevent reactivation of viral particles in neurons. Histone methylation plays a major role in activation of viral early promoters, and this is an interplay between the repressive H3K9 methylation competing with positive H3K4 trimethylation. Lysine-specific demethylase-1 (LSD1) demethylates the repressive H3K9 histones, allowing viral early transcription [33]. LSD1 is inhibited by small molecule monoamine oxidase inhibitors (MAOIs), and treatment of cells with the MAOI inhibitor tranylcypromine resulted in both decreased viral replication due to decreased viral-specific transcription and blocked viral reactivation in murine trigeminal ganglia [33]. Since this finding, other chromatin remodeling factors and HDACs have been found to modulate viral transcription as verified by an RNA interference (RNAi) screen. HDAC inhibitors such as butyrate and trichostatin activate HSV1 virion infection, and this may be of clinical relevance as VA is a structural analog of butyrate which we and others have demonstrated to be an HDAC inhibitor, that is widely used clinically.

15.3 HUMAN PAPILLOMA VIRUS

Human papilloma virus (HPV) is a family of small DNA viruses that are approximately 8 kb in length and normally occur as an extrachromosomal plasmid. Given its small size, it encodes a small number of genes compared with herpesviruses. HPV early genes stimulate cellular replication, while later genes stimulate viral structural proteins.

The two major factors in HPV-induced neoplasia are the ability of viral oncoproteins to bind the tumor suppressors p53 and Rb and whether the virus exists as an episome or integrates into human chromosomal DNA. HPV oncoprotein E6 binds the tumor suppressor protein p53, forms a complex with E6 associated protein (E6AP), and this complex stimulates the ubiquitination and degradation of wild-type p53 [25]. Because the wild-type p53 is successfully inactivated, there is little pressure to mutate p53, so early lesions of HPV rarely contain mutant p53. HPV is subdivided into low-risk and high-risk subtypes based upon their propensity to cause malignancy. Low-risk subtypes include HPV1 and -2, associated with plantar and palmar warts, while high-risk types include HPV16, -18, -31, and -33, and are a major cause of cervical cancer, penile cancer, anal squamous cell carcinoma, and oral carcinoma [42]. High-risk HPV subtypes have the ability to replicate in upper layers of the epidermis compared to low-risk subtypes, which replicate in the basal layer and differentiate in the upper layers of the epidermis. In addition, high-risk subtypes have an increased propensity to integrate into host chromosomal DNA, sometimes interrupting endogenous tumor suppressor genes. E6, in addition to downregulating p53, has the ability to upregulate telomerase, thus further facilitating cellular replication [25].

The ability of E6 to destabilize p53 through proteasomal degradation has been taken advantage of in the use of proteasome inhibitors that have particular efficacy on HPV infected cells. A novel proteasome inhibitor, RA490, was shown to form covalent adducts with adducts to RPN13 cysteine 88, resulting in inhibition of p53 degradation in HPV transformed cells, induction of p53 targets in these cells, and inhibition of tumor growth *in vivo* [25].

Viral promoters are differentially methylated in the HPV genome based upon the location of the infection, with methylation being present on the promoters of structural proteins during times of cellular replication, while being hypomethylated during terminal differentiation. Part of this differential methylation may be due to activity of DNMT1, which is induced by E6 through inactivation of p53 [25].

As in the case of E6, the E7 protein binds the tumor suppressor protein retinoblastoma (Rb), causing its destabilization by proteasomal degradation. In addition, through conserved sequences in what has been termed "the pocket," E7 binds and destabilizes the Rb family proteins RBL1 and -2. More recently, E7 has also been found to bind and activate *FoxM1*, a gene which is amplified in many poor prognosis solid tumors. *FoxM1* causes upregulation of the g2/M genes aurora kinase B (*AURKB*), pololike kinase 4 (*PLK4*), and cyclin B1 (*CCNB1*) [10] (Figure 15.2).

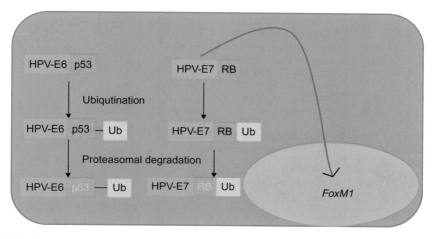

FIGURE 15.2 Common mode of epigenetic regulation in tumorigenic human papillomavirus.

15.4 CONCLUSION

The combination of the E6 and E7 viral oncogenes results in complete carcinogenesis, especially in the presence of high-risk subtypes of HPV which bind host tumor suppressors with high affinity. HPV-induced tumors also evade the immune system by silencing viral capsid antigens and integrate into the genome of solid tumors, rather than persist as an episome. Elimination of HPV-induced tumors will require a combination of immunotherapy, in which the E6 and E7 oncogenes are recognized as tumor antigens. Defects in antigen processing, such as decreased TAP protein and carbonic anhydrase 9, contribute to immune evasion. Finally, class 1 and class II MHC are suppressed in HPV tumors. Effective therapies will induce class 1 and class II MHC, enhance tumor antigen presentation, and activate p53. Interferons and interferon inducers such as imiquimod are only partially effective in this respect, and the addition of adjuvant drugs such as NADPH oxidase inhibitors or honokiol may enhance immune destruction of tumors.

List of Abbreviations

AURKB	aurora kinase B
CCNB1	cyclin B1
DNMT1	DNA methyltransferase 1
E6AP	E6 associated protein
EBER	Epstein–Barr virus-encoded small RNAs
EBNA-1, 2	Epstein–Barr virus nuclear antigen-1, 2

EBV	Epstein−Barr virus
HDACs	histone deacetylases
HHV8, KSHV	Kaposi's sarcoma-associated herpesvirus
HIV	human immunodeficiency virus
HSV1, 2	herpes simplex virus 1, 2
KS	Kaposi's sarcoma
LAT	latency associated transcript
LMP-1, LMP-2A, LMP-2B	Epstein−Barr virus latent membrane protein-1, -2A, -2B
LSD1	lysine-specific demethylase-1
MAOI	monoamine oxidase inhibitors
PLK4	polo-like kinase 4
PTEN	phosphatase and tensin homolog
Rb	retinoblastoma
VA	valproic acid
VZV	varicella zoster virus

References

[1] Ushijima T, Okochi-Takada E. Aberrant methylations in cancer cells: where do they come from? Cancer Sci 2005;96(4):206−11.
[2] Bierne H, Hamon M, Cossart P. Epigenetics and bacterial infections. Cold Spring Harb Perspect Med 2012;2(12):a010272.
[3] Ciarlo E, Savva A, Roger T. Epigenetics in sepsis: targeting histone deacetylases. Int J Antimicrob Agents 2013;42(Suppl.):S8−12.
[4] Costenbader KH, Gay S, Alarcon-Riquelme ME, Iaccarino L, Doria A. Genes, epigenetic regulation and environmental factors: which is the most relevant in developing autoimmune diseases? Autoimmun Rev 2012;11(8):604−9.
[5] Ciuffi A, Telenti A. State of genomics and epigenomics research in the perspective of HIV cure. Curr Opin HIV AIDS 2013;8(3):176−81.
[6] Sun J, Lu H, Wang X, Jin H. MicroRNAs in hepatocellular carcinoma: regulation, function, and clinical implications. Sci World J 2013;2013: 924206.
[7] Grinde B. Herpesviruses: latency and reactivation—viral strategies and host response. J Oral Microbiol 2013;5.
[8] Dittmer DP, Damania B. Kaposi sarcoma associated herpesvirus pathogenesis (KSHV)—an update. Curr Opin Virol 2013;3(3):238−44.
[9] Brandsma JL. The cottontail rabbit papillomavirus model of high-risk HPV-induced disease. Methods Mol Med 2005;119:217−35.
[10] Decaprio JA. Human papillomavirus type 16 E7 perturbs DREAM to promote cellular proliferation and mitotic gene expression. Oncogene 2014;33(31):4036−8.
[11] Steiner I, Kennedy PG, Pachner AR. The neurotropic herpes viruses: herpes simplex and varicella-zoster. Lancet Neurol 2007;6(11):1015−28.
[12] Egan KP, Wu S, Wigdahl B, Jennings SR. Immunological control of herpes simplex virus infections. J Neurovirol 2013;19(4):328−45.
[13] Bhandarkar SS, Mackelfresh J, Fried L, Arbiser JL. Targeted therapy of oral hairy leukoplakia with gentian violet. J Am Acad Dermatol 2008;58(4):711−12.
[14] Tetzlaff MT, Nosek C, Kovarik CL. Epstein−Barr virus-associated leiomyosarcoma with cutaneous involvement in an African child with human immunodeficiency virus: a case report and review of the literature. J Cutan Pathol 2011;38(9):731−9.
[15] Jiang Q, Liu S, Peng J, et al. An extraordinary T/NK lymphoma, nasal type, occurring primarily in the prostate gland with unusual CD30 positivity: case report and review of the literature. Diagn Pathol 2013;8:94.

[16] Oh HS, Bryant KF, Nieland TJ, et al. A targeted RNA interference screen reveals novel epigenetic factors that regulate herpesviral gene expression. MBio 2014;5(1): e01086-13.

[17] Cerimele F, Battle T, Lynch R, et al. Reactive oxygen signaling and MAPK activation distinguish Epstein—Barr virus (EBV)-positive versus EBV-negative Burkitt's lymphoma. Proc Natl Acad Sci USA 2005;102(1):175—9.

[18] Ott G, Rosenwald A, Campo E. Understanding MYC-driven aggressive B-cell lymphomas: pathogenesis and classification. Blood 2013;122(24):3884—91.

[19] Mbulaiteye SM, Pullarkat ST, Nathwani BN, et al. Epstein—Barr virus patterns in US Burkitt lymphoma tumors from the SEER residual tissue repository during 1979—2009. APMIS 2014;122(1):5—15.

[20] Rickinson AB. Co-infections, inflammation and oncogenesis: future directions for EBV research. Semin Cancer Biol 2014;26:99—115.

[21] Iizasa H, Nanbo A, Nishikawa J, Jinushi M, Yoshiyama H. Epstein—Barr virus (EBV)-associated gastric carcinoma. Viruses 2012;4(12):3420—39.

[22] Henkenberens C, Franzke A, Raab P, Oschlies I, Klapper W, Christiansen H. Primary EBV-positive Hodgkin's lymphoma of the CNS under azathioprine treatment: case report and review of the literature. Strahlenther Onkol 2014;190(9):847—52.

[23] Kato N, Yasukawa K, Kimura K, et al. CD2 − CD4+ CD56+ hematodermic/hematolymphoid malignancy. J Am Acad Dermatol 2001;44(2):231—8.

[24] Murata T, Sato Y, Kimura H. Modes of infection and oncogenesis by the Epstein—Barr virus. Rev Med Virol 2014;24(4):242—53.

[25] Au Yeung CL, Tsang WP, Tsang TY, Co NN, Yau PL, Kwok TT. HPV-16 E6 upregulation of DNMT1 through repression of tumor suppressor p53. Oncol Rep 2010;24(6): 1599—604.

[26] Arbiser JL. Implications of Epstein—Barr virus (EBV)-induced carcinogenesis on cutaneous inflammation and carcinogenesis: evidence of recurring patterns of angiogenesis and signal transduction. J Invest Dermatol 2005;124(5):xi—xii.

[27] Klangby U, Okan I, Magnusson KP, Wendland M, Lind P, Wiman KG. p16/INK4a and p15/INK4b gene methylation and absence of p16/INK4a mRNA and protein expression in Burkitt's lymphoma. Blood 1998;91(5):1680—7.

[28] Pise-Masison CA, Radonovich M, Sakaguchi K, Appella E, Brady JN. Phosphorylation of p53: a novel pathway for p53 inactivation in human T-cell lymphotropic virus type 1-transformed cells. J Virol 1998;72(8):6348—55.

[29] Laurie NA, Donovan SL, Shih CS, et al. Inactivation of the p53 pathway in retinoblastoma. Nature 2006;444(7115):61—6.

[30] Jones RG, Saibil SD, Pun JM, et al. NF-kappaB couples protein kinase B/Akt signaling to distinct survival pathways and the regulation of lymphocyte homeostasis *in vivo*. J Immunol 2005;175(6):3790—9.

[31] Parsa AT, Waldron JS, Panner A, et al. Loss of tumor suppressor PTEN function increases B7-H1 expression and immunoresistance in glioma. Nat Med 2007;13(1): 84—8.

[32] Lechowicz M, Dittmer DP, Lee JY, et al. Molecular and clinical assessment in the treatment of AIDS Kaposi sarcoma with valproic acid. Clin Infect Dis 2009;49 (12):1946—9.

[33] Liang Y, Vogel JL, Narayanan A, Peng H, Kristie TM. Inhibition of the histone demethylase LSD1 blocks alpha-herpesvirus lytic replication and reactivation from latency. Nat Med 2009;15(11):1312—17.

[34] Knipe DM, Lieberman PM, Jung JU, et al. Snapshots: chromatin control of viral infection. Virology 2013;435(1):141—56.

[35] Chen CJ, Hsu WL, Yang HI, et al. Epidemiology of virus infection and human cancer. Recent Results Cancer Res 2014;193:11—32.

[36] Nagamachi Y, Takenoshita S. Lymphoid hyperplastic diseases of the intestine, lymphoid hyperplasia of the intestine, malignant lymphoma of the intestine, and Burkitt's lymphoma of the intestine. Ryoikibetsu Shokogun Shirizu 1994;6:595—8.

[37] Geraminejad P, Memar O, Aronson I, Rady PL, Hengge U, Tyring SK. Kaposi's sarcoma and other manifestations of human herpesvirus 8. J Am Acad Dermatol 2002; 47(5):641—55, quiz 656—48.

[38] Giffin L, Damania B. KSHV: pathways to tumorigenesis and persistent infection. Adv Virus Res 2014;88:111—59.

[39] Aguirre AJ, Robertson ES. Epstein—Barr virus recombinants from BC-1 and BC-2 can immortalize human primary B lymphocytes with different levels of efficiency and in the absence of coinfection by Kaposi's sarcoma-associated herpesvirus. J Virol 2000;74(2):735—43.

[40] Chang HH, Ganem D. A unique herpesviral transcriptional program in KSHV-infected lymphatic endothelial cells leads to mTORC1 activation and rapamycin sensitivity. Cell Host Microbe 2013;13(4):429—40.

[41] Wang X, He Z, Xia T, et al. Latency-associated nuclear antigen of Kaposi sarcoma-associated herpesvirus promotes angiogenesis through targeting notch signaling effector Hey1. Cancer Res 2014;74(7):2026—37.

[42] Johannsen E, Lambert PF. Epigenetics of human papillomaviruses. Virology 2013; 445(1—2):205—12.

Epigenetics of Melanoma

Jessica Charlet and Gangning Liang

Department of Urology, Keck School of Medicine,
University of Southern California, Los Angeles, CA

16.1 INTRODUCTION TO THE DISEASE CONDITION

16.1.1 Incidence and Etiology of Melanoma

The human skin is the largest organ of an individual and this is prone to cancer development. Cutaneous malignant melanoma (CMM) is a lethal cancer of the skin that predominantly occurs in fair-skinned people. Compared to the non-melanoma skin cancers basal (BCC) and squamous cell carcinoma (SCC), CMM has a much stronger invasive and metastatic potential [1].

The incidence of CMM is increasing much faster every year than any other human cancer type worldwide [2], and the mortality rate sums up to less than 2 years of survival for individuals presenting with the disease [3]. Below the age of 30, women have a greater incidence of CMM than do men; however, after the age of 30 the incidence is reversed [4].

The etiology of CMM is rather complex—environmental factors, such as the amount of sun (UV) exposure, which is the most common factor (65% of cases) [5], but also different genetic factors, can contribute to the risk of contracting CMM. Thus, Australia with its high level of sun exposure has a higher incidence of CMM [6]. Apart from individuals who already have BCC or SCC, other genetic factors, that increase CMM risk are mostly heritable; they include pigmentation of the skin (fair), eyes (blue and green), and hair (blond and red) [7]. The ability to tan and overall sensitivity to the sun as well as the presence of a large amount of melanocytic or atypical nevi represent other CMM risk factors [8,9].

339
© 2015 Elsevier Inc. All rights reserved.

16.1.2 Origin and Heritability of Melanoma

Melanocytes derive from neural crest cells, which complement the other three germ layers, namely, the ecto-, meso-, and endoderms. Melanocytes are located in the epidermis, eyes, hair bulbs, meninges, and ears. Melanocytes produce the pigment melanin, which is created by the enzyme tyrosinase and is then transferred to keratinocytes through recognition via the PAR-2 receptor. Melanin protects nuclei from harmful UV light through formation of a cap on top of the basal cell nuclei [10].

About 10% of all CMM cases have a familial/inherited background. Familial cases compared to sporadic CMM cannot be distinguished from each other on a clinical level. Both forms of cancer present as large epitheloid cell clusters with no other phenotypical characteristics [11]. The main difference between germ line/familial and sporadic CMM cases is the fact that germ line patients are diagnosed with the disease at a much earlier age and usually have a greater number of tumors, which appear slightly thinner than sporadic/nonfamilial tumors [6].

16.2 DIAGNOSIS OF MELANOMA

In the case of suspicious-looking pigmentation, the skin should be analyzed by dermascopy, which permits even an inexperienced practitioner to make a highly accurate diagnosis. However, a definitive diagnosis of either a melanoma or a benign nevus can only be done after biopsy [12].

CMM usually presents with characteristics referred to as "ABCD," meaning an asymmetric shape, an irregular border, variation in color, and a diameter that is greater than 6 mm.

16.2.1 Melanoma Progression and Subtypes

CMM tumors progress in three different histomorphological steps. The first growth phase is called radial growth phase (RGP)-confined melanoma, where the tumor stays confined to the epidermis. The second phase is the RGP-confined microinvasion, where some cells have infiltrated the superficial papillary dermis. The final phase describes the tumorigenic or mitogenic phase of melanoma and is called the vertical growth phase (VGP) [13].

The current staging of melanoma comprises four clinical subtypes. *Superficial spreading melanoma* (SSM) is the most common form and varies in pigmentation. About 75% of SSM tumors are newly formed [14]. *Nodular melanoma* (NM) has no RGP and is either nodular or pedunculated, or even polyploid [15]. The third variant of melanoma, *lentigo maligna melanoma* (LLM), is found predominantly on the face and upper body extremities of elderly people [16]. The last type of melanoma is *acral lentiginous melanoma* (ALM), which is mostly found on plantar, ungual, and palmar skin of Japanese and black individuals [17].

Some other rare variants of melanoma include myxoid, signet ring, verrucous, small cell, nevoid, and osteogenic melanoma, and childhood melanoma excongenital nevus. Minimal deviation malignant melanoma has a uniform distribution of melanocytes with only few atypia. Animals that produce pigments such as horses are also prone to melanoma [18]. Another type of melanoma that often gets misdiagnosed as a scar or fibroblastic proliferation is desmoplastic melanoma. This tumor has more atypical features than the classical melanomas; it mostly appears in the neck and head region and presents as a plaque or bulky growth [19].

16.3 ETIOLOGY AND PATHOGENESIS OF MELANOMA

16.3.1 Genetic Alterations of Melanoma

16.3.1.1 *Sporadic Mutations*

To date, it appears that CMM might be the cancer with the highest frequency of mutations in a single cell [20]. First screenings for somatic mutations in a metastatic melanoma cell line revealed more than 30,000 mutations, about 300 occurring in the gene's coding sequence [21]. Most nonsynonymous mutations in CMM are C−T transitions, which highlight the correlation between exposure of skin to sunlight and melanoma development [22].

Most nevi that appear after birth have a *BRAF* mutation, which is found in 50−70% of all CMM cases and many other cancers [23]. The *BRAF* gene encodes for the kinase BRAFV600E, which plays a major role in the MAPK signaling pathway [24]. ERK, a downstream target of the latter signaling pathway, is hyperactive in about 90% of all CMM cases [25]. Since most nevi do not develop into CMM, the *BRAF* mutation is necessary for the disease but alone is not sufficient; thus, other genetic mutations and/or epigenetic alterations need to occur for tumorigenesis [26]. Another genetic mutation initiating the formation of CMM occurs in the *NRAS* gene (15−30% of cases), which can also lead to neurological disorders [27]. Nevi do not tend to show chromosomal alterations; however, melanomas gain and lose chromosomal loci. Chromosomes 1q, 6p, 7, 8q, 17, and 20q are frequently gained in CMM, encompassing oncogenes such as *MYC, CDK4, MITF, MDM2,* and *CCND1*, while chromosomes 6q, 8p, 9p, and 10q are usually lost, including the tumor-suppressor genes *P14*, *P15*, *P16*, and *PTEN* [28,29]. Interestingly, the most frequent mutations, *BRAF*, *NRAS*, and *KIT* (<17% of cases) [30], are mutually exclusive for the onset of CMM [31].

16.3.1.2 *Germ Line Mutations*

Family studies of CMM have revealed that predominantly three high-penetrance genes are responsible for CMM, namely, *CDKN2A* [32], *CDK4* [33], and *BAP1* [34]. These three susceptibility genes are inherited in an autosomal dominant pattern, affecting both sexes at the same rate. It is believed that further susceptibility genes in CMM exist; however, they still need to be identified.

The *CDKN2A* locus is on chromosome 9p21 and the gene encodes for the tumor-suppressor gene *p16*, regulating cell growth by arresting the cell cycle at G1; the other transcript produced by *CDKN2A* is p14ARF, inducing apoptosis and cell cycle arrest though *p53* [35,36]. About 20–40% of CMM-prone families have a germ line mutation in *CDKN2A*, which differs by the geographic location. Generally, *p16* has a higher mutation frequency of 38%, compared to *p14*, which is only mutated in 2.5% of cases. The majority of these mutations are missense mutations (65%); deletions/insertions or duplications occur in 23% of cases and only 5% are nonsense mutations [37]. The Swedish and Dutch populations have the same founder mutation, as do British and Australian peoples and French, Spanish, and Italian populations. The North American population mutations can be tracked from different migration patterns. Although *CDKN2A* mutations increase the risk of CMM dramatically, the risk outside high-risk families, meaning the general population, is rather low, ranging from 0.2% to 3% [38].

The *CDK4* gene, located on chromosome 12q14, is an oncogene that is very rarely mutated; however, it also confers an increased risk rate for CMM in melanoma-prone families. So far, only 20 families have been identified worldwide bearing the mutation [33]. To date, two causal mutations have been characterized, occurring in exon 2 of the gene, a region that encodes for the p16 binding site. Thus, because of the rarity of this mutation in CMM-prone families, the *CDK4* mutations in the general population are barely detectable [37].

BRCA1-associated protein 1 or *BAP1* is located on chromosome 3p21 and is considered a tumor-suppressor gene. *BAP1* is considered a rare high-risk factor for melanoma [34]. About 75% of families carrying the mutation have at least one member with cutaneous melanoma. Since *BAP1* germ line mutations were primarily found in patients with cutaneous and ocular melanoma, it is believed that the mutation appears in cutaneous melanoma but primarily in conjunction with uveal melanoma [39].

Other low-risk susceptibility genes in CMM are *MC1R*, which plays a role in normal skin pigmentation variation [40] and the microphthalmia-associated transcription factor (*MITF*). A recent study of proband whole-genome sequencing of melanoma families revealed a novel germ line mutation in *MITF*. The same mutation was also verified in two independent studies of a British and Australian cohort [41].

16.3.2 Epigenetic Alterations of Melanoma

In addition to all the previously mentioned genetic mutations and chromosomal alterations in CMM, epigenetic modifications were detected.

16.3.2.1 DNA Hypomethylation

In many cancer types, global DNA hypomethylation has been reported to primarily occur in repetitive regions of long or short interspersed nuclear elements—at transposons and pericentromeric regions. This causes chromosomal instability, but also the expression of certain oncogenes could be induced by DNA hypomethylation [42,43]. A set of genes that are aberrantly expressed in melanoma cells are, among others, the cancer—testis antigens that are normally repressed through promoter hypermethylation in normal skin cells [44]. Whether the reexpression of these normally silenced genes actively contributes to tumorigenesis is still contested, since this could simply be a secondary effect due to the global chromosomal instability induced by genome-wide DNA hypomethylation. However, the family of *MAGE* genes (Table 16.1) was considered to be a contributor to the cancer phenotype. For example, the testis-specific protein, Y-linked, is considered a putative oncogene, as

TABLE 16.1 Genes Showing DNA Hypomethylation in CMM

Gene name	Common name	Reference
14-3-3σ	Stratifin	[45]
GAGE 1—6	G antigen 1—6	[44]
HMW-MAA	High molecular weight melanoma-associated antigen	[46]
MAGE A1—4 and 6	Melanoma antigen, family A1—4 and 6	[44]
NY-ESO-1	New York esophageal squamous cell carcinoma 1	[47]
PI5	Protease inhibitor 5/maspin	[48]
PRAME	Preferentially expressed antigen of melanoma	[44]
SSX 1—5	Sarcoma, synovial, breakpoint 1—5	[44]
TSPY	Testis-specific protein, Y-linked	[45]

Adapted from Ref. [45].

3. NONIMMUNOLOGIC SKIN DISEASES

it is reexpressed in melanoma through loss of promoter hypermethylation and forces the cell more rapidly through the cell cycle, thus contributing to cell proliferation [49,50]. One theory concerning the loss of DNA methylation and thus reexpression of these cancer germ line genes claims that the transcription factor Brother of the Regulator of Imprinted Sites (BORIS) is involved in this mechanism through its upregulation in CMM. Although exogenous expression of BORIS in a cancer cell line model did not activate the *MAGE* genes, it is believed that the activation of cancer germ line genes is not solely due to BORIS [51].

16.3.2.2 DNA Hypermethylation

As previously mentioned, DNA hypermethylation is another way of gene silencing next to inactivating gene mutations or deletions. Dense CpG regions, also called CpG islands (CGIs), are at least 200 bp long, have a GC content of at least 50% and an observed:expected ratio of more than 0.6. In cancer, these CGI can have aberrant DNA methylation, where transcription factors can no longer bind and thus lead to reduced or complete inhibition of gene expression. This phenomenon can be explained by the expression upregulation of the so-called *de novo* DNA methyltransferases (DNMTs) 3A and 3B in CMM progression [52]. Melanoma growth can be inhibited through knockdown experiments of DNMT3A and DNMT1, and the maintenance DNMT can be suppressed by knocking down BRAF^{V600E} [53,54].

To date, more than 70 different genes have been detected that are aberrantly methylated in CMM (Table 16.2). Several different pathways that are likely to be mutated are also altered through DNA hypermethylation; these include WNT, RB, PI3K, MAPK, apoptosis, cell cycle, DNA repair, and metastasis. The best known and most frequently hypermethylated genes in CMM are *RARB* (59%), *RASSF1A* (28%), and *MGMT* (26%) [61,72]. These three genes can be detected as being aberrantly methylated in circulating tumor cells, which makes them useful diagnostic and prognostic markers for CMM patients [73]. These genes are also frequently disrupted in other human cancer types, and another panel of genes that show aberrant DNA methylation has been discovered by an independent group—namely *DcR1* and *DcR2*, *LOX*, and *TPM1*; these genes have a methylation frequency of 60, 80, 50, and 10%, respectively [45].

Since colorectal cancer is well known for its CpG methylator phenotype (CIMP), which reveals a gradual increase in DNA methylation along with tumor aggressiveness, a similar study has been undertaken in CMM. Although this study is not based on a genome-wide level, several genes have been detected that gain DNA methylation with advanced tumor stage. These genes, *WIF1*, *TFPI2*, *RASSF1A*, *SOCS1*, *MINT17*, and *MINT31*, have been suggested as constituting a CMM

TABLE 16.2   Genes Showing DNA Hypermethylation in CMM

Gene name	Common name	Reference
APC	Adenomatous polyposis coli gene	[55]
ASC	Apoptosis speck-like protein containing a CARD	[56]
BST2	Bone marrow stromal cell antigen 2	[57]
CD10	Membrane metalloendopeptidase	[58]
CDH1	E-cadherin	[59]
CDH8	Cadherin 8	[57]
CDKN1B	Cyclin-dependent kinase inhibitor 1B	[55]
CDKN1C	Cyclin-dependent kinase inhibitor 1C	[58]
CDKN2A	Cyclin-dependent kinase inhibitor 2A	[60]
CIITA-PIV	Class II transactivator promoter IV	[59]
COL1A2	Collagen, type 1, alpha 2	[57]
CYP1B1	Cytochrome P450, subfamily 1, polypeptide 1	[57]
DAL1	Differentially expressed in adenocarcinoma of the lung	[57]
DAPK1	Death-associated protein kinase 1	[61]
DPPIV	Dipeptidyl peptidase IV	[62]
ER-a	Estrogen receptor alpha	[63]
GDF15	Growth/differentiation factor 15	[57]
HOXB13	Homeobox 13	[57]
HSP11	Heat shock protein 11	[64]
KR18	Zinc finger protein 160	[58]
LOX	Lysyl oxidase	[59]
LRRC2	Leucine-rich repeat-containing protein 2	[57]
LXN	Latexin	[57]
MDR1	Multidrug resistance 1	[58]
MEGALIN	Low-density lipoprotein-related protein 2	[58]
MFAP2	Microfibril-associated protein 2	[57]
MGMT	O^6-methylguanine-DNA methyltransferase	[61]
MIB2	Skeletrophin	[65]
P101	Phosphoinositide-3-kinase, regulatory subunit 5	[58]
P73	Tumor protein 73	[58]

(Continued)

3. NONIMMUNOLOGIC SKIN DISEASES

TABLE 16.2 (Continued)

Gene name	Common name	Reference
PCSK1	Proprotein convertase, subtilisin/kexin-type 1	[57]
PRDX2	Peroxiredoxin 2	[66]
PTEN	Phosphatase and tensin homolog	[67]
PTGS2	Prostaglandin-endoperoxide synthase 2	[57]
QPCT	Glutaminyl-peptide cyclotransferase	[57]
RARB2	Retinoic-acid receptor, beta isoform 2	[61]
RASSF1A	RAS association domain family protein 1A	[68]
RIL	PDZ and LIM domain 4	[58]
SOCS-1	Suppressor of cytokine signaling 1	[69]
SOCS-2	Suppressor of cytokine signaling 2	[59]
SOCS-3	Suppressor of cytokine signaling 3	[70]
SYK	Protein-tyrosine kinase SYK	[57]
TFP12	Tissue factor pathway inhibitor 2	[71]
THBS1	Thrombospondin 1	[58]
THBS4	Thrombospondin 4	[58]
TIMP3	Tissue inhibitor of metalloproteinase 3	[59]
TM	Thrombomodulin	[72]
TNFRSF10A	Tumor necrosis factor receptor superfamily, member 10a	[59]
TNFRSF10C	Tumor necrosis factor receptor superfamily, member 10c	[59]
TNFRSF10D	Tumor necrosis factor receptor superfamily, member 10d	[59]
TPM1	Tropomyosin 1	[59]
WFDC1	Wap 4-disulfide core domain 1	[57]

Adapted from Ref. [45].

CIMP [74]. The latter cannot be recognized as CIMP yet, as reproducible studies are lacking. Other studies have attempted microarray analysis of upregulated gene expression after treatment with a DNA methylation inhibitor such as decitabine; however, when it comes to validation of these putative aberrantly methylated genes, the studies are lacking the necessary control material of normal tissue or the number of available samples is too limited to give the findings enough statistical significance to claim new CIMP markers in CMM [75,76].

16.3.2.3 *Histone Posttranslational Modifications and Chromatin Remodeling*

As described earlier, DNA methylation modifications at the promoter of genes influence binding of transcription factors and thus regulate expression of the corresponding gene by interference with RNA polymerase [77]. Another level of gene regulation occurs at histones that directly binds DNA. These histones wrapped around DNA form nucleosomes with stretches of 50 bp of free DNA in between them. The state of the chromatin is controlled by modifications on the out-sticking NH_2-terminal domains of the histone tails. The charge of the tails determines whether the chromatin is in an open (euchromatin) and accessible state for transcription factors and chromatin remodeling complexes or in the opposite scenario, where chromatin becomes more compact thus rendering it inaccessible to regulatory factor binding, then being referred to as heterochromatin [78].

Among the different modifications (histone code, as reviewed in Chapter 1) such as methylation, sumoylation, phosphorylation, and ubiquitination that can take place on the histone tails, the most studied is histone acetylation through histone acetyltransferases (HATs) or deacetylation via histone deacetyltransferases (HDACs). HATs are generally associated with open chromatin, and thus transcriptional activation, while HDACs render the chromatin inaccessible to regulatory factors and thus inhibit transcription [79].

In CMM, changes on the histone tails can lead to aberrant gene expression, in the same manner as do DNA methylation changes and genetic mutations. Although data of aberrant histone modifications in CMM is very limited, it is known that hypoacetylation has the most dramatic effect in CMM. Genes that are downregulated by the latter mechanism are, for example, the cell cycle regulator *CDKN1A* and the pro-apoptotic proteins BAX, caspases, BIM, BID, and APAF1, which are all upregulated [80−82]. Stabilization of oncoproteins in CMM, such as MYC and NFKB, is achieved by TIP60, GCN5, PCAF and TIP60, and HBO1. Also, the transcription coactivators P300 and CBP are involved in apoptosis, differentiation, and proliferation, both proteins associating with the melanocyte lineage survival oncogene *MITF* that is mostly upregulated in metastasis and regulates melanoma proliferation, invasiveness, and apoptosis [83]. The latter correlates with a decreased survival rate in CMM patients [84].

HDACs have not only a repressive effect on gene transcription but are also involved in apoptosis, growth arrest, and differentiation [85]. There are four classes of HDACs: class I, II, III, and IV. These classes have been established based on their homology to yeast HDACs as well as their enzymatic activity and cellular localization. Class I proteins are localized in the nucleus and can be found expressed in all cell types; while class II proteins are moved back and forth between nucleus and cytosol, they are expressed in the brain, muscle, pancreas, and heart. Class IV is composed of a single HDAC (11) and is localized in the same tissues as class II proteins [86]. The remaining class of HDACs, class III, is composed of sirtuin proteins. They are involved in aging in yeast and more recently were identified as playing an important role in cancer. The contribution of these proteins to tumor development has made them a popular target for epigenetic cancer treatment. Whether the designed HDAC inhibitors (HDACis) are selective or of a broad-spectrum nature, they support a new age of cancer therapy [45].

The deregulation of nucleosome sliding or ejection from DNA has been linked to CMM development. One of the main complexes involved in this regulation is the SWI/SNF complex, which is composed of ATP-dependent chromatin remodeling enzymes (Figure 16.1). Tumor-suppressor genes such as *BRG1* and *BRM* of the complex are

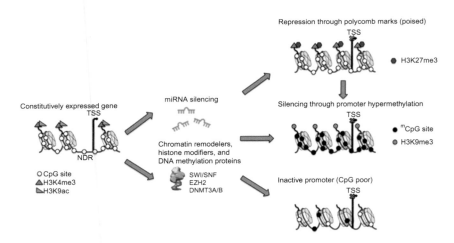

FIGURE 16.1 **Genetic–epigenetic cross talk in tumorigenesis.** This figure highlights examples of aberrant epigenetic gene expression control through genetic modifications during tumorigenesis. A normally expressed gene in a non-malignant cell shows a nucleosome-depleted region (NDR) upstream of its transcription start site (TSS); the promoter region is unmethylated and carries active histone marks (H3K4me3 and H3K9ac) on the surrounding nucleosomes. During tumorigenesis, genes can be silenced by miRNAs, mutated chromatin remodeler proteins, or overexpressed histone modifiers, and proteins involved in DNA methylation. These aberrant epigenetic modifications are illustrated on the right of the figure: from top to bottom: promoter regions carrying both active (H3K4me3) and inactive marks (H3K27me3) are in a poised status that can undergo an epigenetic switch in cancer to permanently silence transcription. This is illustrated in the middle panel, where DNA is highly methylated and histones carry the inactive H3K9me3 mark. The bottom panel shows non-CpG island promoter silencing of a gene.

downregulated in CMM; however, a combined downregulation has not been observed to date [87]. Also, BRG1 binds P16 and thus it is believed that the SWI/SNF complex acts through RB1 and E2F to regulate the cell cycle [88]. In melanoma cells, this complex was also found upstream of *MITF*, increasing the latter's expression, and at the same time *MITF* recruits the complex to melanocyte-specific genes such as *TYRP1*, where chromatin will be remodeled to allow activation of these genes [89,90]. Another component of the SWI/SNF complex that needs to be downregulated in CMM is *SMARCB1*, in order to bypass apoptosis induced by $BRAF^{V600E}$ [91].

16.3.2.4 Polycomb Group Proteins

Polycomb group proteins (PcGs) are involved in gene repression through recognition of histone methylation and ubiquitination. They form the polycomb repressive complexes (PRCs) 1 and 2 that are involved in carcinogenesis. The catalytic subunit of PRC2, EZH2, is gradually upregulated from benign nevi to malignant CMM [92]. EZH2 also interacts with $BRAF^{V600E}$ to induce hypermethylation at target genes, and CDKN1A repression by EZH2 impairs cell senescence [93]. Regarding PRC1, the main acting protein of this complex is BMI1, which suppresses the *INK4B-ARF-INK4A* locus. In CMM, BMI1 expression is lost, which leads to aggressive CMM subgroups and reduced patient survival rates [94]. Neutralizing or counteracting proteins of the PcG proteins are JMJD3 and KDM5B. The first is a H3K27 demethylase and acts in response to RAS signaling, while KDM5B demethylates H3K4 sites and associates with RB1 to control the cell cycle. Both histone demethylases are highly expressed in nonmalignant nevi, and their expression declines with the formation of CMM [95,96].

16.3.2.5 Noncoding RNAs (miRNAs)

Noncoding RNAs or microRNAs (miRNAs) are small sequences of about 22 nucleotides in the human genome that can interact with mRNA, by either causing degradation of the target RNA or inhibition of the translational process [97]. Thousands of miRNAs have been detected in the genome and they play an important role in cancer development by modulating the expression level of an even greater number of target genes (Table 16.3). First, miRNAs are transcribed by RNA polymerase II, and they are then processed by DROSHA, a member of the nuclear RNase III family before they are transported to the cytoplasm by RanGTP/EXPORTIN5 and reduced to their actual size by Dicer. The remaining duplex RNA sequence will contain the one strand that will be the mature miRNA [99].

The enzyme cleaving miRNAs, Dicer, is found upregulated in CMM, indicating a major involvement of miRNAs in this disease [100].

TABLE 16.3   Misregulated miRNAs in CMM

miRNA name		Regulation
Let-7a/b	miR-192-194	
miR-29c	miR-193b	
miR-30b	miR-196a	
miR-31	miR-199a	
miR-34a/b/c	miR-200a/b/c	
miR-125b	miR-203 to 205	Downregulated
miR-148a	miR-211	
miR-149	miR-214	
miR-155	miR-375	
miR-182		
miR-184 to 185		
Let-7i	miR-193b	
miR-9	miR-200a/b/c	
miR-15b	miR-211	
miR-17-5p	miR-214	
miR-20b	miR-221 to 222	Upregulated
miR-21	miR-365	
miR-30b/d	miR-506 to 514	
miR-137		
miR-141		
miR-155		
miR-182		

Adapted from Ref. [98].

In general, upregulated miRNAs silence tumor-suppressor genes, while downregulated ones could induce activation of oncogenes in cancer [101]. In CMM, a large number of the affected target genes are involved in cell cycle control. *CDK4/6, CCND1, CDKN1A/B,* and *MET* but also *MITF* and genes involved in epithelial−mesenchymal transition (EMT) are frequent targets. MITF, the master regulator in CMM development, is regulated by miR-137, -148, and -182, which are all aberrantly transcribed in CMM [102−104]. Inversely, MITF can also regulate certain miRNAs, such as members of the let-7 family that are downregulated in primary melanomas but not in benign tissue. It has been shown that inhibition of let-7a leads to an invasive phenotype of melanocytes and the overexpression of let-7b leads to reduced expression of cyclins and CDK4, which in turn are involved in CMM development as well [105,106].

Certain miRNA expression levels are associated with genetic mutations such as *BRAF^{V600E}*. Patients with a *BRAF* mutation have low levels of miR-193a, -338, and -565 and *NRAS* mutation carriers display low expression of miR-663 [107]. Other miRNAs such as miR-29c can regulate DNMTs and thus change the DNA methylation pattern in CMM.

miR-29c is anticorrelated with DNMT3A/B and CIMP [52]. On the other hand, miRNAs that are considered to have tumor-suppressor functions can be silenced through promoter DNA hypermethylation in CMM [108]. As mentioned before, miRNAs play an important role in EMT; in particular, the miR-200 family is downregulated in CMM, whose members directly target *ZEB1/2*, which are regulators of the repressor *CDH1*. Thus, this family of miRNAs are key initiators of CMM metastasis [109]. See Figure 16.1 for a summary of genetic and epigenetic modification cross-talk in tumorigenesis.

16.4 COMMON TREATMENTS AND EPIGENETIC THERAPY

The management of malignant melanoma consists of a clear excision of the tumor followed by a sentinel lymph node biopsy, which is the first node to pick up ^{99m}Tc among the other nodes that are closest to the area of excision [110]. In the case of in-transit melanomas, which constitute metastasis that appears in between the primary excision site (>2 cm away) and the draining lymph nodes, treatment involves CO_2 laser ablation, hypothermic limb perfusion with melphalan, and intralesional BCG [111]. When adjuvant therapy is required owing to the presence of positive lymph nodes or melanomas thicker than 4 mm, commonly treatment consists of administration of FDA-approved interferon-α 2b [112].

16.4.1 Genetic Drug Treatment in CMM

During the last few years, the standard chemotherapy in metastatic melanoma has been geared toward a more targeted or even immunotherapy. New drugs have been successfully developed to counteract important genetic mutations in CMM. Since 60—70% of melanoma patients have mutations in *CKIT, BRAF*, or *NRAS*, inhibitors have been developed. Three drugs have been approved so far for the treatment of activating *BRAF* mutations: the BRAF inhibitors debrafenib and vemurafenib, and the MEK inhibitor trametinib [113]. Although these new treatments seemed very promising, a large number of patients became resistant to the drugs or the response was only transient. Also, other patients showed the emergence of additional mutations. This led to support for currently in-progress clinical trials, where in patients are treated with a combination of different inhibitors [114]. Small molecules and nucleoside analogs have been designed to, for example, inhibit hTERT, the reverse transcriptase of telomerase [113].

16.4.2 Epigenetic Drug Treatment in CMM

16.4.2.1 *Targeting DNA Methylation*

In tandem with the advances made in epigenetic drug design, DNA methylation changes in cancer have become a popular target for epigenetic drug treatment since these changes, unlike genetic mutations, are reversible. The drugs employed are also being considered in terms of usage as sensitizers in combination treatments with chemotherapy or even immunotherapy.

The most popular DNMT inhibitors are 5-azacytidine (azacytidine) and 5-aza-2′-deoxycytidine (decitabine), which have been approved by the FDA for treatment of myelodysplastic syndrome (Table 16.4). Both interfere with DNMTs and indirectly induce loss of DNA methylation in actively dividing cells. This in turn leads to reactivation of genes that were silenced through promoter DNA hypermethylation to induce cell differentiation or lead to apoptosis [115]. Clinical trials have shown that the efficiency of epigenetic drugs can be increased even more through combination with chemotherapeutic drugs, interferons, or tumor vaccines [116]. Decitabine was shown to be safely combinable with carboplatin-based drugs and interleukin-2. The latter given at high doses with decitabine induced objective responses in 31% of CMM patients [117]. The caveat with those treatments, however, lies in the fact that not only will tumor suppressor genes (TSGs) be reactivated with epigenetic drugs but there is a constant risk of reexpression of tumor-promoting genes. This can only be controlled with the right patient and dosing schedule selections [45].

These frequently hyper- or hypomethylated genes could serve as potential biomarkers in CMM, to help in early diagnosis of the disease or in tumor stratification and treatment, as well as dose selection; and even the treatment response could be monitored by such a panel of genes. Via quantitative methylation-specific PCR (qMSP), certain markers have been detected in different types of cancer, such as *GSTP1* in

TABLE 16.4 Epigenetic Drugs Used in Cutaneous Malignant Melanoma

Chemical structure	Common names	Inhibitor function
5-Azacytidine	Azacytidine	DNA methylation
5-Aza-2′-deoxycytidine	Decitabine	
Short-chain fatty acids	Butyrate and valproic acid (VPA)	
Hydroxamic acids	Trichostatin A (TSA), SAHA, vorinostat and Zolinza™	HDAC inhibition
Benzamides	MS-275 and CI-994	
Cyclic tetrapeptides	Depsipeptide and trapoxin	

Adapted from Ref. [45].

prostate cancer or circulating methylated *RASSF1A* in biochemotherapy responders [73,118]. Specifically in CMM, serum methylated estrogen receptor α serves as an unfavorable prognostic marker that is an indicator of melanoma progression. Other genes have been detected in similar fashion, for which their DNA methylation status informs about the disease progression. These are *RARβ2* and *DCR1*, with both having a methylation frequency of 60% as well as *SOCS1* and *SOCS2* and *DCR2* with methylation frequencies of 90, 80, and 85%, respectively [59]. Since these genes and no other commonly methylated genes in different malignancies are methylated in melanoma, this panel of genes can be considered as true biomarkers for CMM, and one could consider a more targeted treatment approach in order to specifically reactivate these genes.

16.4.2.2 *Targeting Histone Modifications*

Similarly to aberrant DNA methylation in cancer, histone modifications are not permanent and can be reversed. Thus with the same idea, epigenetic drugs against histone modifications, namely, HDACis, were designed to increase tumor suppressor or DNA repair gene expression and lower oncogene expression rates [119]. Since most suppressed genes in cancer carry both layers of silencing, DNA methylation and inactivating histone modifications, such as histone deacetylation, HDACis are currently being extensively studied for their efficiency in melanoma. These drugs are very promising since they act on proteins that are involved in cell death, gene expression, differentiation, and proliferation [85,120].

The designed HDACis used in treatment for cancer patients can be classified into four main categories. The class I and II HDACi are hydroxamic acids, short-chain fatty acids, cyclic tetrapeptides, or benzamides. Class III inhibitors are counteracting sirtuins [121].

In melanoma, most apoptosis with HDACis is induced via the mitochondrial pathway, where normal cells are spared [122]. Although most HDACi-induced apoptosis is independent of the P53 pathway, it was shown that in melanoma cell lines, Na butyrate can induce apoptosis through P53 BAX induction via HDAC1 upregulation [123]. Synergistic effects have been demonstrated on melanoma cells with the combination of hydroxamate and *cis*-retinoic acid, which is a natural and synthetic analog of vitamin A involved in epithelial cell differentiation and growth. The mechanism of action of the combinatorial treatment functions through upregulation of the *RARβ2*, thereby enhancing cancer cell sensitivity to retinoic acid [124].

Other drug combinations revealed to have synergistic effects in clinical trials are xanthine and suramin, which have anti-tumor activity with downregulation of matrix-metalloproteinase 9. Suramin also has inhibitory effects on the proangiogenic factors VEGF and PDGF. Despite its antiproliferative activity, Suramin has a broad range of toxicity and is thus used in combination treatment to chemosensitize cancer cells [125]. Another very important HDACi is VPA. It has been shown to cause dose-dependent growth arrest on melanoma cells, and it sensitizes them to other chemotherapeutic drugs such as cisplatin and etoposide. Largazole is a natural depsipeptide that has an inhibitory effect on HDAC1, 2, and 3 at the picomolar level in melanoma cell lines and thus makes this drug a very interesting candidate, since it is very potent at low dose with, thus, few side effects [126].

16.5 CONCLUSION

In order to completely understand the onset and progression of melanoma, the underlying molecular mechanisms need to be studied in more detail. The genetic alterations in melanoma are better known than the epigenetic ones and their combinatorial effects need to be elucidated in greater depth. One example of this interaction is that of the transcription factor MITF and the *INK4B-ARF-INK4A* locus. Better integration of the techniques specifically designed to investigate genetic mutations, such as exome sequencing and detection of promoter DNA methylation aberrations with DNA methylation arrays, will enable discovery of more melanoma biomarkers or even novel treatment targets. The combination of epigenetic drugs such as decitabine or different HDACi in combination with standard chemotherapeutic drugs might advance the treatment strategies employed for melanoma patients.

The biggest difficulty in melanoma for detecting epigenetic biomarkers is the lack of tumor and, even more often, normal tissue material. Thus, potential biomarkers need to be evaluated in large cohorts with regard to their predictive and prognostic effectiveness.

List of Abbreviations

ALM	acral lentiginous melanoma
BCC	basal cell carcinoma
CGI	CpG island
CIMP	CpG methylator phenotype
CMM	cutaneous malignant melanoma
DNMT	DNA methyltransferase
HAT	histone acetyltransferase
HDAC	histone deacetyltransferase
HDACis	histone deacetylase inhibitors
LLM	lentigo maligna melanoma
miRNA	microRNA
NM	nodular melanoma
PcG	polycomb group protein
PRC	polycomb repressive complex

qMSP	quantitative methylation-specific PCR
RGP	radial growth phase
SCC	squamous cell carcinoma
SSM	superficial spreading melanoma
TSG	tumor suppressor gene
VGP	vertical growth phase

References

[1] Weedon D. Weedon's skin pathology. 3rd ed. London, UK: Churchill Livingstone Elsevier; 2004.

[2] Lens MB, Dawes M. Global perspectives of contemporary epidemiological trends of cutaneous malignant melanoma. Br J Dermatol 2004;150(2):179−85.

[3] Balch CM, Buzaid AC, Atkins MB, et al. A new American Joint Committee on Cancer staging system for cutaneous melanoma. Cancer 2000;88(6):1484−91.

[4] Zell JA, Cinar P, Mobasher M, Ziogas A, Meyskens Jr. FL, Anton-Culver H. Survival for patients with invasive cutaneous melanoma among ethnic groups: the effects of socioeconomic status and treatment. J Clin Oncol 2008;26(1):66−75.

[5] Katsambas A, Nicolaidou E. Cutaneous malignant melanoma and sun exposure. Recent developments in epidemiology. Arch Dermatol 1996;132(4):444−50.

[6] Goldstein AM, Tucker MA. Genetic epidemiology of cutaneous melanoma: a global perspective. Arch Dermatol 2001;137(11):1493−6.

[7] Naldi L, Altieri A, Imberti GL, et al. Cutaneous malignant melanoma in women. Phenotypic characteristics, sun exposure, and hormonal factors: a case-control study from Italy. Ann Epidemiol 2005;15(7):545−50.

[8] Bataille V, Bishop JA, Sasieni P, et al. Risk of cutaneous melanoma in relation to the numbers, types and sites of naevi: a case-control study. Br J Cancer 1996;73 (12):1605−11.

[9] Holly EA, Aston DA, Cress RD, Ahn DK, Kristiansen JJ. Cutaneous melanoma in women. I. Exposure to sunlight, ability to tan, and other risk factors related to ultraviolet light. Am J Epidemiol 1995;141(10):923−33.

[10] Hermanns JF, Petit L, Martalo O, Pierard-Franchimont C, Cauwenbergh G, Pierard GE. Unraveling the patterns of subclinical pheomelanin-enriched facial hyperpigmentation: effect of depigmenting agents. Dermatology 2000;201(2):118−22.

[11] Busam KJ, Sung J, Wiesner T, von Deimling A, Jungbluth A. Combined BRAF (V600E)-positive melanocytic lesions with large epithelioid cells lacking BAP1 expression and conventional nevomelanocytes. Am J Surg Pathol 2013;37(2):193−9.

[12] Clark Jr. WH, From L, Bernardino EA, Mihm MC. The histogenesis and biologic behavior of primary human malignant melanomas of the skin. Cancer Res 1969;29 (3):705−27.

[13] Guerry D, Synnestvedt M, Elder DE, Schultz D. Lessons from tumor progression: the invasive radial growth phase of melanoma is common, incapable of metastasis, and indolent. J Invest Dermatol 1993;100(3):342S−5S.

[14] Demitsu T, Nagato H, Nishimaki K, et al. Melanoma *in situ* of the penis. J Am Acad Dermatol 2000;42(2 Pt 2):386−8.

[15] Kiene P, Petres-Dunsche C, Folster-Holst R. Pigmented pedunculated malignant melanoma. A rare variant of nodular melanoma. Br J Dermatol 1995;133(2):300−2.

[16] Cohen LM. Lentigo maligna and lentigo maligna melanoma. J Am Acad Dermatol 1995;33(6):923−36, quiz 37−40.

[17] Chen YJ, Wu CY, Chen JT, Shen JL, Chen CC, Wang HC. Clinicopathologic analysis of malignant melanoma in Taiwan. J Am Acad Dermatol 1999;41(6):945−9.

[18] Reed RJ, Webb SV, Clark Jr. WH. Minimal deviation melanoma (halo nevus variant). Am J Surg Pathol 1990;14(1):53—68.

[19] Whitaker DC, Argenyi Z, Smith AC. Desmoplastic malignant melanoma: rare and difficult to diagnose. J Am Acad Dermatol 1992;26(5 Pt 1):704—9.

[20] Berger MF, Hodis E, Heffernan TP, et al. Melanoma genome sequencing reveals frequent PREX2 mutations. Nature 2012;485(7399):502—6.

[21] Pleasance ED, Cheetham RK, Stephens PJ, et al. A comprehensive catalogue of somatic mutations from a human cancer genome. Nature 2010;463(7278):191—6.

[22] Krzywinski M, Schein J, Birol I, et al. Circos: an information aesthetic for comparative genomics. Genome Res 2009;19(9):1639—45.

[23] Davies H, Bignell GR, Cox C, et al. Mutations of the BRAF gene in human cancer. Nature 2002;417(6892):949—54.

[24] Pollock PM, Harper UL, Hansen KS, et al. High frequency of BRAF mutations in nevi. Nat Genet 2003;33(1):19—20.

[25] Cohen C, Zavala-Pompa A, Sequeira JH, et al. Mitogen-activated protein kinase activation is an early event in melanoma progression. Clin Cancer Res 2002;8 (12):3728—33.

[26] Maldonado JL, Fridlyand J, Patel H, et al. Determinants of BRAF mutations in primary melanomas. J Natl Cancer Inst 2003;95(24):1878—90.

[27] Kinsler VA, Thomas AC, Ishida M, et al. Multiple congenital melanocytic nevi and neurocutaneous melanosis are caused by postzygotic mutations in codon 61 of NRAS. J Invest Dermatol 2013;133(9):2229—36.

[28] Bastian BC, Olshen AB, LeBoit PE, Pinkel D. Classifying melanocytic tumors based on DNA copy number changes. Am J Pathol 2003;163(5):1765—70.

[29] Cowan JM, Halaban R, Francke U. Cytogenetic analysis of melanocytes from premalignant nevi and melanomas. J Natl Cancer Inst 1988;80(14):1159—64.

[30] Beadling C, Jacobson-Dunlop E, Hodi FS, et al. KIT gene mutations and copy number in melanoma subtypes. Clin Cancer Res 2008;14(21):6821—8.

[31] Banerji U, Affolter A, Judson I, Marais R, Workman P. BRAF and NRAS mutations in melanoma: potential relationships to clinical response to HSP90 inhibitors. Mol Cancer Ther 2008;7(4):737—9.

[32] Goldstein AM, Chan M, Harland M, et al. Features associated with germline CDKN2A mutations: a GenoMEL study of melanoma-prone families from three continents. J Med Genet 2007;44(2):99—106.

[33] Zuo L, Weger J, Yang Q, et al. Germline mutations in the p16INK4a binding domain of CDK4 in familial melanoma. Nat Genet 1996;12(1):97—9.

[34] Wiesner T, Obenauf AC, Murali R, et al. Germline mutations in BAP1 predispose to melanocytic tumors. Nat Genet 2011;43(10):1018—21.

[35] Serrano M, Gomez-Lahoz E, DePinho RA, Beach D, Bar-Sagi D. Inhibition of ras-induced proliferation and cellular transformation by p16INK4. Science 1995;267 (5195):249—52.

[36] Zhang Y, Xiong Y, Yarbrough WG. ARF promotes MDM2 degradation and stabilizes p53: ARF-INK4a locus deletion impairs both the Rb and p53 tumor suppression pathways. Cell 1998;92(6):725—34.

[37] Goldstein AM, Chan M, Harland M, et al. High-risk melanoma susceptibility genes and pancreatic cancer, neural system tumors, and uveal melanoma across GenoMEL. Cancer Res 2006;66(20):9818—28.

[38] Begg CB, Orlow I, Hummer AJ, et al. Lifetime risk of melanoma in CDKN2A mutation carriers in a population-based sample. J Natl Cancer Inst 2005;97(20):1507—15.

[39] Njauw CN, Kim I, Piris A, et al. Germline BAP1 inactivation is preferentially associated with metastatic ocular melanoma and cutaneous-ocular melanoma families. PLOS ONE 2012;7(4):e35295.

[40] Busca R, Ballotti R. Cyclic AMP a key messenger in the regulation of skin pigmentation. Pigment Cell Res 2000;13(2):60—9.

[41] Yokoyama S, Woods SL, Boyle GM, et al. A novel recurrent mutation in MITF predisposes to familial and sporadic melanoma. Nature 2011;480(7375):99—103.

[42] Nishigaki M, Aoyagi K, Danjoh I, et al. Discovery of aberrant expression of R-RAS by cancer-linked DNA hypomethylation in gastric cancer using microarrays. Cancer Res 2005;65(6):2115—24.

[43] Rauch TA, Zhong X, Wu X, et al. High-resolution mapping of DNA hypermethylation and hypomethylation in lung cancer. Proc Natl Acad Sci USA 2008;105(1):252—7.

[44] Sigalotti L, Coral S, Nardi G, et al. Promoter methylation controls the expression of MAGE2, 3 and 4 genes in human cutaneous melanoma. J Immunother 2002;25 (1):16—26.

[45] Howell Jr. PM, Liu S, Ren S, Behlen C, Fodstad O, Riker AI. Epigenetics in human melanoma. Cancer Control 2009;16(3):200—18.

[46] Luo W, Wang X, Kageshita T, Wakasugi S, Karpf AR, Ferrone S. Regulation of high molecular weight-melanoma associated antigen (HMW-MAA) gene expression by promoter DNA methylation in human melanoma cells. Oncogene 2006;25(20):2873—84.

[47] James SR, Link PA, Karpf AR. Epigenetic regulation of X-linked cancer/germline antigen genes by DNMT1 and DNMT3b. Oncogene 2006;25(52):6975—85.

[48] Wada K, Maesawa C, Akasaka T, Masuda T. Aberrant expression of the maspin gene associated with epigenetic modification in melanoma cells. J Invest Dermatol 2004;122(3):805—11.

[49] Schnieders F, Dork T, Arnemann J, Vogel T, Werner M, Schmidtke J. Testis-specific protein, Y-encoded (TSPY) expression in testicular tissues. Hum Mol Genet 1996;5 (11):1801—7.

[50] Simpson AJ, Caballero OL, Jungbluth A, Chen YT, Old LJ. Cancer/testis antigens, gametogenesis and cancer. Nat Rev Cancer 2005;5(8):615—25.

[51] Kholmanskikh O, Loriot A, Brasseur F, De Plaen E, De Smet C. Expression of BORIS in melanoma: lack of association with MAGE-A1 activation. Int J Cancer 2008;122 (4):777—84.

[52] Nguyen T, Kuo C, Nicholl MB, et al. Downregulation of microRNA-29c is associated with hypermethylation of tumor-related genes and disease outcome in cutaneous melanoma. Epigenetics 2011;6(3):388—94.

[53] Deng T, Kuang Y, Wang L, Li J, Wang Z, Fei J. An essential role for DNA methyltransferase 3a in melanoma tumorigenesis. Biochem Biophys Res Commun 2009;387 (3):611—16.

[54] Hou P, Liu D, Dong J, Xing M. The BRAF(V600E) causes widespread alterations in gene methylation in the genome of melanoma cells. Cell Cycle 2012;11(2):286—95.

[55] Worm J, Christensen C, Gronbaek K, Tulchinsky E, Guldberg P. Genetic and epigenetic alterations of the APC gene in malignant melanoma. Oncogene 2004;23(30):5215—26.

[56] Guan X, Sagara J, Yokoyama T, et al. ASC/TMS1, a caspase-1 activating adaptor, is downregulated by aberrant methylation in human melanoma. Int J Cancer 2003;107 (2):202—8.

[57] Muthusamy V, Duraisamy S, Bradbury CM, et al. Epigenetic silencing of novel tumor suppressors in malignant melanoma. Cancer Res 2006;66(23):11187—93.

[58] Shen L, Kondo Y, Guo Y, et al. Genome-wide profiling of DNA methylation reveals a class of normally methylated CpG island promoters. PLOS Genet 2007;3(10):2023—36.

[59] Liu S, Ren S, Howell P, Fodstad O, Riker AI. Identification of novel epigenetically modified genes in human melanoma via promoter methylation gene profiling. Pigment Cell Melanoma Res 2008;21(5):545—58.

[60] Gonzalgo ML, Jones PA. Mutagenic and epigenetic effects of DNA methylation. Mutat Res 1997;386(2):107—18.

[61] Hoon DS, Spugnardi M, Kuo C, Huang SK, Morton DL, Taback B. Profiling epigenetic inactivation of tumor suppressor genes in tumors and plasma from cutaneous melanoma patients. Oncogene 2004;23(22):4014−22.

[62] McGuinness C, Wesley UV. Dipeptidyl peptidase IV (DPPIV), a candidate tumor suppressor gene in melanomas is silenced by promoter methylation. Front Biosci 2008;13:2435−43.

[63] Mori T, Martinez SR, O'Day SJ, et al. Estrogen receptor-alpha methylation predicts melanoma progression. Cancer Res 2006;66(13):6692−8.

[64] Sharma BK, Smith CC, Laing JM, Rucker DA, Burnett JW, Aurelian L. Aberrant DNA methylation silences the novel heat shock protein H11 in melanoma but not benign melanocytic lesions. Dermatology 2006;213(3):192−9.

[65] Takeuchi T, Adachi Y, Sonobe H, Furihata M, Ohtsuki Y. A ubiquitin ligase, skeletrophin, is a negative regulator of melanoma invasion. Oncogene 2006;25(53):7059−69.

[66] Furuta J, Nobeyama Y, Umebayashi Y, Otsuka F, Kikuchi K, Ushijima T. Silencing of Peroxiredoxin 2 and aberrant methylation of 33 CpG islands in putative promoter regions in human malignant melanomas. Cancer Res 2006;66(12):6080−6.

[67] Mirmohammadsadegh A, Marini A, Nambiar S, et al. Epigenetic silencing of the PTEN gene in melanoma. Cancer Res 2006;66(13):6546−52.

[68] Spugnardi M, Tommasi S, Dammann R, Pfeifer GP, Hoon DS. Epigenetic inactivation of RAS association domain family protein 1 (RASSF1A) in malignant cutaneous melanoma. Cancer Res 2003;63(7):1639−43.

[69] Marini A, Mirmohammadsadegh A, Nambiar S, Gustrau A, Ruzicka T, Hengge UR. Epigenetic inactivation of tumor suppressor genes in serum of patients with cutaneous melanoma. J Invest Dermatol 2006;126(2):422−31.

[70] Tokita T, Maesawa C, Kimura T, et al. Methylation status of the SOCS3 gene in human malignant melanomas. Int J Oncol 2007;30(3):689−94.

[71] Nobeyama Y, Okochi-Takada E, Furuta J, et al. Silencing of tissue factor pathway inhibitor-2 gene in malignant melanomas. Int J Cancer 2007;121(2):301−7.

[72] Furuta J, Umebayashi Y, Miyamoto K, et al. Promoter methylation profiling of 30 genes in human malignant melanoma. Cancer Sci 2004;95(12):962−8.

[73] Mori T, O'Day SJ, Umetani N, et al. Predictive utility of circulating methylated DNA in serum of melanoma patients receiving biochemotherapy. J Clin Oncol 2005;23 (36):9351−8.

[74] Hughes LA, Khalid-de Bakker CA, Smits KM, et al. The CpG island methylator phenotype in colorectal cancer: progress and problems. Biochim Biophys Acta 2012;1825 (1):77−85.

[75] Gallagher WM, Bergin OE, Rafferty M, et al. Multiple markers for melanoma progression regulated by DNA methylation: insights from transcriptomic studies. Carcinogenesis 2005;26(11):1856−67.

[76] van Vlodrop IJ, Niessen HE, Derks S, et al. Analysis of promoter CpG island hypermethylation in cancer: location, location, location! Clin Cancer Res 2011;17 (13):4225−31.

[77] Vaissiere T, Sawan C, Herceg Z. Epigenetic interplay between histone modifications and DNA methylation in gene silencing. Mutat Res 2008;659(1−2):40−8.

[78] Verdone L, Agricola E, Caserta M, Di Mauro E. Histone acetylation in gene regulation. Brief Funct Genomic Proteomic 2006;5(3):209−21.

[79] Yang XJ, Seto E. HATs and HDACs: from structure, function and regulation to novel strategies for therapy and prevention. Oncogene 2007;26(37):5310−18.

[80] Katoh M. Network of WNT and other regulatory signaling cascades in pluripotent stem cells and cancer stem cells. Curr Pharm Biotechnol 2011;12(2):160−70.

[81] Shtutman M, Zhurinsky J, Simcha I, et al. The cyclin D1 gene is a target of the beta-catenin/LEF-1 pathway. Proc Natl Acad Sci USA 1999;96(10):5522−7.

[82] Widlund HR, Horstmann MA, Price ER, et al. Beta-catenin-induced melanoma growth requires the downstream target microphthalmia-associated transcription factor. J Cell Biol 2002;158(6):1079−87.

[83] Dynek JN, Chan SM, Liu J, Zha J, Fairbrother WJ, Vucic D. Microphthalmia-associated transcription factor is a critical transcriptional regulator of melanoma inhibitor of apoptosis in melanomas. Cancer Res 2008;68(9):3124−32.

[84] Ugurel S, Houben R, Schrama D, et al. Microphthalmia-associated transcription factor gene amplification in metastatic melanoma is a prognostic marker for patient survival, but not a predictive marker for chemosensitivity and chemotherapy response. Clin Cancer Res 2007;13(21):6344−50.

[85] Minucci S, Pelicci PG. Histone deacetylase inhibitors and the promise of epigenetic (and more) treatments for cancer. Nat Rev Cancer 2006;6(1):38−51.

[86] Gao L, Cueto MA, Asselbergs F, Atadja P. Cloning and functional characterization of HDAC11, a novel member of the human histone deacetylase family. J Biol Chem 2002;277(28):25748−55.

[87] Keenen B, Qi H, Saladi SV, Yeung M, de la Serna IL. Heterogeneous SWI/SNF chromatin remodeling complexes promote expression of microphthalmia-associated transcription factor target genes in melanoma. Oncogene 2010;29(1):81−92.

[88] Dunaief JL, Strober BE, Guha S, et al. The retinoblastoma protein and BRG1 form a complex and cooperate to induce cell cycle arrest. Cell 1994;79(1):119−30.

[89] de la Serna IL, Ohkawa Y, Higashi C, et al. The microphthalmia-associated transcription factor requires SWI/SNF enzymes to activate melanocyte-specific genes. J Biol Chem 2006;281(29):20233−41.

[90] Vachtenheim J, Ondrusova L, Borovansky J. SWI/SNF chromatin remodeling complex is critical for the expression of microphthalmia-associated transcription factor in melanoma cells. Biochem Biophys Res Commun 2010;392(3):454−9.

[91] Wajapeyee N, Serra RW, Zhu X, Mahalingam M, Green MR. Oncogenic BRAF induces senescence and apoptosis through pathways mediated by the secreted protein IGFBP7. Cell 2008;132(3):363−74.

[92] McHugh JB, Fullen DR, Ma L, Kleer CG, Su LD. Expression of polycomb group protein EZH2 in nevi and melanoma. J Cutan Pathol 2007;34(8):597−600.

[93] Fan T, Jiang S, Chung N, et al. EZH2-dependent suppression of a cellular senescence phenotype in melanoma cells by inhibition of p21/CDKN1A expression. Mol Cancer Res 2011;9(4):418−29.

[94] Bachmann IM, Puntervoll HE, Otte AP, Akslen LA. Loss of BMI-1 expression is associated with clinical progress of malignant melanoma. Mod Pathol 2008;21 (5):583−90.

[95] Barradas M, Anderton E, Acosta JC, et al. Histone demethylase JMJD3 contributes to epigenetic control of INK4a/ARF by oncogenic RAS. Genes Dev 2009;23(10):1177−82.

[96] Roesch A, Becker B, Meyer S, et al. Overexpression and hyperphosphorylation of retinoblastoma protein in the progression of malignant melanoma. Mod Pathol 2005;18(4):565−72.

[97] Ambros V. The functions of animal microRNAs. Nature 2004;431(7006):350−5.

[98] van den Hurk K, Niessen HE, Veeck J, et al. Genetics and epigenetics of cutaneous malignant melanoma: a concert out of tune. Biochim Biophys Acta 2012;1826 (1):89−102.

[99] Kosik KS, Krichevsky AM. The elegance of the microRNAs: a neuronal perspective. Neuron 2005;47(6):779−82.

[100] Murakami T, Sato A, Chun NA, et al. Transcriptional modulation using HDACi depsipeptide promotes immune cell-mediated tumor destruction of murine B16 melanoma. J Invest Dermatol 2008;128(6):1506−16.

[101] Croce CM, Calin GA. miRNAs, cancer, and stem cell division. Cell 2005;122(1):6−7.

[102] Bemis LT, Chen R, Amato CM, et al. MicroRNA-137 targets microphthalmia-associated transcription factor in melanoma cell lines. Cancer Res 2008;68(5):1362–8.

[103] Haflidadottir BS, Bergsteinsdottir K, Praetorius C, Steingrimsson E. miR-148 regulates Mitf in melanoma cells. PLOS ONE 2010;5(7):e11574.

[104] Segura MF, Hanniford D, Menendez S, et al. Aberrant miR-182 expression promotes melanoma metastasis by repressing FOXO3 and microphthalmia-associated transcription factor. Proc Natl Acad Sci USA 2009;106(6):1814–19.

[105] Muller DW, Bosserhoff AK. Integrin beta 3 expression is regulated by let-7a miRNA in malignant melanoma. Oncogene 2008;27(52):6698–706.

[106] Schultz J, Lorenz P, Gross G, Ibrahim S, Kunz M. MicroRNA let-7b targets important cell cycle molecules in malignant melanoma cells and interferes with anchorage-independent growth. Cell Res 2008;18(5):549–57.

[107] Caramuta S, Egyhazi S, Rodolfo M, et al. MicroRNA expression profiles associated with mutational status and survival in malignant melanoma. J Invest Dermatol 2010;130(8):2062–70.

[108] Lujambio A, Calin GA, Villanueva A, et al. A microRNA DNA methylation signature for human cancer metastasis. Proc Natl Acad Sci USA 2008;105(36):13556–61.

[109] Bracken CP, Gregory PA, Khew-Goodall Y, Goodall GJ. The role of microRNAs in metastasis and epithelial–mesenchymal transition. Cell Mol Life Sci 2009;66 (10):1682–99.

[110] Walsh P, Gibbs P, Gonzalez R. Newer strategies for effective evaluation of primary melanoma and treatment of stage III and IV disease. J Am Acad Dermatol 2000;42 (3):480–9.

[111] Kandamany N, Mahaffey P. Carbon dioxide laser ablation as first-line management of in-transit cutaneous malignant melanoma metastases. Lasers Med Sci 2009;24 (3):411–14.

[112] Kefford RF. Adjuvant therapy of cutaneous melanoma: the interferon debate. Ann Oncol 2003;14(3):358–65.

[113] Schadendorf D, Hauschild A. Melanoma in 2013: melanoma—the run of success continues. Nat Rev Clin Oncol 2014;11(2):75–6.

[114] Trunzer K, Pavlick AC, Schuchter L, et al. Pharmacodynamic effects and mechanisms of resistance to vemurafenib in patients with metastatic melanoma. J Clin Oncol 2013;31(14):1767–74.

[115] Reu FJ, Bae SI, Cherkassky L, et al. Overcoming resistance to interferon-induced apoptosis of renal carcinoma and melanoma cells by DNA demethylation. J Clin Oncol 2006;24(23):3771–9.

[116] Appleton K, Mackay HJ, Judson I, et al. Phase I and pharmacodynamic trial of the DNA methyltransferase inhibitor decitabine and carboplatin in solid tumors. J Clin Oncol 2007;25(29):4603–9.

[117] Gollob JA, Sciambi CJ, Peterson BL, et al. Phase I trial of sequential low-dose 5-aza-2'-deoxycytidine plus high-dose intravenous bolus interleukin-2 in patients with melanoma or renal cell carcinoma. Clin Cancer Res 2006;12(15):4619–27.

[118] Cairns P, Esteller M, Herman JG, et al. Molecular detection of prostate cancer in urine by GSTP1 hypermethylation. Clin Cancer Res 2001;7(9):2727–30.

[119] Riker AI, Enkemann SA, Fodstad O, et al. The gene expression profiles of primary and metastatic melanoma yields a transition point of tumor progression and metastasis. BMC Med Genomics 2008;1:13.

[120] Xu WS, Parmigiani RB, Marks PA. Histone deacetylase inhibitors: molecular mechanisms of action. Oncogene 2007;26(37):5541–52.

[121] Itoh Y, Suzuki T, Miyata N. Isoform-selective histone deacetylase inhibitors. Curr Pharm Des 2008;14(6):529–44.

[122] Boyle GM, Martyn AC, Parsons PG. Histone deacetylase inhibitors and malignant melanoma. Pigment Cell Res 2005;18(3):160–6.

[123] Bandyopadhyay D, Mishra A, Medrano EE. Overexpression of histone deacetylase 1 confers resistance to sodium butyrate-mediated apoptosis in melanoma cells through a p53-mediated pathway. Cancer Res 2004;64(21):7706–10.

[124] Kato Y, Salumbides BC, Wang XF, et al. Antitumor effect of the histone deacetylase inhibitor LAQ824 in combination with 13-cis-retinoic acid in human malignant melanoma. Mol Cancer Ther 2007;6(1):70–81.

[125] Trapp J, Meier R, Hongwiset D, Kassack MU, Sippl W, Jung M. Structure-activity studies on suramin analogues as inhibitors of NAD+-dependent histone deacetylases (sirtuins). ChemMedChem 2007;2(10):1419–31.

[126] Bowers A, West N, Taunton J, Schreiber SL, Bradner JE, Williams RM. Total synthesis and biological mode of action of largazole: a potent class I histone deacetylase inhibitor. J Am Chem Soc 2008;130(33):11219–22.

17

Cutaneous T-Cell Lymphoma: Mycosis Fungoides and Sézary Syndrome

Henry K. Wong[1], Li Wu[2], Suresh de Silva[2], Pierluigi Porcu[1,3], and Anjali Mishra[1]

[1]Comprehensive Cancer Center and Division of Dermatology, Department of Internal Medicine, The Ohio State University, Columbus, Ohio [2]Center for Retrovirology Research, Department of Veterinary Biosciences, The Ohio State University, Columbus, Ohio [3]Division of Hematology, Department of Internal Medicine, The Ohio State University, Columbus, Ohio

17.1 INTRODUCTION

Cutaneous T-cell lymphomas (CTCLs) are a heterogeneous group of T-cell malignancies that arise from T cells that are home to the skin, without evidence of systemic involvement [1]. The natural history and biologic behavior of CTCL differs among the subtypes, and these differences likely reflect innate properties of the original T cell from which each neoplasm arises. The different clinical and histologic variants of CTCL have been classified by the World Health Organization (WHO)/European Organization for the Treatment of Cancer (EORTC) [1,2]. The most common types of CTCL are mycosis fungoides (MF) and Sézary syndrome (SS), characterized by intermediate aggressiveness with good response to therapy [3]. A more aggressive course, often associated with "large cell transformation," can be observed and associate with CD30[+] expression [4]. Aggressive CTCL variants include extranodal natural killer T-cell lymphoma (ENKTL), primary cutaneous gamma−delta T-cell lymphoma (PCGD-TCL), and other rare types [5]. Their incidence is low and studies focused on them have been hampered by a lack of data leading to useful insight into the pathogenesis and response to treatment, and hence these aggressive variants will not be discussed in this chapter. This chapter will focus on the expression of abnormal genes in MF/SS and the role of epigenetic mechanisms in their expression.

17.2 MF/SS CLINICAL OVERVIEW

Mycosis fungoides generally has an indolent course, with slow incremental clinical progression that can span decades in duration [6]. Sézary syndrome is more aggressive, can present *de novo*, or arises from MF, and has a 5-year survival of less than 50% [3]. MF/SS most frequently represents proliferation of clonal mature $CD4^+$ $CD45RO^+$ effector memory T cells that initially reside in the skin, based on adhesion homing surface markers such as cutaneous leukocyte antigen (CLA) [7]. The visible nature of MF and the involvement of the skin permit longitudinal assessment by direct monitoring and sequential sampling for diagnosis and determination of disease stage and progression. An advantage of treating the skin is the direct accessibility of the malignancy to targeted treatments, such as topical pharmacotherapy or radiation to minimize systemic toxicity, and the ability to monitor clinical response.

The median age at presentation for MF/SS patients is 50–70 years [8]. The epidemiology has been described in many regions around the world and the incidence rate is between 0.3 and 0.6 per 100,000 [3,8–11]. Survival is correlated to the stage with early skin-restricted disease having a life span similar to that of the age-match control [3,12]. Staging is dependent on a thorough clinical examination to assess the degree of skin involvement and palpation for enlarged lymph nodes for evidence of systemic involvement, imaging studies to assess systemic disease when tumors or lymph nodes are appreciated on physical examination, and laboratory studies that may include flow cytometric analysis of peripheral blood [13]. The clinical staging of MF/SS is outlined in Table 17.1; early stage corresponds to stages IA to IIA, while advance stage corresponds to stages IIB to IV [13]. In the early stage of the disease, the neoplastic cells are restricted to the skin, and bone marrow biopsy or imaging studies are unnecessary. When there is extensive skin involvement with extensive plaques (>10% body surface area) and tumor, or lymphadenopathy, the disease stage is higher (stages IIB to IV), more aggressive, and associated with a poorer prognosis [14]. However, in most individuals, MF remains confined to the skin with slow progression over many years. A subset of MF may transform to a more aggressive variant with rapid progression in approximately 5% of the patients [3,15]. SS is a CTCL subtype associated with generalized erythroderma, lymphadenopathy, and peripheral blood involvement,

TABLE 17.1   ISCL/EORTC Revision of MF/SS Staging

Stage	T (Skin)	Node	M	Blood
IA	1 (<10%)	0	0	0, 1
IB	2 (>10%)	0	0	0, 1
IIA	1, 2	1, 2	0	0, 1
IIB	3 (tumors)	0–2	0	0, 1
IIIA	4 (>80%)	0–2	0	0
IIIB	4	0–2	0	1
IVA$_1$	1–4	0–2	0	2
IVA$_2$	1–4	3	0	0–2
IVB	1–4	0–3	1 (involved)	0–2

*Nodes: N1 = Dutch grade 1, N2 = Dutch grade 2, N3 = Dutch grades 3–4. Blood: B1 >5% peripheral blood. B2: >1000/μL Sézary cells.
Olsen et al. [13].

and generally has a more aggressive course. Fortunately, this subset is relatively uncommon, accounting for less than 5% of all CTCLs [3].

17.3 MOLECULAR IMMUNOPATHOLOGY OF MF/SS

The pathogenesis of CTCL remains unclear and, to date, no environmental, infectious, or iatrogenic risk factors have been identified [16,17]. The lack of an infectious basis is supported by whole transcriptome analysis, which did not detect any nonhuman transcripts or viral sequences in CD4$^+$ T cells isolated from SS patients [18].

Since the malignant cells in MF/SS express surface markers found in mature T cells with a memory phenotype (CD45RO), these malignant cells may possess memory functions, which is the ability to express high levels of cytokines [19]. However, when stimulated with phorbol esters, SS T cells poorly express cytokines, indicating that the neoplastic T cells have changed significantly, or originate from cells with low cytokine production [20]. The study by Chong et al. [21] showed that the neoplastic T cells were not activated like normal peripheral T cells. Indeed in SS T cells, multiple cytokine genes are minimally expressed, such as those for IL-2, IL-4, IL-5, IL-10, IL-13, IL-17, and IFN-γ, in comparison to normal memory T cells, and SS T cells show a greater resemblance to regulatory T cells (Tregs), which express only a few cytokines [20].

The original T cell evolving into MF/SS based on the type of T cells defined by cytokine expression remains unclear, and the recent classification of T cells into additional subsets does not facilitate clarification of the initial phenotype in MF/SS. CD4 T cells can be divided into different subtypes, Th1, Th2, Th17, Th22, and Treg, as defined by cytokine profile in response to appropriate stimulus by antigen presenting cell [22−24]. Each T-cell subtype has chemokine receptors to direct trafficking to its target destination [25]. A possible mechanism for development of MF/SS is that during differentiation and development of inflammatory response, clonal expansion occurs, giving rise to MF/SS.

Whether the cell of origin of MF/SS, functionally or immunophenotypically, corresponds to any of the currently known subsets of helper T cells (Th1, Th2, Th17, or Treg) remains unclear. When T-helper cell subsets were divided into two groups prior to the identification of the Th17 subset, the neoplastic cells in MF/SS were favored to be more similar to Th2 T cells, due to high expression of IL-4 and IL-5 [26]. However, studies identifying the expression of addition genes in MF/SS have since showed that the neoplastic T cells in MF/SS express Treg markers, such as CTLA-4, FoxP3 [27,28], and a low level of cytokines in response to stimulation [20]. In the same family as CTLA-4, the expression of PD-1 has been detected in CTCL cells [29]. So far, the number of different genes detected in the neoplastic T cells does not easily permit classification into any of the currently defined T-cell subsets and does not easily hint at one T-cell type being predisposed to transform into MF/SS. Also increasing the difficulty to gain insight into the original clonal T cell is that T cells have recently been shown to demonstrate phenotypic plasticity in cytokine expression under different microenvironments [30,31]. In this scenario, a mechanism to consider is that it is possible that any skin homing T-cell clone has the potential to develop into MF. Thus, the initiating neoplastic clone may best be classified as a mature skin homing $CD4^+$ T cell which might shift its cytokine profile during progression in each patient when subjected to different microenvironments [31].

17.4 MOLECULAR GENE EXPRESSION DIFFERENCES IN MF/SS

The consistent immunophenotype based on surface markers of the neoplastic T cells in MF/SS suggests the expression of a subset of distinctive genes that can provide insight into the development of MF/SS. Disease-specific genes are ideal diagnostic biomarkers to increase the sensitivity to differentiating atypical presentation of clinically and morphologically overlapping but distinct diseases, such as psoriasis. These biomarker genes and the encoded proteins can also provide clues to genetic mechanisms leading to the development and evolution of MF/SS.

17.4.1 Protein Markers in MF/SS

Prior studies of T-cell markers derived from MF/SS patients have identified surface markers in MF/SS, with the atypical MF/SS cell expressing CD3, CD4, CD45RO$^+$, and the skin-homing receptor CLA [32,33]. In addition, the skin-homing malignant T cells have shown the frequent expression of chemokine receptors CCR4 and CCR6 [34]. Loss of surface marker expression has been observed, two markers of T cells lost frequently in MF/SS being CD7 and CD26, which also can be measured by immunohistochemistry and by transcriptional analysis [35,36]. Although this pattern of expression is helpful, it is not sufficiently specific as these markers are present in T cells. Also, studying gene regulation of these common immune markers unfortunately may not yield insight into the mechanism of the disease.

In addition to surface markers, nuclear proteins have been identified to be abnormally expressed in MF/SS. Dysregulation of transcription factors regulating cytokine expression, increased STAT3, and loss of STAT4 have been observed in MF/SS. The increased active STAT3 protein may play a role in directing the Th2 phenotype of MF/SS; however, in our studies, the activity of STAT3 is likely abnormal as SS T cells express low levels of cytokines [37,21]. Genes for nuclear proteins involved in proliferation, such as JunB, JunD, and other cell cycle genes, are increased, but these genes are not specific for MF/SS and have been detected in other cancers, which limits their use as specific biomarkers [38,39]. Another transcription factor, GATA3, is activated in SS from a proteasome defect that contributes to increased ubiquitination of GATA3 [40]. This defect may have profound downstream consequences on cellular targets that contribute to the malignancy, one of which is the upregulation of the *CTLA-4* gene in MF/SS [28].

17.4.2 Novel Genes Expressed in MF/SS

Recently numerous genes expressed specifically in SS have been identified, and these may be valuable to study the mechanism of disease development [6,41]. For example, a gene such as PLS3 (T-plastin), which has been detected in SS and not in psoriasis, could be valuable for studying the disease mechanism [42,43]. Also a surface marker that is a member of the killer receptor, KIR3DL2 or CD158, normally expressed in NK cells, is increased in SS [44]. As a screen for SS, KIR3DL2 was detectable in the malignant clone in the peripheral blood, supporting the use of this molecule as a biomarker for SS [45]. Infrequently, KIR expression is detected in aged T cells and T cells in autoimmune disease, suggesting that KIR detection in combination with assessing *TCR* gene clonality may increase the specificity for accurate diagnosis of MF/SS

[46]. The mechanism for increased KIR receptor expression in aged T cells has been suggested to be epigenetic and may indicate an important mechanism that becomes altered in the development of MF/SS [47]. As these marker genes in SS are also detected among patients with other diseases, a common pathway may become deregulated to alter the expression of these genes in the development of MF/SS.

An unbiased approach to study gene expression globally is to use microarrays. Taking this approach, microarray studies comparing gene expression between normal T cells from SS T cells have identified additional novel genes in MF/SS. Of note is the significant upregulation of genes not normally expressed in T cells, such as TWIST1, DNM3 (dynamin 3), NEDD4L, and PLS3 [42,48]. The functional role and biologic significance of these genes in MF/SS are unclear. While investigations into the function of the proteins encoded by these genes are ongoing, nevertheless these genetic biomarkers can be utilized in the diagnosis of MF/SS. The mechanism for the dysregulation of this number of genes is unlikely to arise solely from mutations affecting each gene individually, but is more likely to derive from epigenetic changes based on the CpG sequence structure of these genes.

17.4.3 Noncoding RNAs in MF/SS

Studies of gene expression have uncovered RNAs corresponding to genomic regions that are not translated into proteins that have a role in regulating gene expression. Analysis of these noncoding RNAs has enabled identification of specific RNAs deregulated in cancer [49]. In MF, studies have described differences in microRNA (miRNA) expression, with miR-155 and miR-92a increased in affected skin lesions [50]. In neoplastic cells from SS patients, there is both a decreased and an increased expression of certain miRNAs [51,52]. In a study of whole transcriptome sequencing of SS T cells, long noncoding RNAs that are uniquely expressed in SS T cells compared to nonneoplastic CD4$^+$ T cells were identified [18]. This study detected transcripts in SS that are also present in MF tumors and supports the hypothesis that these two clinically different types of CTCL may share the same cell of origin. In another deep sequencing study, the gene *DNM3* is increased, and pre-miR-199a2 and pre-miR-214 located on the opposite strand are increased coordinately, suggesting alteration of the chromatin region of *DNM3* to coordinately affect multiple RNA species [53]. These changes in expression of noncoding RNAs suggest altered transcription control as a mechanism that leads to the abnormal gene expression in MF/SS, potentially affecting a common pathway that has consequences for the regulation of many genes.

17.5 POTENTIAL GENETIC AND EPIGENETIC MECHANISMS IN MF/SS

The number of genes altered in MF/SS suggests that multiple genetic mechanisms play an important role in its development, from specific mutations affecting oncogenes and tumor suppressors to epigenetic changes. The specific mechanisms are not completely known. Pathognomonic chromosomal translocations such as those seen in B-cell lymphomas and myeloid leukemias have not been identified in MF/SS. In B cells, the immunoglobulin heavy chain genes undergo recombination during class switching and this mechanism may increase the susceptibility in B cells to aberrant DNA rearrangements and chromosomal translocation. In contrast, genetic maturation by class switching does not occur with the T-cell receptor and the observations of recurrent hallmark translocations are infrequent in CTCL. Nevertheless the inherent developmental program to rearrange antigen receptor genes in T cells may increase susceptibility to genetic mutations and chromosomal alterations. Indeed numerous studies of CTCL show chromosomal alterations, and genomic gains and losses, affecting chromosomes 1, 6–10, and 17 [54,55]. This underlying instability may contribute to the diverse gene expression abnormalities observed in MF/SS.

In addition to the primary DNA sequence, the organization of DNA with histones and modifications of the DNA by methylation can affect gene regulation. Thus genomic changes from modification of DNA or its structural organization can be early steps to alter gene expression programs in the pathogenesis of MF/SS, in altering the phenotype of a normal memory $CD4^+$ T cell to a transformed skin-homing T cell. While accumulation of a series of genetic mutations that increase proto-oncogene function have been observed in MF/SS, such as *c-Jun* and *c-myc*, or inactivation of tumor suppressor genes, epigenetic mechanisms that affect gene regulation at nonsequence level may be important in the development of MF/SS, and it is the totality of gene expression differences that defines MF/SS [54,56].

17.5.1 DNA Methylation

The methylation of DNA at CpG dinucleotides has an important role in controlling gene expression and plays a role in T-cell differentiation. CpG dinucleotides are often located in clusters or "islands" in the promoter region of genes and recruit proteins important in regulating compaction of DNA [57]. Methylated DNA is organized within heterochromatin and restricts access to transcription factors and machinery to block gene expression. Methylated genes are thus silenced while unmethylated genes are expressed. Epigenetic modification at promoters is important in transmitting signals to direct the control of genes in the differentiation of T cells into specific functional subsets [58].

Regulation of DNA methylation has focused on enzymes with the ability to add methyl groups to cytosine residues. DNA methyltransferases (DNMTs) are a family of enzymes that methylate cytosine at CpG dinucleotides [59]. There are three DNMTs that add methyl groups to DNA, DNMT1, DNMT3a, and DNMT3b [60]. DNMT1 is expressed in all cells and plays a role in DNA replication, by adding methyl groups to hemimethylated DNA to preserve the inheritance of the methylation pattern of DNA [60]. DNMT3A and DNMT3b are *de novo* methyltransferases, which add methyl groups to unmethylated cytosine and may play a role in the hypermethylation of CpG islands in transformed cells, particularly at the loci of tumor suppressor genes.

Although cell division has been thought to be required to change the methylation of DNA in genes, dynamic methylation without cell division has now been proposed [61]. Lymphocyte genes undergo changes in DNA methylation during differentiation upon activation and it has been hypothesized that dynamic methylation is an important mechanism in the regulation of genes in T cells [62]. Differentiation-specific genes that are not expressed in naïve T cells, such as cytokine genes, have CpGs that are methylated at the promoter in the naïve state. After expansion with antigen stimulation, cytokine genes (*IFN-γ, IL-2*) are demethylated and expressed highly [63,64]. The mechanism for active demethylation in T cells remains unclear, but the identification of novel enzymes of the TET family of proteins that modify methylated cytosines, suggests that there are mechanisms for removal of methyl groups from DNA without DNA synthesis [65,66]. Abnormal demethylation of genes may contribute to cell transformation in the development of lymphoid malignancies [67].

17.5.2 DNA Methylation and Cancer

DNMTs play a role in autoimmune diseases and this has been best characterized in systemic lupus erythematosus and is reviewed elsewhere in this book. The role of DNMTs in cancer has been revealed by the discovery of mutations in DNMTs in some cancers. The level of DNMTs is elevated in solid tumors such as colon and breast cancers and also hematologic malignancies such as acute myeloid leukemia (AML), specifically DNMT3A [68,69]. Overexpression of DNMT3b has been observed in prostate cells and may have a role in the inactivation of tumor suppressors [70]. The specific role of DNMTs in CTCL remains unclear.

Interestingly, we have identified a gene, *samhd1*, that is regulated by methylation in CTCL [71]. Downregulation of SAMHD1 expression correlates with promoter DNA methylation in SS patients [71]. SAMHD1 is a dNTP triphosphohydrolase involved in nucleotide metabolism and functions as a restriction factor in blocking HIV replication in certain immune cell types [72]. Analysis of the gene shows that there is increased methylation in the promoter region in patients who are in the advanced stages [71]. Thus, SAMHD1 may have a potential role similar to that of a tumor suppressor gene, and repression of its expression by methylation is permissive for T-cell proliferation. These findings indicating methylation affecting *samhd1* expression suggest a possible role for increased DNMT activity; however, further studies in a greater number of patients with MF/SS will be needed.

In contrast to increased methylation from overexpression of DNMTs in cancer, there may be circumstances where there is decreased DNMT in cancer, and, more specifically, T-cell lymphoma. In a murine model targeting the DNMT pathways, cancer develops when DNMT1 is genetically inactivated, and a common type is T-cell lymphoma [73]. In these cancers, an observed consequence in the genome is an increase in DNA hypomethylation [73]. Unlike murine model studies, clinical analysis of human specimens to date has not shown data supporting a consistent loss of DNMTs in MF/SS.

The role of hypomethylation in the development of MF/SS is not clear. DNA hypomethylation is an epigenetic change that has been detected in many types of cancer and may be relevant in MF/SS [74]. In T cells, demethylation of DNA of cytokine genes is observed upon antigen stimulation, and dysregulation of this mechanism can contribute to the development of cancer. Indeed, autoimmune diseases have been described in T cells, such as lupus and rheumatoid arthritis, that have been associated with DNA hypomethylation where T cells are chronically activated [75,76]. A potential role of hypomethylation in MF/SS may be to increase genomic instability, thereby increasing gene expression that promotes cell proliferation.

In MF/SS, there are genes highly expressed in the neoplastic T cells that are not expressed in normal T cells. Specifically, highly expressed genes in SS include *CD158* (*KIR3DL2*), *DNM3*, *PLS3*, and *TWIST1* [6]. Interestingly, these genes have large CpG islands, based on sequence analysis (UCSC Genome Browser: genome.ucsc.edu/), and loss of methylation may be a mechanism for their increased expression. Two of these genes, *PLS3* and *DNM3*, have a structural role based on sequence motif, do not have clear immune functions, and are not normally expressed in T cells or immune cells. Thus the preferential activation of these genes in MF/SS remains an enigma [77]. Some of them, such as T-plastin, are not detected in other hematologic malignancies, suggesting a specific altered regulatory pathway in MF/SS that can reveal mechanisms important for the development of this cancer. The reason for the consistent increased expression of these genes in SS is unclear. Nevertheless, supporting the role of epigenetics in MF/SS is that the *PLS3* gene has been shown to be hypomethylated at CpGs [78]. The role that DNA hypomethylation may play in transformation is to predispose these cells to malignancy by increasing genetic instability [79,80]. Thus increased hypomethylation may play a role in the development of MF/SS.

The mechanism responsible for increasing hypomethylation in different types of cancer is unclear, whether it is from DNMT dysfunction or another mechanism. Interestingly, recent identification of the *TET* gene family has hinted at additional epigenetic proteins regulating CpG DNA methylation to control gene expression [65,81]. Analysis of the *TET2* gene in hematologic malignancies that include AML, chronic myelogenous leukemia, myelodysplastic syndrome, and mastocytosis has found mutations in the *TET* gene [82−85]. Loss-of-function *TET2* mutations lead to DNA hypermethylation, suggesting a role of *TET2* in disruption of DNA demethylation [86]. However, there is one study that shows that *TET2* mutation is associated with an increased hypomethylation in myeloid malignancies and suggests that the pathway for the increased hypomethylation remains unclear. Nevertheless, the increase in hypomethylation observed in MF/SS hints to a potential pathway associated with epigenetic regulation that warrants further studies to determine whether the TET pathway may plays a role in the development and progression of MF/SS.

17.6 SUMMARY

MF/SS is a T-cell malignancy where alteration in the pattern of gene expression suggests a contribution from epigenetic dysregulation. Studies have identified novel biomarkers increased at the transcriptional level, supporting a role for epigenetics in the development of MF/SS. The study of epigenetic mechanisms in MF/SS is nascent, in part from incomplete understanding of the mechanisms regulating epigenetic modification of DNA. Further advances in understanding the steps and mechanisms in the control of DNA modification in gene expression will lead to identification of abnormalities that play a role in the development of MF/SS.

Acknowledgments

This work was supported in part by NIH grants CA164911 to HKW and PP and CA181997 to LW from the National Cancer Institute.

References

[1] Willemze R, Jaffe ES, Burg G, et al. WHO-EORTC classification for cutaneous lymphomas. Blood 2005;105(10):3768–85.

[2] Olsen EA, Rook AH, Zic J, et al. Sezary syndrome: immunopathogenesis, literature review of therapeutic options, and recommendations for therapy by the United States Cutaneous Lymphoma Consortium (USCLC). J Am Acad Dermatol 2011;64 (2):352–404.

[3] Agar NS, Wedgeworth E, Crichton S, et al. Survival outcomes and prognostic factors in mycosis fungoides/Sezary syndrome: validation of the revised International Society for Cutaneous Lymphomas/European Organisation for Research and Treatment of Cancer staging proposal. J Clin Oncol 2010;28(31):4730–9.

[4] Arulogun SO, Prince HM, Ng J, et al. Long-term outcomes of patients with advanced-stage cutaneous T-cell lymphoma and large cell transformation. Blood 2008;112(8):3082–7.

[5] Willemze R, Meijer CJ. Classification of cutaneous T-cell lymphoma: from Alibert to WHO-EORTC. J Cutan Pathol 2006;33(Suppl. 1):18–26.

[6] Wong HK, Mishra A, Hake T, Porcu P. Evolving insights in the pathogenesis and therapy of cutaneous T-cell lymphoma (mycosis fungoides and Sezary syndrome). Br J Haematol 2011;155(2):150–66.

[7] Berger CL, Warburton D, Raafat J, LoGerfo P, Edelson RL. Cutaneous T-cell lymphoma: neoplasm of T cells with helper activity. Blood 1979;53(4):642–51.

[8] Criscione VD, Weinstock MA. Incidence of cutaneous T-cell lymphoma in the United States, 1973–2002. Arch Dermatol 2007;143(7):854–9.

[9] Saunes M, Nilsen TI, Johannesen TB. Incidence of primary cutaneous T-cell lymphoma in Norway. Br J Dermatol 2009;160(2):376–9.

[10] Salehi M, Azimi Z, Fatemi F, Rajabi P, Kazemi M, Amini G. Incidence rate of mycosis fungoides in Isfahan (Iran). J Dermatol 2010;37(8):703–7.

[11] Alsaleh QA, Nanda A, Al-Ajmi H, et al. Clinicoepidemiological features of mycosis fungoides in Kuwait, 1991–2006. Int J Dermatol 2012;49(12):1393–8.

[12] Kim YH, Liu HL, Mraz-Gernhard S, Varghese A, Hoppe RT. Long-term outcome of 525 patients with mycosis fungoides and Sezary syndrome: clinical prognostic factors and risk for disease progression. Arch Dermatol 2003;139(7):857–66.

[13] Olsen E, Vonderheid E, Pimpinelli N, et al. Revisions to the staging and classification of mycosis fungoides and Sezary syndrome: a proposal of the International Society for Cutaneous Lymphomas (ISCL) and the cutaneous lymphoma task force of the European Organization of Research and Treatment of Cancer (EORTC). Blood 2007;110(6):1713–22.

[14] Scarisbrick JJ, Kim YH, Whittaker SJ, et al. Prognostic factors, prognostic indices and staging in mycosis fungoides and Sezary syndrome: Where are we now? Br J Dermatol 2014;170(6):1226–36.

[15] Benner MF, Jansen PM, Vermeer MH, Willemze R. Prognostic factors in transformed mycosis fungoides: a retrospective analysis of 100 cases. Blood 2012;119 (7):1643–9.

[16] Morales Suarez-Varela MM, Llopis Gonzalez A, Marquina Vila A, Bell J. Mycosis fungoides: review of epidemiological observations. Dermatology 2000;201(1):21–8.

[17] Whittemore AS, Holly EA, Lee IM, et al. Mycosis fungoides in relation to environmental exposures and immune response: a case–control study. J Natl Cancer Inst 1989;81(20):1560–7.

[18] Lee CS, Ungewickell A, Bhaduri A, et al. Transcriptome sequencing in Sezary syndrome identifies Sezary cell and mycosis fungoides-associated lncRNAs and novel transcripts. Blood 2012;120(16):3288–97.

[19] Wong HK. Immunopathogenesis of mycosis fungoides/Sezary syndrome (cutaneous T-cell lymphoma). G Ital Dermatol Venereol 2008;143(6):375–83.
[20] Chong BF, Dantzer P, Germeroth T, et al. Induced Sezary syndrome PBMCs poorly express immune response genes up-regulated in stimulated memory T cells. J Dermatol Sci 2010;60(1):8–20.
[21] Chong BF, Wilson AJ, Gibson HM, et al. Immune function abnormalities in peripheral blood mononuclear cell cytokine expression differentiates stages of cutaneous T-cell lymphoma/mycosis fungoides. Clin Cancer Res 2008;14(3):646–53.
[22] Dong C. TH17 cells in development: an updated view of their molecular identity and genetic programming. Nat Rev Immunol 2008;8(5):337–48.
[23] Murphy KM, Reiner SL. The lineage decisions of helper T cells. Nat Rev Immunol 2002;2(12):933–44.
[24] Sallusto F, Lanzavecchia A. Human Th17 cells in infection and autoimmunity. Microbes Infect 2009;11(5):620–4.
[25] Denucci CC, Mitchell JS, Shimizu Y. Integrin function in T-cell homing to lymphoid and nonlymphoid sites: getting there and staying there. Crit Rev Immunol 2009;29 (2):87–109.
[26] Dummer R, Heald PW, Nestle FO, et al. Sezary syndrome T-cell clones display T-helper 2 cytokines and express the accessory factor-1 (interferon-gamma receptor beta-chain). Blood 1996;88(4):1383–9.
[27] Berger CL, Tigelaar R, Cohen J, et al. Cutaneous T-cell lymphoma: malignant proliferation of T-regulatory cells. Blood 2005;105(4):1640–7.
[28] Wong HK, Wilson AJ, Gibson HM, et al. Increased expression of ctla-4 in malignant T-cells from patients with mycosis fungoides—cutaneous T cell lymphoma. J Invest Dermatol 2006;126(1):212–19.
[29] Samimi S, Benoit B, Evans K, et al. Increased programmed death-1 expression on CD4+ T cells in cutaneous T-cell lymphoma: implications for immune suppression. Arch Dermatol 146(12):1382–88.
[30] Hirahara K, Vahedi G, Ghoreschi K, et al. Helper T-cell differentiation and plasticity: insights from epigenetics. Immunology 2011;134(3):235–45.
[31] Zhu J, Paul WE. Heterogeneity and plasticity of T helper cells. Cell Res 2009;20 (1):4–12.
[32] Kim YH, Hoppe RT. Mycosis fungoides and the Sezary syndrome. Semin Oncol 1999;26(3):276–89.
[33] Girardi M, Heald PW, Wilson LD. The pathogenesis of mycosis fungoides. N Engl J Med 2004;350(19):1978–88.
[34] Hwang ST, Janik JE, Jaffe ES, Wilson WH. Mycosis fungoides and Sezary syndrome. Lancet 2008;371(9616):945–57.
[35] Wood GS, Hong SR, Sasaki DT, et al. Leu-8/CD7 antigen expression by CD3+ T cells: comparative analysis of skin and blood in mycosis fungoides/Sezary syndrome relative to normal blood values. J Am Acad Dermatol 1990;22(4):602–7.
[36] Jones D, Dang NH, Duvic M, Washington LT, Huh YO. Absence of CD26 expression is a useful marker for diagnosis of T-cell lymphoma in peripheral blood. Am J Clin Pathol 2001;115(6):885–92.
[37] Sommer VH, Clemmensen OJ, Nielsen O, et al. *In vivo* activation of STAT3 in cutaneous T-cell lymphoma. Evidence for an antiapoptotic function of STAT3. Leukemia 2004;18(7):1288–95.
[38] Mao X, Orchard G, Lillington DM, Russell-Jones R, Young BD, Whittaker SJ. Amplification and overexpression of JUNB is associated with primary cutaneous T-cell lymphomas. Blood 2003;101(4):1513–19.
[39] Mao X, Orchard G, Mitchell TJ, et al. A genomic and expression study of AP-1 in primary cutaneous T-cell lymphoma: evidence for dysregulated expression of JUNB and JUND in MF and SS. J Cutan Pathol 2008;35(10):899–910.

[40] Gibson HM, Mishra A, Chan DV, Hake TS, Porcu P, Wong HK. Impaired proteasome function activates GATA3 in T cells and upregulates CTLA-4: relevance for Sezary syndrome. J Invest Dermatol 2013;133(1):249−57.

[41] Wong HK. Novel biomarkers, dysregulated epigenetics, and therapy in cutaneous T-cell lymphoma. Discov Med 2013;16(87):71−8.

[42] Tang N, Gibson H, Germeroth T, Porcu P, Lim HW, Wong HK. T-plastin (PLS3) gene expression differentiates Sezary syndrome from mycosis fungoides and inflammatory skin diseases and can serve as a biomarker to monitor disease progression. Br J Dermatol 2010;162(2):463−6.

[43] Su MW, Dorocicz I, Dragowska WH, et al. Aberrant expression of T-plastin in Sezary cells. Cancer Res 2003;63(21):7122−7.

[44] Poszepczynska-Guigne E, Schiavon V, D'Incan M, et al. CD158k/KIR3DL2 is a new phenotypic marker of Sezary cells: relevance for the diagnosis and follow-up of Sezary syndrome. J Invest Dermatol 2004;122(3):820−3.

[45] Marie-Cardine A, Huet D, Ortonne N, et al. Killer cell Ig-like receptors CD158a and CD158b display a coactivatory function, involving the c-Jun NH2-terminal protein kinase signaling pathway, when expressed on malignant CD4+ T cells from a patient with Sezary syndrome. Blood 2007;109(11):5064−5.

[46] Li G, Yu M, Weyand CM, Goronzy JJ. Epigenetic regulation of killer immunoglobulin-like receptor expression in T cells. Blood 2009;114(16):3422−30.

[47] Li G, Weyand CM, Goronzy JJ. Epigenetic mechanisms of age-dependent KIR2DL4 expression in T cells. J Leukoc Biol 2008;84(3):824−34.

[48] Booken N, Gratchev A, Utikal J, et al. Sezary syndrome is a unique cutaneous T-cell lymphoma as identified by an expanded gene signature including diagnostic marker molecules CDO1 and DNM3. Leukemia 2008;22(2):393−9.

[49] Cheetham SW, Gruhl F, Mattick JS, Dinger ME. Long noncoding RNAs and the genetics of cancer. Br J Cancer 2013;108(12):2419−25.

[50] van Kester MS, Ballabio E, Benner MF, et al. miRNA expression profiling of mycosis fungoides. Mol Oncol 2011;5(3):273−80.

[51] Ballabio E, Mitchell T, van Kester MS, et al. MicroRNA expression in Sezary syndrome: identification, function, and diagnostic potential. Blood 2010;116(7):1105−13.

[52] Narducci MG, Arcelli D, Picchio MC, et al. MicroRNA profiling reveals that miR-21, miR486 and miR-214 are upregulated and involved in cell survival in Sezary syndrome. Cell Death Dis 2011;2:e151.

[53] Lee YK, Turner H, Maynard CL, et al. Late developmental plasticity in the T helper 17 lineage. Immunity 2009;30(1):92−107.

[54] Izykowska K, Przybylski GK. Genetic alterations in Sezary syndrome. Leuk Lymphoma 2011;52(5):745−53.

[55] Mohr B, Illmer T, Oelschlagel U, et al. Complex cytogenetic and immunophenotypic aberrations in a patient with Sezary syndrome. Cancer Genet Cytogenet 1996;90 (1):33−6.

[56] Garatti SA, Roscetti E, Trecca D, Fracchiolla NS, Neri A, Berti E. bcl-1, bcl-2, p53, c-myc, and lyt-10 analysis in cutaneous lymphomas. Recent Results Cancer Res 1995;139:249−61.

[57] Bird A. DNA methylation patterns and epigenetic memory. Genes Dev 2002;16 (1):6−21.

[58] Russ BE, Prier JE, Rao S, Turner SJ. T cell immunity as a tool for studying epigenetic regulation of cellular differentiation. Front Genet 2013;4:218.

[59] Piccolo FM, Fisher AG. Getting rid of DNA methylation. Trends Cell Biol 2014;24 (2):136−43.

[60] Dawson MA, Kouzarides T. Cancer epigenetics: from mechanism to therapy. Cell 2012;150(1):12−27.

[61] Bhutani N, Burns DM, Blau HM. DNA demethylation dynamics. Cell 2011;146 (6):866—72.

[62] Dong J, Chang HD, Ivascu C, et al. Loss of methylation at the IFNG promoter and CNS-1 is associated with the development of functional IFN-gamma memory in human CD4(+) T lymphocytes. Eur J Immunol 2013;43(3):793—804.

[63] Murayama A, Sakura K, Nakama M, et al. A specific CpG site demethylation in the human interleukin 2 gene promoter is an epigenetic memory. EMBO J 2006;25 (5):1081—92.

[64] Kersh EN, Fitzpatrick DR, Murali-Krishna K, et al. Rapid demethylation of the IFN-gamma gene occurs in memory but not naive CD8 T cells. J Immunol 2006;176 (7):4083—93.

[65] Williams K, Christensen J, Helin K. DNA methylation: TET proteins—guardians of CpG islands? EMBO Rep 2012;13(1):28—35.

[66] Wu SC, Zhang Y. Active DNA demethylation: many roads lead to Rome. Nat Rev Mol Cell Biol 2010;11(9):607—20.

[67] Yang H, Liu Y, Bai F, et al. Tumor development is associated with decrease of TET gene expression and 5-methylcytosine hydroxylation. Oncogene 2013;32(5):663—9.

[68] Ley TJ, Ding L, Walter MJ, et al. DNMT3A mutations in acute myeloid leukemia. N Engl J Med 2010;363(25):2424—33.

[69] Subramaniam D, Thombre R, Dhar A, Anant S. DNA methyltransferases: a novel target for prevention and therapy. Front Oncol 2014;4:80.

[70] Benbrahim-Tallaa L, Waterland RA, Dill AL, Webber MM, Waalkes MP. Tumor suppressor gene inactivation during cadmium-induced malignant transformation of human prostate cells correlates with overexpression of *de novo* DNA methyltransferase. Environ Health Perspect 2007;115(10):1454—9.

[71] de Silva S, Wang F, Hake TS, Porcu P, Wong HK, Wu L. Downregulation of SAMHD1 expression correlates with promoter DNA methylation in Sezary syndrome patients. J Invest Dermatol 2014;134(2):562—5.

[72] de Silva S, Hoy H, Hake TS, Wong HK, Porcu P, Wu L. Promoter methylation regulates SAMHD1 gene expression in human CD4+ T cells. J Biol Chem 2013;288 (13):9284—92.

[73] Gaudet F, Hodgson JG, Eden A, et al. Induction of tumors in mice by genomic hypomethylation. Science 2003;300(5618):489—92.

[74] Feinberg AP, Vogelstein B. Hypomethylation distinguishes genes of some human cancers from their normal counterparts. Nature 1983;301(5895):89—92.

[75] Sunahori K, Juang YT, Kyttaris VC, Tsokos GC. Promoter hypomethylation results in increased expression of protein phosphatase 2A in T cells from patients with systemic lupus erythematosus. J Immunol 2011;186(7):4508—17.

[76] Zufferey F, Williams FM, Spector TD. Epigenetics and methylation in the rheumatic diseases. Semin Arthritis Rheum 2014;43(5):692—700.

[77] von Euw E, Chodon T, Attar N, et al. CTLA4 blockade increases Th17 cells in patients with metastatic melanoma. J Transl Med 2009;7:35.

[78] Jones CL, Ferreira S, McKenzie RC, et al. Regulation of T-plastin expression by promoter hypomethylation in primary cutaneous T-cell lymphoma. J Invest Dermatol 2012;132(8):2042—9.

[79] Eden A, Gaudet F, Waghmare A, Jaenisch R. Chromosomal instability and tumors promoted by DNA hypomethylation. Science 2003;300(5618):455.

[80] Chen RZ, Pettersson U, Beard C, Jackson-Grusby L, Jaenisch R. DNA hypomethylation leads to elevated mutation rates. Nature 1998;395(6697):89—93.

[81] Mohr F, Dohner K, Buske C, Rawat VP. TET genes: new players in DNA demethylation and important determinants for stemness. Exp Hematol 2011;39(3):272—81.

[82] Holmfeldt L, Mullighan CG. The role of TET2 in hematologic neoplasms. Cancer Cell 2011;20(1):1−2.

[83] Mercher T, Quivoron C, Couronne L, Bastard C, Vainchenker W, Bernard OA. TET2, a tumor suppressor in hematological disorders. Biochim Biophys Acta 2012;1825 (2):173−7.

[84] Gaidzik VI, Paschka P, Spath D, et al. TET2 mutations in acute myeloid leukemia (AML): results from a comprehensive genetic and clinical analysis of the AML study group. J Clin Oncol 2012;30(12):1350−7.

[85] Wu YC, Ling ZQ. The role of TET family proteins and 5-hydroxymethylcytosine in human tumors. Histol Histopathol 2014;29(8):991−7.

[86] Figueroa ME, Abdel-Wahab O, Lu C, et al. Leukemic IDH1 and IDH2 mutations result in a hypermethylation phenotype, disrupt TET2 function, and impair hematopoietic differentiation. Cancer Cell 2010;18(6):553−67.

Epigenetics and Aging

Sabita N. Saldanha[1] and Louis Patrick Watanabe[2]

[1]Department of Biological Sciences, Alabama State University,
Montgomery, AL [2]Department of Biology, University of Alabama
at Birmingham, Birmingham, AL

18.1 INTRODUCTION

Aging and its variety of effects are considered as undesirable yet unpreventable phenotypes across cultures around the world to the extent that there is a massive market for global antiaging products. Despite abundant interest in antiaging products, particularly for the skin, the mechanisms behind cutaneous aging are still largely unclear making it difficult to develop treatments or procedures for preventative care.

What we do know is that aging in humans is a combination of genetic and epigenetic changes that occur along the course of time [1]. Classic investigations are purely focused on the genetic component; however, recent findings suggest a role for epigenetic regulatory elements in modulating gene expression involved in the aging of the skin [2]. Although it is still unclear how and why the effects of aging take place, distinct epigenetic aging pathways in epidermal stem cells have been elucidated [2].

Epigenetics is defined as a heritable yet reversible change devoid of alterations in the DNA sequence. DNA methylation, histone amino acid tail modifications, and regulatory activity of microRNAs (miRNAs), a group of noncoding RNAs (ncRNAs), are the most established epigenetic processes known to influence mammalian development and aging [3]. Of these, DNA methylation is an extensively studied process and is regulated by three enzymes: DNA methyltransferase 1 (DNMT1), DNA methyltransferase 3a (DNMT3a), and DNA methyltransferase 3b (DNMT3b) [4]. DNMT1 ensures that methylation patterns are faithfully

© 2015 Elsevier Inc. All rights reserved.

copied in subsequent generations regulating cellular lineage, identity, and functions [5]. DNMT3a and 3b are *de novo* enzymes responsible for laying down new methylation patterns and are necessary during reprogramming events of the embryonic epigenome [6]. Methyltransferases catalyze the transfer of methyl groups from *S*-adenosylmethionine to the 5′ position of cytosine residues that are presented in the CpG dinucleotide configuration [7]. Normally, gene transcript inhibition and induction are associated with hypermethylation and hypomethylation states, respectively. However, there are exceptions to the norm where, in a few genes, the observed methylated state to transcript expression effect is reversed [8]. The plasticity of epigenetic changes allows for reversal of cellular or metabolic processes and therefore, though complex, it is feasible to rejuvenate aged tissue to a younger phenotype by targeting such epigenetic modifiers.

Regulation of chromatin introduces another layer of epigenetic complexity and involves covalent modifications of histone amino acid tails. Of these modifications, histone acetylation and methylation are well-established and understood processes [9]. Other modifications that influence chromatin structure impacting gene expression are histone phosphorylation, ubiquitination, biotinylation, sumoylation, and ADP-ribosylation [10]. Histone acetylation patterns are catalyzed by histone acetyltransferases and histone deacetylases (HDACs) [10]. These two enzymes maintain the dynamic equilibrium of acetylated chromatin states contributing to changes in gene function.

Since the discovery of miRNAs in 1993, research has shed light on the influence of miRNAs in the regulation of the epigenetic landscape of various human tissues [11,12]. Unlike DNA methylation and chromatin alterations, these ncRNAs, which are about 22−23 nucleotides in length, regulate mammalian systems by affecting the transcriptome through increased/decreased transcript stability or degradation [13]. miRNAs play critical roles in various skin pathobiologies but the understanding of the mechanics involved in these roles in normal cutaneous homeostasis and aging is limited [14,15].

Aging and age-associated changes are difficult to assess due to the heterogeneity of tissues and variations among individuals. It is also relatively hard to sift the causes of aging from the changes that arise due to aging. Given these problems, the skin serves as an ideal model to assess age-associated changes due to its minimal structural complexity. In this chapter, age-associated epigenetic changes that govern the transformation of cutaneous tissue from a youthful appearance to an aged state will be discussed. How these epigenetic changes can be potentially reversed to attain rejuvenation of the skin and therapies that target epigenetic factors to mediate the same will also be covered.

18.2 SENESCENCE AND AGING

The hallmark of aging is the accumulation of damage to cellular macromolecules such as DNA, protein, and lipids. The damage is mediated by free radicals generated as by-products of metabolic activity [16,17]. With age, quenchers of free radicals decline and the reactive oxygen species (ROS) increase, parallel DNA damage, and contribute to cellular and mitochondrial dysfunction [17]. When innate cellular repair machinery fail to limit and repair the damage the cells are forced into a phase of senescence where they remain viable but lack replicative potential. Therefore, with time, mitotic cells undergo the process of replicative senescence [18]. In mammals, in general, when the replicative potential of stem cells declines, specific age-associated changes occur.

Since the establishment of newer techniques such as pyrosequencing, bisulfite sequencing, and genome-wide microarray technology, epigenetic factors have gained prominence in age-related gene regulation. When epigenetic drifts target the replicative potentials of adult stem cells, the sustenance of which are crucial to tissue integrity it can trigger the onset of age-related changes. Tissues with high turnover rates that rely on the maintenance of stem cell populations are susceptible to this type of epigenetic regulation. It is plausible that the skin, being a highly proliferative organ, is affected similarly with age. As yet, the complete epigenetic basis of aging is not very clear and is currently being investigated through various model systems such as yeast, worms, and flies [19].

18.3 CHANGES ASSOCIATED WITH AGING SKIN

Epidermal cells, melanocytes, basal cells, keratinocytes, fibroblasts, collagen fibers, and adipose cells found within various layers of the skin undergo dramatic structural or functional changes with age (Figure 18.1). Wrinkles, sagging skin, and loss of pigmentation are visible signs of aged skin. Environmental factors such as exposure to sunlight also contribute to these changes [20]. Photoaging of the skin arises due to the loss of pigment producing cells and from damage by UVB radiations of the sun [21]. With age, the epidermal layer flattens out and the pigment containing cells decrease in number, and therefore aged skin appears thinner, paler, and more translucent [22,23]. Age spots begin to appear in sun-exposed areas [24]. Loss of collagen reduces the strength and flexibility of skin and loss of elastic fibers reduces the elasticity, and the combination of both in sun-exposed areas causes solar elastosis [25]. The hypodermis thins leading to lowered insulation and padding, and sweat production decreases, and taken together, the ability of the body to regulate normal

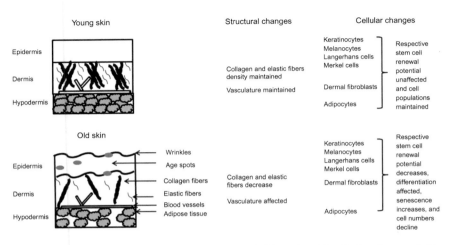

FIGURE 18.1   **Changes associated with normal skin aging.** The distinct layers of skin populated with different cell types and structural components dictate cutaneous functions in young versus old skin. With age, collagen ( ) and elastic fibers ( ) of the dermal region decrease and enhance wrinkle formation. Sun damage induced decline in melanin production by the loss in number and proliferation of melanocytes initiates age spots ( ). Blood vessels ( ) thin out and adipocyte numbers ( ) decrease reducing insulation and homeostasis of body temperature. Although the immune-related functions and epigenetic regulation of Langerhans cells in skin are not clearly known, the cells undergo a fate similar to that of melanocytes. The decrease in Langerhans cells increases the risk of pathogen invasion often seen in aged populations. The loss of Merkel cells in aging contributes to the loss of touch sensation. However, the age-related Merkel cell dysfunction needs to be validated by further investigations. Taken together, age-associated skin damage is increased by the thinning of the epidermal layer and weakening of the dermal layer resulting in easy tears and cuts, increased infections, and reduced wound healing.

temperature declines. In addition to senescence, DNA damage also contributes to these observed changes. However, epigenetic events that target the expression of key genes in skin homeostasis are also postulated to be involved in the aging process.

18.4 EPIGENETIC CHANGES IN CELLULAR AND STRUCTURAL COMPONENTS OF SKIN

Mammalian systems are exposed to various environmental stimuli and thus the epigenetic landscapes are often altered. Several human diseases associated with age, and cancer, have an epigenetic bearing, and epigenetic factors that change with time and influence the aging process are currently being investigated. The structural integrity of skin is highly dependent on the functions of its epidermal and dermal layers. The epidermal layer is involved in protection functions and consists of various cell types including keratinocytes and melanocytes [26,27]. The dermal region is highly vascular, and interspersed in between are collagen and elastic fibers that impart strength and elasticity to the tissue [28]. The junction where the two layers meet consists of basal epidermal keratinocytes firmly anchored to the collagen-rich basement membrane by hemidesmosomes with the dermal region supported by collagen VII fibrils [29,30]. Dermal fibroblasts are involved in the synthesis of dermal extracellular matrix (ECM) proteins that dominate the structure and function of the dermis [31].

In aged skin, reduction and disorganization of its major ECM components, such as collagen and other elastic fibers, are observed (Figure 18.1). Loss of dermal integrity due to ECM aberrant changes alters fibroblast behavior, preventing their association with integrin-collagen fibers and thus hindering mechanical stretching [32]. In aged skin, changes in morphology, metabolic activity, decline in the synthesis of ECM proteins mainly collagens, and overexpression of proteases in fibroblasts are observed. [33]. Some of the genes that regulate the structural and cellular components of skin and which are affected by DNA methylation, histone modifications, and miRNAs are discussed herein.

18.4.1 Tet Methylcytosine Dioxygenase 2 (TET2) and Epidermis

TET enzymatic proteins catalyze the reaction of 5-methylcytosine (5mC) to 5-hydroxymethylcytosine (5hmC), a process considered to be involved in passive as well as active demethylation mechanisms [34]. The subsequent dilution of 5hmC after series of replicative rounds contributes to a possible epigenetic reprogramming event. Imprinting of genes occurs early in development and DNA methylation patterns contributed individually from paternal and maternal genomes are erased prior to new patterns being laid down in the zygote. The role of TET in these events is now slowly emerging through developmental studies [34]. It has been observed that TET proteins are expressed at specific stages of embryonal development. For example, at embryo stage E0.5 paternal genome demethylation occurs in the presence of TET3, at embryo stage E8.5 global DNA demethylation I occurs in the presence of TET 1/2 proteins, and at embryo stage E11.5, global DNA demethylation II imprint erasure occurs [34]. There exists a possibility that TET proteins can be considered as potential targets for rejuvenation therapy of skin through epigenetic reprogramming of epidermal cells, however, this prospect requires further exploration and experimental support.

Currently, knowledge of the role of TET2 in skin is limited to the understanding of its functions in maintaining active chromatin states through the conversion of 5mC−5hmC. In epidermal and hair follicle cells, the DNA is 5-hydroxymethylated and structures of skin have been shown to contain adult stem cells in which *TET* genes are expressed. Conversion of pluripotent and multipotent stem cells into specific-cell lineages requires extensive epigenomic reprogramming. Active chromatin states mediated by TET2 are observed at such lineage-specific loci [35]. Ironically, in aged skin, *TET2* is downregulated through methylation. Often hypermethylation of *TET2* accounts for poor turnover of epidermal cells from adult stem cells leading to epidermal thinning [36] (Table 18.1).

TABLE 18.1   Epigenetic Regulation of Genes in Aged Skin

Gene/locus	Epigenetic change	Type of epigenetic change	Effect	Type of tissue analyzed	Reference
CDKN2B	DNA methylation	Hypermethylation	–	Human fibroblasts	[37]
SEC31L2	DNA methylation	Hypermethylation	Important for collagen secretion	Epidermis/dermis	[36]
TET2	DNA methylation	Hypermethylation	Tumor suppressor/demethylator	Epidermis/dermis	[36]
DDAH2	DNA methylation	Hypermethylation	Marker for differentiation	Epidermis/dermis	[36]
COL1A1	DNA methylation	Hypermethylation	Decreased expression of collagen	Epidermis	[38]
INK4b/ARF/INK4a	Histone methylation and ANRIL RNA	H3K27me3 and ANRIL binding	Repress *INK4b/ARF/INK4a*	MEFs, IMR90 cells	[39]
			Activate p53-dependent cell cycle arrest, deplete sebaceous gland cells		

18.4.2 Regulation of Dimethylarginine Dimethylaminohydrolase 2 (DDAH2) and Keratinocytes

DDAH2 encodes a key enzyme in the nitric oxide pathway [40]. The pathway is necessary for the regulation of keratinocyte proliferation as well as the development of skin cancer [36]. Bisulfite sequence analysis of aged skin specimens has revealed age-associated hypermethylation of *DDAH2* (Table 18.1) [36]. Such epigenetic changes may contribute to the deterioration of the epidermal layers by lack of maintenance of sufficient keratinocyte populations. This may also explain that, with age, the skin becomes fragile, and tears easily with poor wound healing capabilities. Together, the hypermethylated status of *TET2* and *DDAH2* as observed in aged skin samples may account for the loss of structural and functional integrity of the epidermal layers [36].

18.4.3 Collagen

Type 1 and type III collagens are the most prevalent collagen fibers in skin tissue with type I being more predominant of the two. The proteins impart structural strength to human skin. Progressive collagen loss had been associated with aging resulting in the thinning and fragility of elderly skin. The reasons behind the decrease or loss of collagen in aging are not well understood. However, two important genes, connective tissue growth factor (*CTGF*) and transforming growth factor (*TGF*)-β have been implicated owing to their role in the decrease of collagen synthesis in the dermis of aged skin [41,42]. Reports have shown that *CTGF* is constitutively expressed in normal human dermis and that the levels of CTGF/TGF-β are highly reduced in dermal fibroblasts of aged human skin samples *in vivo* [41]. Dermal fibroblast cells are also the major collagen producing cells and the populations of these cell types decrease with age.

CTGF is a multifunctional matrix cellular protein and is also a direct target of TGF-β. Conversely, CTGF mediates the downstream functions of *TGF-β* [43]. It is possible that these genes are regulated by chromatin remodeling or miRNA target regulation, which may account for the observed decrease in expression in aged skin cells. Reports have shown that the *CTGF* gene is repressed by miR-18a, and by miR-214 based on cell type and the source of tissue origin [44,45]. TSA, an HDAC inhibitor, regulates TGF-β function. In the presence of the HDAC inhibitor, TGF-β fails to induce collagen alpha-2(I) (*COL1A2*) promoter activity and is found to be associated with reduced specificity protein 1 (Sp1) levels [46]. Probably, in skin fibroblasts, HDAC inhibition may induce gene-specific regulation inhibiting TGF-β-induced collagen synthesis.

18.4.4 Melanocytes

Loss of pigmentation is a prominent histological age-associated change observed in skin. Oxidative stress mediated by external stimuli such as sunlight may contribute to the loss of pigmentation. This loss has been reported to result from the decrease in melanocytes from human hair follicle [47]. The decrease in melanocyte production is the direct culmination of the decrease in the number of viable melanocytes accompanied by increased melanocyte apoptosis in the aging hair follicle bulge and bulb.

The role of microphthalmia-associated transcription factor (*MITF*) in melanocyte biology has been well documented [48]. It has also been reported that factors such as Paired box 3 (*PAX3*), Sry-related HMG box 10 (*SOX10*), CREB-binding protein (*CBP*), and β-catenin/lymphoid enhancer-binding factor 1 (*LEF1*) control MITF expression in melanocyte development in the mouse model [49]. However, the mechanisms regulating *MITF* expression in human melanocytes and melanomas still remain obscure. B-cell lymphoma 2 (*Bcl2*) is a direct target of MITF and modulates lineage survival and melanoma cell viability [50]. *MITF* acts as a master regulator of melanocyte biology by regulating various differentiation and cell cycle progression genes. It is also an oncogene and has been implicated in a fraction of human melanomas [47]. Therefore, MITF functions in instructing melanocytes towards terminal differentiation and/or pigmentation or, alternatively, promotes malignant behavior. Understanding the molecular function of MITF and its associated pathways will hopefully shed light on strategies for improving therapeutic approaches for these diseases as well in aging.

Although presently, no direct evidence is available that lends support to an epigenetic basis of *MITF* regulation, MITF regulation of DICER, an important component of miRNA biogenesis has been reported [51]. Studies that have analyzed the miRNA expression profiles in differentiating melanocytes have observed two populations, one with upregulated pre-miRNAs and the other upregulated as mature miRNAs. Conversion of pre-miRNAs to fully processed miRNAs is mediated by the induction of DICER, transcriptionally regulated by MITF. The DICER-dependent processing of the pre-miRNA-17 targets B-cell lymphoma 2 interacting mediator (*BIM*). BIM is a proapoptotic regulator of melanocyte survival [51]. These observations highlight a central mechanism underlying the influence of miRNA in lineage-specific melanocyte development.

18.4.5 Mesenchymal Stem Cells and Dermal Fibroblasts

Mesenchymal stem cells (MSCs) have multipotent differential potential such as adipogenic, osteogenic, and chondrogenic differentiation [52,53]. Similarly, dermal fibroblasts that are necessary for maintaining skin architecture and ECM synthesis have also been described as having multipotent properties comparable to those of MSCs [53]. However, variability in differential potential between the two cell types is also strongly argued [37,52,53]. Human dermal fibroblasts are shown to contain distinct cell subtype populations based on the anatomical site of origin. A study conducted by the Koch group isolated, cultured, and analyzed fibroblasts from various dermal sites to evaluate the epigenetic age-associated changes that occur in comparison to MSCs [37]. The site of origin of the dermal region governs the functionality of the fibroblast subtypes and epigenetic changes in these cells govern positional effects through reciprocal epithelial—mesenchymal interactions.

MSCs are characterized by their ability to have remarkable differential potential but it is relatively hard to discern MSCs from dermal fibroblasts as they appear to be morphologically similar. Both are also comprised of multipotent cells and exhibit differential potential as well. Nonetheless, differential potential analysis of MSCs versus dermal fibroblast in the Koch study revealed that fibroblast subtypes had limited differentiation potential. MSCs are able to attain adipogenic and otogenic states whereas fibroblasts do not [25]. The limited differential potential in fibroblasts in comparison to MSCs may be attributed to differences in DNA methylation profiles between the two and also to the varied epigenetic make-up of the subtypes within MSCs [37].

Epithelial—mesenchymal interactions dictate positional identity of epidermal differentiation. Experiments with chick models have shown that transplantation of wing epithelium to leg mesenchyme induces scale development of the transplanted epithelium [54]. Global gene expression profiles of fibroblasts in conjunction with their positional variation contribute to the anatomical demarcation such as anterior—posterior, proximal—distal, and dermal versus nondermal [55]. Upon aging, the two cell populations, MSCs and fibroblasts, display DNA methylation changes at similar CpG sites but are differently regulated, suggestive of the fundamental functional difference between the cell types and the involvement of different mechanisms for cellular aging [37]. As seen in previous studies where the cellular milieu influences phenotypic outcome, it is plausible that epigenetic modifications occurring in these two populations are differentially regulated by the microenvironments in the dermis and the bone marrow [56].

18.5 PHOTOAGING OF SKIN

p53 and retinoblastoma (Rb) are tumor suppressor pathways that are essential to the regulation of the cellular senescence process. p53 activation is initiated in the presence of DNA damage and includes activation of the gene transcriptome involved in growth arrest such as cyclin-dependent kinase (*p21WAF1*). Simultaneously, the upregulation of inhibitor of kinase 4a (*p16INK4A*) hinders the phosphorylation of the retinoblastoma protein (pRb) sustaining cell cycle arrest [57]. Clearly, p16INK4A appears to be an important marker for senescence. The simultaneous activation of *p21WAF1* and inhibition of *p16INK4A* by p53 and pRb, respectively, initiate and reinforce senescence. These observations are, however, cell-type specific.

Stress-induced premature senescence (SIPS) occurs in the presence of oxidative stress, irradiation, or replicative stress. This type of premature senescence is independent of telomere damage. Normally, human diploid fibroblasts (HDFs) serve as the experimental model for determining UVB-induced extrinsic skin aging. The accumulation of senescent fibroblast cells in HDFs determines the extent of photoaging by UVB [58]. UVB irradiation is known to induce a DNA damage response that triggers the p53 response pathway, but p53-independent pathways may also be involved and have been supported by knock-down studies [59]. miRNAs are important transcriptome regulators and changes in miRNA expression contribute to cellular senescence and organismic aging [60]. A small set of miRNAs that are differentially regulated by UVB-mediated skin damage have been identified as miR-20a, -20b, -15a, -93, and -101 [58].

18.6 DNA METHYLATION AND OVERALL AGING IN SKIN

Genome-wide investigations of methylation patterns in aging skin reveal differences in global versus loci-specific methylation states [36,61]. In some instances, the observed increase in loci-specific methylation changes is not found to be accompanied by concomitant changes in DNMT1, DNMT3a, and DNMT3b expression [61]. How changes in methylation patterns occur devoid of the changes in the expressed state of the modifying enzyme remains unclear and it is plausible that chromatin modifications work in concordance with DNA methylation. Contradictory to this observation, the work presented in another study showed that the expression levels of epigenetic enzymes DNMT3a, DNMT3b, and TET2 in aged skin decline significantly [36]. The disparities in the observations between the two studies probably stem from the different model systems used. The Lyko group study involved the comparison of young versus old human epidermal tissue whereas the work conducted by the Xu group was on a mouse model [36,62]. Functional analysis of genes altered by the activity of these enzymes reveals that inhibition of cell proliferation and activation of the immune response are contributing factors to the development of aged skin [62].

Global hypomethylation with regional or gene-specific hypermethylation in different human tissues is generally the well-established epigenetic norm. Global hypomethylation may be the result of passive demethylation due to progressive inefficiencies of DNMT1 with age. In aging, hypermethylation of specific loci, gene clusters, receptors, and growth factors can occur. However, it is uncertain which cues regulate site-specific methylation. The HumanMethylation27 BeadChip microarray analysis of mesenchymal stromal cells which assessed 27,578 unique CpG sites within more than 14,000 promoter regions reveals that the methylation patterns in long-term cultures and aging are maintained, but exhibit significant differences at loci-specific CpG sites [37,63].

In the skin, *DDAH2* and *TET2* are hypermethylated and thus transcriptionally silenced [36] (Table 18.1). Only a small percentage of genes have been found to be affected by *de novo* methylation in aging skin [36]. These observations also support CpG hypermethylation being able to affect DNA binding of transcription factors and may be age dependent [64]. Altered gene expression can have manifold effects in biological pathways imposing a myriad of phenotypic alterations. Some of these regulate senescence and apoptosis [65]. In most cancers, the deregulation of tumor suppressors in addition to other factors account for the development of tumors, and transcriptional silencing through hypermethylation is one such process. Contrarily, in aging of skin, these genes appear to undergo *de novo* methylation [65,66]. TET2 that is actively involved in melanomas and in the onset of healthy aged skin is an example of this observation.

18.7 CHROMATIN AND AGING SKIN

Chromatin fibers in eukaryotic systems comprise DNA associated with histone as well as nonhistone proteins [67]. Histones exist in an octameric configuration consisting of two copies of each of the four core histone proteins H2A, H2B, H3, and H4. Together they form the nucleosome responsible for orchestrating DNA accessibility [68]. Posttranslational modifications of the histone core via the amino acid tails of the respective subunits help to maintain the dynamic equilibrium of chromatin and are largely responsible for chromatin regulation [68]. The myriad of modifications that the histone tails are subjected to are well reviewed elsewhere [68]. Of these, histone methylation and acetylation are the known established regulators of the epigenome. In general, the combinations of specific modifications result in a code termed the histone code. The histone code serves as a special epigenetic signature that dictates the conformational changes of chromatin structure, thereby regulating DNA accessibility [69]. Therefore chromatin and its codes are able to regulate several biological processes.

Evidence supporting the role of chromatin modifications in the senescent phenotype leading to organismal aging is on the rise. Previous experiments in human skin fibroblasts have demonstrated that with age the nucleosomal organization is altered and the linker DNA that tethers the nucleosomal core tends to become more heterogeneous [70]. The dynamic nature of chromatin makes it an ideal target from the perspective of treatment of human diseases. Chromatin modifications control a large number of cellular processes and include gene transcription, DNA stability, and chromatin organization [71]. Chromatin remodeling occurs in mammalian cells with age and the changes are observed both globally and at specific loci. With increased improvement in bioinformatics and whole-genome analysis, it has now become possible to determine the relationship between chromatin structure and the aging phenotype.

18.7.1 Histone Methylation

The specific age-associated histone codes are yet to be ascertained. To add to the complexity in most cases, inactivating the enzymes that are responsible for the specific modification leads to cell lethality limiting experimental contributions in the understanding of these changes. The pleiotropic effects induced by the enzymes regulating chromatin modifications also hinder the process of deciphering their role in organismal aging. Nonetheless, there are histone modifications found to be associated with the aging process.

The trimethylation of histone H4 at lysine 20 (H4K20me3) has been found to be upregulated *in vitro* [72]. Similarly, altered histone H3 lysine 27 trimethylation (H3K27me3) has been observed in primary Hutchinson—Gilford progeria syndrome (HGPS) skin fibroblasts [73] and Suv4-20, a histone methyltransferase thought to be responsible for the modification. However, comparative studies assessing the activity and expression of the enzyme in various aging tissue (young vs. old) are yet to be performed. The inactivation of Suv4-20 has been shown to lead to proliferation defects induced by DNA damage. Another histone modifier, histone methyltransferase, enhancer of zeste homologue 2 (EZH2), is better assessed in relation to its role in cellular senescence.

18.7.1.1 INK4 Loci

Under cellular duress, alternate reading frame (p19ARF) and p16INK4A proteins accumulate in the cell and are products of the *INK4A* locus. The *INK4B-ARF-INK4A* (*INK4* box) is epigenetically regulated and codes for the expression of *INK4A*, *INK4B*, and *ARF*, all of which show increased expression during senescence (Table 18.1) [74]. Another study from 2010 took a more in-depth approach to understanding the mechanisms controlling *INK4* and showed that the transcription is controlled by two epigenetic regulatory elements, ncRNAs and histone modifications [39]. The proposed mechanism is that ncRNAs at *INK4A* recruit polycomb-repressive complex 1 and 2 (PRC1/2). PRC1 contains chromobox 7 (CBX7), which directly inhibits *INK4A* expression, while PRC2 induces histone 3 lysine 27 methylation (H3K27me), a marker indicating repressive chromatin structure.

In proliferating cells, EZH2 induces *INK4A* locus repression. Decrease in EZH2 levels and the loss of H3K27me3 mark at the *INK4A* locus leads to loci-specific hypomethylation [75,76]. ARF in particular, which activates p53-dependent cell cycle arrest, is regulated in such a manner and this is indicative of a potential mechanism for the contribution of *INK4A* to aging [74]. Specifically, pyrosequencing of human dermal fibroblasts has shown that the *INK4A* locus is hypermethylated in aged fibroblasts [37].

18.7.1.2 P53 and Rb

Ongoing research has shown that p53 plays a critical role in the aging process [77]. In relation to the aging of the epidermis in particular, one study has found that p53 is responsible for skin aging through the depletion of sebaceous gland cells in mice [78]. These cells are responsible for controlling hydration in the outer layer of the epidermis, as well as providing the lipid mixture sebum to the skin, two traits that are often lacking in aging skin. The study showed that p53 overexpression in mutant mice has sebaceous gland cells that are not maintained, which draws comparisons to the inactivation of sebaceous gland cells in humans. This could indicate a signaling pathway involving the methylation of *INK4* at H3K27 resulting in chromatin alterations, which subsequently lead to the activation of p53-dependent cell cycle arrest, thus inactivating the sebaceous gland cells and ultimately leading to epidermal senescence.

p53 and Rb pathways are also required for the induction of senescence processes in the presence of oxidative stress and DNA damage [79]. Histone demethylases, lysine demethylase 2a (KDM2a), and lysine demethylase 2b (KDM2b), target methylated histone H3 lysine 36 (H3K36). The demethylation of H3K36me histones occurs at the *p15INK4B* locus, resulting in repression [80]. Thus, demethylation activity by demethylases mediates senescence inhibitory action via the modulation of p53 and Rb pathways [80]. Independent of the epigenetic code, histone modifiers either induce or prevent cellular senescence and this regulation is found to be dependent on the code prevalent in the tissue.

These observations have been supported by those for gene-specific chromatin changes described above. However, genome-wide approaches are becoming necessary for the analysis of the same functions.

18.7.1.3 p63

Transcription factor such as p63 regulates the biogenesis of epidermal stem cells by interacting with histone modifiers [81]. Human fibroblast specification is hampered and epidermal stratification is affected in the absence of HDAC1 and HDAC2. In addition to the absence of HDAC1 and HDAC2, this cell-specific affected phenotype is also observed in parallel with the downregulation of transcription factor p63 [2]. p63 interaction with chromatin modifiers is crucial for epidermal embryogenesis and adult epidermal homeostasis as well. In the presence of HDAC1/2 and methyl-CpG (meCpG) stabilizer lymphoid-specific helicase/helicase lymphoid specific (Lsh/HELLS), p63 inhibits the expression of antiproliferative genes [2,81].

18.7.2 Histone Acetylation

Of the covalent modifications governing histone epigenetic alterations, acetylation and methylation appear to have a pertinent role in aged human tissue. HDACs represent a class of enzymes that catalyze the removal of acetyl groups and affect acetylation patterns of the genome. Of the several classes of HDACs known, class III HDACs have a pivotal role in aging [82]. Decreased expression of Sirtuin 1 (SIRT1) induces chromatin instability that is found to be associated with aging in mammals [82]. Enhancing the expression of silent mating type information regulation 2 homolog (sirtuins) has been described to have beneficial antiapoptotic effects and thus serves as a target for antiaging therapies [83].

Changes in global histone acetylation levels occur as the organism ages. It has been reported that the levels of the histone deacetylase HDAC1 decrease in cultured primary human fibroblasts upon serial passaging and that the use of specific class I and II HDAC inhibitors induces a senescent-like phenotype [82,84]. Therefore, histone acetylation levels are important in establishing a senescent state but it remains unclear whether global changes in histone acetylation or loci-specific changes are necessary for the process. Overexpression of HDAC1 has also been shown to induce an irreversible senescent program but the changes observed may be cell-type specific [85]. Another study in human fibroblasts found that the presence of histone deacetylase inhibitors (HDACIs) is able to induce early senescence by the induction of loss of silent heterochromatin [86].

Researchers have observed that in the presence of HDACIs, *p16INK4A* is upregulated, which has led to the development of biomarkers indicating senescence [87]. More work is necessary to understand the exact mechanism by which HDACIs induce senescence. The study also identified p53 in the aging pathway and has found that *p53* null mutant mice show resistance to senescence induced by HDACI. The contribution of HDACs to the senescence program would benefit from genome-wide approaches that could determine direct HDAC targets involved in age-associated changes.

18.7.3 Polycomb Complex Group

Recently, the polycomb complex group (PcG) complex of proteins have been shown to have a pivotal role in transcriptional repression by their epigenetic interaction with chromatin. They are responsible for establishing and maintaining cell fates and are important to stem cell self-renewal and in cancer development. Their epigenetic roles in controlling skin stem cell renewal and differentiation have been determined [88]. The targets of PcG complex proteins are susceptible to methylation with age [89]. In some cases, methylation and chromatin modifications act in concert and dictate epigenetic outcomes. It has been observed that hypermethylation also tends to occur predominantly at bivalent chromatin domain promoters and are involved in developmental regulation. Bivalent domains are often seen in embryonic stem cells and consist of activating histone 3 lysine 4 trimethylation (H3K4me3) and inactivating histone 3 lysine 27 trimethylation (H3K27me3) histone marks [90,91]. The bivalent epigenetic modifications can present genes in a bipotential state contributing to differential expression in various cell lineages. Such factors can also contribute to varied DNA methylation changes in different aged tissue.

Polycomb complexes cooperate to create a network of stable gene silencing through the following mechanism. First, PRC2 induces the trimethylation of histone 3 lysine 27 (H3K27me3) through the EZH2 protein. PRC1 is then able to bind to H3K27me3 through CBX7, a protein on PRC1. Following this binding, the ubiquitination of histone 2A at lysine 119 by PRC1 induces chromatin compaction thereby directly inhibiting the expression of genes such as *INK4* [39,92]. As the functions and effects of PRC1 are elucidated, the research focus has shifted to the roles of its individual components, and in particular, *Bmi1*. Studies on Bmi1 in keratinocytes have found that it localizes in many layers of the epidermis, including the stem cells [92]. It is used as a biomarker for aging and is the obverse of *INK4* expression, in that Bmi1 is abundant in keratinocytes taken from younger individuals and lacking in those collected from older individuals [93]. This dichotomy has led to several studies that have implied that Bmi1 may be a modulator of *INK4A/ARF* expression [92]. Interestingly, increased expression of *Bmi1* results in an increase of EZH2, the protein responsible for H3K27me3, further solidifying the cooperation between the PRC complexes [94].

18.8 MicroRNAs

RNA molecules that do not encode a protein are commonly termed ncRNAs [95]. However, these ncRNAs do contain information (probably hidden signals) and have various regulatory functions [95]. They undergo further processing through splicing events to yield smaller products which include epigenetic regulators such as miRNAs.

The most recently discovered of the major mechanisms regulating cellular senescence are those associated with the miRNAs (Table 18.2). These are short 22-bp RNA sequences that are regulated at least partly by epigenetic mechanisms and can affect gene expression posttranscriptionally [29]. A large cohort of miRNAs are expressed in the skin, and particularly strong expression in the epidermis comes from those in the

TABLE 18.2   miRNAs Associated in the Aging of Skin

miRNA	Expression	Interaction with	Effect	References
miR-130b, -138, -181a, -181b	Increased	p63 (represses)	Delay aging	[98]
miR-137	Increased	p53 (concomitantly induced)	Induces senescence	[97]
miR-668	Increased	p16^{INK4A} (concomitantly induced)	Induces senescence	[97]
miR-191	Increased	CDK6	Commences senescence	[99]
miR-152	Increased	ITGA5 (repressed)	Inhibits fibroblast adhesion	[29]
miR-29a, -30	Increased	BMYB (repressed)	Induce senescence	[100]
miR-29a	Increased	Collagens	Senescence	[96]
miR-203	Increased	p63 (repressed)	Induces senescence	[98,101]

miR-200 and 19/20 families [96]. It is interesting to note that a study in 2011 has shown that certain miRNAs (miR-137 and miR-668) are concomitantly upregulated in aging skin cells with the previously discussed *p53* and *INK4* loci [97].

A host of miRNAs have been studied in the context of aging since their rise to prominence, and many such studies have focused on keratinocyte senescence [98,99]. For example, p63, a transcription factor from the p53 family that is heavily involved in counteracting aging has been identified as a target for miR-130b silencing [98] (Table 18.2). Interestingly, there are miRNAs that facilitate aging which are repressed by p63, indicative that both the inhibition and induction of senescence are controlled by miRNAs [78].

The examination of the miRNA profile in aged tissue has shown more upregulation than downregulation of miRNAs [102]. The genes controlled by these miRNAs are associated with mitochondrial function, oxidative stress, and proliferation [102]. miRNAs have varied roles in aging and some have been found to regulate senescence-dependent growth arrest [103], skeletal muscle aging [104], and aging delay [105]. Mouse models representing premature aging conditions reveal that the deregulation of miRNA-29 is associated with DNA damage and the p53 pathway [106].

Specific sets of miRNAs are upregulated in the replicative senescence of fibroblasts and keratinocytes [29]. The upregulation of miR-152 and miR-181a has been found to be associated with the induction of senescent markers, reduced expression of *p16INK4A* in young fibroblasts and a decrease in cell proliferation [29]. These observations have been backed by studies which have shown an increase in miR-138, -181a, -181b, and -130b levels in passaged human primary keratinocytes [98].

p63 transcription factor regulates keratinocyte proliferative potential by maintaining the turnover of basal keratinocytes in developing as well as mature epidermal layers [98,107]. miRNAs which are upregulated in keratinocyte replicative senescence and which target Sirt1 and p63 are miR-138, -181a, -181b, and -130b [98]. miRNAs that target p63 in aged skin may therefore have a bearing on the differentiation process of epidermal stem cells resulting in a weaker and flattened epidermal layer. Studies have shown that miR-138 that is significantly upregulated in aged skin, targets p63 [98]. In addition, p63 is also a direct target of miR-203, which is found active in terminally differentiating cells and not in proliferative progenitor populations [108]. The roles of miR-203 and -138 in the replicative senescence biology of fibroblasts have been determined [98].

ITAG5 and CO16A1 have been identified as targets of these miRNAs and the reduction in expression of the gene transcripts is thought to be involved in ECM aberrant remodeling of aged skin (Table 18.2). In skin, integrins convey signals from the matrix to cells, help in the organization and remodeling of the matrix via ECM. These proteins are severely affected upon aging partly due to the upregulation of miR-152 followed by a decrease in ITGA5 expression [109]. Overexpression of miR-152 prevents fibroblast cell adhesion. miR-181a targets collagen XVI. Although this type of collagen is a minor component of ECM its presence in the dermal epidermal junction (DEJ) zone is required for mechanical anchorage of the cell and conduction of signals both in and out [110]. Therefore, the upregulation of miR-152 and miR-181a, which individually target respective integrin and collagen fibers, maintains a senescent phenotype, and plays a complex role in ECM remodeling characteristics of aged skin [29].

18.9 REVERSING AGING

Studies in various model systems have confirmed that the aging clock can be reversed through epigenetic reprogramming, termed rejuvenation [111]. Accumulation of senescent cells enhances the increased imbalance in homeostatic mechanisms caused by the alteration of tissue structure and functions. The advances in aging research and epigenetics have shown promising insights to reverse the aging clock. The erasure of programmed memory in gametes at the time of fertilization to a reconfigured phenotype in the zygote differs from the reversal of the aging clock in differentiated cells. Rejuvenation of aged cells constitutes the reversal of two systems—aging and differentiation [111]. Uncoupling the two poses a challenge. Studies have shown that exposing aged cells to the milieu of young mice causes cells to revert to a youthful phenotype without altering the differentiated state of the cells in question [112].

Information from such preliminary experiments is useful as they contribute to the plausible mechanisms of reversing aging. Some of these mechanisms could include a change in the cellular milieu or inhibition of regulatory genes that affect enzymes, protein metabolism, and in turn several pathways. A complication presented here is that unlike pluripotency where a few transcription factors are necessary for the reprogramming event, aging is a multifactorial process, and therefore it remains unclear whether only a few genes may be necessary to reverse aging. Attainment of youthful cellular phenotypes from aged cells without dedifferentiation retains the structural strata of the tissue and, therefore, rejuvenation presents a useful therapeutic strategy against many diseases.

18.10 EPIGENETIC THERAPEUTICS IN PREVENTION OR TREATMENT OF AGED SKIN

With the rapid progress in our understanding of the epigenetic mechanisms involved in skin aging, possible therapeutic strategies for treatment/prevention of epidermal senescence are now being explored [111]. As described above, altered chromatin states based on a multitude of epigenetic modifications are a hallmark of aging skin, thus simply reversing these modifications is an attractive model for rejuvenating the chromatin states of youthful skin [111]. Potential targets thus include DNMTs, hyper- or hypomethylated DNA, histone modifications, novel epigenetic modifiers, and miRNAs [113]. Other more macro-scale epigenetic influencers include caloric restriction (CR) or the consumption of folate, among other methylation-modifying foods [114,115].

The human tripeptide glycyl-L-histidyl-L-lysine (GHK) has been used for the treatment of wounds as well as for its antiaging benefits since its discovery in 1973 [113]. Recent studies have shown that, in epidermal stem cells, GHK upregulates an antisenescence transcription factor p63 and furthermore increases the expression of other proteins abundant in youthful skin such as collagen, glycosaminoglycans, and decorin [29]. The gene regulatory elements observed in GHK suggest that it is a potential epigenetic modulator; however, more experimental data is required to determine its mechanism of action.

Another aging-prevention and lifespan-increasing treatment is caloric restriction (CR). The benefits of CR were first discovered in rats, and since then it has been studied in relation to a variety of diseases and genetic loci [115]. Epigenetically regulated genes such as *p16INK4a* and *p53*, which both modulate aging in epidermal (and other) cells, exhibit downregulation as a result of CR. A general proposed mechanism shows that CR influences DNA methylation and histone modifications, which in turn control the expression of key genes *p16INK4a, p53, hTERT*, and so on in aging and lead to the delay of aging and increased longevity [77].

Accompanying our improved understanding of epigenetic influences in aging as well as the slow introduction of the concept of epigenetics to the mainstream, cosmetic care companies have introduced products claiming to reduce the effects of senescence through epigenetics. It is important to note that the research for these products has been done in-house and thus independent *in vitro* and clinical studies are required to validate their effects. One such compound was developed by Sederma, a cosmetics ingredient company, and is called Senestem [116]. This ingredient is claimed to target miRNAs and inhibit the decrease in protein synthesis that occurs with age. However, no publicly available manuscripts are available on Senestem and thus the mechanistic action and the targeted miRNAs need to be ascertained.

Another product called ReGenistem Red Rice (R3) from Lonza Personal Care also claims to alleviate the effects of aging skin through epigenetic mechanisms [38]. The effectiveness of the product depends on the secondary metabolites from the meristem cells of Himalayan Red Rice. The meristems are cultured and then exposed to ozone stress for the production of secondary metabolites. The researchers applied a 2% R3 preparation on epidermal tissues from old and young subjects and measured CpG methylation. They found that, in aged human fibroblasts, the hypermethylation at promoters genome-wide was decreased after treatment with 2% R3. Treatment with 2% R3 also enhanced collagen expression, most likely due to the demethylation at *COL1A1*, a gene involved in collagen biosynthesis [117]. Expression of dermatopontin, a protein responsible for keratinocyte adhesion, significantly increased following treatment with 2% R3. The research suggests that R3 may be able to tighten skin and alleviate some of the phenotypes of senescence through demethylating key promoters, thus returning chromatin to a youthful state. It is appreciated that Lonza have publicly provided their research; however, third party studies are needed to exclude bias and uncover the potential detrimental effects of ReGeniStem.

Generation of reactive oxidative species (ROS) is another major mechanism that accelerates skin aging. ROS increase with age and antioxidants, which are quenchers of free radicals, decrease. The increase in ROS with the concomitant decrease in endogenous mechanisms that counteract them leads to changes in and damage of cellular structures. Antioxidant compounds can provide protection from oxidative stress by scavenging free radicals and the effects of antioxidant extracts on skin aging markers have been evaluated [118]. The transcriptional effect of the antioxidant extracts were evaluated in normal human dermal fibroblasts (NHDFs), a cellular model that represents the replicative senescence process. Aging markers that were assessed included dermal structure, cell renewal, inflammatory response, and oxidative stress mechanisms. The extracts tested significantly regulated five genes with roles in the inflammatory response, cell renewal, and antioxidant defenses. Therefore such genes can serve as targets for antiaging nutraceutical therapeutic strategies.

miRNA research in the field of dermatology is in its early phase, but the early finds are substantial, pointing toward a vast opportunity for developing effective therapies for treatment of skin diseases and wounds.

18.11 CONCLUSION

Aging is a multifactorial phenomenon and therefore very complex. Even with the current existing epigenetic techniques, specific skin senescence markers are difficult to assess and further research in these areas is necessary. Most studies analyzing the epigenetic factors involved in the aging process of skin are recent (within approximately the last 3–4 years). Therefore, epigenetics and the aging of skin is still considered to be in its infancy and incompletely explored.

Investigations conducted thus far have analyzed epigenetic marks in specific skin cell types or genes and in limited model systems. Compiling the epigenetic information obtained from the data of such investigations and correlating it as epigenetic changes observed in aging populations in general may be erroneous. Therefore, epidemiological studies are warranted for the same. Should epigenetic codes be used as markers to achieve prevention, treatment, or rejuvenation of aged skin, concomitant assessment of epigenetic marks such as DNA methylation, histone modifications, and miRNA changes in conjunction with comparative studies between young and old aged human skin tissue is paramount.

Glossary

Epigenetics is defined as a heritable yet reversible change devoid of alterations in the DNA sequence.
miRNAs are short 22–23 nucleotide sequences that target specific transcripts inhibiting or inducing expression of the gene.
Senescence is a process whereby cells lose their replicative potential due either to damage or to the internal mitotic clock but remain viable and metabolically active.

List of Abbreviations

BIM	B-cell lymphoma 2 interacting mediator
CBP	CREB-binding protein
CBX7	chromobox 7
CDKN2B	cyclin-dependent kinase inhibitor 2B
COL1A2	collagen alpha-2(I)
CR	caloric restriction
CTGF	connective tissue growth factor
DDHA2	dimethylarginine dimethylaminohydrolase 2
DNMT	DNA methyltransferase
DNMT1	DNA methyltransferase 1
DNMT3a	DNA methyltransferase 3a
DNMT3b	DNA methyltransferase 3b
ECM	extracellular matrix
EZH2	enhancer of zeste homologue 2
H3K27me3	histone H3 lysine 27 trimethylation
H3K36	histone H3 lysine 36
H4K20me3	histone H4 lysine 20 trimethylation
HAT	histone acetyltransferase
HDAC	histone deacetylase
HDACI	histone deacetylase inhibitor
HDF	human dermal fibroblast
hTERT	human telomerase reverse transcriptase

ITAG5	integrin alpha 5
KDM2a	lysine demethylase 2a
KDM2b	lysine demethylase 2b
LEF1	lymphoid enhancer-binding factor 1
Lsh/HELLS	lymphoid-specific helicase/helicase lymphoid specific
meCpG	methyl-CpG (meCpG)
MITF	microphthalmia-associated transcription factor
ncRNA	noncoding RNA
p15INK4B	inhibitor of kinase 4B (15 kDa)
p16INK4A	inhibitor of kinase 4A (16 kDa)
p53	tumor protein p53 (53 kDa)
PAX3	Paired box 3
PcG	polycomb group
PCR2	polycomb repressor complex 2
pRb	phosphorylated retinoblastoma
Rb	retinoblastoma
ROS	reactive oxygen species
SIPS	stress-induced premature senescence
Sirtuin	silent mating type information regulation 2 homolog
SIRT1	sirtuin 1
SOX10	Sry-related HMG box 10 (SOX10)
Sp1	specificity protein 1
TET2	Tet methylcytosine dioxygenase 2
TGF-β	transforming growth factor (TGF)-β

References

[1] Gonzalo S. Epigenetic alterations in aging. J Appl Physiol 2010;109:586–97.
[2] Shen Q, Jin H, Wang X. Epidermal stem cells and their epigenetic regulation. Int J Mol Sci 2013;14:17861–80.
[3] Brunet A, Berger SL. Epigenetics of aging and aging-related disease. J Gerontol A Biol Sci Med Sci 2014;69(Suppl. 1):S17–20.
[4] Casillas Jr. MA, Lopatina N, Andrews LG, Tollefsbol TO. Transcriptional control of the DNA methyltransferases is altered in aging and neoplastically-transformed human fibroblasts. Mol Cell Biochem 2003;252:33–43.
[5] Lopatina N, Haskell JF, Andrews LG, Poole JC, Saldanha S, Tollefsbol T. Differential maintenance and de novo methylating activity by three DNA methyltransferases in aging and immortalized fibroblasts. J Cell Biochem 2002;84:324–34.
[6] Rivera RM, Ross JW. Epigenetics in fertilization and preimplantation embryo development. Prog Biophys Mol Biol 2013;113:423–32.
[7] Detich N, Hamm S, Just G, Knox JD, Szyf M. The methyl donor S-adenosylmethionine inhibits active demethylation of DNA: a candidate novel mechanism for the pharmacological effects of S-adenosylmethionine. J Biol Chem 2003;278:20812–20.
[8] Guilleret I, Yan P, Grange F, Braunschweig R, Bosman FT, Benhattar J. Hypermethylation of the human telomerase catalytic subunit (hTERT) gene correlates with telomerase activity. Int J Cancer 2002;101:335–41.
[9] Wolffe AP. Packaging principle: how DNA methylation and histone acetylation control the transcriptional activity of chromatin. J Exp Zool 1998;282:239–44.
[10] Jenuwein T, Allis CD. Translating the histone code. Science 2001;293:1074–80.
[11] Persengiev S, Kondova I, Otting N, Koeppen AH, Bontrop RE. Genome-wide analysis of miRNA expression reveals a potential role for miR-144 in brain aging and spinocerebellar ataxia pathogenesis. Neurobiol Aging 2011;32(2316):e17–27.

[12] Lee RC, Feinbaum RL, Ambros V. The *C. elegans* heterochronic gene lin-4 encodes small RNAs with antisense complementarity to lin-14. Cell 1993;75:843—54.

[13] Ying SY, Chang DC, Lin SL. The microRNA (miRNA): overview of the RNA genes that modulate gene function. Mol Biotechnol 2008;38:257—68.

[14] Lerman G, Avivi C, Mardoukh C, et al. MiRNA expression in psoriatic skin: reciprocal regulation of hsa-miR-99a and IGF-1R. PLOS ONE 2011;6:e20916.

[15] Liu Y, Yang D, Xiao Z, Zhang M. miRNA expression profiles in keloid tissue and corresponding normal skin tissue. Aesthetic Plast Surg 2012;36:193—201.

[16] Poljsak B, Dahmane R. Free radicals and extrinsic skin aging. Dermatol Res Pract 2012;2012:135206.

[17] Hekimi S, Lapointe J, Wen Y. Taking a "good" look at free radicals in the aging process. Trends Cell Biol 2011;21:569—76.

[18] Linskens MH, Harley CB, West MD, Campisi J, Hayflick L. Replicative senescence and cell death. Science 1995;267:17.

[19] Warner HR. Subfield history: use of model organisms in the search for human aging genes. Sci Aging Knowledge Environ 2003;2003(6):RE1.

[20] Vierkotter A, Krutmann J. Environmental influences on skin aging and ethnic-specific manifestations. Dermatoendocrinology 2012;4:227—31.

[21] Helfrich YR, Sachs DL, Voorhees JJ. Overview of skin aging and photoaging. Dermatol Nurs 2008;20:177—83 [quiz 84].

[22] Kurban RS, Bhawan J. Histologic changes in skin associated with aging. J Dermatol Surg Oncol 1990;16:908—14.

[23] Branchet MC, Boisnic S, Frances C, Robert AM. Skin thickness changes in normal aging skin. Gerontology 1990;36:28—35.

[24] Goyarts E, Muizzuddin N, Maes D, Giacomoni PU. Morphological changes associated with aging: age spots and the microinflammatory model of skin aging. Ann NY Acad Sci 2007;1119:32—9.

[25] Thomas NE, Kricker A, From L, et al. Associations of cumulative sun exposure and phenotypic characteristics with histologic solar elastosis. Cancer Epidemiol Biomarkers Prev 2010;19:2932—41.

[26] Cichorek M, Wachulska M, Stasiewicz A, Tyminska A. Skin melanocytes: biology and development. Postepy Dermatol Alergol 2013;30:30—41.

[27] Hirobe T, Furuya R, Akiu S, Ifuku O, Fukuda M. Keratinocytes control the proliferation and differentiation of cultured epidermal melanocytes from ultraviolet radiation B-induced pigmented spots in the dorsal skin of hairless mice. Pigment Cell Res 2002;15:391—9.

[28] Robert J, Hartmann DJ, Sengel P. Production of fibronectin and collagen types I and III by chick embryo dermal cells cultured on extracellular matrix substrates. Int J Dev Biol 1989;33:267—75.

[29] Mancini M, Saintigny G, Mahe C, Annicchiarico-Petruzzelli M, Melino G, Candi E. MicroRNA-152 and -181a participate in human dermal fibroblasts senescence acting on cell adhesion and remodeling of the extra-cellular matrix. Aging 2012;4:843—53.

[30] Steplewski A, Kasinskas A, Fertala A. Remodeling of the dermal—epidermal junction in bilayered skin constructs after silencing the expression of the p.R2622Q and p.G2623C collagen VII mutants. Connect Tissue Res 2012;53:379—89.

[31] Lindner D, Zietsch C, Becher PM, et al. Differential expression of matrix metalloproteases in human fibroblasts with different origins. Biochem Res Int 2012;2012:875742.

[32] Sahuc F, Nakazawa K, Berthod F, Collombel C, Damour O. Mesenchymal—epithelial interactions regulate gene expression of type VII collagen and kalinin in keratinocytes and dermal—epidermal junction formation in a skin equivalent model. Wound Repair Regen 1996;4:93—102.

[33] Daly CH, Odland GF. Age-related changes in the mechanical properties of human skin. J Invest Dermatol 1979;73:84−7.

[34] Hill PW, Amouroux R, Hajkova P. DNA demethylation, Tet proteins and 5-hydroxymethylcytosine in epigenetic reprogramming: an emerging complex story. Genomics 2014;104:324−33.

[35] Bocker MT, Tuorto F, Raddatz G, et al. Hydroxylation of 5-methylcytosine by TET2 maintains the active state of the mammalian HOXA cluster. Nat Commun 2012;3:818.

[36] Gronniger E, Weber B, Heil O, et al. Aging and chronic sun exposure cause distinct epigenetic changes in human skin. PLOS Genet 2010;6:e1000971.

[37] Koch CM, Suschek CV, Lin Q, et al. Specific age-associated DNA methylation changes in human dermal fibroblasts. PLOS ONE 2011;6:e16679.

[38] Ludwig P, Bennet S, Gruber JV. Rice meristems stimulate epigenetic rejuvenation. L. P. Care. Lonza (Personal Care) 2011;1−6.

[39] Yap KL, Li S, Munoz-Cabello AM, et al. Molecular interplay of the noncoding RNA ANRIL and methylated histone H3 lysine 27 by polycomb CBX7 in transcriptional silencing of INK4a. Mol Cell 2010;38:662−74.

[40] Breckenridge RA, Kelly P, Nandi M, Vallance PJ, Ohun TJ, Leiper J. A role for dimethylarginine dimethylaminohydrolase 1 (DDAH1) in mammalian development. Int J Dev Biol 2010;54:215−20.

[41] Quan T, Shao Y, He T, Voorhees JJ, Fisher GJ. Reduced expression of connective tissue growth factor (CTGF/CCN2) mediates collagen loss in chronologically aged human skin. J Invest Dermatol 2010;130:415−24.

[42] Igarashi A, Okochi H, Bradham DM, Grotendorst GR. Regulation of connective tissue growth factor gene expression in human skin fibroblasts and during wound repair. Mol Biol Cell 1993;4:637−45.

[43] Leask A, Abraham DJ. The role of connective tissue growth factor, a multifunctional matricellular protein, in fibroblast biology. Biochem Cell Biol 2003;81:355−63.

[44] Ohgawara T, Kubota S, Kawaki H, et al. Regulation of chondrocytic phenotype by micro RNA 18a: involvement of Ccn2/Ctgf as a major target gene. FEBS Lett 2009;583:1006−10.

[45] Chen L, Charrier A, Zhou Y, et al. Epigenetic regulation of connective tissue growth factor by MicroRNA-214 delivery in exosomes from mouse or human hepatic stellate cells. Hepatology 2014;59:1118−29.

[46] Ghosh AK, Mori Y, Dowling E, Varga J. Trichostatin A blocks TGF-beta-induced collagen gene expression in skin fibroblasts: involvement of Sp1. Biochem Biophys Res Commun 2007;354:420−6.

[47] Arck PC, Overall R, Spatz K, et al. Towards a "free radical theory of graying": melanocyte apoptosis in the aging human hair follicle is an indicator of oxidative stress induced tissue damage. FASEB J 2006;20:1567−9.

[48] Levy C, Khaled M, Fisher DE. MITF: master regulator of melanocyte development and melanoma oncogene. Trends Mol Med 2006;12:406−14.

[49] Cimadamore F, Shah M, Amador-Arjona A, et al. SOX2 modulates levels of MITF in normal human melanocytes, and melanoma lines in vitro. Pigment Cell Melanoma Res 2012;25:533−6.

[50] McGill GG, Horstmann M, Widlund HR, et al. Bcl2 regulation by the melanocyte master regulator Mitf modulates lineage survival and melanoma cell viability. Cell 2002;109:707−18.

[51] Levy C, Khaled M, Robinson KC, et al. Lineage-specific transcriptional regulation of DICER by MITF in melanocytes. Cell 2010;141:994−1005.

[52] Junker JP, Sommar P, Skog M, Johnson H, Kratz G. Adipogenic, chondrogenic and osteogenic differentiation of clonally derived human dermal fibroblasts. Cells Tissues Organs 2010;191:105−18.

[53] Haniffa MA, Collin MP, Buckley CD, Dazzi F. Mesenchymal stem cells: the fibroblasts' new clothes? Haematologica 2009;94:258–63.

[54] Rinn JL, Wang JK, Liu H, Montgomery K, van de Rijn M, Chang HY. A systems biology approach to anatomic diversity of skin. J Invest Dermatol 2008;128:776–82.

[55] Rinn JL, Bondre C, Gladstone HB, Brown PO, Chang HY. Anatomic demarcation by positional variation in fibroblast gene expression programs. PLOS Genet 2006;2:e119.

[56] Yu KR, Kang KS. Aging-related genes in mesenchymal stem cells: a mini-review. Gerontology 2013;59:557–63.

[57] Rayess H, Wang MB, Srivatsan ES. Cellular senescence and tumor suppressor gene p16. Int J Cancer 2012;130:1715–25.

[58] Greussing R, Hackl M, Charoentong P, et al. Identification of microRNA–mRNA functional interactions in UVB-induced senescence of human diploid fibroblasts. BMC Genomics 2013;14:224.

[59] Ha L, Ichikawa T, Anver M, et al. ARF functions as a melanoma tumor suppressor by inducing p53-independent senescence. Proc Natl Acad Sci USA 2007;104:10968–73.

[60] Grillari J, Grillari-Voglauer R. Novel modulators of senescence, aging, and longevity: small non-coding RNAs enter the stage. Exp Gerontol 2010;45:302–11.

[61] Raddatz G, Hagemann S, Aran D, et al. Aging is associated with highly defined epigenetic changes in the human epidermis. Epigenetics Chromatin 2013;6:36.

[62] Qian H, Xu X. Reduction in DNA methyltransferases and alteration of DNA methylation pattern associate with mouse skin ageing. Exp Dermatol 2014;23:357–9.

[63] Karymov MA, Tomschik M, Leuba SH, Caiafa P, Zlatanova J. DNA methylation-dependent chromatin fiber compaction in vivo and in vitro: requirement for linker histone. FASEB J 2001;15:2631–41.

[64] Harries LW, Hernandez D, Henley W, et al. Human aging is characterized by focused changes in gene expression and deregulation of alternative splicing. Aging Cell 2011;10:868–78.

[65] Salminen A, Ojala J, Kaarniranta K. Apoptosis and aging: increased resistance to apoptosis enhances the aging process. Cell Mol Life Sci 2011;68:1021–31.

[66] Salminen A, Kaarniranta K. Control of p53 and NF-kappaB signaling by WIP1 and MIF: role in cellular senescence and organismal aging. Cell Signal 2011;23:747–52.

[67] Geiman TM, Robertson KD. Chromatin remodeling, histone modifications, and DNA methylation—how does it all fit together? J Cell Biochem 2002;87:117–25.

[68] Peterson CL, Laniel MA. Histones and histone modifications. Curr Biol 2004;14:R546–51.

[69] Strahl BD, Allis CD. The language of covalent histone modifications. Nature 2000;403:41–5.

[70] Ishimi Y, Kojima M, Takeuchi F, Miyamoto T, Yamada M, Hanaoka F. Changes in chromatin structure during aging of human skin fibroblasts. Exp Cell Res 1987;169:458–67.

[71] Zhang Y, Reinberg D. Transcription regulation by histone methylation: interplay between different covalent modifications of the core histone tails. Genes Dev 2001;15:2343–60.

[72] Sarg B, Koutzamani E, Helliger W, Rundquist I, Lindner HH. Postsynthetic trimethylation of histone H4 at lysine 20 in mammalian tissues is associated with aging. J Biol Chem 2002;277:39195–201.

[73] McCord RP, Nazario-Toole A, Zhang H, et al. Correlated alterations in genome organization, histone methylation, and DNA-lamin A/C interactions in Hutchinson-Gilford progeria syndrome. Genome Res 2013;23:260–9.

[74] Salminen A, Kaarniranta K, Hiltunen M, Kauppinen A. Histone demethylase Jumonji D3 (JMJD3/KDM6B) at the nexus of epigenetic regulation of inflammation and the aging process. J Mol Med 2014;92:1035–43.

[75] Stepanik VA, Harte PJ. A mutation in the E(Z) methyltransferase that increases tri-methylation of histone H3 lysine 27 and causes inappropriate silencing of active Polycomb target genes. Dev Biol 2012;364:249–58.

[76] Ezhkova E, Lien WH, Stokes N, Pasolli HA, Silva JM, Fuchs E. EZH1 and EZH2 cogovern histone H3K27 trimethylation and are essential for hair follicle homeostasis and wound repair. Genes Dev 2011;25:485–98.

[77] Rufini A, Tucci P, Celardo I, Melino G. Senescence and aging: the critical roles of p53. Oncogene 2013;32:5129–43.

[78] Kim J, Nakasaki M, Todorova D, et al. p53 induces skin aging by depleting Blimp1 + sebaceous gland cells. Cell Death Dis 2014;5:e1141.

[79] Shay JW, Pereira-Smith OM, Wright WE. A role for both RB and p53 in the regulation of human cellular senescence. Exp Cell Res 1991;196:33–9.

[80] He J, Kallin EM, Tsukada Y, Zhang Y. The H3K36 demethylase Jhdm1b/Kdm2b reg-ulates cell proliferation and senescence through p15(Ink4b). Nat Struct Mol Biol 2008;15:1169–75.

[81] Wu N, Rollin J, Masse I, Lamartine J, Gidrol X. p63 regulates human keratinocyte proliferation via MYC-regulated gene network and differentiation commitment through cell adhesion-related gene network. J Biol Chem 2012;287:5627–38.

[82] Willis-Martinez D, Richards HW, Timchenko NA, Medrano EE. Role of HDAC1 in senescence, aging, and cancer. Exp Gerontol 2010;45:279–85.

[83] Dali-Youcef N, Lagouge M, Froelich S, Koehl C, Schoonjans K, Auwerx J. Sirtuins: the 'magnificent seven', function, metabolism and longevity. Ann Med 2007;39:335–45.

[84] Chuang JY, Hung JJ. Overexpression of HDAC1 induces cellular senescence by Sp1/PP2A/pRb pathway. Biochem Biophys Res Commun 2011;407:587–92.

[85] Goding CR. Melanoma senescence: HDAC1 in focus. Pigment Cell Res 2007;20:336–8.

[86] Ogryzko VV, Hirai TH, Russanova VR, Barbie DA, Howard BH. Human fibroblast commitment to a senescence-like state in response to histone deacetylase inhibitors is cell cycle dependent. Mol Cell Biol 1996;16:5210–18.

[87] Munro J, Barr NI, Ireland H, Morrison V, Parkinson EK. Histone deacetylase inhibi-tors induce a senescence-like state in human cells by a p16-dependent mechanism that is independent of a mitotic clock. Exp Cell Res 2004;295:525–38.

[88] Zhang J, Bardot E, Ezhkova E. Epigenetic regulation of skin: focus on the Polycomb complex. Cell Mol Life Sci 2012;69:2161–72.

[89] Teschendorff AE, Menon U, Gentry-Maharaj A, et al. Age-dependent DNA methyla-tion of genes that are suppressed in stem cells is a hallmark of cancer. Genome Res 2010;20:440–6.

[90] Azuara V, Perry P, Sauer S, et al. Chromatin signatures of pluripotent cell lines. Nat Cell Biol 2006;8:532–8.

[91] Zhao XD, Han X, Chew JL, et al. Whole-genome mapping of histone H3 Lys4 and 27 trimethylations reveals distinct genomic compartments in human embryonic stem cells. Cell Stem Cell 2007;1:286–98.

[92] Eckert RL, Adhikary G, Rorke EA, Chew YC, Balasubramanian S. Polycomb group proteins are key regulators of keratinocyte function. J Invest Dermatol 2011;131:295–301.

[93] Cordisco S, Maurelli R, Bondanza S, et al. Bmi-1 reduction plays a key role in physio-logical and premature aging of primary human keratinocytes. J Invest Dermatol 2010;130:1048–62.

[94] Balasubramanian S, Adhikary G, Eckert RL. The Bmi-1 polycomb protein antagonizes the (−)-epigallocatechin-3-gallate-dependent suppression of skin cancer cell survival. Carcinogenesis 2010;31:496–503.

[95] Mattick JS, Makunin IV. Non-coding RNA. Hum Mol Genet 2006;15 Spec No. 1: R17–29.

[96] Banerjee J, Chan YC, Sen CK. MicroRNAs in skin and wound healing. Physiol Genomics 2011;43:543–56.

[97] Shin KH, Pucar A, Kim RH, et al. Identification of senescence-inducing microRNAs in normal human keratinocytes. Int J Oncol 2011;39:1205–11.

[98] Rivetti di Val Cervo P, Lena AM, Nicoloso M, et al. p63-microRNA feedback in keratinocyte senescence. Proc Natl Acad Sci USA 2012;109:1133–8.

[99] Lena AM, Mancini M, Rivetti di Val Cervo P, et al. MicroRNA-191 triggers keratinocytes senescence by SATB1 and CDK6 downregulation. Biochem Biophys Res Commun 2012;423:509–14.

[100] Martinez I, Dimaio D. B-Myb, cancer, senescence, and microRNAs. Cancer Res 2011;71:5370–3.

[101] Koster MI, Kim S, Mills AA, DeMayo FJ, Roop DR. p63 is the molecular switch for initiation of an epithelial stratification program. Genes Dev 2004;18:126–31.

[102] Maes OC, An J, Sarojini H, Wu H, Wang E. Changes in MicroRNA expression patterns in human fibroblasts after low-LET radiation. J Cell Biochem 2008;105: 824–34.

[103] Bonifacio LN, Jarstfer MB. MiRNA profile associated with replicative senescence, extended cell culture, and ectopic telomerase expression in human foreskin fibroblasts. PLOS ONE 2010;5:pii. e12519.

[104] Drummond MJ, McCarthy JJ, Sinha M, et al. Aging and microRNA expression in human skeletal muscle: a microarray and bioinformatics analysis. Physiol Genomics 2011;43:595–603.

[105] Bates DJ, Li N, Liang R, et al. MicroRNA regulation in Ames dwarf mouse liver may contribute to delayed aging. Aging Cell 2010;9:1–18.

[106] Ugalde AP, Ramsay AJ, de la Rosa J, et al. Aging and chronic DNA damage response activate a regulatory pathway involving miR-29 and p53. EMBO J 2011;30:2219–32.

[107] Chakrabarti SK, Francis J, Ziesmann SM, Garmey JC, Mirmira RG. Covalent histone modifications underlie the developmental regulation of insulin gene transcription in pancreatic beta cells. J Biol Chem 2003;278:23617–23.

[108] Viticchie G, Lena AM, Cianfarani F, et al. MicroRNA-203 contributes to skin re-epithelialization. Cell Death Dis 2012;3:e435.

[109] Leiss M, Beckmann K, Giros A, Costell M, Fassler R. The role of integrin binding sites in fibronectin matrix assembly in vivo. Curr Opin Cell Biol 2008;20:502–7.

[110] Grassel S, Unsold C, Schacke H, Bruckner-Tuderman L, Bruckner P. Collagen XVI is expressed by human dermal fibroblasts and keratinocytes and is associated with the microfibrillar apparatus in the upper papillary dermis. Matrix Biol 1999;18:309–17.

[111] Rando TA, Chang HY. Aging, rejuvenation, and epigenetic reprogramming: resetting the aging clock. Cell 2012;148:46–57.

[112] Villeda SA, Plambeck KE, Middeldorp J, et al. Young blood reverses age-related impairments in cognitive function and synaptic plasticity in mice. Nat Med 2014;20:659–63.

[113] Pickart L, Vasquez-Soltero JM, Margolina A. The human tripeptide GHK-Cu in prevention of oxidative stress and degenerative conditions of aging: implications for cognitive health. Oxid Med Cell Longev 2012;2012:324832.

[114] Crider KS, Yang TP, Berry RJ, Bailey LB. Folate and DNA methylation: a review of molecular mechanisms and the evidence for folate's role. Adv Nutr 2012;3: 21–38.

[115] Li Y, Daniel M, Tollefsbol TO. Epigenetic regulation of caloric restriction in aging. BMC Med 2011;9:98.

[116] Cosmetics & Toiletries. Sederma targets MicroRNAs to fade senescence. May 29, 2013.

[117] Nazaruk J, Galicka A. The influence of selected flavonoids from the leaves of *Cirsium palustre* (L.) Scop. on collagen expression in human skin fibroblasts. Phytother Res 2014;28:1399–405.

[118] Dudonne S, Coutiere P, Woillez M, Merillon JM, Vitrac X. DNA macroarray study of skin aging-related genes expression modulation by antioxidant plant extracts on a replicative senescence model of human dermal fibroblasts. Phytother Res 2011;25:686–93.

APPLICATIONS OF EPIGENETICS

Targeting Epigenetics in the Development of New Diagnostic Applications—Lessons from Autoimmune Diseases

Hui-Min Chen[1,2], Frederic L. Chedin[2], Christopher Chang[1], and Patrick S.C. Leung[1]

[1]Division of Rheumatology, Allergy and Clinical Immunology,
University of California, Davis, CA [2]Department of Molecular
and Cellular Biology, University of California, Davis, CA

19.1 INTRODUCTION

Epigenetics was first defined by Conrad Waddington in the 1940s as a branch of biology that studies the causal interactions between genes and their products which bring the phenotype into being. He also coined the term "epigenetic landscape" for the molecular mechanisms that convert this genetic information into observable traits or phenotypes [1]. A popular, and much-debated, modern definition of epigenetics has defined it as "stable and heritable patterns of gene expression and cellular function that do not involve any alterations to the original DNA sequence." A more recent and minimalist definition of epigenetics refers to "the structural adaption of chromosomal regions in order to register, signal, or perpetuate altered activity states" [2]. Epigenetic information consists of a series of posttranslational modifications, often referred to as "marks," added to histone proteins—the components of nucleosomes, and of cytosine DNA methylation. These marks, in their multiple

© 2015 Elsevier Inc. All rights reserved.

combinatorial patterns, create binding sites for epigenetic "readers" which tie the presence of particular marks to a specific biological output [3]. Altogether, epigenetic mechanisms account for a highly sophisticated regulatory system that plays critical roles in regulating nearly all nuclear processes including the precise and timely expression of genes, mRNA splicing, DNA recombination, genome stability, the suppression of transposable element mobility, and chromosome segregation. Not surprisingly, epigenetics plays fundamental roles during cellular differentiation and is necessary for X-chromosome inactivation and genomic imprinting [4–7].

One key difference between genetic and epigenetic information relates to the stability and the heritability of this information from cell to cell or generation to generation. It is now clear that epigenetic information is intrinsically reversible. Histone marks can be added and removed through the action of numerous "writers" and "erasers." In addition, the nucleosomes themselves only have a limited half-life on the chromatin template and undergo frequent turnover [8]. Chromatin-based epigenetic modifications are therefore best thought of as long-lived, but reversible, metastable states that can undergo dynamic fluctuations [9]. This concept has important implications for our understanding of the "heritability" of epigenetic information. Faithful propagation of epigenetic patterns over multiple cell divisions in somatic cells likely implies the existence of mechanisms that can instruct the reconstruction of these patterns post replication. Disruption of these mechanisms through genetic polymorphisms, inadequate supplies of epigenetic building blocks such as methyl groups and acetyl groups, and exposure to epigenetic disruptors, may sensitize cells and patients to epigenetic abnormalities.

The transgenerational inheritance of epigenetic information, and in particular of altered epigenetic states, remains a controversial area and is often ill-defined [10]. Early development, characterized by rapid cellular expansion and the differentiation of multiple cell types, is accompanied by clear epigenomic changes. In that context, it is conceivable that the maternal metabolic state, taken here to include diet, exposure to pollutants, and other stresses, can influence the epigenomic trajectory of the developing fetus, including its germ line and hence potentially the next generation [11,12]. In this sense, the human epigenome can be regarded as a biochemical record of past life events and exposures, linking them to present and future health outcomes and disease risks. In recent years, accumulating evidence has clearly demonstrated that, in addition to genetics, other complementary mechanisms such as epigenetics are involved in the pathogenesis of autoimmunity [1,7,13–33].

19.2 THE MOLECULAR BASIS OF EPIGENETICS

Epigenetic mechanisms primarily consist of DNA methylation and histone modifications. These epigenetic mechanisms are not mutually exclusive, and in many cases are mechanistically linked to each other, thereby synergistically affecting gene expression.

19.2.1 DNA Methylation

DNA methylation is currently the best studied epigenetic mechanism. It plays a key role in genomic imprinting, embryogenesis, cellular differentiation, tissue-specific development, and genome stability [34]. DNA methylation involves the transfer of a methyl group to the fifth position of a cytosine to form 5-methylcytosine (5mC). In mammals, DNA methylation occurs predominantly at CpG dinucleotides. DNA methylation can also exist in a CHG and CHH (H = A, C, or T) context, particularly in stem cells [35] and in neuronal tissues [36], but the function of non-CpG methylation, if any, remains unclear.

In general, DNA methylation is associated with long-term gene inactivation. DNA methylation can inhibit gene expression by various mechanisms. Methylated DNA can promote the recruitment of methyl-CpG-binding domain (MBD) proteins; MBD proteins in turn recruit histone-modifying and chromatin-remodeling complexes to methylated sites to alter the chromatin structure toward a compact, silent state [34,37]. It should be noted, however, that knocking out MBD protein function, through single and multiple gene knockouts, does not lead to the reactivation of gene expression from DNA methylated targets [38]. This suggests that MBD proteins may not be directly responsible for enforcing gene silencing in response to the presence of methylation marks. The factor(s) mediating gene repression in response to DNA methylation remain(s) to be fully identified. DNA methylation can in some instances directly inhibit transcription by precluding the recruitment of DNA-binding proteins from their target sites [39].

DNA methylation is catalyzed by DNA methyltransferase (DNMT) enzymes. This family is composed of four members: DNMT1, DNMT3A, DNMT3B, and DNMT3L [40,41]. The catalytic members of the DNMT family are classified into *de novo* and maintenance enzymes. DNMT3A and DNMT3B are involved in the *de novo* methylation of unmethylated and hemimethylated sites that occurs during embryonic development and germ line reprogramming. DNMT3L is expressed during gametogenesis when genomic imprinting takes place and is required for establishing maternal genomic imprinting. DNMT3L acts as a general stimulatory factor for DNMT3A and DNMT3B and interacts with them in the nucleus [40]. DNMT1 is the most abundant DNMT enzyme within the cell and is transcribed mostly during the S phase in the cell cycle. DNMT1, together with its conserved partner UHRF1 [42], localizes to the DNA replication fork, where they ensure that hemimethylated sites generated during semiconservative DNA replication are restored to full methylation on both DNA strands [43,44]. The combined action of the *de novo* and maintenance proteins ensures that DNA methylation patterns are faithfully copied from mother to daughter cells at each generation, therefore representing a form of mitotically stable epigenetic memory [45–47].

The most widely studied type of DNA methylation affects CpG islands (CGIs) promoter regions. CGIs are regions of more than 200 base pairs of a G + C content of at least 50% and a ratio of observed to statistically expected CpG frequencies of at least 0.6. About 60% of human genes are transcribed from CGI promoters. Despite the fact that CGIs are ideal substrates for the DNMT enzymes, they remain for the most part refractory to DNA methylation. This basal unmethylated state renders these regions transcriptionally permissive. Indeed, methylation of CGI promoters is associated with long-term gene silencing, as observed for instance during X-chromosome inactivation. Some CGIs (~6%) become methylated in a tissue-specific manner during early development or in differentiated tissues [16,48]. CGI shores, which are regions of lower CpG density located in close proximity (~2 kb) to CGIs, also undergo variable DNA methylation. As observed for CGI promoters, methylation of CGI shores is closely associated with transcriptional inactivation [49]. It is important to note that most tissue-specific DNA methylation occurs at CGI shores, but not at CGIs.

DNA methylation also occurs in gene bodies, where it appears to be positively correlated with gene expression, with the exception of highly transcribed genes which tend to be less methylated [50,51]. This apparent positive correlation may be explained by the fact that the DNMT3A *de novo* DNMT recognizes the H3K36 trimethylation mark [52] deposited along gene bodies during transcription elongation [53]. Extensive gene body methylation is thought to prevent spurious transcription initiation even when the chromatin fiber is undergoing dynamic transcription and transcription-coupled remodeling [53].

Finally, DNA methylation is most prevalent over repetitive DNA sequences. This type of DNA methylation is suggested to maintain chromosomal integrity, in particular over pericentromeric regions characterized by the accumulation of large arrays of satellite DNA repeats. DNMT3B appears to play a particular role there, and mutations in *DNMT3B* cause ICF syndrome, an immunodeficiency disorder associated with DNA hypomethylation and pericentromeric decondensation and instability [54,55]. DNA methylation also targets dispersed repeats

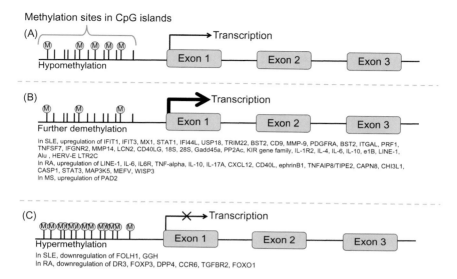

FIGURE 19.1 **DNA methylation regulates gene expression.** In general, DNA methylation leads to gene inactivation. In contrast, unmethylated DNA generates a chromatin structure favorable for gene expression. Aberrant DNA methylation has been reported in several autoimmune diseases. (A) In basal states, promoter regions are hypomethylated to allow gene expression. (B) Further promoter demethylation leads to gene transactivation and expression. (C) On the other hand, hypermethylation of promoter regions represses gene expression.

such as retrotransposons, thereby preventing the reactivation of endogenous retroelements [56,57]. Notably, in some autoimmune diseases such as systemic lupus erythematosus (SLE) and rheumatoid arthritis (RA), reactivation of endogenous retroelements has been observed. This phenomenon might play a role as the initial trigger of these diseases (see below) [58,59] and several perturbations in DNA methylation patterns have been reported in some autoimmune diseases (Figure 19.1).

DNA methylation can, in principle, be counterbalanced by DNA demethylation. DNA demethylation can be passive, that is, generated upon DNA replication in the absence of DNA maintenance methylation and can be active as well, which is thought to be enzyme-mediated DNA demethylation. Multiple mechanisms involving an oxidative pathway mediated by the ten eleven translocation (TET) family of enzymes and/or the deamination of methylated residues, followed by the repair of these modified bases have been proposed [45–47].

19.2.2 Histone Modifications

Modification of histone proteins plays an important role in transcriptional regulation, DNA repair [60], DNA replication, alternative splicing

[61], and chromosome condensation [62]. Histones are conserved proteins that package and organize DNA to form chromatin, the packaged form of DNA. Histone proteins can be differentiated into core histones (H2A, H2B, H3, and H4) and linker histones (H1 and H5). The core histones aggregate into two H2A−H2B dimers and one H3−H4 tetramer to form a histone octamer. A 147 bp segment of DNA is wrapped in 1.65 turns around the histone octamer to make up one nucleosome, the basic unit of chromatin. Neighboring nucleosomes are separated by ∼50 bp of linker DNA. The linker histones do not form part of the nucleosome; instead they bind to linker DNA by sealing off the nucleosome at the location where DNA enters and leaves [62,63]. The core histones are predominantly globular except for their N-terminal tails, which are unstructured [62].

Numerous posttranslational modifications occur in histone tails, such as acetylation, methylation, phosphorylation, ubiquitination, SUMOylation, and ADP-ribosylation [62,64]. These modifications induce changes to the structure of chromatin and thereby affect the accessibility of DNA to transcription factors and other enzymes, resulting in gene activation or repression (see below) [65]. The best understood histone modification is acetylation. Histone acetylation neutralizes positive charges on lysine and weakens the interactions between DNA and histone. As a result, histone acetylation is usually associated with upregulated transcriptional activity of the associated gene. Contrarily, histone deacetylation by histone deacetylases (HDACs) is thought to stabilize the local chromatin structure by restoring the positive charges in lysine residues. Therefore, HDACs are mainly considered to be transcriptional repressors [66].

Histone acetylation and deacetylation are catalyzed by histone acetyltransferases (HATs) and HDACs, respectively [67,68]. HATs promote gene expression by transferring an acetyl group to lysine, whereas HDACs contribute to gene repression by removing an acetyl group from the lysine tail. Notably, many transcriptional coactivators such as GCN5, PCAF, CBP, p300, Tip60, and MOF possess intrinsic HAT activity, whereas many transcriptional corepressor complexes such as mSin3a, NCoR/SMRT, and Mi-2/NuRD contain subunits with HDAC activity [69]. Unlike acetylation, histone methylation does not alter the charge of histone proteins. Histone methylation mainly occurs at lysine and arginine residues of histones H3 and H4 [66]. Depending on the position and the degree of methylation, histone methylation can be associated with gene activation or repression.

Histone modifications can occur within the same nucleosome [69], in the same histone tail [70], and among different histone tails [71], giving rise to a crosstalk among different signals. Notably, a single histone mark does not determine the outcome; instead, it is the combination of all marks that leads to a specific result. As an example, it has been shown that there are up to 51 distinct chromatin states existing based on combinations of histone modifications, suggesting that distinct biological roles are determined by the sum of different chromatin states [72]. Furthermore, euchromatin is characterized by high levels of acetylation and trimethylated H3K4, H3K36, and H3K79, whereas heterochromatin is characterized by low levels of acetylation and high levels of H3K9, H3K27, and H4K20 methylation [73]. Histone modification levels are predictive of gene expression: actively transcribed genes are characterized by high levels of H3K4me3, H3K27ac, H2BK5ac, and H4K20me1 in the promoter and H3K79me1 and H4K20me1 along the gene body [74]. Some perturbations in histone modifications are exemplified in several autoimmune diseases (Figure 19.2).

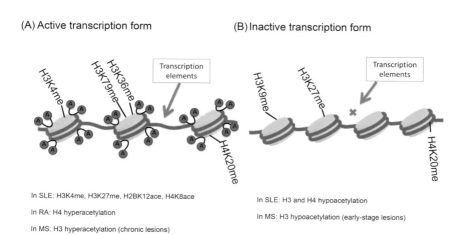

(A) Active transcription form

(B) Inactive transcription form

In SLE: H3K4me, H3K27me, H2BK12ace, H4K8ace

In RA: H4 hyperacetylation

In MS: H3 hyperacetylation (chronic lesions)

In SLE: H3 and H4 hypoacetylation

In MS: H3 hypoacetylation (early-stage lesions)

FIGURE 19.2 **Histone modifications.** Several posttranslational modifications occur in histone tails. These modifications induce changes of the chromatin structure and thereby affect the accessibility of the DNA to transcriptional factors and enzymes, resulting in gene activation or repression. (A) Histone acetylation neutralizes positive charges on lysine and then weakens the interactions between DNA and histone. As a result, histone acetylation is usually associated with upregulated transcriptional activity of the associated gene. Unlike acetylation, histone methylation does not alter the charge of histone proteins. Euchromatin (active transactivation) is characterized by high levels of acetylation and H3K4me, H3K36me, H3K79me, and H4K20me in the promoter, and H3K79me and H4K20me along the gene body (not shown). (B) In contrast, histone deacetylation is thought to stabilize the local chromatin structure by restoring the positive charges in lysine residues. Therefore, HDACs are mainly considered to be transcriptional repressors. Heterochromatin (inactive transactivation) is characterized by low levels of acetylation and high levels of H3K9me, H3K27me, and H4K20me.

19.3 EPIGENETIC PERTURBATIONS IN AUTOIMMUNE DISEASES—POTENTIAL TARGETS FOR THE DEVELOPMENT OF DIAGNOSTIC MARKERS AND NOVEL THERAPEUTIC INTERVENTIONS

19.3.1 Systemic Lupus Erythematosus

SLE is a systemic autoimmune disease manifested clinically by extreme fatigue, joint pain and/or swelling (arthritis), unexplained fever, and skin rash. SLE can affect many parts of the body, including the joints, skin, kidneys, heart, lungs, blood vessels, and brain. Immunologically, SLE is characterized by the presence of autoantibodies against nuclear and/or cytoplasmic antigens [75–78].

SLE is thought to result from a combination of genetic susceptibility and environmental exposures, leading to a breakdown in immune tolerance mechanisms [79]. Despite rigorous research to elucidate the etiology of SLE, the pathogenesis of SLE is still poorly understood. There have been extensive studies focusing on DNA methylation patterns in immune cell subsets from patients with SLE. For example, it has been reported that DNA from $CD4^+$ T cells in patients with SLE and subacute cutaneous SLE is predominantly hypomethylated. Moreover, promoter regions display trends toward DNA hypomethylation in $CD4^+$ T cells from SLE patients [80]. Lupus T cells showed a $15-20\%$ decrease in global DNA methylation levels. The percentage of 5mC was also inversely correlated with the SLE disease activity [81,82]. Furthermore, DNMT mRNA levels in T cells from acute SLE patients showed a 50% reduction compared with healthy controls, consistent with a $\sim 50\%$ decrease in DNMT enzyme activity found in active lupus T cells [82]. An array-based DNA methylation analysis identified 166 CpG sites in $CD19^+$ B cells, 97 CpG sites in $CD14^+$ monocytes, and 1033 CpG sites in $CD4^+$ T cells with significant changes in DNA methylation levels. Since type I IFN is often upregulated in SLE, it is not surprising that IFN-regulated genes in $CD4^+$ T cells from SLE patients, including *IFIT1*, *IFIT3*, *MX1*, *STAT1*, *IFI44L*, *USP18*, *TRIM22*, and *BST2*, are also hypomethylated [83]. It has also been reported that there are widespread and severe hypomethylated signatures in type I IFN signaling in active and quiescent SLE patients [84].

Furthermore, a widespread shift in methylation near genes involved in cell division and MAPK signaling was also observed in lupus $CD4^+$ T cells [84]. In another comparative DNA methylation study on $CD4^+$ T cells between lupus patients and healthy controls, 236 hypomethylated CpG sites and 105 hypermethylated CpG sites in lupus $CD4^+$ T cells were identified, including hypomethylation in a number of genes related to immune functions including *CD9*, *MMP-9*, *PDGFRA*, and *BST2*. At the same time, hypermethylation was reported in the genes *FOLH1* and *GGH*, which are involved in folate biosynthesis, and also in the transcription factor RUNX3 [19,76]. In this study, researchers also identified the transcription factor HNF4a as a regulatory hub affecting differentially methylated genes through protein—protein interaction maps. Interestingly, the methylation status of *RAB22A, STX1B2, LGALS3BP, DNASE1L1,* and *PREX1* also correlated with disease activity in lupus patients [76].

The expression of a number of methylation-sensitive autoimmune-related genes including *ITGAL (CD11a)*, *PRF1 (perforin)*, *TNFSF7 (CD70)*, and *CD40LG* was elevated in CD4$^+$T cells from patients with SLE [33,85]. In lupus CD4$^+$ T cells, the increased expression of CD11a mRNA was inversely correlated with DNA methylation. CD4$^+$T cells from active, but not inactive, SLE patients overexpressed perforin. This overexpression was related to demethylation of the promoter regulatory element, which normally suppresses perforin transcription in primary CD4$^+$ T cells [86]. HMGB1 mRNA and protein expression were significantly increased in lupus CD4$^+$ T cells and positively correlated with CD11a and CD70 mRNA expression and SLE clinical activity. HMGB1 binds to Gadd45a and may be involved in DNA demethylation in CD4$^+$T cells during lupus flares [87]. Lupus CD4$^+$T cells also displayed increased levels of Gadd45a proportional to the decreased global methylation [88].

Reduced ERK (extracellular signal-regulated kinases) signaling was responsible for the overexpression of CD11a and CD70 through decreased DNMT expression and correlated with SLE disease activity [89,90]. Deregulation of X-chromosome-located CD40LG may contribute to the striking female predilection of SLE. The X-chromosome-located CD40LG is unmethylated in men, but in healthy women, one allele is methylated and the other allele is unmethylated. In lupus patients, demethylated CD40LG in CD4$^+$T cells from women, but not men, led to the overexpression of CD40LG in lupus CD4$^+$ T cells and subsequent overstimulation of B cells [91—93].

Expression levels of PP2Ac were significantly increased in lupus T cells as a result of DNA demethylation in its promoter, and it was shown that the methylation intensity of the PP2Ac promoter correlated inversely with SLE disease activity [94]. *KIR* genes, normally suppressed by DNA methylation in CD4$^+$ T cells, were expressed in lupus T cells in proportion to disease activity. Abnormal KIR expression in lupus T cells was linked to IFN-gamma production and macrophage killing [95,96]. DNA hypomethylation was also reported in *IL-10* and *IL-1R2* in SLE patients. Furthermore, expression levels of IL-4 and IL-6 were significantly increased in lupus T cells as a result of demethylated CGIs in the promoter regions and was closely related to disease severity [97,98].

The clinical significance of DNA hypomethylation in reactivation of normally silent endogenous retroelements was also investigated in SLE. Hypomethylation of LINE-1 has been detected in lupus CD4$^+$ T cells,

$CD8^+$ T cells, and B cells, especially in $CD4^+$ and $CD8^+$ T cells from patients with active SLE. On the other hand, Alu hypomethylation was observed in $CD8^+$ T cells from an inactive-SLE group [59]. HERV-E LTR2C methylation levels in $CD4^+$ T cells of active SLE were significantly lower than those from inactive-SLE patients and healthy controls. Hypomethylation of HERV-E LTR2C was positively correlated with lymphopenia in active SLE. However, HERV-K LTR5_Hs hypomethylation was detected in $CD4^+$ T cells from patients with inactive SLE when compared with active-SLE patients and normal controls and correlated with complement activity and SLE disease activity [99]. Other than lupus $CD4^+$ T cells, CD5-E1B promoter was demethylated in B cells from SLE patients but not in healthy controls, and this differential methylation is more pronounced following B-cell antigen receptor (BCR) engagement [100] (Table 19.1, Figure 19.1).

Other studies have examined altered histone modifications in SLE. Nucleosomes, which are the primary sources of autoantigens in SLE, are released as a result of disturbed cell apoptosis and/or insufficient clearance of apoptotic debris. During apoptosis, nucleosomes are modified, creating more immunogenic epitopes. Subsequently epitopes spreading leads to the production of autoantibodies against nucleosomes. Histone modifications such as H3K4me3, H4K8 triacetylation, H3K27me3, and H2BK12 acetylation are known to cause increased apoptotic nucleosomes and further generate autoimmunogenicity [7,101–104]. Global histone H3 and H4 acetylation is reduced in lupus $CD4^+$ T cells. Levels of histone H3 acetylation are negatively correlated with SLE disease activity index [102].

Global histone H3K9 hypomethylation was observed in both active and inactive lupus $CD4^+$ T cells. The same study also found that SIRT1 mRNA levels were significantly increased in active lupus $CD4^+$ T cells, whereas mRNA levels of CREBBP, P300, HDAC2, HDAC7, SUV39H2, and EZH2 were significantly downregulated [102]. Expression levels of transcription regulatory factor cAMP-responsive element modulator alpha (CREMα) were increased in SLE. CREMα enhanced histone H3 methylation and CpG DNA methylation of human Notch-1 promoter, which contributed to decreased Notch-1 expression in various T-cell populations in SLE patients. Notch-1 expression levels inversely correlated with SLE disease activity and decreased Notch-1 levels were associated with elevated IL-17A levels [105]. Furthermore, CREMα bound to an as yet unidentified CRE site within the proximal promoter, leading to reduced expression of IL-17F in SLE lymphocytes. Overall, CREMα disrupted the balance between IL-17A and IL-17F in SLE T cells, in favor of IL-17A. An increased IL-17A/IL-17F ratio may aggravate the proinflammatory phenotype in SLE [106].

TABLE 19.1   Summary of Epigenetic Perturbations Involved in Human Autoimmune Diseases

	SYSTEMIC LUPUS ERYTHEMATOSUS
DNA methylation	DNA hypomethylation in the promoter regions of IFIT1, IFIT3, MX1, STAT1, IFI44L, USP18, TRIM22, BST2 [83], CD9, MMP-9, PDGFRA, BST2 [76], ITGAL [85], PRF1 [86], TNFSF7 [192], IFGNR2 [193], MMP14 [193], LCN2 [193], CD40LG [91], ribosomal RNA promoter regions (18S and 28S) [193], Gadd45a [88], PP2Ac [94], KIR gene family [95,96], IL-1R2, IL-4, IL-6, IL-10 [97,98], e1B [100], LINE-1, Alu [59], and HERV-E LTR2C [99]
	DNA hypermethylation in the promoter regions of FOLH1, GGH [76]
	~50% decrease of DNMT mRNA levels in lupus T cells, consistent with the ~50% decrease in DNMT enzyme activity [82]
Histone modifications	H3K4me3, H4K8 triacetylation, and H3K27me3, and H2BK12 acetylation cause increased apoptotic nucleosomes and generate autoimmunogenicity in SLE [7,101−104]
	Global histone H3 and H4 hypoacetylation in CD4$^+$ T cells; negatively correlated with SLE disease activity index (SLEDAI) [102]
	Global histone H3K9 hypomethylation in CD4$^+$ T cells; mRNA levels of SIRT1 were increased, whereas CREBBP, P300, HDAC2, HDAC7, SUV39H2, and EZH2 were downregulated [102]
	H3K27me3 was increased at the HPK1 promoter in lupus CD4$^+$ T cells [78]
	RHEUMATOID ARTHRITIS
DNA methylation	DNA hypomethylation in the promoter regions of LINE-1 [116,117], IL-6 [122,123], IL6-R [110], TNF-alpha [119], IL-10 [120], IL-17A [121], CXCL12 [124], CD40L [125], ephrinB1 [122,123], TNFAIP8/TIPE2 [110], CAPN8 [110], CHI3L1 [112], CASP1 [112], STAT3 [112], MAP3K5 [112], MEFV [112], and WISP3 [112]
	DNA hypermethylation in the promoter regions of DR3 [113−115], FOXP3 [121], DPP4 [110], CCR6 [110], TGFBR2 [112], and FOXO1 [112]
	Expression levels of SSAT1, AMD, PMFBP1, SLC3A2, and DASp are increased in RA synovial fibroblasts (RASFs), whereas SAM and DNMT1 are decreased [58,111,117,118]
Histone modifications	Strong histone acetylation in macrophages, RASFs, and RA synovial tissues [130]
	Increased histone H4 acetylation in MMP-1 promoter region in RASFs [23]
	Upregulated EZH2, a histone methyltransferase (HMT) enhancer in RASFs [132]
	Upregulated aurora kinase A and B accompanied by increased phosphorylation of histone H3 [133]
	MULTIPLE SCLEROSIS
DNA methylation	DNA hypomethylation in the promoter regions of PAD2 [145]
Histone modifications	Histone H3 hyperacetylation in chronic MS lesions and older patients, whereas early-stage MS lesions showed remarkable histone H3 deacetylation [150]

The expression and activity of the transcription factor RFX1 were decreased in lupus CD4$^+$T cells. RFX1 binds to promoter regions of CD11a and CD70, and then recruits DNMT1, SUV39H1, and HDAC1 to the CD11a and CD70 promoters, thereby repressing their expression. Downregulation of *RFX1* causes DNA hypomethylation, reduced H3K9 trimethylation, and histone H3 hyperacetylation, and subsequently contributes to CD11a and CD70 overexpression in lupus CD4$^+$ T cells [32,107]. Expression of E4BP4 was increased in lupus CD4$^+$T cells. E4BP4 directly regulated CD40L expression by binding to its promoter region and altering histone acetylation and methylation of *CD40L* loci. E4BP4 can negatively regulate self-reactivity of lupus CD4$^+$T cells and initiate a protective mechanism in these T cells [108]. H3K27me3 was increased at the HPK1 promoter in lupus CD4$^+$ T cells, consistent with significantly decreased HPK1 mRNA and protein levels. Downregulation of *HPK1* accelerated T-cell proliferation and production of IFN-gamma and IgG. A striking decrease in JMJD3 binding, but no marked change in EZH2 binding, was seen at the HPK1 promoter region in lupus CD4$^+$T cells, suggesting that inhibited HPK1 expression in lupus CD4$^+$T cells was associated with loss of JMJD3 binding and H3K27me3 enrichment at the HPK1 promoter, which contributed to T-cell overactivation and B-cell overstimulation in SLE [78] (Table 19.1, Figure 19.2).

19.3.2 Rheumatoid Arthritis

RA is an autoimmune disease characterized by the progressive destruction of peripheral joints, as is associated with chronic inflammation and pain in the joints. RA affects around 1% of the population. Although familial studies have established genetic risk factors for most phenotypes of RA, genetic variants identified so far explain only a small proportion of the total phenotypic variance. The low concordance rate of RA in monozygotic twins (12−22% [7]) suggests that environmental factors and epigenetics may also contribute to the etiopathogenesis of RA. Data from global methylation analysis showed that DNA methylation levels of peripheral blood mononuclear cells were significantly lower in patients with RA [109].

In RA, several cell types are involved in the destruction of the joints, whereas synovial fibroblasts are thought to be the most important among all [110]. The RASFs play a major role in the initiation and perpetuation of the disease. In addition to macrophages and lymphocytes, RASFs are also capable of producing a multitude of inflammatory cytokines and chemokines and thereby actively contribute to the inflammatory state in RA. Their intrinsic activation, inhibition of apoptosis, and recruitment of inflammatory cells including macrophages and lymphocytes are the main reasons for the excessive hyperplasia of the synovial tissue in RA joints [18]. Analysis of spermidine/spermine N1-acetyltransferase (SSAT1), S-adenosyl methionine decarboxylase (AMD), and polyamine-modulated factor 1-binding protein 1 (PMFBP1) expression levels in RASFs by flow cytometry revealed that their levels were significantly increased when compared with controls. Solute carrier family 3 member 2 (SLC3A2) and diacetylspermine (DASp) in cell culture supernatant of RASFs were also upregulated.

The expression of S-adenosyl methionine (SAM), DNMT1 protein, and 5-MeC was decreased but the parameters of polyamine metabolism were increased at the same time, which suggested that high consumption of SAM may contribute to global DNA hypomethylation in RASFs; global DNA hypomethylation has a crucial role in intrinsic activation of RASFs [111]. A genome-wide DNA methylation loci analysis in RASFs isolated from the site of disease in RA identified 1859 differentially methylated loci. Hypomethylated loci were identified in key genes relevant to RA, such as *CHI3L1*, *CASP1*, *STAT3*, *MAP3K5*, *MEFV*, and *WISP3*, and were associated with increased gene expression. Hypermethylation was also observed, including that of *TGFBR2* and *FOXO1*. Grouped analysis identified 207 hypermethylated or hypomethylated genes with multiple differentially methylated loci. Hypomethylation was increased in multiple pathways related to cell migration, including focal adhesion, cell adhesion, transendothelial migration, and extracellular matrix interactions [112].

The methylation status of CGIs within the promoter region of death receptor 3 gene *DR3*, a member of the apoptosis-inducing *Fas* gene family, was examined. The expression of DR3 protein was downregulated and the *DR3* gene promoter was specifically methylated in RASFs. This change may provide resistance to the apoptosis of RASFs [113−115]. Furthermore, it is also postulated that RASFs may play a critical role in the onset and development of RA through global DNA hypomethylation or CGIs hypomethylation at the endogenous retroelement LINE-1 promoter regions [116,117]. There is a 30- to 300-fold increase of LINE-1 in RASFs; there are fewer 5mC and less methylated CG sites upstream of the LINE-1 ORF in RASFs. Some reports showed that DNMT1 is deficient in proliferating RASFs [58,117,118].

The methylation status of cytokines and chemokines in RA has also drawn much attention and has yielded some interesting observations. For example, a high expression level of TNF-alpha and a low methylation status within the promoter of the gene was observed in RA [119]. Methylation of the CpG motif may also regulate expression of IL-10 in RA [120].

The *IL-17A* gene may also be regulated by promoter methylation. Synovial-infiltrating CD4$^+$ T cells displayed a demethylation state in IL-17A loci. *FOXP3* was significantly hypermethylated in RA peripheral blood CD4$^+$ T cells, whereas the methylation status of *IFN-gamma*, *IL-13*, and *IL-17* was similar to that in healthy controls [121]. In RA PBMCs, loss of methylation in the promoter regions of *IL-6* and *ephrinB1* is associated with increased expression of these genes in RA [122,123]. One DNA methylation profiling system using universal bead array has indicated that *IL-6R* is hypomethylated in RASFs [110]. The percentage of CpG methylation in the *CXCL12* promoter was lower in RASFs with a negative correlation between mRNA expression and *CXCL12* promoter methylation [124].

Interestingly, a comparison of DNA methylation of the *CD40L* gene on the X chromosome in CD4$^+$ T cells between female and male RA patients and controls showed that the *CD40L* promoter region in CD4$^+$ T cells has a higher degree of demethylation in female RA patients when compared with male RA patients and controls [125]. This phenomenon may be one explanation for the female predisposition to RA disease.

Using universal bead array, analysis of DNA methylation profiles of RASFs revealed a number of hypomethylated genes, including *TNFAIP8/TIPE2*, which is a negative mediator of apoptosis that plays a role in inflammation. Another candidate epigenetic gene is *CAPN8*, which has not previously been associated with RA, although it is involved in some inflammatory processes such as irritable bowel syndrome. In addition, other hypermethylated genes including *DPP4* and *CCR6* were detected. *DDP4* encodes a serine protease, which cleaves a number of regulatory factors, including chemokines and growth factors [110] (Table 19.1, Figure 19.1).

A shift toward histone hyperacetylation owing to the decreased HDAC activity and protein expression, particularly of HDAC1 and HDAC2, has been observed in RA synovial tissue [126]. However, this observation has not been consistent, as one study reported that HDAC activity was increased in RA synovial tissue [127], while other studies demonstrated that the expression of HDACs was not significantly different, with the exception of higher HDAC1 and lower HDAC4 [128,129]. Acetylation has also been reported to occur in macrophages and RASFs in RA synovial tissues [130]. The activity and expression of HDAC1 are significantly induced by TNF-alpha stimulation in RA synovium [131]. Acetylation of histone H4 was increased in the distal region of the *MMP-1* promoter in RASFs. However, the overexpression of SENP1 resulted in decreased acetylation, with accumulation of histone HDAC4 on the *MMP-1* promoter, leading to downregulation of MMP-1 [23].

Zest homologue 2 (EZH2), an HMT enhancer, is overexpressed in RASFs. Expression of secreted frizzled-related protein 1 (*SFRP1*), a target gene of EZH2, correlates with the occupation of its promoter with activating and silencing histone markers [132]. Aurora kinases A and B are highly expressed in treatment-naïve patients. Elevated aurora kinase expression was accompanied by increased phosphorylation of histone H3, which promotes proliferation of T cells [133] (Table 19.1, Figure 19.2).

19.3.3 Multiple Sclerosis

Multiple sclerosis (MS) is a chronic inflammatory and neurodegenerative autoimmune disease of the brain and spinal cord, characterized by myelin destruction followed by a progressive neurodegeneration [7,134]. Monozygotic twins have a concordance rate of around 6−31% in MS, despite sharing 100% of their genes [135−139]. These studies suggest that the development of MS is not determined by genetic factors alone. Furthermore, MS susceptibility is characterized by a maternal parent-of-origin effect with an increased female penetrance. Enhanced maternal transmission of risk alleles suggests that inheritable epigenetic patterns may influence MS susceptibility [140]. Although research focusing on epigenetic patterns in MS only began in the past decade, a growing body of literature suggests that epigenetic perturbations may be involved in the development and progression of MS.

It is known that epigenetic patterns are highly sensitive to environmental influences, indicating that the effects of known environmental risk factors on MS, such as smoking, vitamin D deficiency, and Epstein−Barr virus infection, may contribute to perturbations in patients' epigenetic profiles [134]. Epigenetic studies on MS are beginning to deliver important insights and to delineate the underlying pathophysiology of MS. Inflammation and demyelination in relapsing−remitting MS (RRMS) are related to macrophage activation and differentiation of a T-cell lineage toward a Th17 phenotype, which are both known to be epigenetically regulated mechanisms [121,141−143]. Moreover, citrullination of myelin basic protein (MBP) and CNS neurosteroid synthesis are critical neurodegenerative mechanisms regulated epigenetically [144,145].

It has been suggested that posttranslational citrullination of MBP, which is regulated by epigenetic mechanisms, plays a critical role in the pathogenesis of MS [146]. MBP is a major component of myelin in the CNS, and MBP can be modified in several ways after translation. In biopsy samples from 13 patients with MS, normal-appearing white matter (NAWM) revealed an increased level of citrullinated MBP compared with MBP levels in 18 normal controls and levels in six patients with Alzheimer's disease [147]. Citrullinated MBP is less stable than unmodified MBP, which may lead to myelin breakdown and eventually result in the development of autoimmune responses against MBP [145].

It is known that MBP is citrullinated by an enzyme named peptidyl arginine deiminase type II (PAD2) [148]. One study of NAWM in biopsy samples collected from 15 patients with MS (among them, 13 patients with progressive disease) demonstrated increased levels of both citrullinated MBP and PAD2 enzyme compared with healthy controls. DNA methylation of the promoter region of *PAD2* in MS patients was decreased to one-third compared with control groups. DNA demethylase activity in supernatants collected from MS patients was twofold higher than normal [145]. These findings suggest that hypomethylation in the *PAD2* promoter region leads to the overexpression of PAD2 enzyme, and consequently an increase in citrullinated MBP. Citrullination of MBP is regulated by DNA methylation in the promoter region and may play a role in the pathogenesis of MS.

A preliminary study was conducted to investigate DNA methylation patterns in relation to MS disease activity. The investigators included a panel of 56 gene promoter regions to study DNA methylation status using cell-free plasma DNA collected from 29 MS patients during relapse and 30 patients during remission. The relapse and remission states can be distinguished from each other with a sensitivity of 70.8% and specificity of 71.2%. This study suggests that samples from cell-free plasma DNA have the potential to be used as a diagnostic tool for MS [149] (Table 19.1, Figure 19.1).

With regard to the role of histone modifications in MS pathogenesis, one study has demonstrated changed histone acetylation patterns in early-stage MS [150]. Brain biopsy samples from 47 patients with progressive MS compared with those of 69 controls without neurological disorders, displayed a shift toward histone acetylation in the white matter. There was a trend of an increase in histone H3 acetylation in oligodendrocytes within the chronic MS lesions, whereas oligodendrocytes within early-stage MS lesions showed remarkable histone H3 deacetylation. Increased histone H3 acetylation was also seen in samples from older patients and the extent of acetylation correlated with disease duration. Increased histone H3 acetylation in oligodendrocytes was associated with impaired differentiation, as shown by high levels of transcriptional inhibitors for oligodendrocyte differentiation (such as TCF7L2, ID2, and SOX2) and higher HAT transcript levels (such as CBP and P300) in MS patients and might lead to impaired remyelination in patients with MS. Samples from MS patients also revealed an enrichment of acetyl-histone H3 at the promoter regions of target genes, such as *TCF7L2*. Altogether, these data suggest that histone deacetylation is a process that occurs in the early stages of the disease and whose efficiency decreases with the disease duration [150]. These data indicate that there may be a role for the evaluation of histone modifications as a diagnostic tool in MS (Table 19.1, Figure 19.2).

19.4 CLINICAL APPLICATIONS

Epigenetic mechanisms regulate gene functions in a manner that is maintained over multiple cell divisions but flexible enough to reflect environmental changes. This dual property of stability and plasticity renders epigenetic information highly useful for monitoring cellular and disease states. Recent rigorous studies on epigenetics and autoimmune diseases have generated a wealth of data [7,20,21,33,44,134,151,152] (Table 19.1) and given rise to much excitement and anticipation in how these data can be translated to clinical applications, with a focus on biomarker development and novel treatment.

19.4.1 Development of Diagnostic Tools

The goal of epigenetic biomarker development is to design standardized assays that can be applied to disease-specific diagnosis. Specifically, biomarkers provide physicians with relevant information about the presence or absence of a disease (diagnostic biomarkers) as well as about patient and disease characteristics that influence treatment decisions (prognostic and therapy-optimization biomarkers) (see Chapter 20). Biomarkers may also help to screen high-risk populations long before the onset of disease and provide biomarker-based selection of the patients that should receive treatment with epigenetic drugs. Before biomarkers can be utilized in these roles, they must be identified. Comparison of asymptomatic patients with specific autoimmune antibodies with those who are clinically affected may help identify biomarkers. Other strategies may include comparing the clinical, genetic, and immunologic characteristics of responders and nonresponders to a specific medication or treatment.

At the present time, the diagnosis of autoimmune diseases is frequently costly and often lengthy, as it is based on a combination of multiple clinical symptoms and laboratory tests. With the growing body of knowledge on epigenetics and autoimmune diseases, the ability to evaluate disease-specific epigenetic signatures using noninvasive methods, such as sampling from peripheral blood, may facilitate the diagnostic processes. Furthermore, the ability to analyze DNA methylation profiles from blood samples in high-risk populations before the onset of clinical symptoms may lead to early diagnosis and improved preventive medicine. The diagnostic potential for histone modifications currently faces more technical challenges compared with DNA methylation. Histone modifications are less stable and have been shown to be more dynamic in nature. The assessment of histone modifications is also more difficult to standardize and to apply in regular laboratory settings.

Candidate gene studies have identified a set of genes that undergo aberrant DNA hypomethylation in the promoter regions and consequently gene overexpression in SLE and RA. Classic methylation-sensitive autoimmune-related genes have been identified in lupus CD4$^+$ T cells, including *ITGAL (CD11a)*, *PRF1* (perforin), *TNFSF7 (CD70)*, *IFGNR2*, *MMP14*, *LCN2*, *PP2Ac*, *IL-4*, *IL-6*, *IL-10*, *IL-1R2*, and *CD40LG* [7,33,85]. In RA, several DNA methylation profiling studies have also indentified loci with a potential for diagnostic biomarkers, for example, hypomethylation of *IL-6R* in RASFs [110]; a high expression level of TNF-alpha and a low methylation status in its promoter region [119]; upregulation of *IL-17A* by promoter hypomethylation; hypermethylation of the *FOXP3* locus [121]; loss of methylation in the promoter regions of *IL-6* and *ephrinB1* [122,123]; hypomethylation of the *CXCL12* promoter [124]; DNA hypomethylation in *CD40L* [125]; hypermethylation of the *DR3* gene promoter relative to resistance to apoptosis [113–115], and global DNA hypomethylation at the LINE-1 promoter regions [58,116–118]. Notably, many of these epigenetic perturbations in SLE and RA have been shown to be correlated with disease activity and/or clinical scores. Therefore, they may be potential candidates for diagnostic and/or prognostic biomarkers and provide crucial information for treatment selection dependent on which epigenetic perturbations may be observed in individual patients. However, "driver or passenger," causality or effect of these epigenetic perturbations on disease onset is not fully confirmed. Few replication or reproducibility studies of these results and lack of systemic analysis render these data currently preliminary.

A comparative study of DNA methylation patterns between patients with MS, patients with RRMS, and healthy controls has shown that healthy controls can be distinguished from patients with RRMS with a sensitivity of 79.2% (remission) or 91.5% (exacerbation) and with a specificity of 92.9% (remission) or 91.5% (exacerbation) [149]. However, the clinical and pathological significance of DNA methylation profiles in MS is unclear.

19.4.2 Treatment of Diseases

Owing to the increased resolution of epigenetic profiling technologies, targeted epigenetic treatment, such as specific enzyme inhibitors, are undergoing extensive research. Epigenetically mediated changes in gene function might be reversible by drug treatments that target epigenetic mechanisms; for example, DNA hypermethylation may be modulated using DNMT inhibitors, and histone deacetylation can be counteracted with HDAC inhibitors. If administered appropriately and in timely fashion, the process of epigenetic perturbations is potential amenable to correction by targeting the enzymes involved. This approach will be particularly attractive if epigenetic perturbations are clearly identified as causal in the pathogenesis of the disease. Specific treatments that target epigenetic modifiers are the subject of current investigation.

There are two major classes of DNA methylation inhibitors, nucleoside analogs and nonnucleoside analogs. Nucleoside analogs consist of ribose or deoxyribose fused to a modified cytosine ring. Phosphorylation of the modified nucleoside into a nucleotide and subsequent incorporation into

DNA prevents DNA methylation. Non-nucleoside analogs are small molecules that bind to the active site of DNMTs or prevent the expression of DNMTs. A wide variety of drugs including short chain fatty acids, hydroxamic acids, benzamides, and cyclic peptides have been found to possess HDAC inhibition activity.

To date, five epigenetic drugs have been approved by the FDA for the treatment of cancer, including two DNMT inhibitors, two HDAC inhibitors, and one JAK1/2 inhibitor. Many more are in the pipeline for preclinical studies and clinical trials. More than 100 agents are in various stages of development, and the field of epigenetics holds exciting promise for new drug development. 5-Azacytidine (5-aza-CR; manufactured by Celgene) was approved in 2004 to inhibit DNA methylation. Two years later its variant 5-aza-2′-deoxycytidine was also approved (5-aza-CdR; manufactured by Eisai). Both are approved for the treatment of higher-risk myelodysplastic syndromes. In addition, S110, a dinucleotide containing 5-aza-CdR with enhanced stability and efficiency, has been added to the list of DNMT inhibitors. Vorinostat (manufactured by Merck), a pan-HDAC inhibitor for the treatment of cutaneous T-cell lymphoma, was approved in 2006. Romidepsin (manufactured by Celgene), a class-I HDAC inhibitor, also revealed remarkable efficacy for the treatment of cutaneous T-cell lymphoma and was approved in 2009. Two additional HDAC inhibitors, panobinostat (manufactured by Novartis) and CI-994 (manufactured by Pfizer), are currently undergoing phase III clinical trials for the treatment of lymphomas and non-small-cell lung cancer, respectively [151]. Ruxolitinib (manufactured by Incyte) is a Janus kinase 2 (JAK2) inhibitor that blocks histone H3Y41 phosphorylation. It was approved by FDA in 2011 for the treatment of intermediate- or high-risk myelofibrosis [153].

The old adage "first do no harm" should be mentioned in the development of epigenetic drugs and must be an important consideration. The concern surrounding use of DNMT inhibitors and HDAC inhibitors is the lack of specificity of these agents: treatment of the relevant autoimmune diseases would inhibit the target enzymes globally and, therefore, potentially lead to adverse events and toxic effects. New medications with well-defined target specificity, prolonged activity, and optimum delivery systems are required in the future.

The properties of HDAC inhibitors that can be targeted for treatment of RA include the effects on cytokine production, T-cell differentiation, and the function of macrophages, dendritic cells, osteoblasts, osteoclasts, and synovial fibroblasts. Clinical data have so far been consistent with those from animal models of RA and suggest that HDAC inhibitors are promising as a safe oral option as an alternative for RA treatment [154]. Another target of HDAC inhibitors is related to their ability to suppress bone destruction in chronic inflammatory diseases such as RA [155].

Although there are so far no clinical data or animal studies to substantiate the direct contribution of DNA methylation and/or histone modifications in the pathogenesis of MS, preclinical studies in experimental autoimmune encephalomyelitis (EAE) and clinical trials in other settings nonetheless suggest that this concept has some merit. *In vitro* studies have demonstrated the association of histone deacetylation in the major immunopathological mechanisms involved in MS, such as the differentiation and immunostimulatory capacity of dendritic cells [156,157] as well as oligodendrocyte maturation [158–160], suggesting that HDAC inhibitors may be of therapeutic potential in MS. Studies of mice with EAE also demonstrated that the HDAC inhibitors sodium phenylacetate and trichostatin A can ameliorate the symptoms of EAE [159–162]. Valproate is a powerful HDAC inhibitor. No studies have investigated the effects of valproate either in patients with MS or in animals with EAE, although one animal study showed that valproate treatment was associated with the preservation of axons and oligodendrocytes after spinal cord injury [163]. One important advantage of using valproate to inhibit HDACs is that it is a well-established generic drug with known tolerability and adverse-effect profile, rendering clinical trials comparatively potentially easier and more cost-effective. Further preclinical and clinical research on the effects of HDAC inhibitors in patients with MS and animals with EAE are certainly warranted. A more detailed discussion of the therapeutic applications of epigenetics can be found in Chapter 20.

19.5 FUTURE PERSPECTIVES

To date, research into the potential epigenetic mechanisms underlying autoimmune diseases has been dominated by relatively small cross-sectional studies that have either focused on a few genes or loci, or have utilized methods that provide limited information. Therefore, epigenetic perturbations in autoimmune diseases have not been fully characterized in a genome-wide quantitative manner. Genome-wide epigenetic studies in autoimmune diseases are few. These studies have identified epigenomic signatures in health and in disease, clustering subgroups as known and as new epigenetic correlations [33]. Epigenome-wide association studies (EWAS), which is analogous to the better known genome-wide association studies (GWAS), aim at identifying epigenetic markers associated with disease by performing hypothesis-free testing across the entire genome [164]. EWAS, if sufficiently powered with large sample sizes or twin designs with epigenome-wide coverage, will have great potential for the identification of biomarkers for disease onset, progression, or treatment response.

Notably, current studies have been focusing on identifying differences in epigenetic profiles between patients with autoimmune diseases and healthy controls. Epigenetic changes that are found associated with a particular phenotype may take place prior to the presentation of the phenotype, but even then may not necessarily be causally linked. In addition, epigenetic changes may occur as a consequence of diseases or interventions and not necessarily be the triggering step in disease susceptibility. Conclusions on causality cannot be drawn from retrospective data. Longitudinal cohorts will allow the sequence of change to be determined and point to underlying causality; therefore, these will be advantageous in teasing apart cause and effect. Furthermore, collecting samples at diagnosis before medication starts and then longitudinally comparing the epigenetic profiles between responders versus nonresponders for a specific medication may show epigenomic differences induced by treatment response or that previous differences preclude drug effects. These studies may help to guide treatment decisions and identify prognostic biomarkers.

Future studies will be informative if they have a prospective rather than cross-sectional design. Other factors, including methodological design, small sample sizes, lack of replicable data, less integrative validation with other biological-omics, and reduced genomic coverage of assays, render many results preliminary [33]. Further studies aimed at investigating potential epigenetic mechanisms underlying autoimmune diseases, utilizing rigorous study design, larger sample sizes, and more homogeneous populations, such as twin pairs, are clearly warranted.

With the Human Genome Project completed, the NIH Roadmap Epigenomics Mapping Consortium was launched in 2008 with the goal of providing a resource of human epigenomic data to catalyze basic biology and disease-oriented research. The Consortium leverages experimental pipelines built around next-generation sequencing technologies to map DNA methylation and histone modifications. By examining healthy and diseased tissues, specific genomic regions involved in disease development, tissue-specific expression, environmental susceptibility, and pathogenesis will be identified. Moreover, several international projects have also been established such as the ENCODE Project, the AHEAD Project, and the Epigenomics NCBI browser. The detailed study of epigenetic maps would be of enormous advantage in basic and applied research and would be relevant for focusing pharmacological research on the most promising epigenetic targets. These projects expect to deliver a collection of epigenomes that will provide a framework and reference for comparison and integration within a broad array of future studies.

Recently, microRNAs (miRNAs) have been shown as biomarkers for diagnosis in various diseases, including autoimmunity [110,165–169]. The biogenesis and mechanisms of action of miRNA have been reviewed extensively [170,171]. miRNAs are single-stranded, noncoding RNAs around 18–23 nucleotides in length that can negatively control their target gene expression posttranscriptionally by binding to cognate mRNA to repress translation. They are encoded in the genome and are generally transcribed by RNA polymerase II [172]. Aberrant miRNA expression has been extensively studied in cancers, inherited diseases, cardiovascular diseases, obesity, neurodegenerative diseases, and in autoimmune diseases [173]. In autoimmune diseases discussed herein, it is already known that miR-21, miR-29b, miR-126, miR-128a, miR-148a, miR-126, miR-155, miR-31, miR-182, and miR-96-183 cluster are upregulated, and miR-125a and miR-146a are downregulated in SLE [174–180]; miR-10b, miR-16, miR-18a, miR-137, miR-146a, miR-155, miR-203, miR-223, miR-335, miR-346, miR-454, miR-550, and miR-551b are upregulated and miR-124a, miR-204, miR-219-5p, miR-363, miR-498, miR-503, miR-542-5p, miR-596, miR-625, and miR-708 are downregulated in RA [110,181–188]; miRNA-34a, miRNA-155, miRNA-326, miRNA-17-5p, miR-155, miR-338, and miR-491 are upregulated and miR-17 and miR-20a are downregulated in MS [141–144,189]. These miRNA data undoubtedly provide a window for future development of diagnostic and prognostic biomarkers and hold promise for novel treatments for autoimmune diseases [190,191].

Research into the roles of epigenetic changes in autoimmune diseases is still in its infancy, and broad, comprehensive and well-designed studies of epigenetic mechanisms in patients with autoimmune diseases are scarce. Epigenomic information will be necessary to provide a comprehensive framework for understanding the pathogenesis underlying autoimmune diseases at the molecular level. Epigenomics can also offer novel opportunities for the development of diagnostic tests, such as DNA methylation biomarkers and histone modification profiles, as well as therapeutic interventions, such as correction of epigenetic defects and inhibition or activation of signal transduction pathways, aimed directly at blunting the aberrant reactions driving autoimmune diseases. The primary goals over the next decade will include improving our understanding of the interplay between epigenetic mechanisms, gene expression, and the environmental factors, and moving from animal models to clinical trials of novel epigenetic applications.

19.6 TAKE-HOME MESSAGES

1. Epigenetic perturbations, which may reflect environmental factors, play a critical role in the pathogenesis of autoimmune diseases. Epigenetics may also provide an insight into the female predominance seen in some autoimmune diseases.

2. Deregulation of epigenetic mechanisms, including DNA methylation and histone modifications, has been shown in many studies to correlate with the pathogenesis of autoimmune diseases. Many disease-related epigenetic perturbations are also linked to disease activity.

3. The data discussed above are critical to understand the full story of the pathogenesis of autoimmune diseases and can be utilized for future development of biomarkers for diseases, providing additional tools for diagnosis, disease monitoring, optimization of treatment, and prognostic prediction. They are also important in preventive medicine, personalized medicine, and new drug development.

4. Epigenome-wide approaches are necessary to develop clinical biomarkers for autoimmune diseases and other human diseases in the future.

List of Abbreviations

5-aza-CdR	5-aza-2′-deoxycytidine
5-aza-CR	5-azacytidine
5mC	5-methylcytosine
AMD	S-adenosyl methionine decarboxylase
CGI	CpG island
CREMα	cAMP-responsive element modulator alpha
DASp	diacetylspermine
DNMT	DNA methyltransferase
ERK	extracellular signal-regulated kinase
EWAS	epigenome-wide association studies
EZH2	zest homolog 2
GWAS	genome-wide association studies
HAT	histone acetyltransferase
HDAC	histone deacetylase
HMT	histone methyltransferase
JAK2	Janus kinase 2
MBD	methyl-CpG-binding domain
MBP	myelin basic protein
miRNA	microRNA
MS	multiple sclerosis
NAWM	normal-appearing white matter
ORF	open reading frame
PAD2	peptidyl arginine deaminase type II
PMFBP1	polyamine-modulated factor 1-binding protein 1
PRF1	perforin1
RA	rheumatoid arthritis
RASF	rheumatoid arthritis synovial fibroblast
RRMS	relapsing–remitting MS
SAM	S-adenosyl methionine
SFRP1	secreted frizzled-related protein 1
SLC3A2	solute carrier family 3 member 2
SLE	systemic lupus erythematosus

SLEDAI	SLE disease activity index
SSAT1	spermidine/spermine N1-acetyltransferase
TET	ten−eleven translocation
UHRF1	ubiquitin-like, containing PHD and RING finger domain1

References

[1] Dupont C, Armant DR, Brenner CA. Epigenetics: definition, mechanisms and clinical perspective. Semin Reprod Med 2009;27:351−7.
[2] Bird A. Perceptions of epigenetics. Nature 2007;447:396−8.
[3] Musselman CA, Lalonde ME, Cote J, Kutateladze TG. Perceiving the epigenetic landscape through histone readers. Nat Struct Mol Biol 2012;19:1218−27.
[4] Smith ZD, Meissner A. DNA methylation: roles in mammalian development. Nat Rev Genet 2013;14:204−20.
[5] Hemberger M, Dean W, Reik W. Epigenetic dynamics of stem cells and cell lineage commitment: digging Waddington's canal. Nat Rev Mol Cell Biol 2009;10:526−37.
[6] Reik W. Stability and flexibility of epigenetic gene regulation in mammalian development. Nature 2007;447:425−32.
[7] Quintero-Ronderos P, Montoya-Ortiz G. Epigenetics and autoimmune diseases. Autoimmune Dis 2012;2012:593720.
[8] Deal RB, Henikoff JG, Henikoff S. Genome-wide kinetics of nucleosome turnover determined by metabolic labeling of histones. Science 2010;328:1161−4.
[9] LaSalle JM, Powell WT, Yasui DH. Epigenetic layers and players underlying neurodevelopment. Trends Neurosci 2013;36:460−70.
[10] Heard E, Martienssen RA. Transgenerational epigenetic inheritance: myths and mechanisms. Cell 2014;157:95−109.
[11] Feil R, Fraga MF. Epigenetics and the environment: emerging patterns and implications. Nat Rev Genet 2011;13:97−109.
[12] Kaelin Jr. WG, McKnight SL. Influence of metabolism on epigenetics and disease. Cell 2013;153:56−69.
[13] Brooks WH, Le Dantec C, Pers JO, Youinou P, Renaudineau Y. Epigenetics and autoimmunity. J Autoimmun 2010;34:J207−19.
[14] Cooper ME, El-Osta A. Epigenetics: mechanisms and implications for diabetic complications. Circ Res 2010;107:1403−13.
[15] Costa-Reis P, Sullivan KE. Genetics and epigenetics of systemic lupus erythematosus. Curr Rheumatol Rep 2013;15:369.
[16] Esteller M. Epigenetics in cancer. N Engl J Med 2008;358:1148−59.
[17] Feinberg AP. Genome-scale approaches to the epigenetics of common human disease. Virchows Arch 2010;456:13−21.
[18] Gay S, Wilson AG. The emerging role of epigenetics in rheumatic diseases. Rheumatology (Oxford) 2014;53:406−14.
[19] Jeffries MA, Sawalha AH. Epigenetics in systemic lupus erythematosus: leading the way for specific therapeutic agents. Int J Clin Rheumtol 2011;6:423−39.
[20] Koch MW, Metz LM, Kovalchuk O. Epigenetics and miRNAs in the diagnosis and treatment of multiple sclerosis. Trends Mol Med 2013;19:23−30.
[21] Lu Q. The critical importance of epigenetics in autoimmunity. J Autoimmun 2013; 41:1−5.
[22] Luo Y, Wang Y, Wang Q, Xiao R, Lu Q. Systemic sclerosis: genetics and epigenetics. J Autoimmun 2013;41:161−7.
[23] Maciejewska-Rodrigues H, Karouzakis E, Strietholt S, Hemmatazad H, Neidhart M, Ospelt C, et al. Epigenetics and rheumatoid arthritis: the role of SENP1 in the regulation of MMP-1 expression. J Autoimmun 2010;35:15−22.

[24] Millington GW. Epigenetics and dermatological disease. Pharmacogenomics 2008; 9:1835–50.

[25] Ngalamika O, Zhang Y, Yin H, Zhao M, Gershwin ME, Lu Q. Epigenetics, autoimmunity and hematologic malignancies: a comprehensive review. J Autoimmun 2012; 39:451–65.

[26] Ospelt C, Reedquist KA, Gay S, Tak PP. Inflammatory memories: is epigenetics the missing link to persistent stromal cell activation in rheumatoid arthritis? Autoimmun Rev 2011;10:519–24.

[27] Renaudineau Y, Youinou P. Epigenetics and autoimmunity, with special emphasis on methylation. Keio J Med 2011;60:10–16.

[28] Richardson B. Primer: epigenetics of autoimmunity. Nat Clin Pract Rheumatol 2007; 3:521–7.

[29] Rodenhiser D, Mann M. Epigenetics and human disease: translating basic biology into clinical applications. CMAJ 2006;174:341–8.

[30] Selmi C, Mayo MJ, Bach N, Ishibashi H, Invernizzi P, Gish RG, et al. Primary biliary cirrhosis in monozygotic and dizygotic twins: genetics, epigenetics, and environment. Gastroenterology 2004;127:485–92.

[31] Wu C, Morris JR. Genes, genetics, and epigenetics: a correspondence. Science 2001; 293:1103–5.

[32] Zhao M, Sun Y, Gao F, Wu X, Tang J, Yin H, et al. Epigenetics and SLE: RFX1 down-regulation causes CD11a and CD70 overexpression by altering epigenetic modifications in lupus CD4+ T cells. J Autoimmun 2010;35:58–69.

[33] Zufferey F, Williams FM, Spector TD. Epigenetics and methylation in the rheumatic diseases. Semin Arthritis Rheum 2014;43:692–700.

[34] Esteller M. Epigenetic gene silencing in cancer: the DNA hypermethylome. Hum Mol Genet 2007;16 Spec No 1:R50–9.

[35] Lister R, Pelizzola M, Dowen RH, Hawkins RD, Hon G, Tonti-Filippini J, et al. Human DNA methylomes at base resolution show widespread epigenomic differences. Nature 2009;462:315–22.

[36] Lister R, Mukamel EA, Nery JR, Urich M, Puddifoot CA, Johnson ND, et al. Global epigenomic reconfiguration during mammalian brain development. Science 2013; 341:1237905.

[37] Lopez-Serra L, Esteller M. Proteins that bind methylated DNA and human cancer: reading the wrong words. Br J Cancer 2008;98:1881–5.

[38] Baubec T, Schubeler D. Genomic patterns and context specific interpretation of DNA methylation. Curr Opin Genet Dev 2014;25C:85–92.

[39] Kuroda A, Rauch TA, Todorov I, Ku HT, Al-Abdullah IH, Kandeel F, et al. Insulin gene expression is regulated by DNA methylation. PLOS ONE 2009;4:e6953.

[40] Chedin F. The DNMT3 family of mammalian de novo DNA methyltransferases. Prog Mol Biol Transl Sci 2011;101:255–85.

[41] Goll MG, Bestor TH. Eukaryotic cytosine methyltransferases. Annu Rev Biochem 2004; 74:481–574.

[42] Sharif J, Muto M, Takebayashi S, Suetake I, Iwamatsu A, Endo TA, et al. The SRA protein Np95 mediates epigenetic inheritance by recruiting Dnmt1 to methylated DNA. Nature 2007;450:908–12.

[43] Law JA, Jacobsen SE. Establishing, maintaining and modifying DNA methylation patterns in plants and animals. Nat Rev Genet 2010;11:204–20.

[44] Portela A, Esteller M. Epigenetic modifications and human disease. Nat Biotechnol 2010;28:1057–68.

[45] Kohli RM, Zhang Y. TET enzymes, TDG and the dynamics of DNA demethylation. Nature 2013;502:472–9.

[46] Pastor WA, Aravind L, Rao A. TETonic shift: biological roles of TET proteins in DNA demethylation and transcription. Nat Rev Mol Cell Biol 2013;14:341–56.

[47] Fritz EL, Papavasiliou FN. Cytidine deaminases: AIDing DNA demethylation? Genes Dev 2010;24:2107–14.

[48] Straussman R, Nejman D, Roberts D, Steinfeld I, Blum B, Benvenisty N, et al. Developmental programming of CpG island methylation profiles in the human genome. Nat Struct Mol Biol 2009;16:564–71.

[49] Irizarry RA, Ladd-Acosta C, Wen B, Wu Z, Montano C, Onyango P, et al. The human colon cancer methylome shows similar hypo- and hypermethylation at conserved tissue-specific CpG island shores. Nat Genet 2009;41:178–86.

[50] Jjingo D, Conley AB, Yi SV, Lunyak VV, Jordan IK. On the presence and role of human gene-body DNA methylation. Oncotarget 2012;3:462–74.

[51] Rauch TA, Wu X, Zhong X, Riggs AD, Pfeifer GP. A human B cell methylome at 100-base pair resolution. Proc Natl Acad Sci USA 2009;106:671–8.

[52] Dhayalan A, Rajavelu A, Rathert P, Tamas R, Jurkowska RZ, Ragozin S, et al. The Dnmt3a PWWP domain reads histone 3 lysine 36 trimethylation and guides DNA methylation. J Biol Chem 2010;285:26114–20.

[53] Wagner EJ, Carpenter PB. Understanding the language of Lys36 methylation at histone H3. Nat Rev Mol Cell Biol 2012;13:115–26.

[54] Ehrlich M, Sanchez C, Shao C, Nishiyama R, Kehrl J, Kuick R, et al. ICF, an immuno-deficiency syndrome: DNA methyltransferase 3B involvement, chromosome anomalies, and gene dysregulation. Autoimmunity 2008;41:253–71.

[55] Xu GL, Bestor TH, Bourc'his D, Hsieh CL, Tommerup N, Bugge M, et al. Chromosome instability and immunodeficiency syndrome caused by mutations in a DNA methyltransferase gene. Nature 1999;402:187–91.

[56] Bourc'his D, Bestor TH. Meiotic catastrophe and retrotransposon reactivation in male germ cells lacking Dnmt3L. Nature 2004;431:96–9.

[57] Yoder JA, Walsh CP, Bestor TH. Cytosine methylation and the ecology of intragenomic parasites. Trends Genet 1997;13:335–40.

[58] Ali M, Veale DJ, Reece RJ, Quinn M, Henshaw K, Zanders ED, et al. Overexpression of transcripts containing LINE-1 in the synovia of patients with rheumatoid arthritis. Ann Rheum Dis 2003;62:663–6.

[59] Nakkuntod J, Avihingsanon Y, Mutirangura A, Hirankarn N. Hypomethylation of LINE-1 but not Alu in lymphocyte subsets of systemic lupus erythematosus patients. Clin Chim Acta 2011;412:1457–61.

[60] Huertas D, Sendra R, Munoz P. Chromatin dynamics coupled to DNA repair. Epigenetics 2009;4:31–42.

[61] Luco RF, Pan Q, Tominaga K, Blencowe BJ, Pereira-Smith OM, Misteli T. Regulation of alternative splicing by histone modifications. Science 2010;327:996–1000.

[62] Kouzarides T. Chromatin modifications and their function. Cell 2007;128:693–705.

[63] Daujat S, Zeissler U, Waldmann T, Happel N, Schneider R. HP1 binds specifically to Lys26-methylated histone H1.4, whereas simultaneous Ser27 phosphorylation blocks HP1 binding. J Biol Chem 2005;280:38090–5.

[64] Rando OJ, Chang HY. Genome-wide views of chromatin structure. Annu Rev Biochem 2009;78:245–71.

[65] Dieker J, Muller S. Epigenetic histone code and autoimmunity. Clin Rev Allergy Immunol 2010;39:78–84.

[66] Bannister AJ, Kouzarides T. Regulation of chromatin by histone modifications. Cell Res 2011;21:381–95.

[67] de Ruijter AJ, van Gennip AH, Caron HN, Kemp S, van Kuilenburg AB. Histone deacetylases (HDACs): characterization of the classical HDAC family. Biochem J 2003;370:737–49.

[68] Roth SY, Denu JM, Allis CD. Histone acetyltransferases. Annu Rev Biochem 2001; 70:81−120.

[69] Wang Z, Zang C, Rosenfeld JA, Schones DE, Barski A, Cuddapah S, et al. Combinatorial patterns of histone acetylations and methylations in the human genome. Nat Genet 2008;40:897−903.

[70] Duan Q, Chen H, Costa M, Dai W. Phosphorylation of H3S10 blocks the access of H3K9 by specific antibodies and histone methyltransferase. Implication in regulating chromatin dynamics and epigenetic inheritance during mitosis. J Biol Chem 2008;283:33585−90.

[71] Nakanishi S, Lee JS, Gardner KE, Gardner JM, Takahashi YH, Chandrasekharan MB, et al. Histone H2BK123 monoubiquitination is the critical determinant for H3K4 and H3K79 trimethylation by COMPASS and Dot1. J Cell Biol 2009;186:371−7.

[72] Ernst J, Kellis M. Discovery and characterization of chromatin states for systematic annotation of the human genome. Nat Biotechnol 2010;28:817−25.

[73] Li B, Carey M, Workman JL. The role of chromatin during transcription. Cell 2007; 128:707−19.

[74] Karlic R, Chung HR, Lasserre J, Vlahovicek K, Vingron M. Histone modification levels are predictive for gene expression. Proc Natl Acad Sci USA 2010;107:2926−31.

[75] Deng Y, Zhao J, Sakurai D, Kaufman KM, Edberg JC, Kimberly RP, et al. MicroRNA-3148 modulates allelic expression of toll-like receptor 7 variant associated with systemic lupus erythematosus. PLOS Genet 2013;9:e1003336.

[76] Jeffries MA, Dozmorov M, Tang Y, Merrill JT, Wren JD, Sawalha AH. Genome-wide DNA methylation patterns in CD4+ T cells from patients with systemic lupus erythematosus. Epigenetics 2011;6:593−601.

[77] Yu C, Gershwin ME, Chang C. Diagnostic criteria for systemic lupus erythematosus: a critical review. J Autoimmun 2014;48−49:10−13.

[78] Zhang Q, Long H, Liao J, Zhao M, Liang G, Wu X, et al. Inhibited expression of hematopoietic progenitor kinase 1 associated with loss of Jumonji domain containing 3 promoter binding contributes to autoimmunity in systemic lupus erythematosus. J Autoimmun 2011;37:180−9.

[79] Pan Y, Sawalha AH. Epigenetic regulation and the pathogenesis of systemic lupus erythematosus. Transl Res 2009;153:4−10.

[80] Luo Y, Li Y, Su Y, Yin H, Hu N, Wang S, et al. Abnormal DNA methylation in T cells from patients with subacute cutaneous lupus erythematosus. Br J Dermatol 2008; 159:827−33.

[81] Corvetta A, Della Bitta R, Luchetti MM, Pomponio G. 5-Methylcytosine content of DNA in blood, synovial mononuclear cells and synovial tissue from patients affected by autoimmune rheumatic diseases. J Chromatogr 1991;566:481−91.

[82] Richardson B, Scheinbart L, Strahler J, Gross L, Hanash S, Johnson M. Evidence for impaired T cell DNA methylation in systemic lupus erythematosus and rheumatoid arthritis. Arthritis Rheum 1990;33:1665−73.

[83] Coit P, Jeffries M, Altorok N, Dozmorov MG, Koelsch KA, Wren JD, et al. Genome-wide DNA methylation study suggests epigenetic accessibility and transcriptional poising of interferon-regulated genes in naive CD4+ T cells from lupus patients. J Autoimmun 2013;43:78−84.

[84] Absher DM, Li X, Waite LL, Gibson A, Roberts K, Edberg J, et al. Genome-wide DNA methylation analysis of systemic lupus erythematosus reveals persistent hypomethylation of interferon genes and compositional changes to CD4+ T-cell populations. PLOS Genet 2013;9:e1003678.

[85] Lu Q, Kaplan M, Ray D, Ray D, Zacharek S, Gutsch D, et al. Demethylation of ITGAL (CD11a) regulatory sequences in systemic lupus erythematosus. Arthritis Rheum 2002;46:1282−91.

[86] Kaplan MJ, Lu Q, Wu A, Attwood J, Richardson B. Demethylation of promoter regulatory elements contributes to perforin overexpression in CD4+ lupus T cells. J Immunol 2004;172:3652−61.

[87] Li Y, Huang C, Zhao M, Liang G, Xiao R, Yung S, et al. A possible role of HMGB1 in DNA demethylation in CD4+ T cells from patients with systemic lupus erythematosus. Clin Dev Immunol 2013;2013:206298.

[88] Li Y, Zhao M, Yin H, Gao F, Wu X, Luo Y, et al. Overexpression of the growth arrest and DNA damage-induced 45alpha gene contributes to autoimmunity by promoting DNA demethylation in lupus T cells. Arthritis Rheum 2010;62:1438−47.

[89] Deng C, Kaplan MJ, Yang J, Ray D, Zhang Z, McCune WJ, et al. Decreased Ras-mitogen-activated protein kinase signaling may cause DNA hypomethylation in T lymphocytes from lupus patients. Arthritis Rheum 2001;44:397−407.

[90] Miyamoto A, Nakayama K, Imaki H, Hirose S, Jiang Y, Abe M, et al. Increased proliferation of B cells and auto-immunity in mice lacking protein kinase Cdelta. Nature 2002;416:865−9.

[91] Lu Q, Wu A, Tesmer L, Ray D, Yousif N, Richardson B. Demethylation of CD40LG on the inactive X in T cells from women with lupus. J Immunol 2007;179:6352−8.

[92] Sawalha AH, Wang L, Nadig A, Somers EC, McCune WJ; Michigan Lupus Cohort, et al. Sex-specific differences in the relationship between genetic susceptibility, T cell DNA demethylation and lupus flare severity. J Autoimmun 2012;38:J216−22.

[93] Zhou Y, Yuan J, Pan Y, Fei Y, Qiu X, Hu N, et al. T cell CD40LG gene expression and the production of IgG by autologous B cells in systemic lupus erythematosus. Clin Immunol 2009;132:362−70.

[94] Sunahori K, Juang YT, Kyttaris VC, Tsokos GC. Promoter hypomethylation results in increased expression of protein phosphatase 2A in T cells from patients with systemic lupus erythematosus. J Immunol 2011;186:4508−17.

[95] Liu Y, Kuick R, Hanash S, Richardson B. DNA methylation inhibition increases T cell KIR expression through effects on both promoter methylation and transcription factors. Clin Immunol 2009;130:213−24.

[96] Basu D, Liu Y, Wu A, Yarlagadda S, Gorelik GJ, Kaplan MJ, et al. Stimulatory and inhibitory killer Ig-like receptor molecules are expressed and functional on lupus T cells. J Immunol 2009;183:3481−7.

[97] Mi XB, Zeng FQ. Hypomethylation of interleukin-4 and -6 promoters in T cells from systemic lupus erythematosus patients. Acta Pharmacol Sin 2008;29:105−12.

[98] Lin SY, Hsieh SC, Lin YC, Lee CN, Tsai MH, Lai LC, et al. A whole genome methylation analysis of systemic lupus erythematosus: hypomethylation of the IL10 and IL1R2 promoters is associated with disease activity. Genes Immun 2012;13:214−20.

[99] Nakkuntod J, Sukkapan P, Avihingsanon Y, Mutirangura A, Hirankarn N. DNA methylation of human endogenous retrovirus in systemic lupus erythematosus. J Hum Genet 2013;58:241−9.

[100] Garaud S, Le Dantec C, Jousse-Joulin S, Hanrotel-Saliou C, Saraux A, Mageed RA, et al. IL-6 modulates CD5 expression in B cells from patients with lupus by regulating DNA methylation. J Immunol 2009;182:5623−32.

[101] van Bavel CC, Dieker JW, Tamboer WP, van der Vlag J, Berden JH. Lupus-derived monoclonal autoantibodies against apoptotic chromatin recognize acetylated conformational epitopes. Mol Immunol 2010;48:248−56.

[102] Hu N, Qiu X, Luo Y, Yuan J, Li Y, Lei W, et al. Abnormal histone modification patterns in lupus CD4+ T cells. J Rheumatol 2008;35:804−10.

[103] Amoura Z, Koutouzov S, Piette JC. The role of nucleosomes in lupus. Curr Opin Rheumatol 2000;12:369−73.

[104] Koutouzov S, Jeronimo AL, Campos H, Amoura Z. Nucleosomes in the pathogenesis of systemic lupus erythematosus. Rheum Dis Clin North Am 2004;30:529−58, ix.

[105] Rauen T, Grammatikos AP, Hedrich CM, Floege J, Tenbrock K, Ohl K, et al. cAMP-responsive element modulator alpha (CREMalpha) contributes to decreased Notch-1 expression in T cells from patients with active systemic lupus erythematosus (SLE). J Biol Chem 2012;287:42525–32.

[106] Hedrich CM, Rauen T, Kis-Toth K, Kyttaris VC, Tsokos GC. cAMP-responsive element modulator alpha (CREMalpha) suppresses IL-17F protein expression in T lymphocytes from patients with systemic lupus erythematosus (SLE). J Biol Chem 2012;287:4715–25.

[107] Zhao M, Wu X, Zhang Q, Luo S, Liang G, Su Y, et al. RFX1 regulates CD70 and CD11a expression in lupus T cells by recruiting the histone methyltransferase SUV39H1. Arthritis Res Ther 2010;12:R227.

[108] Zhao M, Liu Q, Liang G, Wang L, Luo S, Tang Q, et al. E4BP4 overexpression: a protective mechanism in CD4+ T cells from SLE patients. J Autoimmun 2013; 41:152–60.

[109] Liu CC, Fang TJ, Ou TT, Wu CC, Li RN, Lin YC, et al. Global DNA methylation, DNMT1, and MBD2 in patients with rheumatoid arthritis. Immunol Lett 2011; 135:96–9.

[110] de la Rica L, Urquiza JM, Gomez-Cabrero D, Islam AB, Lopez-Bigas N, Tegner J, et al. Identification of novel markers in rheumatoid arthritis through integrated analysis of DNA methylation and microRNA expression. J Autoimmun 2013;41:6–16.

[111] Karouzakis E, Gay RE, Gay S, Neidhart M. Increased recycling of polyamines is associated with global DNA hypomethylation in rheumatoid arthritis synovial fibroblasts. Arthritis Rheum 2012;64:1809–17.

[112] Nakano K, Whitaker JW, Boyle DL, Wang W, Firestein GS. DNA methylome signature in rheumatoid arthritis. Ann Rheum Dis 2013;72:110–17.

[113] Osawa K, Takami N, Shiozawa K, Hashiramoto A, Shiozawa S. Death receptor 3 (DR3) gene duplication in a chromosome region 1p36.3: gene duplication is more prevalent in rheumatoid arthritis. Genes Immun 2004;5:439–43.

[114] Fas SC, Fritzsching B, Suri-Payer E, Krammer PH. Death receptor signaling and its function in the immune system. Curr Dir Autoimmun 2006;9:1–17.

[115] Takami N, Osawa K, Miura Y, Komai K, Taniguchi M, Shiraishi M, et al. Hypermethylated promoter region of DR3, the death receptor 3 gene, in rheumatoid arthritis synovial cells. Arthritis Rheum 2006;54:779–87.

[116] Karouzakis E, Gay RE, Michel BA, Gay S, Neidhart M. DNA hypomethylation in rheumatoid arthritis synovial fibroblasts. Arthritis Rheum 2009;60:3613–22.

[117] Neidhart M, Rethage J, Kuchen S, Kunzler P, Crowl RM, Billingham ME, et al. Retrotransposable L1 elements expressed in rheumatoid arthritis synovial tissue: association with genomic DNA hypomethylation and influence on gene expression. Arthritis Rheum 2000;43:2634–47.

[118] Karouzakis E, Gay RE, Gay S, Neidhart M. Epigenetic control in rheumatoid arthritis synovial fibroblasts. Nat Rev Rheumatol 2009;5:266–72.

[119] Sullivan KE, Reddy AB, Dietzmann K, Suriano AR, Kocieda VP, Stewart M, et al. Epigenetic regulation of tumor necrosis factor alpha. Mol Cell Biol 2007;27:5147–60.

[120] Fu LH, Ma CL, Cong B, Li SJ, Chen HY, Zhang JG. Hypomethylation of proximal CpG motif of interleukin-10 promoter regulates its expression in human rheumatoid arthritis. Acta Pharmacol Sin 2011;32:1373–80.

[121] Janson PC, Linton LB, Bergman EA, Marits P, Eberhardson M, Piehl F, et al. Profiling of CD4+ T cells with epigenetic immune lineage analysis. J Immunol 2011;186:92–102.

[122] Nile CJ, Read RC, Akil M, Duff GW, Wilson AG. Methylation status of a single CpG site in the IL6 promoter is related to IL6 messenger RNA levels and rheumatoid arthritis. Arthritis Rheum 2008;58:2686–93.

[123] Kitamura T, Kabuyama Y, Kamataki A, Homma MK, Kobayashi H, Aota S, et al. Enhancement of lymphocyte migration and cytokine production by ephrinB1 system in rheumatoid arthritis. Am J Physiol Cell Physiol 2008;294:C189−96.

[124] Karouzakis E, Rengel Y, Jungel A, Kolling C, Gay RE, Michel BA, et al. DNA methylation regulates the expression of CXCL12 in rheumatoid arthritis synovial fibroblasts. Genes Immun 2011;12:643−52.

[125] Liao J, Liang G, Xie S, Zhao H, Zuo X, Li F, et al. CD40L demethylation in CD4(+) T cells from women with rheumatoid arthritis. Clin Immunol 2012;145:13−18.

[126] Huber LC, Brock M, Hemmatazad H, Giger OT, Moritz F, Trenkmann M, et al. Histone deacetylase/acetylase activity in total synovial tissue derived from rheumatoid arthritis and osteoarthritis patients. Arthritis Rheum 2007;56:1087−93.

[127] Smeets TJ, Barg EC, Kraan MC, Smith MD, Breedveld FC, Tak PP. Analysis of the cell infiltrate and expression of proinflammatory cytokines and matrix metalloproteinases in arthroscopic synovial biopsies: comparison with synovial samples from patients with end stage, destructive rheumatoid arthritis. Ann Rheum Dis 2003;62:635−8.

[128] Kawabata T, Nishida K, Takasugi K, Ogawa H, Sada K, Kadota Y, et al. Increased activity and expression of histone deacetylase 1 in relation to tumor necrosis factor-alpha in synovial tissue of rheumatoid arthritis. Arthritis Res Ther 2010;12:R133.

[129] Horiuchi M, Morinobu A, Chin T, Sakai Y, Kurosaka M, Kumagai S. Expression and function of histone deacetylases in rheumatoid arthritis synovial fibroblasts. J Rheumatol 2009;36:1580−9.

[130] Grabiec AM, Tak PP, Reedquist KA. Targeting histone deacetylase activity in rheumatoid arthritis and asthma as prototypes of inflammatory disease: Should we keep our HATs on? Arthritis Res Ther 2008;10:226.

[131] Grabiec AM, Tak PP, Reedquist KA. Function of histone deacetylase inhibitors in inflammation. Crit Rev Immunol 2011;31:233−63.

[132] Trenkmann M, Brock M, Gay RE, Kolling C, Speich R, Michel BA, et al. Expression and function of EZH2 in synovial fibroblasts: epigenetic repression of the Wnt inhibitor SFRP1 in rheumatoid arthritis. Ann Rheum Dis 2011;70:1482−8.

[133] Glant TT, Besenyei T, Kadar A, Kurko J, Tryniszewska B, Gal J, et al. Differentially expressed epigenome modifiers, including aurora kinases A and B, in immune cells in rheumatoid arthritis in humans and mouse models. Arthritis Rheum 2013;65: 1725−35.

[134] Koch MW, Metz LM, Kovalchuk O. Epigenetic changes in patients with multiple sclerosis. Nat Rev Neurol 2013;9:35−43.

[135] Ebers GC, Bulman DE, Sadovnick AD, Paty DW, Warren S, Hader W, et al. A population-based study of multiple sclerosis in twins. N Engl J Med 1986;315: 1638−42.

[136] Kuusisto H, Kaprio J, Kinnunen E, Luukkaala T, Koskenvuo M, Elovaara I. Concordance and heritability of multiple sclerosis in Finland: study on a nationwide series of twins. Eur J Neurol 2008;15:1106−10.

[137] Ristori G, Cannoni S, Stazi MA, Vanacore N, Cotichini R, Alfo M, et al. Multiple sclerosis in twins from continental Italy and Sardinia: a nationwide study. Ann Neurol 2006;59:27−34.

[138] French Research Group on Multiple Sclerosis. Multiple sclerosis in 54 twinships: concordance rate is independent of zygosity. Ann Neurol 1992;32:724−7.

[139] Hansen T, Skythe A, Stenager E, Petersen HC, Bronnum-Hansen H, Kyvik KO. Concordance for multiple sclerosis in Danish twins: an update of a nationwide study. Mult Scler 2005;11:504−10.

[140] Chao MJ, Herrera BM, Ramagopalan SV, Deluca G, Handunnetthi L, Orton SM, et al. Parent-of-origin effects at the major histocompatibility complex in multiple sclerosis. Hum Mol Genet 2010;19:3679−89.

[141] Du C, Liu C, Kang J, Zhao G, Ye Z, Huang S, et al. MicroRNA miR-326 regulates TH-17 differentiation and is associated with the pathogenesis of multiple sclerosis. Nat Immunol 2009;10:1252−9.

[142] Cox MB, Cairns MJ, Gandhi KS, Carroll AP, Moscovis S, Stewart GJ, et al. MicroRNAs miR-17 and miR-20a inhibit T cell activation genes and are underexpressed in MS whole blood. PLOS ONE 2010;5:e12132.

[143] Junker A, Krumbholz M, Eisele S, Mohan H, Augstein F, Bittner R, et al. MicroRNA profiling of multiple sclerosis lesions identifies modulators of the regulatory protein CD47. Brain 2009;132:3342−52.

[144] Noorbakhsh F, Ellestad KK, Maingat F, Warren KG, Han MH, Steinman L, et al. Impaired neurosteroid synthesis in multiple sclerosis. Brain 2011;134:2703−21.

[145] Mastronardi FG, Noor A, Wood DD, Paton T, Moscarello MA. Peptidyl argininedeiminase 2 CpG island in multiple sclerosis white matter is hypomethylated. J Neurosci Res 2007;85:2006−16.

[146] Moscarello MA, Mastronardi FG, Wood DD. The role of citrullinated proteins suggests a novel mechanism in the pathogenesis of multiple sclerosis. Neurochem Res 2007;32:251−6.

[147] Moscarello MA, Wood DD, Ackerley C, Boulias C. Myelin in multiple sclerosis is developmentally immature. J Clin Invest 1994;94:146−54.

[148] Lamensa JW, Moscarello MA. Deimination of human myelin basic protein by a peptidylarginine deiminase from bovine brain. J Neurochem 1993;61:987−96.

[149] Liggett T, Melnikov A, Tilwalli S, Yi Q, Chen H, Replogle C, et al. Methylation patterns of cell-free plasma DNA in relapsing−remitting multiple sclerosis. J Neurol Sci 2010;290:16−21.

[150] Pedre X, Mastronardi F, Bruck W, Lopez-Rodas G, Kuhlmann T, Casaccia P. Changed histone acetylation patterns in normal-appearing white matter and early multiple sclerosis lesions. J Neurosci 2011;31:3435−45.

[151] Heyn H, Esteller M. DNA methylation profiling in the clinic: applications and challenges. Nat Rev Genet 2012;13:679−92.

[152] Klein K, Gay S. Epigenetic modifications in rheumatoid arthritis, a review. Curr Opin Pharmacol 2013;13:420−5.

[153] Ho AS, Turcan S, Chan TA. Epigenetic therapy: use of agents targeting deacetylation and methylation in cancer management. Onco Targets Ther 2013;6:223−32.

[154] Vojinovic J, Damjanov N. HDAC inhibition in rheumatoid arthritis and juvenile idiopathic arthritis. Mol Med 2011;17:397−403.

[155] Cantley MD, Bartold PM, Fairlie DP, Rainsford KD, Haynes DR. Histone deacetylase inhibitors as suppressors of bone destruction in inflammatory diseases. J Pharm Pharmacol 2012;64:763−74.

[156] Reddy P, Sun Y, Toubai T, Duran-Struuck R, Clouthier SG, Weisiger E, et al. Histone deacetylase inhibition modulates indoleamine 2,3-dioxygenase-dependent DC functions and regulates experimental graft-versus-host disease in mice. J Clin Invest 2008;118:2562−73.

[157] Nencioni A, Beck J, Werth D, Grunebach F, Patrone F, Ballestrero A, et al. Histone deacetylase inhibitors affect dendritic cell differentiation and immunogenicity. Clin Cancer Res 2007;13:3933−41.

[158] Shen S, Li J, Casaccia-Bonnefil P. Histone modifications affect timing of oligodendrocyte progenitor differentiation in the developing rat brain. J Cell Biol 2005; 169:577−89.

[159] Faraco G, Cavone L, Chiarugi A. The therapeutic potential of HDAC inhibitors in the treatment of multiple sclerosis. Mol Med 2011;17:442−7.

[160] Gray SG, Dangond F. Rationale for the use of histone deacetylase inhibitors as a dual therapeutic modality in multiple sclerosis. Epigenetics 2006;1:67−75.

[161] Camelo S, Iglesias AH, Hwang D, Due B, Ryu H, Smith K, et al. Transcriptional therapy with the histone deacetylase inhibitor trichostatin A ameliorates experimental autoimmune encephalomyelitis. J Neuroimmunol 2005;164:10−21.

[162] Dasgupta S, Zhou Y, Jana M, Banik NL, Pahan K. Sodium phenylacetate inhibits adoptive transfer of experimental allergic encephalomyelitis in SJL/J mice at multiple steps. J Immunol 2003;170:3874−82.

[163] Penas C, Verdu E, Asensio-Pinilla E, Guzman-Lenis MS, Herrando-Grabulosa M, Navarro X, et al. Valproate reduces CHOP levels and preserves oligodendrocytes and axons after spinal cord injury. Neuroscience 2011;178:33−44.

[164] Rakyan VK, Down TA, Balding DJ, Beck S. Epigenome-wide association studies for common human diseases. Nat Rev Genet 2011;12:529−41.

[165] de Candia P, Torri A, Pagani M, Abrignani S. Serum microRNAs as biomarkers of human lymphocyte activation in health and disease. Front Immunol 2014;5:43.

[166] Hayes J, Peruzzi PP, Lawler S. MicroRNAs in cancer: biomarkers, functions and therapy. Trends Mol Med 2014;20:460−9.

[167] Jimenez SA, Piera-Velazquez S. Potential role of human-specific genes, human-specific microRNAs and human-specific non-coding regulatory RNAs in the pathogenesis of systemic sclerosis and Sjogren's syndrome. Autoimmun Rev 2013; 12:1046−51.

[168] Zhang Y, Zhao M, Sawalha AH, Richardson B, Lu Q. Impaired DNA methylation and its mechanisms in CD4(+)T cells of systemic lupus erythematosus. J Autoimmun 2013;41:92−9.

[169] Zhao M, Liu S, Luo S, Wu H, Tang M, Cheng W, et al. DNA methylation and mRNA and microRNA expression of SLE CD4+ T cells correlate with disease phenotype. J Autoimmun 2014;54:127−36.

[170] Schraivogel D, Meister G. Import routes and nuclear functions of Argonaute and other small RNA-silencing proteins. Trends Biochem Sci 2014;39:420−31.

[171] Tufekci KU, Meuwissen RL, Genc S. The role of microRNAs in biological processes. Methods Mol Biol 2014;1107:15−31.

[172] Chuang JC, Jones PA. Epigenetics and microRNAs. Pediatr Res 2007;61:24R−9R.

[173] Maqbool R, Ul Hussain M. MicroRNAs and human diseases: diagnostic and therapeutic potential. Cell Tissue Res 2014;358:1−15.

[174] Pan W, Zhu S, Yuan M, Cui H, Wang L, Luo X, et al. MicroRNA-21 and microRNA-148a contribute to DNA hypomethylation in lupus CD4+ T cells by directly and indirectly targeting DNA methyltransferase 1. J Immunol 2010;184:6773−81.

[175] Qin H, Zhu X, Liang J, Wu J, Yang Y, Wang S, et al. MicroRNA-29b contributes to DNA hypomethylation of CD4+ T cells in systemic lupus erythematosus by indirectly targeting DNA methyltransferase 1. J Dermatol Sci 2013;69:61−7.

[176] Zhao S, Wang Y, Liang Y, Zhao M, Long H, Ding S, et al. MicroRNA-126 regulates DNA methylation in CD4+ T cells and contributes to systemic lupus erythematosus by targeting DNA methyltransferase 1. Arthritis Rheum 2011;63: 1376−86.

[177] Zhao X, Tang Y, Qu B, Cui H, Wang S, Wang L, et al. MicroRNA-125a contributes to elevated inflammatory chemokine RANTES levels via targeting KLF13 in systemic lupus erythematosus. Arthritis Rheum 2010;62:3425−35.

[178] Dai R, Ahmed SA. MicroRNA, a new paradigm for understanding immunoregulation, inflammation, and autoimmune diseases. Transl Res 2011;157:163−79.

[179] Amarilyo G, La Cava A. miRNA in systemic lupus erythematosus. Clin Immunol 2012;144:26−31.

[180] Tang Y, Luo X, Cui H, Ni X, Yuan M, Guo Y, et al. MicroRNA-146A contributes to abnormal activation of the type I interferon pathway in human lupus by targeting the key signaling proteins. Arthritis Rheum 2009;60:1065−75.

[181] Stanczyk J, Pedrioli DM, Brentano F, Sanchez-Pernaute O, Kolling C, Gay RE, et al. Altered expression of microRNA in synovial fibroblasts and synovial tissue in rheumatoid arthritis. Arthritis Rheum 2008;58:1001–9.

[182] Nakamachi Y, Kawano S, Takenokuchi M, Nishimura K, Sakai Y, Chin T, et al. MicroRNA-124a is a key regulator of proliferation and monocyte chemoattractant protein 1 secretion in fibroblast-like synoviocytes from patients with rheumatoid arthritis. Arthritis Rheum 2009;60:1294–304.

[183] Murata K, Yoshitomi H, Tanida S, Ishikawa M, Nishitani K, Ito H, et al. Plasma and synovial fluid microRNAs as potential biomarkers of rheumatoid arthritis and osteoarthritis. Arthritis Res Ther 2010;12:R86.

[184] Nakasa T, Miyaki S, Okubo A, Hashimoto M, Nishida K, Ochi M, et al. Expression of microRNA-146 in rheumatoid arthritis synovial tissue. Arthritis Rheum 2008;58:1284–92.

[185] Pauley KM, Satoh M, Chan AL, Bubb MR, Reeves WH, Chan EK. Upregulated miR-146a expression in peripheral blood mononuclear cells from rheumatoid arthritis patients. Arthritis Res Ther 2008;10:R101.

[186] Stanczyk J, Ospelt C, Karouzakis E, Filer A, Raza K, Kolling C, et al. Altered expression of microRNA-203 in rheumatoid arthritis synovial fibroblasts and its role in fibroblast activation. Arthritis Rheum 2011;63:373–81.

[187] Alsaleh G, Suffert G, Semaan N, Juncker T, Frenzel L, Gottenberg JE, et al. Bruton's tyrosine kinase is involved in miR-346-related regulation of IL-18 release by lipopolysaccharide-activated rheumatoid fibroblast-like synoviocytes. J Immunol 2009;182:5088–97.

[188] Li J, Wan Y, Guo Q, Zou L, Zhang J, Fang Y, et al. Altered microRNA expression profile with miR-146a upregulation in CD4+ T cells from patients with rheumatoid arthritis. Arthritis Res Ther 2010;12:R81.

[189] Lindberg RL, Hoffmann F, Mehling M, Kuhle J, Kappos L. Altered expression of miR-17-5p in CD4+ lymphocytes of relapsing–remitting multiple sclerosis patients. Eur J Immunol 2010;40:888–98.

[190] Chen CZ, Schaffert S, Fragoso R, Loh C. Regulation of immune responses and tolerance: the microRNA perspective. Immunol Rev 2013;253:112–28.

[191] Singh RP, Massachi I, Manickavel S, Singh S, Rao NP, Hasan S, et al. The role of miRNA in inflammation and autoimmunity. Autoimmun Rev 2013;12:1160–5.

[192] Oelke K, Lu Q, Richardson D, Wu A, Deng C, Hanash S, et al. Overexpression of CD70 and overstimulation of IgG synthesis by lupus T cells and T cells treated with DNA methylation inhibitors. Arthritis Rheum 2004;50:1850–60.

[193] Javierre BM, Fernandez AF, Richter J, Al-Shahrour F, Martin-Subero JI, Rodriguez-Ubreva J, et al. Changes in the pattern of DNA methylation associate with twin discordance in systemic lupus erythematosus. Genome Res 2010;20:170–9.

Principles of Epigenetic Treatment

Shannon Doyle Tiedeken[1] and Christopher Chang[2]

[1]Department of Pediatrics, Thomas Jefferson University, Nemours/A.I. duPont Hospital for Children, Wilmington, DE [2]Division of Rheumatology, Allergy and Clinical Immunology, University of California, Davis, CA

20.1 INTRODUCTION

The use of medications and treatment modalities that involve epigenetic modifications existed long before the concept of epigenetics was proposed. DNA methylating agents have been used in the treatment of cancer for many years, but the true impact as an epigenetic mechanism was not fully appreciated until much later. The two primary modes of epigenetic regulation of gene expression involve DNA methylation and histone modification. Therefore, the epigenetic therapies used in treating cancer and autoimmune and other diseases involve agents that can affect these two critical processes of gene expression. There is considerable debate regarding the role of a third potential mode of therapy that may also affect gene expression, namely, the microRNAs. Some say that microRNAs are not strictly epigenetic and are simply molecules that happen to bind to portions of DNA that affect gene transcription. However, simplistically based on this characteristic, microRNAs may actually satisfy the definition of an epigenetic phenomenon.

The premise behind epigenetic transformation of gene expression embraces two additional features, namely, that the changes are hereditable and that they are reversible. It is because they are reversible that they are rendered susceptible to manipulation by currently used and potential future medications that affect the processes of epigenetics. Drugs that have DNA methylation activity or involve use of histone acetylase or deacetylase enzymes (HAT and HDAT, respectively), have already had

© 2015 Elsevier Inc. All rights reserved.

widespread use in the treatment of cancers. Other drugs that have been used, particularly some derived from plant products, have subsequently been found to have varying degrees of these activities as well.

20.2 DNA METHYLATION

Of the many promising targets that epigenetic therapies aim toward, DNA methylation is at the forefront. DNA methylation is described as the addition of a methyl group to coupled cytosine and guanine nucleotides known as CpG dinucleotides. Many promoter regions in the human genome are saturated with CpG dinucleotides, which makes DNA methylation of cytosine and guanine couplets positioned upstream of gene promoters promising marks for changing the expression of genes [1]. DNA methylation can silence cells directly by binding transcriptional activators dependent on methylation or indirectly by changing the affinity of proteins pertinent in chromatin remodeling [2]. Abnormal DNA methylation patterns have been more commonly seen in disease processes as compared to healthy cells [3]. In some studies, DNA methylation has been present in nearly every type of cancer [4]. Decreased methylation, also known as hypomethylation, prevalent in repetitive genomic elements, has been associated with chromatin that is active, while increased methylation, or hypermethylation, noted in the promoter regions of tumor suppressor genes, regulates and has the potential to silence genes. The hyper- and hypomethylation leads to disorganized chromatin and abnormal DNA replication [5]. These notable changes with methylation of DNA makes it a target for epigenetic therapy in multiple disease processes. (Figure 20.1).

20.2.1 Histone Modification

Histone modification and chromatin decompaction have also been noted to have huge potential in epigenetic therapies, since these markers are located all throughout the genome. Alterations, including phosphorylation, acetylation, methylation, ribosylation, and ubiquitination aimed at the core histones, can change the genetic expression of DNA [6]. The primary target of these potential changes, particularly acetylation, deacetylation, and methylation, is the N terminal of the histone tail. Specific lysine residues are necessary for the formation of chromatin domains [6]. Removing acetyl groups from the lysine residues by enzymes such as histone deacetylases (HDACs) makes the chromatin condensed and transcriptionally silent. The addition of acetyl groups to lysine residues by

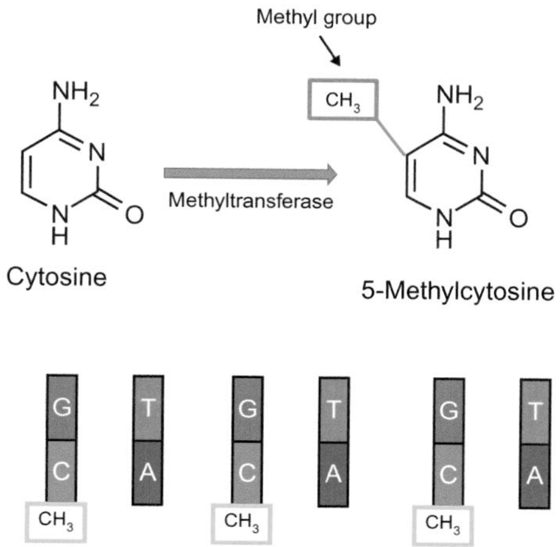

FIGURE 20.1 DNA methylation: addition of methyl groups to DNA bases, forming CpG dinucleotides.

enzymes known as histone acetyltransferases (HATs) makes the chromatin less condensed and more active. Histone modification allows for genome compartmentalization into transcriptionally active versus silent chromatin [7].

20.2.2 Histone Acetylation Enzymes/Histone Deacetylation Enzymes

Aside from DNA methylation, histone modification through acetylation, deacetylation, and methylation are some of the most well-studied epigenetic therapies. As stated above, histone acetyltransferases (HATs) mediate histone acetylation making euchromatin less condensed and more active. Working against HATs are HDACs whose role is to remove the acetyl group, making chromatin more compact and transcriptionally silenced. There are four different classes of HDACs, named classes I, II, III, and IV. Zinc ions have been noted to be necessary for HDAC classes I, II, and IV to function while class III's role is primarily in the area of immunity [8]. The interaction between histone acetylation and deacetylation within the immune system is very complex. HDACs have been noted to play a role in Toll-like receptor signaling, proinflammatory mediation through the inhibition of NF-κB, regulatory T-cell homeostasis, type I interferon production, and suppression of activated T-cell cytokine production. Given the complex role that HDACs have in the immune system as either proinflammatory activators or immunosuppressors, inhibition of the enzymes by HDAC inhibitors may be a critical factor in withdrawing and reversing the changes set forth by HDACs (Table 20.1).

TABLE 20.1   Histone Modifications and Their Effects on Chromatin's Shape and Function

Histone modification	Target of action	Chromatin shape	Function
Histone acetylation enzyme	Lysine residue, N terminal of histone tail	Less condensed	Transcriptionally more active
Histone deacetylation enzyme	N terminal of histone tail	More condensed	Transcriptionally silent
Histone deacetylation enzyme inhibitor	N terminal of histone tail	Less condensed	Transcriptionally more active
Histone methyltransferases	Lysine or arginine residues, N terminal of histone tail	More condensed	Transcriptionally silent
Histone demethylases	N terminal of histone tail	Less condensed	Transcriptionally more active

20.2.3 Histone Deacetylation Enzyme Inhibitors

The majority of HDACis have been studied with regard to their role in cancer research. Some others include butyrates, trichostatin A (TSA) used in treating fungal infections, and valproic acid used in treating seizure disorders—all of which can induce hyperacetylation leading to decondensation of chromatin [9]. One newer HDACi known as suberoylanilide hydroxamic acid (SAHA) has been made available in the treatment of different cancers including T-cell lymphomas and gliomas [10]. Another promising target for HDACis is in the treatment of rheumatoid arthritis (RA). TSA and phenylbutazone (PB) were reviewed regarding their utility in halting disease progression, and studies showed that targeting cell cycle inhibitors including p16^{INK4} and p21^{Cip1} can inhibit proliferation of synovial fibroblasts. Rat studies have shown decreased joint swelling in rodents that received TSA or topical PB [11]. An additional pearl noted in these findings was that the effects of TSA and PB persisted even after they were removed from the recipient [12]. Again these studies reinforced the belief of the important role that HDACis can play in eliminating or lessening disease severity.

20.2.4 Histone Methyltransferases/Histone Demethylases

Histone methyltransferases (HMTs), like HATs and HDACs, act on the histone tails, particularly amino acid side chains containing lysine and arginine [13]. The donor methyl group for HMTs comes from

S-adenosyl-methionine. Arginine is usually mono- or dimethylated while lysine can be mono-, di-, or trimethylated, which can vary during transcriptional activation [14]. Methylating one or more of the amino acids within the histone tail leads to transcriptional silencing similar to that with DNA methylation. It is believed that histone methylation is an irreversible process. Early studies showed that this process was steadfast because histones and methyl-lysines have the same half-lives [15]. In addition to these earlier studies, there is more recent evidence that methyl groups can undergo turnover, but at very low rates, through an active process known as demethylation. Certain demethylase enzymes have been named including *N*-methyl-tryptophan oxidase, sarcosine demethylase, and Elp3 [16]. Additional investigation is required to identify other possible demethylase enzymes and their future role in epigenetic therapies.

20.2.5 The Use of Epigenetic Therapy in Cancer

There has been much research surrounding the role of epigenetics in cancer given the crosstalk between the human genome and the epigenome [17]. A more current belief is that cancer, in part, is caused by the disruption of epigenetic regulatory mechanisms that are normally present as a function of immune surveillance [18]. The role of epigenetics in cancer is not straightforward but may be simultaneously dependent on DNA methylation, histone modifications, and nucleosome remodeling that affects chromatin structure and regulatory RNAs [19]. As tumors progress, the epigenome is altered by diffuse hypomethylation, increased methylation of CpG couplets in promoter regions, and changes in nucleosomes [17]. Recent evidence has shown that the mechanisms of genetics and epigenetics in cancer are not as separate as previously believed. They work together in a symbiotic relationship to accelerate tumorigenesis [17].

There are several epigenetic disruptions that cause key mutations in genes and signaling pathways that are responsible for cancer development. DNA methylation is one of those key disruptions. Tumor suppressor genes exhibit hypermethylation in the promoter region [18]. Genes that aid in DNA repair and cell cycle control including *RB, BRCA 1* and *2,* and *PTEN* are all mutated in cancer through hypermethylation [20]. The predominant mechanism through which genes are silenced in cancer is via promoter hypermethylation leading to loss of function [18]. In addition to tumor suppressor genes that are silenced, DNA repair genes are also subject to hypermethylation. These include hypermethylated *MGMT, KRAS* and *p53,* and *MLH1* seen in endometrial and colorectal cancer [21,22]. Epigenetics has also been shown to play a large role in ovarian carcinoma revealing a predominance of *p53* mutations and DNA methylation as well as additional mutations in genes such as *BRCA 1, BRCA 2,* and *RB* according to the Cancer Genome Atlas [23]. The silencing caused by epigenetics leads to abnormal proliferation pathways and causes an increased likelihood of mutations, leading to an increased risk of cancer [18].

In the 1990s, DNA hypermethylation and CpG couplets in promoter regions were identified as potential marks to be used in assessing cancer risk as well as early detection of cancer, prognosis, and response to therapy [24]. More research has been focused over the last decade on genes, their role in cancer progression and developing detection strategies of these genes [25,26]. Recent developments in colon cancer risk and detection include the detection of hypermethylated genes in blood and stool DNA and have proven to be highly specific and sensitive markers [27,28]. Urine samples and biopsy samples positive for glutathione S-transferase PI *(GSTP 1)* hypermethylation have been used in the detection of prostate cancer [29,30].

In addition to detection of hypermethylated genes, mapping of DNA methylation patterns has been deemed a possible way to identify cancers without a known primary site [4]. Determining DNA methylation patterns can also help predict the cancers' response to chemotherapy. In patients with glioma who have received treatment with an alkylating therapy such as temozolomide, the best survival time and response is predicted by hypermethylation that silences the DNA repair gene *MGMT* [22,31]. These data have recently been documented in an international phase III trial which should lead to Food and Drug Administration (FDA) approval and the ability to use this marker in clinical practice. DNA methylation markers have also been studied in nonsmall cell lung cancer and can be used to determine molecular prognosis and potential for cure [32]. As previously discussed, other potential biomarkers include chromatin abnormalities including patterns of histone acetylation and methylation, which can be used to determine the risk of tumor recurrence [33,34]. The future utility of these markers will depend on the future research in the field.

There has been much excitement surrounding the role of epigenetics in reversing mutations noted in different cancers [10,35]. 5-Azanucleosides which can demethylate DNA methyltransferases and make them inactive were previously felt to be too toxic for clinical use. But there has been recent work that has shown promise in therapeutic efficacy at very low drug doses [36,37]. 5-Azanucleosides at lower doses have been used in a trial of patients with myelodysplastic syndrome (MDS) and showed a delay in progression to leukemia and increased survival overall [37]. Two additional inhibitors, azacytidine and decitabine have been FDA approved for the treatment of MDS and show promise in treating leukemia as well as solid tumors [18] (Table 20.2).

TABLE 20.2   Potential Alternative Therapies in Treating and Preventing Cancer

Therapies in cancer	Trade name	Class	Mechanism of action
Azacitidine	Vidaza	5-Azanucleoside	Demethylates DNA methyltransferases
Decitabine	Dacogen	5-Azanucleoside	Demethylates DNA methyltransferases
Vorinostat	Zolinza	Histone Deacetylation Enzyme Inhibitor (HDACi)	Inhibits histone deacetylation enzymes
Romidepsin	Istodax	Histone Deacetylation Enzyme Inhibitor (HDACi)	Inhibits histone deacetylation enzymes

A second target that is currently under investigation is the class of HDACi [38]. Two FDA-approved HDACis include vorinostat and romidepsin for their use and efficacy against T-cell lymphoma [39,40]. HDACis have also shown promise in the major problem of resistance to different cancer therapies [41]. They have been shown to have the capability to reverse therapeutic resistance in select populations of stem-like cells in culture of different cancers including melanoma. The target has been the overexpression of *JARID1A*, which is a histone demethylase [41,42]. While HDACis have been shown to reverse the cancer process, epigenetic therapy alone has not proven to be curative. Therefore, a combination of epigenetic therapies and standard chemotherapeutic agents may offer the best strategy in fighting cancer [18].

Over the past decade there have been vast increases in progress and interest in the field of cancer epigenetics. There have been advances in using epigenetics as biomarkers to track cancer and in the actual treatment of different malignant processes. There is still much to learn surrounding the best combinations of therapies to use and how to tie them to current therapies in practice.

20.2.6 Epigenetic Therapy in Autoimmune Diseases

More recently, additional research has been done looking at the role of epigenetic therapy in autoimmune diseases such as systemic lupus erythematosus (SLE), multiple sclerosis (MS), inflammatory bowel disease (IBD), and RA. Autoimmune diseases occur due to an imbalance in the immune system and the inability to tolerate self antigens. For each specific disease state it has been noted that particular cell types may be involved including lymphocytes in SLE, neurons in MS, and synoviocytes in RA. Cytokines have also been noted to play a role in autoimmune disease and their inhibition may be a potential target for epigenetic therapy (Table 20.3).

TABLE 20.3   Potential Epigenetic Targets in Treating Autoimmune Diseases

Autoimmune disease	Potential targets	Potential drug name	Mechanism of action
Inflammatory bowel disease	HDACis	1. Butyrate	1. Inhibition of NF-κB-mediated inflammation; inhibits inducers of proinflammatory cytokines, lipopolysaccharide (LPS) and tumor necrosis factor (TNF); prevents movement of NF-κB from the cytoplasm to the nucleus
Multiple sclerosis	HDACis	1. TSA 2. Valproate 3. TSA; phenylacetate	1. Inhibition of *IL-2* gene expression, decreasing inflammation 2. Preservation of axons and oligodendrocytes 3. Unknown
Systemic lupus erythematosus	Overexpressed microRNAs	1. miR-126 inhibitor	1. Inhibition of miR-126 restores DNMT1 function; increased methylation of promoter regions (TNFSF7 and ITGAL); decreased production of CD11a and CD70; and decreased T- and B-cell autoreactivity
Rheumatoid arthritis	Overexpressed microRNAs, HDACis	1. miR-155 inhibitor 2. HDACi	1. Inhibition of miR-155 decreases collagen-induced arthritis, decreases expression of IL-6, IL-22, and IL-17, decreases pathogenic T and B cells; decreasing the number of osteoclasts leads to decreased bone destruction 2. Affect T-cell differentiation, cytokine production, and various cells: macrophages, dendritic cells, osteoclasts, osteoblasts, and synovial fibroblasts leading to decreased bone destruction

IBD is a complex disease but links have been made between the genome and epigenome. The vast majority of single-nucleotide polymorphisms (SNPs) associated with IBD are in noncoding regions, which suggests that they may play a role in gene expression regulation [43].

Epigenetics has played a role in learning how the environment regulates phenotype, which likely makes it critical in mediating effects of the microbiota and in how diet affects homeostasis in the intestine. Experiments conducted in mice have revealed that supplementing a maternal diet with methyl donors was linked to altered DNA methylation in the offspring and they had an increased risk of developing dextran sodium sulfate-induced colitis [44].

MicroRNA has been another potential area in epigenetic therapy that has been of great interest. MicroRNAs are 21−23 nucleotides long and function as noncoding RNAs whose role is to regulate posttranscriptional gene expression in nearly half of all protein coding genes in mammals. They can therefore alter gene expression. Thus, microRNAs have often been considered to have epigenetic properties, although whether or not they are truly epigenetic agents is still under considerable debate. MicroRNAs have also been shown to play a role in DNA methylation. They generally bind to nucleotides in the untranslated area of messenger RNA and inhibit translation of RNA. The up- and downregulation of microRNAs in gene expression has proven to play a pivotal role in the maintenance of immune homeostasis. The role that microRNAs play in the reversible modification of gene expression makes them a promising target for epigenetic therapy.

Further epigenetic studies have also shown that patients with ulcerative colitis as compared to control patients have variable expressions of microRNAs [45]. There have also been reports that analysis of a sample of peripheral blood can be used to differentiate the different subtypes of IBD by identifying microRNAs, making this a possible new biomarker [45,46].

HDACis have also shown promise in treating IBD. When the HDACi butyrate was applied to peripheral blood mononuclear cell (PBMC) cultures from biopsy specimens obtained from patients with Crohn's disease, NF-κB-mediated inflammation was noted to be inhibited. Butyrate was also able to affect mRNA expression of proinflammatory cytokines by inhibiting the production of the inducers LPS and TNF. Butyrate can also prevent the movement of NF-κB from the cytoplasm to the nucleus through inhibiting the degradation of I-κB, which is an NF-κB inhibitory protein [47].

HDACis have also been studied with regard to MS and there have been several promising mechanisms identified. Interferon gamma has been noted to play a role in the pathogenesis of MS due to its ability to induce cells, particularly antigen presenting cells, to secrete IL-12. IL-12 then aides in the process of differentiation of naïve Th0 cells to Th1 cells [48]. Th1 cells play a role in autoimmune diseases through the secretion of IL-2, which is a proinflammatory cytokine. The role of HDACis, particularly TSA, in MS is through the inhibition of *IL-2* gene expression, therefore decreasing inflammation [49,50]. Murine studies have also shown that HDACis, including TSA and phenylacetate, may have potential in treating patients with MS. Another potent HDACi is valproate. While no studies have directly looked at the effect of valproate in animals or humans with MS, one animal study has in fact revealed that after spinal cord injury, valproate is associated with preservation of axons and oligodendrocytes.

Research has expanded our knowledge of the role of microRNAs in autoimmune diseases [51], by virtue of their ability to regulate translation [52]. Several different microRNAs have been identified as being overexpressed in SLE but miR-126 was noted to be the most predominant. Increased microRNA levels led to decreased DNMT1 levels through inhibition of DNMT1 translation. In patients with SLE, increased miR-126 in CD4[+] T cells was associated with decreased activity of DNMT1, leading to decreased methylation of particular immune-related genes, *TNFSF7* and *ITGAL*, which are responsible for the coding of CD70 and CD11a [53]. This leads to overexpression of CD70 and CD11a, which results in increased B- and T-cell autoreactivity [54]. Through inhibiting miR-126, DNMT1 function is restored, resulting in increased methylation of the promoter regions of TNFSF7 and ITGAL. This causes decreased production of CD11a and CD70 which then in turn decreases T- and B-cell autoreactivity [53]. This has been one of the targets of epigenetic therapy in SLE. Additional microRNAs have been studied including miR-146a and miR-125a, which are both downregulated in SLE and potential markers of epigenetic therapy [55,56].

MicroRNAs are also promising in the treatment of RA, including miR-155 [57]. Murine studies have shown that knocking out miR-155 leads to decreased collagen-induced arthritis. Inhibiting miR-155 also leads to decreased expression of IL-6, IL-22, and IL-17, which leads to decreased numbers of pathogenic T and B cells [58]. By decreasing the number of osteoclasts, bone destruction is decreased, which makes miR-155 a prime target for epigenetic therapy in RA. HDACis have also shown promise in treating RA due to their effect on T-cell differentiation, cytokine production, function of various cells including macrophages, dendritic cells, osteoclasts, osteoblasts, and synovial fibroblasts, and with their ability to dampen bone destruction. HDACis have shown to be a safe and effective alternative oral therapy for RA.

In scleroderma, disease occurs due to aberrant fibroblasts that secrete extracellular matrix rich with collagen in place of normal tissue. It has been noted that various cytokines, including IL-4, platelet-derived growth factor beta, and transforming growth factor, inversely correlate with miR-29a levels. In polymyositis and dermatomyositis, microRNAs including miR-146b, miR-155, miR-214, miR-221, and miR-22 have been noted to be important [59]. MicroRNAs are already being used as biomarkers in Sjogren's syndrome including miR-17-92 cluster and miR-146a as previously discussed [60].

20.3 EPIGENETIC THERAPY IN DERMATOLOGY

In more recent studies, different dermatologic diseases have been associated with epigenetic variations. For example, in one variant of cutaneous T-cell lymphoma, Sezary syndrome, the pathogenesis is due to the loss of expression of the Fas receptor [59]. It is believed that the most prevalent cause of FAS inactivation is due to DNA hypermethylation, which is associated with epigenetics [60]. The pathogenesis behind melanoma has also been associated with an aberrant display of DNA methylation in genes such as *PTEN* and *CDKN2A* [61]. Histone modifications have also been noted to possibly enhance dermatologic malignancies as well including melanoma through the overexpression of *EZH2* and *SETDB1*, which both code for methyltransferases that can accelerate the development of melanoma [62].

In dermatopathology, epigenetics has shown promise in serving as a biomarker for the diagnostic process, prognostic evaluation, and response to therapy of cancers. Methylation patterns have been able to discriminate a nevus from melanoma [63]. This has also aided in refining the prognosis process. For example, in 60% of melanomas, *PTEN* is hypermethylated and associated with a poorer patient survival [64].

The use of epigenetic activity has also been used to treated different dermatopathologies. HDACis such as vorinostat and romidepsin have recently been approved by the FDA in the treatment of cutaneous T-cell lymphomas [65]. DNA methyltransferase (DNMT) inhibitors have been used in the treatment of melanoma and have had mixed results [66]. More recent data has also shown that HDACis at very low concentrations may have some therapeutic benefit [67]. Both HDACis and DNMT inhibitors upregulate the expression of surface molecules in melanoma cells including the melanoma antigen coding gene (*MAGE-1*) and major histocompatibility complexes [68–70]. Their use, in addition to other regimens including chemotherapy, radiotherapy, and immunotherapy, has shown promise. MicroRNAs have also been studied but there must be further differentiation of epigenetic regulation (Table 20.4).

20.3.1 Epigenetics in Neurological Diseases

The primary mechanisms behind regulating gene expression for memory formation and brain development are DNA modifications and

TABLE 20.4 Potential Epigenetic Therapies in Dermatologic Diseases

Dermatologic disease	Class	Drug name	Mechanism of action
Cutaneous T-cell lymphoma	HDACi	Vorinostat, Romidepsin	Inhibit histone deacetylation enzymes
Melanoma	DNMT inhibitor, HDACi	–	Upregulate the expression of melanoma antigen coding gene (*MAGE-1*) and major histocompatibility complexes

TABLE 20.5 Potential Epigenetic Therapies in Neurologic Diseases

Neurologic diseases	Cause	Potential treatment	Drug name	Mechanism of action
Huntington's disease	Histone deacetylation leads to neuronal death	HDAC inhibitors	Butyrate and phenylbutyrate	Rescue photoreceptors from neurodegeneration; decrease neuronal loss and increase motor function
Parkinson's disease	Alpha-synuclein binds to histones decreasing the levels of acetylated histones; causes death of dopaminergic neurons	HDAC inhibitors	–	Prevent toxicity induced by alpha-synuclein
Alzheimer's disease	Transmembrane amyloid precursor proteins form beta-amyloid peptides that join together forming plaques that disrupt transcription regulation	HDAC inhibitors	–	Restore histone acetylation, leading to learning and memory restoration

chromatin remodeling [71]. HDACis have been shown to improve symptoms in various neurologic disorders including Huntington's disease (HD), Parkinson's disease (PD), Alzheimer's disease (AD), and mood disorders [71] (Table 20.5).

HD is an inherited, autosomal dominant neurodegenerative disorder that is described as motor, cognitive, and psychiatric decline [72]. The neuronal loss that is seen is due to polyglutamine expansion in CAG, in the Huntingtin gene, leading to transcriptional dysregulation [72,73]. There is histone deacetylation that leads to neuronal death. In a *Drosophila* model, HDACis have been shown to rescue photoreceptors from neurodegeneration [74]. Mouse models were then completed that showed that HDACis, including butyrate and phenylbutyrate, decrease neuronal loss and increase motor function [75,76]. These studies have shown that epigenetic therapy may be efficacious in altering the course of the disease.

PD is also a progressive neurodegenerative disease caused by death of dopaminergic neurons. There is currently no therapy that dampens the death of dopaminergic neurons. Again, HDACis have shown promise in ameliorating the progressive degeneration seen in this disease. Through the use of a *Drosophila* model, it was shown that sequestering alpha-synuclein in the cytoplasm may have a protective role [77]. Alpha-synuclein directly binds to histones, decreasing the levels of acetylated histones. HDACis prevent toxicity induced by alpha-synuclein, making them a potential successful intervention in the cognitive and neurodegenerative processes seen in PD.

A third progressive neurodegenerative disease that epigenetic therapy may benefit is AD. In the AD disease process, transmembrane amyloid precursor proteins form beta-amyloid peptides that join together and form plaques. These plaques then disrupt transcription regulation [78]. Mouse models have shown that HDACis restore histone acetylation, which can lead to improved learning and memory [79].

Epigenetics has also been shown to be effective in treating various psychiatric and mood disorders. There is strong evidence to support that histone acetylation can be effective in treating depression [80]. Current therapies in treating depression target serotonin and norepinephrine levels but it takes several weeks to see an effect. Certain tricyclic antidepressants, such as imipramine, cause increased histone acetylation at particular promoter genes, leading to decreased HDAC5 levels in the hippocampus. Long-term use of tricyclic antidepressants has been shown to reverse and overcome hypermethylation, revealing that increased histone acetylation can have potential effects in reversing symptoms of depression. This information supports the idea that HDACis play a role in the mechanism of action of antidepressants [71].

Epigenetic therapy has been studied in the treatment of schizophrenia, a diagnosis made based on positive symptoms such as delusions and hallucinations or negative symptoms such as social withdrawal and apathy [81]. Recent studies have shown that epigenetics may play a role in the etiology of schizophrenia. Postmortem brains studied from patients with a diagnosis of schizophrenia revealed decreased levels of reelin, an extracellular matrix protein [82]. There are multiple DNA methylation sites on the reelin promoter region. HDACis and DNA methyltransferase inhibitors have been shown to increase reelin expression. Medications that target epigenetics are potential therapies for treating patients with mental disorders.

20.3.2 Epigenetics in Human Imprinting Disorders

Genes responsible for human imprinting play a large role in regulating growth, development, and behavior [83]. The expression of different genes is controlled by imprinting control regions (ICRs) that are inherited by the mother or father. DNA methylation controls ICRs. Disruption of human imprinting can be seen in different diseases including Beckwith—Wiedemann syndrome, Angelman syndrome, hydatidiform mole, and Prader—Willi syndrome [83]. Imprinted genes are usually clustered across the genome. Deletions occurring at the imprinting centers cause abnormal expression of genes, resulting in abnormal phenotypes. The targets of imprinting centers are allele-specific modifications seen in chromatin [84]. The main epigenetic modifications seen in imprinting are methylation of cytosines and histone acetylation. Many DNA-binding proteins and modifying enzymes have been linked to human epigenetic diseases. In Rett syndrome, mutations seen at the methylated CpG-binding protein, MECP2, did not lead to any changes in chromatin [85]; however, in mice, mutations in *Dnmt1* and *Dnmt31* have an effect on imprinting [86]. These studies showed that certain mechanisms that control epigenetic effects at imprinting centers may be different from mechanisms in other epigenetic diseases.

20.3.3 Epigenetics in Allergy and Allergic Dermatological Diseases

Recent research has revealed the role of epigenetics in asthma and allergic disease. Histone modifications and DNA methylation have been shown to regulate T-cell activation which leads toward allergic phenotypes [87]. Both Th1 and Th2 cells are stimulated through activation of their respective receptors, TCR/IL-4 and TCR/IL-12. Activation of these receptors leads to phosphorylation of STAT6 and STAT4, respectively, which leads to further expression of Th2 and Th1 master regulators [87]. The master regulator of Th2 cells is GATA3, involving cytokines including IL-4, IL-5, and IL-13; for Th1 cells the master regulator is T-BET, involving the signature cytokine IFN-gamma; the silencing of Th2 cytokines is the result. Both IL-4 and IFN-γ are methylated in CD4$^+$ T cells that are at rest [88]. In T cells that are allergen specific, the IL-4 promoter region is demethylated after allergic sensitization [88]. These histone modifications are also absent in naïve T cells, as compared to Th2 cells [89].

Th2 genes are located at the Th2 locus control region (LCR). The LCR interacts with GATA3, which leads to chromatin remodeling. This GATA3 and chromatin complex then binds to HDACs and leads to activation of Th1 genes and suppression of Th2 genes [90]. Th1 cytokines are further suppressed through DNA methylation; Th2 genes have exhibited a DNA demethylation mechanism which is not fully understood [87].

Allergies and asthma have been believed to be caused by a dual interaction between environmental and inherited factors [91]. However, genetics fails to account for the increase in the prevalence of allergies and asthma. Alternatively, it is suspected that epigenetics plays a role in the development of allergies and asthma, with epigenetic changes being induced by the environment and being passed on from parents to offspring [87]. There have been multiple loci suggested to play a role in both allergic phenotypes and environmental risk factors. Some of the selected loci associated with asthma have included FOXP3, IFN-γ, STAT5a, ARG2, ALOX12, ASCL3, and ADRB2 [87]. Other loci that have been linked to affecting the atopic phenotype include thymic stromal lymphopoietin (TSLP), prostaglandin D2 receptor (PTGDR), and cytochrome P450, family 26, subfamily A, polypeptide 1 (CYP26A1) [87]. Many loci from previous DNA methylation studies have been associated with environmental exposures such as tobacco smoke, farming, pollution, and pet keeping [87]. Studies have been completed to provide a description of epigenetic landscapes responsible for a particular phenotype which may be targets for future therapies and to serve as potential biomarkers [87]. However, since different cell types have different methylation patterns, any change in the ratio or proportion of Th1 and Th2 cells will affect the methylation pattern [87]. Alternatively, on a more exciting note, epigenetic marks have shown potential in silencing genes which can alter the effect of different polymorphisms on the risk of developing disease [87]. The role of epigenetics in asthma and allergy has only just begun to be studied but does shown future promise in modifying disease. Further studies to determine the underlying mechanisms and role epigenetics plays are needed.

20.3.4 Epigenetics in Other Diseases (Cardiovascular, Obesity, Renal Disease)

Epigenetics has also showed promise in the pathogenesis and treatment of other diseases. One such disease is cardiovascular disease (CVD) which has remained one of the leading causes of death worldwide. CVD has been associated with several modifiable and genetic risk factors yet still much of the disease is still unaccounted for. Some studies have shown promise in epigenetics having the ability to fill that void. Through histone modifications, microRNA changes, and DNA methylation, many risk factors for CVD including nutrition, stress, and smoking may be finally linked through modifications of epigenetic marks [92]. The role of epigenetic therapy is through altering and reversing inherited epigenetic modifications through pharmacologic agents. HDACs and HMTs are the targets for transcriptional regulation and modulation of genes that are associated with causes of CVD including inflammation, atherogenesis, and smooth muscle proliferation [92]. An additional target of epigenetic therapy is RNA interference which limits the transcription of genes that tend to be overexpressed in particular endothelial cells that leads to atherosclerosis [93]. Additional studies are needed to explore further pharmacologic therapeutics targeted specifically at epigenetic mechanisms affecting CVD.

Epigenetics has also exhibited a role in obesity and has been posited as a possible alternative etiology of metabolic disorders [94]. Many studies are exploring how the environment and one's diet affect the epigenome and risk of disease. The mechanism of epigenetics is believed to operate through regulation of energy balance, yet this is still not clear. Epigenetic modifications through DNA methylation and histone modifications are suspected to affect the expression of imprinted and nonimprinted metabolism genes [95]. Understanding the genome-wide mosaicisms and epigenetic differences in genes may help to understand the variation in fat mass seen in different people. Much future research is needed to identify specific genomic loci that house obesity-specific epigenetic modifications which can serve as marks for epigenetic therapy [95].

Epigenetics has been shown to play a role in renal disease as well, primarily through RNA interference. MicroRNAs are necessary to maintain glomerular homeostasis and any disruption thereof can lead to renal disease. Epigenetic modifications in the kidney have been shown to cause fibrosis in renal tissue [96]. In mouse podocyte studies, when an enzyme, Dicer, which aids in the production of microRNAs, was inactivated, the mice developed proteinuria and podocyte dedifferentiation which progressed to chronic kidney disease [97]. The mice died secondarily to renal failure [98]. Additional studies have shown that epigenetic modifications, such as DNA methylation, can lead to fibrosis within the kidney parenchyma. Hypermethylation of the RAS protein activator like-1 gene (RASAL1) led to the persistent activation of fibroblasts, resulting in fibrogenesis of the kidney [99]. These studies have provided preliminary evidence that epigenetic modifications through transcriptional regulation may lead to glomerular and interstitial fibrosis [100]. Many epigenetic therapeutic agents are in development but at this point they are lacking specific targets. Drugs include DNA methyltransferase inhibitors and HDACis [101]. Again, further studies are needed to enable elaboration on the targets and efficacy of these drugs in renal disease.

20.4 DISCUSSION

The main targets of epigenetic modification are DNA methylation and histone modifications via acetylation, deacetylation, and methylation, which can all be targeted to alter gene expression. DNA methylation has been well studied for decades pertaining to its role in cancer. CpG couplets in promoter regions have been used as early markers for detection of cancer as well as prognosis and potential response to chemotherapy. Epigenetics, particularly DNA methylation, have been studied in both colon cancer and prostate cancer but most recently have been highlighted as a possible avenue to identify cancers when the primary site is unknown. HDACis have also been targeted in different cancers which have shown resistance to known therapies and have the capacity to reverse resistance. Despite all of this progress, epigenetic modifications have not yet been shown to be curative and treatments still require standard chemotherapeutic agents. While there has been much progress surrounding epigenetics and cancer, still more research needs to be done specifically looking at the future potential of treating cancer with epigenetic modifications as the sole therapy by determining the exact underlying cause. In autoimmune diseases, more specific cell types have been identified as potential targets in epigenetic therapy. One newer area in the field of epigenetics is microRNAs, which have shown promise in changing gene expression in diseases such as IBD, SLE, and RA. MicroRNAs have also been noted to have an effect on DNA methylation. Given their potential reversible modification of gene expression, microRNAs have proven to be a new and promising target in epigenetic therapy.

Of the many fields potentially affected by epigenetic modifications, one that may benefit the most is neurology. Many devastating diseases including HD and AD have the potential to be treated by epigenetic modifications, which can have significant effects on the lifespan and lifestyle of many patients. Also, epigenetics has proved to be a possible intervention in treating and preventing obesity. Obesity is the starting point for many other comorbidities including metabolic syndrome, hypertension, hypercholesterolemia, diabetes, and coronary artery disease. If epigenetics can be targeted to change people's body habitus, the number of comorbid conditions can be significantly decreased, which can lead to a decrease in medical healthcare costs. Among all of the epigenetic modifications discovered, there may be new ways of thinking about diseases and choices of treatment on the horizon. Many more studies are still needed to further fine-tune all of the modifications and potential therapeutic agents.

Despite the above accolades spouting the potential promise and benefits of epigenetic therapies, one should still proceed with a degree of caution. Every drug or treatment modality, however successful, has its other "dark" side. Some drugs are associated with severe side effects but may have life-saving benefits. Others may relieve symptoms with very few side effects. Whenever we administer a drug or treatment to a patient, we must always consider benefits and risks. Sometimes, the risk is not always completely known. Clinical trials are performed on a limited number of subjects, and long-term side effects are often not recognized until years of use of the drug in the real world. This may be particularly true with epigenetic therapies, where we are making significant changes in the basic inherited genetic makeup of individuals. The lesson is to proceed with caution and humility. It is safe to say that we know more about the universe than we do about how the human body works. As we develop more and more promising therapies based on new research and knowledge, we must always consider the fact that we can truly never completely understand the immune and biological mechanisms that affect how we function, how we differ from each other, and how we may respond differently to various treatments for the diseases that plague us during our lifetimes.

20.5 CONCLUSION

While epigenetic modifications have been used as targets, biomarkers, and therapies in cancer for many decades, the mechanisms underlying these processes have only been more recently studied and understood. However, more recently epigenetics has proven successful in targeting and treating other disease processes including autoimmune diseases, neurologic diseases, dermatologic diseases, asthma, and allergies. From an autoimmune standpoint, there has been success in treating RA, IBD, SLE, and MS. Advances have also been made in originating new promising treatment strategies for anxiety disorders and other devastating neurologic disorders including HD, PD, and AD. Newer studies have also linked epigenetics with asthma, allergies, and even obesity which can play a huge role in how we treat obese patients who are at increased risk for further diseases including metabolic syndrome, hypertension, and heart disease. This is an exciting field with many promising prospects on the horizon. While much more research is needed to further understand and pinpoint specific targets and biomarkers and to refine therapeutic agents, epigenetics can serve as the next large area of medicine with the potential to treat and possibly prevent many future morbidities and mortalities.

References

[1] Saxonov S, Berg P, Brutlag DL. A genome-wide analysis of CpG dinucleotides in the human genome distinguishes two distinct classes of promoters. Proc Natl Acad Sci USA 2006;103(5):1412–17.

[2] Suzuki MM, Bird A. DNA methylation landscapes: provocative insights from epigenomics. Nat Rev Genet 2008;9(6):465–76.

[3] Feinberg AP, Vogelstein B. Hypomethylation of ras oncogenes in primary human cancers. Biochem Biophys Res Commun 1983;111(1):47–54.

[4] Fernandez AF, et al. A DNA methylation fingerprint of 1628 human samples. Genome Res 2012;22(2):407–19.

[5] Berman BP, Weisenberger DJ, Aman JF, Hinoue T, Ramjan Z, Liu Y, et al. Regions of focal DNA hypermethylation and long-range hypomethylation in colorectal cancer coincide with nuclear lamina-associated domains. Nat Genet 2012;44:40–6.

[6] Bartova E, et al. Histone modifications and nuclear architecture: a review. J Histochem Cytochem 2008;56(8):711–21.

[7] Martin C, Zhang Y. The diverse functions of histone lysine methylation. Nat Rev Mol Cell Biol 2005;6(11):838–49.

[8] Stunkel W, Campbell RM. Sirtuin 1 (SIRT1): the misunderstood HDAC. J Biomol Screen 2011;16(10):1153–69.

[9] Toth KF, et al. Trichostatin A-induced histone acetylation causes decondensation of interphase chromatin. J Cell Sci 2004;117(Pt 18):4277–87.

[10] Kelly WK, Marks PA. Drug insight: histone deacetylase inhibitors—development of the new targeted anticancer agent suberoylanilide hydroxamic acid. Nat Clin Pract Oncol 2005;2(3):150–7.

[11] Chung YL, et al. A therapeutic strategy uses histone deacetylase inhibitors to modulate the expression of genes involved in the pathogenesis of rheumatoid arthritis. Mol Ther 2003;8(5):707–17.

[12] Taddei A, et al. Reversible disruption of pericentric heterochromatin and centromere function by inhibiting deacetylases. Nat Cell Biol 2001;3(2):114–20.

[13] Kouzarides T. Histone methylation in transcriptional control. Curr Opin Genet Dev 2002;12(2):198–209.

[14] Zhang Y, Reinberg D. Transcription regulation by histone methylation: interplay between different covalent modifications of the core histone tails. Genes Dev 2001;15 (18):2343–60.

[15] Byvoet P. In vivo turnover and distribution of radio-N-methyl in arginine-rich histones from rat tissues. Arch Biochem Biophys 1972;152(2):887–8.

[16] Khanna P, Jorns MS. Characterization of the FAD-containing N-methyltryptophan oxidase from Escherichia coli. Biochemistry 2001;40(5):1441–50.

[17] You JS, Jones PA. Cancer genetics and epigenetics: two sides of the same coin? Cancer Cell 2012;22(1):9–20.

[18] Baylin SB, Jones PA. A decade of exploring the cancer epigenome-biological and translational implications. Nat Rev Cancer 2011;11(10):726–34.

[19] Sharma S, Kelly TK, Jones PA. Epigenetics in cancer. Carcinogenesis 2010;31 (1):27–36.

[20] Hatziapostolou M, Iliopoulos D. Epigenetic aberrations during oncogenesis. Cell Mol Life Sci 2011;68(10):1681−702.

[21] Krivtsov AV, Armstrong SA. MLL translocations, histone modifications and leukaemia stem-cell development. Nat Rev Cancer 2007;7(11):823−33.

[22] Esteller M, et al. Inactivation of the DNA-repair gene MGMT and the clinical response of gliomas to alkylating agents. N Engl J Med 2000;343(19):1350−4.

[23] Cancer Genome Atlas Research Network. Integrated genomic analyses of ovarian carcinoma. Nature 2011;474(7353):609−15.

[24] Laird PW. The power and the promise of DNA methylation markers. Nat Rev Cancer 2003;3(4):253−66.

[25] Bailey VJ, et al. MS-qFRET: a quantum dot-based method for analysis of DNA methylation. Genome Res 2009;19(8):1455−61.

[26] Li M, et al. Sensitive digital quantification of DNA methylation in clinical samples. Nat Biotechnol 2009;27(9):858−63.

[27] Glockner SC, et al. Methylation of TFPI2 in stool DNA: a potential novel biomarker for the detection of colorectal cancer. Cancer Res 2009;69(11):4691−9.

[28] Lofton-Day C, et al. DNA methylation biomarkers for blood-based colorectal cancer screening. Clin Chem 2008;54(2):414−23.

[29] Rosenbaum E, et al. Promoter hypermethylation as an independent prognostic factor for relapse in patients with prostate cancer following radical prostatectomy. Clin Cancer Res 2005;11(23):8321−5.

[30] Cairns P, et al. Molecular detection of prostate cancer in urine by GSTP1 hypermethylation. Clin Cancer Res 2001;7(9):2727−30.

[31] Hegi ME, et al. MGMT gene silencing and benefit from temozolomide in glioblastoma. N Engl J Med 2005;352(10):997−1003.

[32] Brock MV, et al. DNA methylation markers and early recurrence in stage I lung cancer. N Engl J Med 2008;358(11):1118−28.

[33] Seligson DB, et al. Global histone modification patterns predict risk of prostate cancer recurrence. Nature 2005;435(7046):1262−6.

[34] Fraga MF, et al. Loss of acetylation at Lys16 and trimethylation at Lys20 of histone H4 is a common hallmark of human cancer. Nat Genet 2005;37(4):391−400.

[35] Issa JP, Kantarjian HM. Introduction: emerging role of epigenetic therapy: focus on decitabine. Semin Hematol 2005;42(3 Suppl. 2):S1−2.

[36] Silverman LR, Mufti GJ. Methylation inhibitor therapy in the treatment of myelodysplastic syndrome. Nat Clin Pract Oncol 2005;2(Suppl. 1):S12−23.

[37] Fenaux P, et al. Efficacy of azacitidine compared with that of conventional care regimens in the treatment of higher-risk myelodysplastic syndromes: a randomised, open-label, phase III study. Lancet Oncol 2009;10(3):223−32.

[38] Yoo CB, Jones PA. Epigenetic therapy of cancer: past, present and future. Nat Rev Drug Discov 2006;5(1):37−50.

[39] Duvic M, et al. Phase 2 trial of oral vorinostat (suberoylanilide hydroxamic acid, SAHA) for refractory cutaneous T-cell lymphoma (CTCL). Blood 2007;109(1):31−9.

[40] Olsen EA, et al. Phase IIb multicenter trial of vorinostat in patients with persistent, progressive, or treatment refractory cutaneous T-cell lymphoma. J Clin Oncol 2007;25(21):3109−15.

[41] Sharma SV, et al. A chromatin-mediated reversible drug-tolerant state in cancer cell subpopulations. Cell 2010;141(1):69−80.

[42] Roesch A, et al. A temporarily distinct subpopulation of slow-cycling melanoma cells is required for continuous tumor growth. Cell 2010;141(4):583−94.

[43] Hardison RC. Genome-wide epigenetic data facilitate understanding of disease susceptibility association studies. J Biol Chem 2012;287(37):30932−40.

[44] Schaible TD, et al. Maternal methyl-donor supplementation induces prolonged murine offspring colitis susceptibility in association with mucosal epigenetic and microbiomic changes. Hum Mol Genet 2011;20(9):1687–96.

[45] Wu F, et al. Peripheral blood microRNAs distinguish active ulcerative colitis and Crohn's disease. Inflamm Bowel Dis 2011;17(1):241–50.

[46] Chen Y, et al. miR-200b is involved in intestinal fibrosis of Crohn's disease. Int J Mol Med 2012;29(4):601–6.

[47] Segain JP, et al. Butyrate inhibits inflammatory responses through NFkappaB inhibition: implications for Crohn's disease. Gut 2000;47(3):397–403.

[48] Bright JJ, et al. Expression of IL-12 in CNS and lymphoid organs of mice with experimental allergic encephalitis. J Neuroimmunol 1998;82(1):22–30.

[49] Takahashi I, et al. Selective inhibition of IL-2 gene expression by trichostatin A, a potent inhibitor of mammalian histone deacetylase. J Antibiot (Tokyo) 1996;49(5):453–7.

[50] Camelo S, et al. Transcriptional therapy with the histone deacetylase inhibitor trichostatin A ameliorates experimental autoimmune encephalomyelitis. J Neuroimmunol 2005;164(1–2):10–21.

[51] Krol J, Loedige I, Filipowicz W. The widespread regulation of microRNA biogenesis, function and decay. Nat Rev Genet 2010;11(9):597–610.

[52] Ceribelli A, et al. Lupus T cells switched on by DNA hypomethylation via microRNA? Arthritis Rheum 2011;63(5):1177–81.

[53] Zhao S, et al. MicroRNA-126 regulates DNA methylation in CD4+ T cells and contributes to systemic lupus erythematosus by targeting DNA methyltransferase 1. Arthritis Rheum 2011;63(5):1376–86.

[54] Lu Q, et al. Demethylation of ITGAL (CD11a) regulatory sequences in systemic lupus erythematosus. Arthritis Rheum 2002;46(5):1282–91.

[55] Higgs BW, et al. Patients with systemic lupus erythematosus, myositis, rheumatoid arthritis and scleroderma share activation of a common type I interferon pathway. Ann Rheum Dis 2011;70(11):2029–36.

[56] Zhao X, et al. MicroRNA-125a contributes to elevated inflammatory chemokine RANTES levels via targeting KLF13 in systemic lupus erythematosus. Arthritis Rheum 2010;62(11):3425–35.

[57] Stanczyk J, et al. Altered expression of microRNA in synovial fibroblasts and synovial tissue in rheumatoid arthritis. Arthritis Rheum 2008;58(4):1001–9.

[58] Leng RX, et al. Role of microRNA-155 in autoimmunity. Cytokine Growth Factor Rev 2011;22(3):141–7.

[59] Contassot E, French LE. Epigenetic causes of apoptosis resistance in cutaneous T-cell lymphomas. J Invest Dermatol 2010;130(4):922–4.

[60] Jones CL, et al. Downregulation of Fas gene expression in Sezary syndrome is associated with promoter hypermethylation. J Invest Dermatol 2010;130(4):1116–25.

[61] Sigalotti L, et al. Epigenetics of human cutaneous melanoma: setting the stage for new therapeutic strategies. J Transl Med 2010;8:56.

[62] Ceol CJ, et al. The histone methyltransferase SETDB1 is recurrently amplified in melanoma and accelerates its onset. Nature 2011;471(7339):513–17.

[63] Conway K, et al. DNA-methylation profiling distinguishes malignant melanomas from benign nevi. Pigment Cell Melanoma Res 2011;24(2):352–60.

[64] Lahtz C, et al. Methylation of PTEN as a prognostic factor in malignant melanoma of the skin. J Invest Dermatol 2010;130(2):620–2.

[65] Boumber Y, Issa JP. Epigenetics in cancer: what's the future? Oncology (Williston Park) 2011;25(3):220–6, 228.

[66] McMillan TJ, Hart IR. Enhanced experimental metastatic capacity of a murine melanoma following pre-treatment with anticancer drugs. Clin Exp Metastasis 1986;4(4):285–92.

[67] Woods DM, et al. The antimelanoma activity of the histone deacetylase inhibitor panobinostat (LBH589) is mediated by direct tumor cytotoxicity and increased tumor immunogenicity. Melanoma Res 2013. [Epub ahead of print] PMID: 23963286.

[68] Weber J, et al. Expression of the MAGE-1 tumor antigen is up-regulated by the demethylating agent 5-aza-2'-deoxycytidine. Cancer Res 1994;54(7):1766–71.

[69] Serrano A, et al. Methylated CpG points identified within MAGE-1 promoter are involved in gene repression. Int J Cancer 1996;68(4):464–70.

[70] De Smet C, et al. The activation of human gene MAGE-1 in tumor cells is correlated with genome-wide demethylation. Proc Natl Acad Sci USA 1996;93(14):7149–53.

[71] Abel T, Zukin RS. Epigenetic targets of HDAC inhibition in neurodegenerative and psychiatric disorders. Curr Opin Pharmacol 2008;8(1):57–64.

[72] Bates GP. Huntington's disease. Exploiting expression. Nature 2001;413(6857):691, 693–694.

[73] Sugars KL, Rubinsztein DC. Transcriptional abnormalities in Huntington disease. Trends Genet 2003;19(5):233–8.

[74] Steffan JS, et al. Histone deacetylase inhibitors arrest polyglutamine-dependent neurodegeneration in *Drosophila*. Nature 2001;413(6857):739–43.

[75] Ferrante RJ, et al. Histone deacetylase inhibition by sodium butyrate chemotherapy ameliorates the neurodegenerative phenotype in Huntington's disease mice. J Neurosci 2003;23(28):9418–27.

[76] Gardian G, et al. Neuroprotective effects of phenylbutyrate in the N171-82Q transgenic mouse model of Huntington's disease. J Biol Chem 2005;280(1):556–63.

[77] Kontopoulos E, Parvin JD, Feany MB. Alpha-synuclein acts in the nucleus to inhibit histone acetylation and promote neurotoxicity. Hum Mol Genet 2006;15(20):3012–23.

[78] Mattson MP. Pathways towards and away from Alzheimer's disease. Nature 2004;430(7000):631–9.

[79] Fischer A, et al. Recovery of learning and memory is associated with chromatin remodelling. Nature 2007;447(7141):178–82.

[80] Tsankova N, et al. Epigenetic regulation in psychiatric disorders. Nat Rev Neurosci 2007;8(5):355–67.

[81] Sawa A, Snyder SH. Schizophrenia: diverse approaches to a complex disease. Science 2002;296(5568):692–5.

[82] Chen Y, et al. On the epigenetic regulation of the human reelin promoter. Nucleic Acids Res 2002;30(13):2930–9.

[83] Arnaud P, Feil R. Epigenetic deregulation of genomic imprinting in human disorders and following assisted reproduction. Birth Defects Res C Embryo Today 2005;75(2):81–97.

[84] Paulsen M, Ferguson-Smith AC. DNA methylation in genomic imprinting, development, and disease. J Pathol 2001;195(1):97–110.

[85] Balmer D, et al. MECP2 mutations in Rett syndrome adversely affect lymphocyte growth, but do not affect imprinted gene expression in blood or brain. Hum Genet 2002;110(6):545–52.

[86] Howell CY, et al. Genomic imprinting disrupted by a maternal effect mutation in the Dnmt1 gene. Cell 2001;104(6):829–38.

[87] Begin P, Nadeau KC. Epigenetic regulation of asthma and allergic disease. Allergy Asthma Clin Immunol 2014;10(1):27.

[88] Kwon NH, et al. DNA methylation and the expression of IL-4 and IFN-gamma promoter genes in patients with bronchial asthma. J Clin Immunol 2008;28(2):139–46.

[89] Wei G, et al. Global mapping of H3K4me3 and H3K27me3 reveals specificity and plasticity in lineage fate determination of differentiating CD4+ T cells. Immunity 2009;30(1):155–67.

[90] Hosokawa H, et al. Gata3/Ruvbl2 complex regulates T helper 2 cell proliferation via repression of Cdkn2c expression. Proc Natl Acad Sci USA 2013;110(46):18626–31.

[91] Palmer LJ, et al. Independent inheritance of serum immunoglobulin E concentrations and airway responsiveness. Am J Respir Crit Care Med 2000;161(6):1836–43.

[92] Ordovas JM, Smith CE. Epigenetics and cardiovascular disease. Nat Rev Cardiol 2010;7(9):510–19.

[93] Zhang J, Burridge KA, Friedman MH. *In vivo* differences between endothelial transcriptional profiles of coronary and iliac arteries revealed by microarray analysis. Am J Physiol Heart Circ Physiol 2008;295(4):H1556–61.

[94] Pembrey ME. Time to take epigenetic inheritance seriously. Eur J Hum Genet 2002;10(11):669–71.

[95] Stoger R. Epigenetics and obesity. Pharmacogenomics 2008;9(12):1851–60.

[96] Dwivedi RS, et al. Beyond genetics: epigenetic code in chronic kidney disease. Kidney Int 2011;79(1):23–32.

[97] Harvey SJ, et al. Podocyte-specific deletion of dicer alters cytoskeletal dynamics and causes glomerular disease. J Am Soc Nephrol 2008;19(11):2150–8.

[98] Shi S, et al. Podocyte-selective deletion of dicer induces proteinuria and glomerulosclerosis. J Am Soc Nephrol 2008;19(11):2159–69.

[99] Bechtel W, et al. Methylation determines fibroblast activation and fibrogenesis in the kidney. Nat Med 2010;16(5):544–50.

[100] Liu Y. New insights into epithelial-mesenchymal transition in kidney fibrosis. J Am Soc Nephrol 2010;21(2):212–22.

[101] Ptak C, Petronis A. Epigenetics and complex disease: from etiology to new therapeutics. Annu Rev Pharmacol Toxicol 2008;48:257–76.

How the Environment Influences Epigenetics, DNA Methylation, and Autoimmune Diseases

Christelle Le Dantec[1], Pierre Gazeau[1,2], Sreya Mukherjee[3], Wesley H. Brooks[3], and Yves Renaudineau[1,2]

[1]EA2216, INSERM ESPRI, ERI29, European University of Brittany and Brest University, Brest, France; SFR ScInBioS, LabEx IGO "Immunotherapy Graft Oncology," and "Réseau Épigénétique du Cancéropole Grand Ouest," France [2]Laboratory of Immunology and Immunotherapy, CHU Morvan, Brest, France [3]Department of Chemistry, University of South Florida, Tampa, FL

21.1 INTRODUCTION

Several lines of evidence, when taken together (Table 21.1), strongly support a critical and pathogenic role for environmental factors in autoimmune disease development. These associations can be established by testing whether or not the environmental diseases definition criteria are respected: (i) disease development after exposure (challenge), (ii) disease resolution when the exposure is withdrawn (dechallenge), and (iii) increased disease activity after reintroduction of the same exposure (rechallenge). According to these criteria, several infectious agents, drugs, silica, ultraviolet rays (sun), vaccines, and medical implants, as well as chemical molecules are suspected to contribute to autoimmune diseases.

© 2015 Elsevier Inc. All rights reserved.

TABLE 21.1   Arguments to Support the Involvement of Environmental Factors in the Development of Autoimmune Diseases

1. Epidemiologic associations between environmental exposures, geographic clustering, seasonality, and autoimmune diseases
2. Disease improvement resolution after removal
3. Disease recurrence/worsening after reexposure
4. Animal models supporting the direct relationship between environmental factor exposure and autoimmune disease development
5. Low disease concordance between monozygotic twins (MTs) (usually <50%)
6. Changes in disease frequency when genetically predisposed cohorts move to different geographic areas

Adapted from Ref. [1].

TABLE 21.2   Concordance Rates in MT and DT Pairs for Autoimmune Diseases [3]

Disease	Monozygotic	Dizygotic
Ankylosing spondylitis	50%	20%
Celiac disease	75–83%	11%
Crohn's disease	25%	3–5%
Graves' disease	17–31%	1.9–4.7%
Multiple sclerosis	25–31%	7%
Primary biliary cirrhosis	60%	Not available
Psoriasis	67%	15%
Rheumatoid arthritis	12–15%	3.5%
Systemic lupus erythematosus	33%	2%
Type 1 diabetes	21–70%	0–13%
Ulcerative colitis	18.7%	3%

Other arguments supporting the contribution of environmental factors to the development of autoimmune diseases are related to studies testing disease frequency according to the geographic area, season, latitude, time, and the genetic background of the individual [2]. Comparison of the disease concordance rate between monozygotic (MTs) and dizygotic twins (DTs) is another elegant way to evaluate the contribution of environmental factors in genetically identical individuals (Table 21.2). An illustrative example is systemic lupus erythematosus (SLE), which is significantly affected by environmental factors; not surprisingly, the concordance rate for SLE in MTs is less than 50%. Next, with regard to the worldwide incidence of SLE, variations are reported with differences from country to country (>10-fold), geographical aggregates, and seasonal variations [4].

The increased exposure to chemicals and industrial pollutants in the last few decades may contribute to the ever-increasing prevalence observed for SLE [5]. Infections are often suspected to be the source of seasonal or geographic associations; however, other exposures are seasonal, such as exposure to UV rays, which have been implicated in SLE [6]. It is also important to note that most of the epidemiologic studies conducted so far are limited and need to be confirmed using large, well-designed international studies to minimize bias and to more thoroughly define the specific environmental risk factors for each autoimmune disease. Another bias is the genetic background of the patients, which influences the incidence of the disease as well as its clinical expression. Indeed, Europeans and their descendants are protected from SLE, and renal involvement is less severe in SLE patients with European ancestry than in those from African-American, Asian, and Hispanic populations [7].

21.2 ENVIRONMENTAL FACTORS

21.2.1 Infectious Agents

Bacteria, viruses, fungi, and parasites are all suspected of contributing to the development of autoimmune diseases through their actions on the immune system. Such hypotheses are reinforced by the observations that antibodies (Ab) to viral components (Epstein–Barr virus, cytomegalovirus, hepatitis C, parvovirus-19, and others) are more frequently detected in the sera of patients with autoimmune diseases [8]. Included among the autoimmune diseases having enough data to support an infectious origin are SLE, rheumatoid arthritis (RA), Sjögren's syndrome (SS), antiphospholipid syndrome, dermatopolymyositis, autoimmune thyroid diseases, autoimmune liver diseases, vasculitis, and multiple sclerosis [9].

21.2.2 Drugs and Diet

Drug-induced lupus (DIL) disease development appears to be the most common autoimmune condition associated with drug intake and hundreds of drugs have been associated with DIL, in case reports or published case series [10]. Procainamide, used to treat cardiac arrhythmias, and hydralazine, used to treat hypertension, are considered to be associated with the highest risk for developing DIL (5% and 20% incidence in the first year, respectively), whereas quinidine and the newer biological modulators (tumor necrosis factor (TNF)-alpha inhibitors and cytokines) have a moderate risk (<1% incidence), and all other drugs have low or very low risk. Unlike idiopathic SLE that presents Ab to double-stranded DNA, in DIL double-stranded DNA Ab are rarely detected in patients. After removing the causative medication, drug-induced antihistone Ab and antisingle-stranded DNA-Ab gradually disappear, and reappear with reintroduction of the drug [11]. DIL represents approximately 10% of SLE cases, and differs from idiopathic SLE in the clinical presentation with more frequent arthritis, cutaneous manifestations, and less frequent neurologic and renal involvement [12]. Drugs have also been implicated in other autoimmune diseases, including RA, SS, scleroderma, dermatopolymyositis, myasthenia gravis, membranous glomerulonephritis, autoimmune hepatitis, autoimmune thyroiditis, and autoimmune hemolytic anemia [13].

Hydralazine is referred to as the prototype of DIL [14−17]. Cannat and Seligmann developed a rodent animal model to test the impact of orally administrated hydralazine on normal mice [18]. After several weeks, mice developed an SLE/SS-like disease with antinuclear antibodies (ANAs) thus confirming the contribution of hydralazine in the autoimmune process. It is important to note that variations were observed according to the animal strain, age, and sex. Similar studies were performed with procainamide, revealing a more pronounced effect [19]. Both hydralazine and procainamide have been demonstrated to interfere with the epigenetic machinery [20,21].

Abnormal dietary components can induce lupus such as L-canavanine, an amino acid similar to L-arginine, present in certain leguminous plants, which has been reported to induce diet-induced lupus in humans and monkeys when ingested as a component of alfalfa sprouts [22,23]. As observed with DIL, symptoms disappeared when L-canavanine consumption was stopped and reappeared upon rechallenge [24]. In mammals, L-canavanine is incorporated into proteins, in place of L-arginine [25]. DNA−histone interactions in the nucleosome are fairly dependent on arginine residues in the histones to coordinate DNA methylation/demethylation with histone deacetylation/acetylation, and substitution of L-canavanine is suspected to interfere with this coordination.

21.2.3 Respiratory Exposures

Tobacco smoke has been reported as a major risk factor in neoplasms, in respiratory and heart diseases, and in autoimmune diseases such as RA, autoimmune thyroid disease, and Crohn's disease [26]. Inconsistent results were reported in SLE, and smoking appears to be associated with a reduced risk of ulcerative colitis [27]. In RA, smoking is also a predictor for a poorer prognosis in anticitrullinated protein antibody (ACPA)-positive patients.

Strong associations have been reported for silica dust exposure with SLE, RA, scleroderma, and vasculitis associated with antineutrophil cytoplasmic autoantibodies (ANCA) [28,29]. Less strong associations are reported for solvent exposures with scleroderma, and for farming-associated chemicals or pesticide exposures with RA [30].

21.2.4 Other Factors

A number of autoimmune diseases have been reported to develop following vaccinations but only a few of them have shown significant associations in well-controlled epidemiologic studies. These studies include Guillain—Barre syndrome after influenza and polio vaccines [31], thrombocytopenic purpura after measles vaccines [32], and chronic arthritis after rubella vaccines [33]. Because vaccines are often injected with adjuvants into muscle, adjuvants such as aluminum are suspected of contributing to the development of adjuvant-mediated autoimmune diseases that has been recently referred to as "ASIA—Autoimmune/ Inflammatory Syndrome Induced by Adjuvants" [34].

Heavy metals such as mercury, cadmium, and gold salts have been associated with autoimmunity [35]. In animal models, chronic introduction of heavy metals have confirmed their contribution in the development of autoimmunity [36]. Again, differences with these compounds were observed according to the genetic background in both humans and mice.

Medical devices, particularly silicone implants, are suspected of associations with development of cancers and autoimmune disorders such as SLE, RA, SS, systemic sclerosis, and mixed connective tissue diseases as well as other connective tissue diseases [37]. Patients with severe complications attributed to implanted silicone devices were found to have increased antisilicone IgG-specific antibody [38].

Finally, ultraviolet radiation exposure appeared as an emerging environmental factor through its action on the innate and adaptive immune system in the skin. Opposite effects may be observed with UV light that can either turn on the pathogenic inflammatory pathways as observed in SLE and atopic disorders, or switch pathogenic immune aggression off in other individuals.

21.3 EPIGENETICS

21.3.1 DNA Methylation and Histone Modifications

Epigenetics is defined as mechanisms by which stable and heritable changes in transcriptional control take place. Further, these changes are, for the most part, independent of the underlying DNA sequence, reversible, cell-type specific, and they are impacted by age, sex, and environmental factors. The epigenetic machinery includes the control of DNA methylation and histone modifications as well as transcript regulation by splicing and microRNAs [39].

DNA methylation involves addition of the methyl group (CH_3) at position 5 of the cytosine (C) pyrimidine ring to form 5-methylcytosine (5mC) within CpG base pairs (Figure 21.1). The DNA methylation reaction is

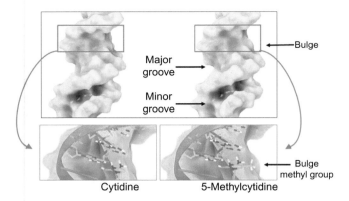

Cytidine   5-Methylcytidine

FIGURE 21.1   **DNA methylation of cytidine.** The methyl group added to cytidine creates a hindrance to polymerases and transcription factors such that they fail to bind their recognition sequences in the major groove of the underlying DNA. The "bulge" from the methyl group is subtle but is sufficient, especially in a "CpG island" in a promoter region where there are typically many methylated cytidines with the additional suppression from binding of methyl-binding proteins. The DNA structure is based on 4GLG.pdb [40] available from the Protein Data Bank (www.rcsb.org).

mediated by DNA methyltransferases (DNMTs) that use *S*-adenosyl-methionine (SAM) as the methyl donor compound. Such addition occurs predominantly in the CpG islands that have been conserved throughout evolution within the regulatory regions in approximately half of human genes. Outside these active genomic areas, CpG motifs have been progressively lost due to transition mutations from cytosine to thymine. Within the promoter region of a gene, CpG islands control the formation of the transcriptional machinery, and recruit specific proteins called methyl-CpG binding proteins (MBPs), which also help to inhibit transcription.

Whereas DNA methylation has been extensively studied in eukaryotes, the demethylation process is complex and a multistep model is actually proposed based on observations performed in zebrafish and mice embryos [41,42]. In the first step, the ten−eleven translocation (TET) proteins convert 5mC to 5-hydroxy-mC (5hmC). In the second step, the activation-induced cytidine deaminase (AICDA) converts 5hmC to 5-hydroxyuracil (5hmU), and then (third step) an apyrimidic acid residue is generated from 5hmU after glycosylation in the presence of the thymidine DNA glycosylase (TDG) MBD4. Finally (fourth step), the T:G mismatch is converted to C:G through standard base excision repair (BER) that involves the UV radiation stress sensor Gadd45-alpha. Whereas 5mC is generally associated with the inhibition of gene expression, 5hmC has been associated with increased gene expression and is involved in cellular processes such as differentiation, development, and aging [43].

Nucleosomes constitute the functional unit of chromatin, which consists of approximately 145 DNA base pairs wrapped in 1.5 turns around a histone octamer core that contains two copies each of histones H2A, H2B, H3, and H4 (Figure 21.2). In humans, nucleosomes are separated by approximately 55 DNA base pairs referred to as linker DNA. Thus, the nucleosome and linker average 200 base pairs in humans but have some positioning preferences based on the underlying DNA sequences. Chromatin can exist in a noncondensed and transcriptionally active state (euchromatin) or in an inactive state (heterochromatin). Linkers, such as histone H1, regulate the transition from euchromatin to heterochromatin. Chromatin compaction is also under the control of posttranslational modifications that occur in the protruding tails of the histones present in the nucleosome [46]. Numerous posttranslational modifications have been reported that include histone acetylation, methylation, phosphorylation, ubiquitination, ribosylation, citrullination, and sumoylation [47]. The combination of histone modifications and their crosstalk with DNA methylation and chromatin remodeling proteins is an active and dynamic process that regulates chromatin structure and transcription.

21.3.2 Epigenetics and Environmental Factors

Fraga et al. were pioneers who demonstrated that DNA methylation patterns and histone acetylation profiles diverge in peripheral blood mononuclear cells (PBMCs) isolated from younger and older healthy MTs [48]. Furthermore, it was also noted that MTs with the greatest differences were those who have different lifestyles, and different medical histories, thus providing evidence that changes may be related to differences in environmental exposures and/or diet. Such observations were reproduced by performing longitudinal studies and variations were more important when buccal cells were used instead of PBMCs [49].

The demonstration that exposure to tobacco smoking influences DNA methylation was recently provided [50]. Moreover, the effect is reversible when quitting tobacco smoking, and DNA methylation changes are reported at birth in relation to maternal smoking [51]. Other studies have examined the effects of pesticides and there is accumulating data that suggests a role of arsenic, organic pollutants, and endocrine disruptors in DNA methylation [52]. The presence of aberrant methylation is also reported in relation to bacterial infection (*Helicobacter pylori*) [53], and viral infections (hepatitis, Epstein—Barr virus (EBV)) [54].

21.3.3 Epigenetics and Autoimmune Diseases

Taking inspiration from genome-wide association studies (GWAS), epigenome-wide association studies (EWAS) were developed to compare

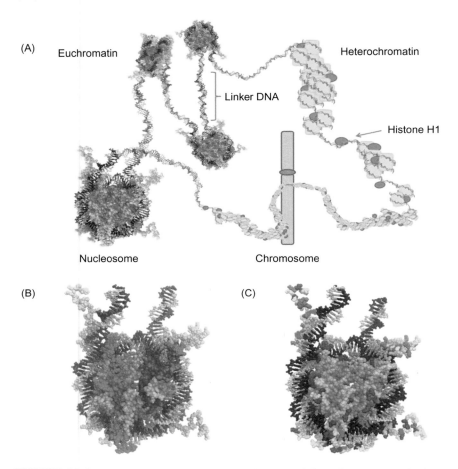

FIGURE 21.2   **Chromatin states and the nucleosome.** (A) Nucleosomes are the basic repeated unit of chromatin. Nucleosomes consist of a core of histones (two copies each of H2A, H2B, H3, and H4) which are well-conserved across species due to their need to fold compactly in order to maximize the organization and reduction of space required by the nuclear DNA. The DNA length is reduced sevenfold as it wraps around the histone core approximately 1.5 supercoil turns, with typically 145 base pairs in contact with the DNA. There is a linker DNA of approximately 55 base pairs between core particles. Nucleosomes occur on average every 200 base pairs in human chromatin. Euchromatin consists of DNA and nucleosomes in an open format, sometimes referred to as "beads on a string." The DNA is accessible for transcription, repair, and modification (such as methylation). Heterochromatin is condensed DNA and nucleosomes, typically with histone H1, binding the linker DNA and the core histones. Loops containing hundreds of nucleosomes can be coiled up in the layers of a chromosome when the underlying genes are not needed. (B) The nucleosome has a core of eight histones. A heterotetramer of H3/H4 forms a central unit with heterodimers of H2A/H2B on either side. The tails of the histones extend out where they can bind DNA and/or histone H1 depending on the modifications on key residues in the tails. H2A = blue; H2B = purple; H3 = yellow; H4 = red. DNA is black and gray strands. (C) Modification sites

TABLE 21.3 DNA Methylation and Autoimmune Diseases

Disease	Target cell	DNA methylation
Systemic lupus erythematosus	Lymphocytes	Demethylation ($\downarrow$ DNTM1, $\uparrow$ Gadd45α, $\uparrow$ MBD4, $\uparrow$ TET1/2)
Sjögren's syndrome	Epithelial cells	Demethylation ($\downarrow$ DNMT1, $\uparrow$ Gadd45α)
Rheumatoid arthritis	Synoviocytes	Demethylation ($\downarrow$ DNMT1, $\downarrow$SAM, $\downarrow$ MBD2)
Multiple sclerosis	White matter	Demethylation
	PBMCs	Demethylation ($\downarrow$ DNMT1, $\downarrow$TET2, $\downarrow$ 5hmC)
Scleroderma	Endothelial cells	Hypermethylation ($\uparrow$ DNMT1)
	Fibroblasts	Hypermethylation ($\uparrow$ DNMT1)
	Lymphocytes	Demethylation ($\downarrow$ DNMT1, $\downarrow$ MBD4)

Adapted from Ref. [57].

methylation CpG sites across the genome, and to capture variations associated with disease and/or environmental risk factors. Applied to autoimmune and rheumatic diseases these studies have already identified methylation signatures within subgroups of patients, and associations with therapeutic responses [55]. However, to minimize bias it is recommended that, as far as is possible, homogeneous cell populations and independent replication cohorts are used [56].

Epigenetic changes in autoimmune diseases are supported by studies in patients (Table 21.3), revealing tissue-specific epigenetic modifications associated with DNA methylation (usually DNA demethylation), The contribution of epigenetic mechanisms in development, acceleration, or repression of the autoimmune disease was further supported from animal studies using mice, including autoimmune-prone mice, treated with demethylating drugs and histone deacetylase (HDAC) inhibitors [58,59].

◀ on the core histones are shown. Note the high concentration of modifiable residues in the H3 tail especially. Positively charged arginine (green) and lysine (blue) can interact with the phosphates of DNA to help hold the compaction. However, methylation and acetylation of these residues can reduce the hold, loosening the DNA to make it accessible to transcription factors and other activities. The methylation of DNA can aid recruitment of the appropriate histone modifying enzymes so that there is coordination in the changes in the DNA and histones to enable the desired state to be reached for the activity of the underlying genes. Modifiable sites: lysine (blue); arginine (green); serine (red); tyrosine (purple); threonine (pink, but not seen in this view). The "histone" code comprises the various combinations of sites and modifications that can occur to give a specific epigenetic state. Some of the histone modification types are methylation, acetylation, phosphorylation, ubiquitination, ribosylation, citrullination, and sumoylation [44]. Nucleosome structure shown is based on 1KX5 pdb [45].

Another argument suggesting epigenetic defects in autoimmune diseases is related to the detection of human endogenous retroviruses (HERVs) in several chronic diseases, including autoimmunity, cancer, and neurological dysfunction [60,61]. HERVs are retroviral elements that have been incorporated in, and represent up to 7% of, the human genome [62]. For the most part, retroviral HERV proteins are not expressed except when the epigenetic machinery is defective [63−65]. As a consequence of this expression, retroviral antigens are produced, and autoantibodies to Gag and Env regions of HERVs have been reported in patients with autoimmune diseases [66,67]. Environmental factors such as viruses, UV rays, and drugs are known to dysregulate HERV expression in susceptible cells, suggesting a role for HERV elements in the initiation and/or evolution of autoimmune diseases [68,69].

21.4 MECHANISM OF ACTION

Most studies performed to explore epigenetic dysregulations in autoimmunity have been performed on CD4$^+$ T cells and B cells from SLE patients [70−72], synovial fibroblasts from RA patients [73], and salivary gland epithelial cells from SS patients [74]. From these studies (Figure 21.3), it appears that modifications in DNA methylation may

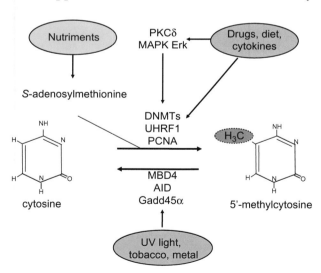

FIGURE 21.3 **Environmental factors influence DNA methylation through different pathways.** First, through an effect on the PKC-delta/Erk/DNMTs pathway. Second, through an effect on the DNA demethylation partners MBD4 and Gadd45-alpha. Third, through an effect on SAM production and/or degradation.

result (i) from a direct or an indirect effect on DNMTs, (ii) from an effect on the DNA demethylation partners MBD4 and Gadd45-alpha, and (iii) from an effect on SAM production and/or degradation. Furthermore, results from animal models suggest that blocking HDAC activity with drugs such as trichostatin A prevents lupus development in the Mrl/lpr lupus-prone mouse through an action on neutrophils, dendritic cells, and regulatory T cells [58,75].

21.4.1 DNMTs and Environmental Factors

A large body of data points toward DNMT1 as a critical component leading to autoimmunity and several environmental factors have indicated direct or indirect DNMT1 inhibitory activity (Table 21.4; Figure 21.4). Indeed, direct DNMT1 inhibitors such as 5-azacytidine and hydralazine are sufficient to induce DIL in animal models. In addition, blocking protein kinase C (PKC)-delta activation (procainamide) or downstream mitogen-activated protein kinase kinase (MEK or MAP2K)/extracellular

TABLE 21.4   Selected Products and Drugs Reported to Inhibit DNMT1 [16,76]

Name	Drug/origin	Direct/indirect inhibition	Autoimmunity
5-Azacytidine and analogs	Anticancer drug	Direct	DIL in mice
Hydralazine	Antihypertensive drug	Direct	DIL
Procainamide	Antiarrhythmic drug	Indirect	DIL
Procaine	Local anesthetic drug		
Disulfiram	Treat alcohol dependence	Direct	Unknown
Curcumin	Curry	Direct and indirect	Prevention?
Epigallocatechin-3-gallate	Green tea	Direct and indirect	Prevention?
Genistein	Soy	Indirect	Unknown
Withaferin A	Plant	Indirect	Unknown
Resveratrol	Plant	Indirect	Unknown
Guggulsterone	Guggul	Indirect	Unknown
Laccaic acid A (LCA)	Plant	Direct	Unknown
Ultraviolet B	Sun	Indirect	Unknown

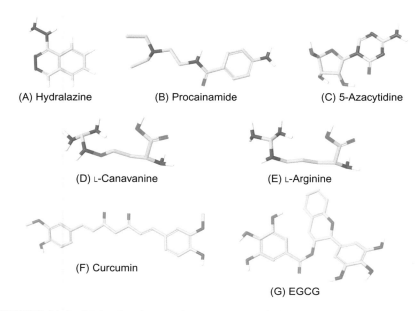

(A) Hydralazine   (B) Procainamide   (C) 5-Azacytidine

(D) L-Canavanine   (E) L-Arginine

(F) Curcumin

(G) EGCG

FIGURE 21.4   **Molecules that can impact DNA methylation.** (A) Hydralazine [IUPAC nomenclature: 1-hydrazinylphthalazine; CAS number: 86-54-4; molecular weight (MW) = 160.176 g/mol; formula: $C_8H_8N_4$]. Second line treatment for hypertension. (B) Procainamide [IUPAC: 4-amino-N-(2-diethylaminoethyl)benzamide; CAS: 51-06-9; MW = 235.325 g/mol; formula: $C_{13}H_{21}N_3O$]. Used to treat cardiac arrhythmia. (C) 5-Azacytidine [IUPAC: 4-amino-1-β-D-ribofuranosyl-1,3,5-triazin-2(1H)-one; CAS: 320-67-2; MW = 244.205 g/mol; formula: $C_8H_{12}N_4O_5$]. Analog of cytosine that is believed to act by inhibition of DNMTs and/or incorporation into DNA and RNA resulting in cell death [77]. (D) L-Canavanine [IUPAC: (2S)-2-amino-4{[(diaminomethylidene)amino]oxy}butanoic acid; CAS: 543-38-4; MW = 176.17 g/mol; formula: $C_5H_{12}N_4O_3$]. Amino acid found in seeds and sprouts where it is a source of nitrogen during early growth. (E) L-Arginine [IUPAC: 2-amino-5-guanidinopentaenoic acid; CAS: 74-79-3 (S); MW = 174.20 g/mol; formula: $C_6H_{14}N_4O_2$]. Considered to be semiessential, it is one of the 20 common amino acids in humans. It has a key role in the urea cycle and, as peptidyl arginine, is important in protein—nucleic acid interactions. (F) Curcumin [IUPAC: (1E,6E)-1,7-bis(4-hydroxy-3-methoxyphenyl)-1,6-heptadiene-3,5-dione; CAS: 458-37-7; MW = 368.38 g/mol; formula: $C_{21}H_{20}O_6$]. A component of the spice turmeric. (G) EGCG (green tea catechin, epigallocatechin-3-gallate) [IUPAC: (2R,3R)-5,7-dihydroxy-2-(3,4,5-trihydroxyphenyl)chroman-3-yl 3,4,5-trihydroxybenzoate; CAS: 989-51-5; MW = 458.372 g/mol; formula: $C_{22}H_{18}O_{11}$]. An antioxidant found in white and green tea, it is an inhibitor of DNMTs, histone acetyltransferases, topoisomerases, and several other enzymes [78]. Key: carbon = turquoise; oxygen = red; nitrogen = blue; hydrogen = white. Nonpolar hydrogens are not shown.

signal-regulated kinase (Erk) activation (PD98059) also causes DIL in mice. In cancer, natural products with anti-DNMT1 activity have received increasing attention revealing that these compounds can be found in many fruits and vegetables including turmeric (curcumin), green tea (epigallocatechin-3-gallate or EGCG), soy (genistein), and guggul (guggulsterone) [16]. Most of these products indirectly inhibit DNA methylation by controlling

DNMT1 expression and to our knowledge none of these products has been associated with autoimmunity. Others like curcumin and EGCG have been proposed to inhibit DNMT1 by alkylating the catalytic thiolate of DNMT1 and by blocking the active site of DNMT1, respectively [79,80]. Interestingly, both curcumin and EGCG have been cited as ameliorating SLE, RA, multiple sclerosis, psoriasis, and Crohn's disease in humans or animal models [81,82]. One plausible explanation for this paradoxical effect on autoimmunity is that curcumin and EGCG have a weak effect on DNMT activity, while another explanation may be related to the inhibitory effect of these two compounds on HDAC-1 [76].

The ubiquitin-like, containing plant homeodomain (PHD) and ring finger domains protein 1 (UHRF1) which contributes to recruitment of DNMT1 on hemimethylated DNA, is another target for environmental factors that is downregulated again by curcumin and EGCG [83,84]. The example of these two compounds therefore highlights the complexity of the effects of the environmental factors on epigenetic control.

21.4.2 DNA Demethylation and Environmental Factors

Active DNA demethylation is mainly achieved through an enzymatic complex that involves the TET dioxygenases, the AICDA deaminase, the MBD4 glycosylase, and the GADD45-alpha BER. Among them, GADD45-alpha is inducible in response to oxidative stress such as UV light, metal nanoparticles, tobacco smoke, bacterial lipopolysaccharides, and cytokines [85,86]. Both salivary gland epithelial cells from SS patients and T cells from SLE patients have increased levels of GADD45-alpha [74,87], and it was elegantly demonstrated in SLE T cells that GADD45-alpha expression and DNA demethylation increased following ultraviolet B irradiation. The TET activity in human cells, rats, and zebrafish is induced by hyperglycemia, and blocking the TET-dependent oxidative pathway restores DNA methylation, thus opening new perspectives in diabetes [88]. Ascorbate (vitamin C) has been proposed to participate in DNA demethylation as a cofactor of TET [89].

21.4.3 SAM and Environmental Factors

A sufficient level of the methyl group donor SAM is required for an effective DNA methylation process. As a consequence, decreased SAM production and/or increased SAM degradation, such as conversion to decarboxylated SAM during cellular stress, impacts intracellular DNA methylation.

Through the one-carbon metabolic pathway, folic acid, B2, B6, and B12 vitamins are the source of coenzymes necessary to methylate homocysteine and to form methionine, the specific precursor for SAM. Dietary folate supplementations are available in North America as dietary supplements and they are suspected to have a beneficial effect for patients with RA and juvenile idiopathic arthritis receiving the anti-folic acid drug methotrexate to prevent mucosal, gastrointestinal, hepatic, or hematologic side effects. By analogy, SAM or a SAM ana-logue might be a candidate agent for treating autoimmune diseases and it has shown some efficacy in clinical studies in osteoarthritis [90].

In addition to its action on DNA methylation, SAM is also a common precursor of nicotinamide/NAD$^+$ (using nicotinamide N-methyltransfer-ase, NNMT) and decarboxylated SAM (using SAM decarboxylase) in the polyamine pathway controlled by the rate-limiting enzymes ornithine decarboxylase (ODC), spermidine−spermine $N(1)$-acetyltransferase (SSAT; encoded by Sat1), and polyamine oxidase (PAO) [91]. As a conse-quence, increased nicotinamide and polyamine pathway activity would affect SAM levels and, in turn, DNA methylation [92], a hypothesis that was tested and validated in type 2 diabetes through the demonstration that NNMT overexpression controls SAM levels [93], and in synovial fibroblasts from RA patients revealing an increased expression of SSAT in these cells [73]. More interestingly, blocking SSAT restored DNA methyla-tion and reduced synovial fibroblast invasiveness [94]. As a consequence, blocking the polyamine pathway and/or SAM decarboxylase may reveal a new epigenetics-based therapeutic treatment [95].

21.5 CONCLUSION

Understanding the interactions between environmental factors and the epigenetic machinery offers the promise of preventing or treating autoim-mune diseases in novel ways. To accomplish this, however, critical questions remain to be answered. What are the detailed epigenetic mechanisms altered in autoimmune diseases? How do environmental fac-tors interact with the epigenetic machinery? Answer to these questions may lead to better treatment and cures, and even prevention in patients with autoimmune diseases. The typical onset of most autoimmune dis-eases in adults, as opposed to early in life, suggests an accumulation of abnormal epigenetic changes due to cellular stress events that lower DNMT1, SAM, and/or MBPs and which can eventually lead to altered expression of underlying genes due to changes in the DNA methylation patterns. This suggests research on epigenetics and autoimmune diseases is important. In addition, since the innate immune response is first to encounter many of the environmental factors when they initially induce cellular stress, such as viral invasion, the involvement of the innate immune response with regard to epigenetic states appears to have an important place in future autoimmune disease research [96,97].

References

[1] Miller FW. Environmental agents and autoimmune diseases. Adv Exp Med Biol 2011;711:61–81.

[2] Youinou P, Pers JO, Gershwin ME, Shoenfeld Y. Geo-epidemiology and autoimmunity. J Autoimmun 2010;34:J163–7.

[3] Brooks WH, Le Dantec C, Pers JO, Youinou P, Renaudineau Y. Epigenetics and autoimmunity. J Autoimmun 2010;34:J207–19.

[4] Tsokos GC. Systemic lupus erythematosus. N Engl J Med 2011;365:2110–21.

[5] Parks C, De Roos A. Pesticides, chemical and industrial exposures in relation to systemic lupus erythematosus. Lupus 2014;23:527–36.

[6] Barbhaiya M, Costenbader K. Ultraviolet radiation and systemic lupus erythematosus. Lupus 2014;23:588–95.

[7] Sanchez E, Nadig A, Richardson BC, Freedman BI, Kaufman KM, Kelly JA, et al. Phenotypic associations of genetic susceptibility loci in systemic lupus erythematosus. Ann Rheum Dis 2011;70:1752–7.

[8] Shapira Y, Agmon-Levin N, Renaudineau Y, Porat-Katz BS, Barzilai O, Ram M, et al. Serum markers of infections in patients with primary biliary cirrhosis: evidence of infection burden. Exp Mol Pathol 2012;93:386–90.

[9] Vojdani A. A potential link between environmental triggers and autoimmunity. Autoimmun Dis 2014;2014:437231.

[10] Xiao X, Miao Q, Chang C, Gershwin ME, Ma X. Common variable immunodeficiency and autoimmunity—an inconvenient truth. Autoimmun Rev 2014;13:858–64.

[11] Araujo-Fernandez S, Ahijon-Lana M, Isenberg D. Drug-induced lupus: including anti-tumour necrosis factor and interferon induced. Lupus 2014;23:545–53.

[12] Vedove CD, Del Giglio M, Schena D, Girolomoni G. Drug-induced lupus erythematosus. Arch Dermatol Res 2009;301:99–105.

[13] Chang C, Gershwin ME. Drug-induced lupus erythematosus: incidence, management and prevention. Drug Saf 2011;34:357–74.

[14] Irias JJ. Hydralazine-induced lupus erythematosus-like syndrome. Am J Dis Child 1975;129:862–4.

[15] Darwaza A, Lamey PJ, Connell JM. Hydralazine-induced Sjogren's syndrome. Int J Oral Maxillofac Surg 1988;17:92–3.

[16] Medina-Franco JL, Lopez-Vallejo F, Kuck D, Lyko F. Natural products as DNA methyltransferase inhibitors: a computer-aided discovery approach. Mol Divers 2011;15:293–304.

[17] Borchers AT, Keen CL, Gershwin ME. Drug-induced lupus. Ann NY Acad Sci 2007;1108:166–82.

[18] Cannat A, Seligmann M. Induction by isoniazid and hydralazine of antinuclear factors in mice. Clin Exp Immunol 1968;3:99–105.

[19] Richardson B, Sawalha AH, Ray D, Yung R. Murine models of lupus induced by hypomethylated T cells (DNA hypomethylation and lupus…). Methods Mol Biol 2012;900:169–80.

[20] Deng C, Lu Q, Zhang Z, Rao T, Attwood J, Yung R, et al. Hydralazine may induce autoimmunity by inhibiting extracellular signal-regulated kinase pathway signaling. Arthritis Rheum 2003;48:746–56.

[21] Lee BH, Yegnasubramanian S, Lin X, Nelson WG. Procainamide is a specific inhibitor of DNA methyltransferase 1. J Biol Chem 2005;280:40749–56.

[22] Akaogi J, Barker T, Kuroda Y, Nacionales DC, Yamasaki Y, Stevens BR, et al. Role of non-protein amino acid L-canavanine in autoimmunity. Autoimmun Rev 2006;5:429–35.

[23] Malinow MR, Bardana Jr. EJ, Pirofsky B, Craig S, McLaughlin P. Systemic lupus erythematosus-like syndrome in monkeys fed alfalfa sprouts: role of a nonprotein amino acid. Science 1982;216:415–17.

[24] Akaogi J, Nozaki T, Satoh M, Yamada H. Role of PGE2 and EP receptors in the pathogenesis of rheumatoid arthritis and as a novel therapeutic strategy. Endocr Metab Immune Disord Drug Targets 2006;6:383–94.

[25] Stasyuk NE, Gaida GZ, Gonchar MV. L-Arginine assay with the use of arginase I. Appl Biochem Microbiol 2013;49:529–34.

[26] Farhat SC, Silva CA, Orione MA, Campos LM, Sallum AM, Braga AL. Air pollution in autoimmune rheumatic diseases: a review. Autoimmun Rev 2011;11:14–21.

[27] Nunes T, Etchevers MJ, Merino O, Gallego S, Garcia-Sanchez V, Marin-Jimenez I, et al. Does smoking influence Crohn's disease in the biologic era? The TABACROHN study. Inflamm Bowel Dis 2013;19:23–9.

[28] Speck-Hernandez CA, Montoya-Ortiz G. Silicon, a possible link between environmental exposure and autoimmune diseases: the case of rheumatoid arthritis. Arthritis 2012;2012:604187.

[29] Gomez-Puerta JA, Gedmintas L, Costenbader KH. The association between silica exposure and development of ANCA-associated vasculitis: systematic review and meta-analysis. Autoimmun Rev 2013;12:1129–35.

[30] Saevarsdottir S, Rezaei H, Geborek P, Petersson I, Ernestam S, Albertsson K, et al. Current smoking status is a strong predictor of radiographic progression in early rheumatoid arthritis: results from the SWEFOT trial. Ann Rheum Dis 2014. Available from: http://dx.doi.org/doi:10.1136/annrheumdis-2013-204601. PMID: 24706006.

[31] Israeli E, Agmon-Levin N, Blank M, Chapman J, Shoenfeld Y. Guillain–Barre syndrome—a classical autoimmune disease triggered by infection or vaccination. Clin Rev Allergy Immunol 2012;42:121–30.

[32] Miller E, Waight P, Farrington CP, Andrews N, Stowe J, Taylor B. Idiopathic thrombocytopenic purpura and MMR vaccine. Arch Dis Child 2001;84:227–9.

[33] Perricone C, Colafrancesco S, Mazor RD, Soriano A, Agmon-Levin N, Shoenfeld Y. Autoimmune/inflammatory syndrome induced by adjuvants (ASIA) 2013: unveiling the pathogenic, clinical and diagnostic aspects. J Autoimmun 2013;47:1–16.

[34] Shoenfeld Y, Agmon-Levin N. 'ASIA'—autoimmune/inflammatory syndrome induced by adjuvants. J Autoimmun 2011;36:4–8.

[35] Afridi HI, Kazi TG, Kazi N, Talpur FN, Shah F, Naeemullah, Arain SS, et al. Evaluation of status of arsenic, cadmium, lead and zinc levels in biological samples of normal and arthritis patients of age groups (46–60) and (61–75) years. Clin Lab 2013;59:143–53.

[36] Leffel EK, Wolf C, Poklis A, White Jr. KL. Drinking water exposure to cadmium, an environmental contaminant, results in the exacerbation of autoimmune disease in the murine model. Toxicology 2003;188:233–50.

[37] Hajdu SD, Agmon-Levin N, Shoenfeld Y. Silicone and autoimmunity. Eur J Clin Invest 2011;41:203–11.

[38] Goldblum RM, Pelley RP, O'Donell AA, Pyron D, Heggers JP. Antibodies to silicone elastomers and reactions to ventriculoperitoneal shunts. Lancet 1992;340:510–13.

[39] Renaudineau Y, Youinou P. Epigenetics and autoimmunity, with special emphasis on methylation. Keio J Med 2011;60:10–16.

[40] Renciuk D, Blacque O, Vorlickova M, Spingler B. Crystal structures of B-DNA dodecamer containing the epigenetic modifications 5-hydroxymethylcytosine or 5-methylcytosine. Nucleic Acids Res 2013;41:9891–900.

[41] Rai K, Huggins IJ, James SR, Karpf AR, Jones DA, Cairns BR. DNA demethylation in zebrafish involves the coupling of a deaminase, a glycosylase, and gadd45. Cell 2008;135:1201−12.

[42] Guo JU, Su Y, Zhong C, Ming GL, Song H. Emerging roles of TET proteins and 5-hydroxymethylcytosines in active DNA demethylation and beyond. Cell Cycle 2011;10:2662−8.

[43] Colquitt BM, Allen WE, Barnea G, Lomvardas S. Alteration of genic 5-hydroxymethylcytosine patterning in olfactory neurons correlates with changes in gene expression and cell identity. Proc Natl Acad Sci USA 2013;110:14682−7.

[44] Strahl BD, Allis CD. The language of covalent histone modifications. Nature 2000;403:41−5.

[45] Davey CA, Sargent DF, Luger K, Maeder AW, Richmond TJ. Solvent mediated interactions in the structure of the nucleosome core particle at 1.9 a resolution. J Mol Biol 2002;319:1097−113.

[46] Lu Q, Renaudineau Y, Cha S, Ilei G, Brooks WH, Selmi C, et al. Epigenetics in autoimmune disorders: highlights of the 10th Sjogren's syndrome symposium. Autoimmun Rev 2010;9:627−30.

[47] Dieker J, Muller S. Epigenetic histone code and autoimmunity. Clin Rev Allergy Immunol 2010;39:78−84.

[48] Fraga MF, Ballestar E, Paz MF, Ropero S, Setien F, Ballestar ML, et al. Epigenetic differences arise during the lifetime of monozygotic twins. Proc Natl Acad Sci USA 2005;102:10604−9.

[49] Martino D, Loke YJ, Gordon L, Ollikainen M, Cruickshank MN, Saffery R, et al. Longitudinal, genome-scale analysis of DNA methylation in twins from birth to 18 months of age reveals rapid epigenetic change in early life and pair-specific effects of discordance. Genome Biol 2013;14:R42.

[50] Zeilinger S, Kuhnel B, Klopp N, Baurecht H, Kleinschmidt A, Gieger C, et al. Tobacco smoking leads to extensive genome-wide changes in DNA methylation. PLOS ONE 2013;8:e63812.

[51] Joubert BR, Haberg SE, Nilsen RM, Wang X, Vollset SE, Murphy SK, et al. 450 K epigenome-wide scan identifies differential DNA methylation in newborns related to maternal smoking during pregnancy. Environ Health Perspect 2012;120:1425−31.

[52] Collotta M, Bertazzi PA, Bollati V. Epigenetics and pesticides. Toxicology 2013;307:35−41.

[53] Niwa T, Tsukamoto T, Toyoda T, Mori A, Tanaka H, Maekita T, et al. Inflammatory processes triggered by *Helicobacter pylori* infection cause aberrant DNA methylation in gastric epithelial cells. Cancer Res 2010;70:1430−40.

[54] Hernando H, Shannon-Lowe C, Islam AB, Al-Shahrour F, Rodriguez-Ubreva J, Rodriguez-Cortez VC, et al. The B cell transcription program mediates hypomethylation and overexpression of key genes in Epstein−Barr virus-associated proliferative conversion. Genome Biol 2013;14:R3.

[55] Zufferey F, Williams FM, Spector TD. Epigenetics and methylation in the rheumatic diseases. Semin Arthritis Rheum 2014;43:692−700.

[56] Michels KB, Binder AM, Dedeurwaerder S, Epstein CB, Greally JM, Gut I, et al. Recommendations for the design and analysis of epigenome-wide association studies. Nat Methods 2013;10:949−55.

[57] Le Dantec C, Chevailler A, Renaudineau Y. Epigénétique et auto-immunité. Revue Francophone des Laboratoires 2013;457:67−73.

[58] Mishra N, Reilly CM, Brown DR, Ruiz P, Gilkeson GS. Histone deacetylase inhibitors modulate renal disease in the MRL-lpr/lpr mouse. J Clin Invest 2003;111:539−52.

[59] Garaud S, Youinou P, Renaudineau Y. DNA methylation and B-cell autoreactivity. Adv Exp Med Biol 2011;711:50−60.

[60] Le Dantec C, Varin MM, Brooks WH, Pers JO, Youinou P, Renaudineau Y. Epigenetics and Sjogren's syndrome. Curr Pharm Biotechnol 2012;13:2046−53.

[61] Konsta OD, Thabet Y, Le Dantec C, Brooks WH, Tzioufas AG, Pers JO, et al. The contribution of epigenetics in Sjogren's syndrome. Front Genet 2014;5:71.

[62] Balada E, Ordi-Ros J, Vilardell-Tarres M. Molecular mechanisms mediated by human endogenous retroviruses (HERVs) in autoimmunity. Rev Med Virol 2009;19:273–86.

[63] Renaudineau Y, Hillion S, Saraux A, Mageed RA, Youinou P. An alternative exon 1 of the CD5 gene regulates CD5 expression in human B lymphocytes. Blood 2005;106:2781–9.

[64] Renaudineau Y, Vallet S, Le Dantec C, Hillion S, Saraux A, Youinou P. Characterization of the human CD5 endogenous retrovirus-E in B lymphocytes. Genes Immun 2005;6:663–71.

[65] Garaud S, Le Dantec C, Berthou C, Lydyard PM, Youinou P, Renaudineau Y. Selection of the alternative exon 1 from the cd5 gene down-regulates membrane level of the protein in B lymphocytes. J Immunol 2008;181:2010–18.

[66] Brookes SM, Pandolfino YA, Mitchell TJ, Venables PJ, Shattles WG, Clark DA, et al. The immune response to and expression of cross-reactive retroviral gag sequences in autoimmune disease. Br J Rheumatol 1992;31:735–42.

[67] Hishikawa T, Ogasawara H, Kaneko H, Shirasawa T, Matsuura Y, Sekigawa I, et al. Detection of antibodies to a recombinant gag protein derived from human endogenous retrovirus clone 4-1 in autoimmune diseases. Viral Immunol 1997;10:137–47.

[68] Perron H, Lang A. The human endogenous retrovirus link between genes and environment in multiple sclerosis and in multifactorial diseases associating neuroinflammation. Clin Rev Allergy Immunol 2010;39:51–61.

[69] Thabet Y, Canas F, Ghedira I, Youinou P, Mageed RA, Renaudineau Y. Altered patterns of epigenic changes in systemic lupus erythematosus and auto-antibody production: is there a link? J Autoimmun 2012;39:154–60.

[70] Zhang Y, Zhao M, Sawalha AH, Richardson B, Lu Q. Impaired DNA methylation and its mechanisms in CD4(+)T cells of systemic lupus erythematosus. J Autoimmun 2013;41:92–9.

[71] Fali T, Le Dantec C, Thabet Y, Jousse S, Hanrotel C, Youinou P, et al. DNA methylation modulates HRES1/p28 expression in B cells from patients with lupus. Autoimmunity 2014;47:265–71.

[72] Garaud S, Le Dantec C, Jousse-Joulin S, Hanrotel-Saliou C, Saraux A, Mageed RA, et al. IL-6 modulates CD5 expression in B cells from patients with lupus by regulating DNA methylation. J Immunol 2009;182:5623–32.

[73] Karouzakis E, Gay RE, Gay S, Neidhart M. Increased recycling of polyamines is associated with global DNA hypomethylation in rheumatoid arthritis synovial fibroblasts. Arthritis Rheum 2012;64:1809–17.

[74] Thabet Y, Le Dantec C, Ghedira I, Devauchelle V, Cornec D, Pers JO, et al. Epigenetic dysregulation in salivary glands from patients with primary Sjogren's syndrome may be ascribed to infiltrating B cells. J Autoimmun 2013;41:175–81.

[75] Pieterse E, Hofstra J, Berden J, Herrmann M, Dieker J, van der Vlag J. Acetylated histones contribute to the immunostimulatory potential of neutrophil extracellular traps in systemic lupus erythematosus. Clin Exp Immunol 2015;179:68–74.

[76] Mirza S, Sharma G, Parshad R, Gupta SD, Pandya P, Ralhan R. Expression of DNA methyltransferases in breast cancer patients and to analyze the effect of natural compounds on DNA methyltransferases and associated proteins. J Breast Cancer 2013;16:23–31.

[77] Derissen EJ, Beijnen JH, Schellens JH. Concise drug review: azacitidine and decitabine. Oncologist 2013;18:619–24.

[78] Singh BN, Shankar S, Srivastava RK. Green tea catechin, epigallocatechin-3-gallate (EGCG): mechanisms, perspectives and clinical applications. Biochem Pharmacol 2011;82:1807–21.

[79] Liu Z, Xie Z, Jones W, Pavlovicz RE, Liu S, Yu J, et al. Curcumin is a potent DNA hypomethylation agent. Bioorg Med Chem Lett 2009;19:706–9.

[80] Fang MZ, Wang Y, Ai N, Hou Z, Sun Y, Lu H, et al. Tea polyphenol (-)-epigallocate-chin-3-gallate inhibits DNA methyltransferase and reactivates methylation-silenced genes in cancer cell lines. Cancer Res 2003;63:7563—70.

[81] Peairs A, Dai R, Gan L, Shimp S, Rylander MN, Li L, et al. Epigallocatechin-3-gallate (EGCG) attenuates inflammation in MRL/lpr mouse mesangial cells. Cell Mol Immunol 2010;7:123—32.

[82] Bright JJ. Curcumin and autoimmune disease. Adv Exp Med Biol 2007;595:425—51.

[83] Abusnina A, Keravis T, Yougbare I, Bronner C, Lugnier C. Anti-proliferative effect of curcumin on melanoma cells is mediated by PDE1A inhibition that regulates the epi-genetic integrator UHRF1. Mol Nutr Food Res 2011;55:1677—89.

[84] Achour M, Mousli M, Alhosin M, Ibrahim A, Peluso J, Muller CD, et al. Epigallocatechin-3-gallate up-regulates tumor suppressor gene expression via a reac-tive oxygen species-dependent down-regulation of UHRF1. Biochem Biophys Res Commun 2013;430:208—12.

[85] Feng L, Zhang Y, Jiang M, Mo Y, Wan R, Jia Z, et al. Up-regulation of Gadd45alpha after exposure to metal nanoparticles: the role of hypoxia inducible factor 1alpha. Environ Toxicol 2013. Available from: http://dx.doi.org/doi:10.1002/tox.21926.

[86] Yu Y, Li J, Wan Y, Lu J, Gao J, Huang C. GADD45alpha induction by nickel nega-tively regulates JNKs/p38 activation via promoting PP2Calpha expression. PLOS ONE 2013;8:e57185.

[87] Li Y, Zhao M, Yin H, Gao F, Wu X, Luo Y, et al. Overexpression of the growth arrest and DNA damage-induced 45alpha gene contributes to autoimmunity by promoting DNA demethylation in lupus T cells. Arthritis Rheum 2010;62:1438—47.

[88] Dhliwayo N, Sarras Jr. MP, Luczkowski E, Mason S, Intine RV. Parp inhibition pre-vents ten eleven translocase enzyme activation and hyperglycemia induced DNA demethylation. Diabetes 2014;63:3069—76.

[89] Dickson KM, Gustafson CB, Young JI, Zuchner S, Wang G. Ascorbate-induced gener-ation of 5-hydroxymethylcytosine is unaffected by varying levels of iron and 2-oxoglutarate. Biochem Biophys Res Commun 2013;439:522—7.

[90] Gregory PJ, Sperry M, Wilson AF. Dietary supplements for osteoarthritis. Am Fam Physician 2008;77:177—84.

[91] Brooks WH. Increased polyamines alter chromatin and stabilize autoantigens in auto-immune diseases. Front Immunol 2013;4:91.

[92] Brooks WH. Autoimmune diseases and polyamines. Clin Rev Allergy Immunol 2012;42:58—70.

[93] Kraus D, Yang Q, Kong D, Banks AS, Zhang L, Rodgers JT, et al. Nicotinamide N-methyltransferase knockdown protects against diet-induced obesity. Nature 2014;508:258—62.

[94] Neidhart M, Karouzakis E, Jungel A, Gay RE, Gay S. Inhibition of spermidine/ spermine N1-acetyltransferase (SSAT1) activity—a new therapeutical concept in rheu-matoid arthritis. Arthritis Rheumatol 2014;66:1723—33.

[95] Brooks WH, McCloskey DE, Daniel KG, Ealick SE, Secrist III JA, Waud WR, et al. In silico chemical library screening and experimental validation of a novel 9-aminoacridine based lead-inhibitor of human S-adenosylmethionine decarboxylase. J Chem Inf Model 2007;47:1897—905.

[96] Renaudineau Y. The revolution of epigenetics in the field of autoimmunity. Clin Rev Allergy Immunol 2010;39:1—2.

[97] Renaudineau Y, Garaud S, Le Dantec C, Alonso-Ramirez R, Daridon C, Youinou P. Autoreactive B cells and epigenetics. Clin Rev Allergy Immunol 2010;39:85—94.

Index

Note: Page numbers followed by *"f"* and *"t"* refers to figures and tables, respectively.

Printed in the United States
By Bookmasters